DATA STRUCTURES

Theory and Practice

Computer Science and Applied Mathematics

A SERIES OF MONOGRAPHS AND TEXTBOOKS

Editor

Werner Rheinboldt

University of Maryland

Hans P. Künzi, H. G. Tzschach, and C. A. Zehnder
NUMERICAL METHODS OF MATHEMATICAL OPTIMIZATION: WITH ALGOL AND FORTRAN PROGRAMS, 1968

Azriel Rosenfeld
PICTURE PROCESSING BY COMPUTER, 1969

James Ortega and Werner Rheinboldt
ITERATIVE SOLUTION OF NONLINEAR EQUATIONS IN SEVERAL VARIABLES, 1970

A. T. Berztiss
DATA STRUCTURES: THEORY AND PRACTICE, 1971

Azaria Paz
INTRODUCTION TO PROBABILISTIC AUTOMATA, 1971

In preparation

David Young
ITERATIVE SOLUTION OF LARGE LINEAR SYSTEMS

Ann Yasuhara
RECURSIVE FUNCTION THEORY AND LOGIC

DATA STRUCTURES

Theory and Practice

A. T. Berztiss

University of Pittsburgh

Academic Press New York and London

ACADEMIC PRESS, INC.
111 Fifth Avenue, New York, New York 10003

United Kingdom Edition published by
ACADEMIC PRESS, INC. (LONDON) LTD.
Berkeley Square House, London W1X 6BA

LIBRARY OF CONGRESS CATALOG CARD NUMBER: 75-137603

AMS (MOS) 1970 Subject Classifications: 00-01; 04-01, 04-04, 04A05, 05-01; 05-04, 05A10, OSC20; 68-01, 68A15; 90-01, 90-04, 90B10, 90B35

Third Printing, 1972

PRINTED IN THE UNITED STATES OF AMERICA

Contents

Chapter 6. **Paths and Cycles in Digraphs**

Chapter 7. **Digraphs of Programs**

Chapter 8. **Other Applications of Graphs**

Part III

COMPUTER REPRESENTATION OF STRUCTURES

Chapter 9. **Arrays**

Chapter 10. **Pushdown Stores, Lists and List Structures**

Preface

A man looking at a road map wishes to go from place X to place Y. His problem: Which of the numerous possible routes should he follow? Let us interpret the road map as the picture of a mathematical object, namely of a graph. Then the choosing of the shortest route translates into the well-known shortest path problem of graph theory, and an algorithm for solving this problem exists. The remaining task is the computer implementation of the algorithm, and this requires a suitable computer representation of the graph. Here we can identify three distinct phases in the complete solution process. First, a mathematical model has to be found. Second, an algorithm has to be developed within the mathematical model (we hope, however, that somebody has already done this for us). Third, a computer representation has to be selected for the data on which the algorithm is to operate.

The three parts of this book correspond more or less to the three phases. The selection of a mathematical model requires knowledge of the basic properties of some mathematical systems. These fundamentals form the subject matter of Part I, Discrete Structures in Mathematics. Part II, Applications of Structures, deals mainly with algorithms, but, in addition, the examples should help one gain experience in the selection of mathematical models for real life problems. Part III is precisely described by its title, Computer Representation of Structures.

In broad terms, this book aims at developing a productive *attitude* to the solution of problems. Of course, the main purpose is sometimes obscured by the details.

Part I, taken as a whole, is a selection of mathematical topics that are essential background knowledge for anyone who wishes to study data structures in depth. Despite appearances, the selection is highly pragmatic. Some of the topics may be irrelevant to an understanding of later parts of *this book*, but they have direct relevance to the proper understanding of journal articles dealing with mathematical models of computer data structures. The order of courses in a computer science curriculum varies so much from place to place that I could not take any specific item of background knowledge for granted. This practical consideration has prevented inclusion of more examples from computer science, which would have made the general relevance of discrete mathematics more apparent. The instructor, being familiar with the background of his students, can, and should, enliven the presentation with appropriate examples.

The first two chapters deal with topics in set theory, and most readers will find them familiar. I feel, however, that a student having just a knowledge of set theory is inadequately prepared. One must be able to think in abstract terms. That is why these chapters belabor set theory in great detail. They aim at dispelling false conceptions that a student may have and at improving his intuitive understanding of the nature of a mathematical proof. By the end of Chapter 2 the student should be able to digest an abstract argument fairly rapidly.

Chapter 3, which introduces graph theory, is the core of the book in that nearly all of Part II is based on this chapter and the various list structures of Chapter 10 are merely computer representations of directed graphs. Chapter 4 is needed for background to the applications discussed in Part II.

Part II establishes links with a number of disciplines in which mathematical structures can be put to good use. This way the mathematical structures are transformed into data structures, but the actual computer representation is understated. The object here is to unify various disciplines by the use of similar data structures.

In Part III efficiency is considered. There, for example, it is shown that the designer of a representation of a data structure must consider the peculiarities of his computer and of the operations that are to be performed on the data structure. Because of the rapid changes in computer science, one of the hardest tasks in writing a text is to decide which parts of the subject are basic. I hope to have avoided much that is ephemeral, but this means that the study of Part III should be supplemented with other material. Detailed suggestions are given below.

The material in this book covers much of what the ACM Curriculum Committee on Computer Science recommends for course B3 (Introduction to Discrete Structures) and I1 (Data Structures) [see *CACM* **11**, 151–197 (1968)]. The book is intended for the following classes of readers.

(a) Students taking a single course that combines the study of mathematical structures and data structures. Many colleges still oppose an undergraduate degree program in Computer Science for a variety of philosophical and practical reasons. At many of these schools just one combined course might be offered. The course may have Modern Algebra as taught by the Mathematics Department as a prerequisite. If it does, then Part I can be covered in two weeks by means of reading assignments, and the other parts can be dealt with almost in their entirety in the remaining time. If the course has no such prerequisite, Chapter 1 is still a reading assignment, but the instructor may have to set aside four weeks or so for a pruned down version of Chapters 2–4. He would have to be very selective in Part II, and perhaps half the available time would be spent on Part III.

(b) Students taking a course on discrete mathematical structures (at the sophomore level) and a course on data structures (at junior level). Here Parts I and II are envisaged as the material for the first course and Part III as the basis of the second course. I have tried to avoid Part III becoming a technological handbook. Therefore, this part of the book does not contain enough material for it to stand on its own as a text for a whole course. The course could be supplemented with the study of a programming language for nonnumerical computation, the manual for this language serving as a supporting text. Alternatively, the students can themselves extend an existing general purpose programming language to enable it to deal with some class of nonnumerical problems. If the course is taught to seniors, reading assignments covering journal articles can be set as well. Bibliographical notes are appended to each chapter; in Part III they are sufficiently detailed for this purpose. The very detailed development in Chapters 1 and 2 is not to be taken as a suggestion that the instructor should spend much time on these chapters. On the contrary, the detail is there to enable the instructor to adopt a much more rapid pace than he could if he himself had to supply the detail in class.

(c) Practicing programmers and systems analysts. They will find the book a balanced mixture of background mathematics and matter having immediate relevance. The very detailed presentation in the initial chapters should make the book an adequate self-study text.

Although the relation between mathematical structures and computing is not stressed in the earlier chapters, nearly all chapters contain some algorithms put in the form of computer programs and some programming exercises. Algol and Fortran are the only general purpose languages for which standards have been published. Because Fortran is the more primitive of the two, translation of Fortran into Algol is easier than the reverse. Moreover,

Fortran is better known. Therefore, Fortran is the general purpose language used here. Algorithms that cannot be implemented in Fortran could be presented as assembler language programs, but, since there is no standard assembler language, one would then have to play at being a computer manufacturer and design one's own computer with its assembler languge. Alternatively, the algorithms could be left in the form of flowcharts or English sentences. Since translation of an algorithm from one assembler language into another is at least as difficult as translation of a flowchart, the latter course is followed here.

Most of the material consists of reasonably well-established results and procedures. Exceptions are to be found in Chapter 7 and in the development of a theory of linear formulas for the representation of directed graphs (the K-formulas of Section 3e and later sections). In the discussion of program segmentation and automatic flowcharting in Chapter 7 I have tried to give a few pointers to possible lines of research in a very difficult area that has been little explored, and I hope that the more adventurous students will be stimulated into giving some thought to these problems. Similarly, the theory of K-formulas has been included in the hope that it may lead to the development of new algorithms.

Acknowledgments

The text grew out of lecture notes for a course entitled Information Structures, which I taught at the University of Pittsburgh while on sabbatical leave. I am indebted to my students there and at the University of Melbourne for their questions, comments, and corrections. I am very grateful to Rex Harris, Dale Isner, Francis Sullivan, Antons Susts, and Richard Watkins for suggestions that have improved the presentation of quite a number of sections. Richard Watkins tested the Comit examples of Section 12b. The use that has been made of contributions published in the literature will be acknowledged at the appropriate places. In selecting references my main criterion was the suitability of the reference in a program of further study. I apologize to those concerned for having given little attention to priorities of discovery.

PART I

DISCRETE STRUCTURES IN MATHEMATICS

CHAPTER 1

Set Theory

1a. Basic Definitions

Mathematics investigates relations between abstract objects. The objects might be natural numbers, or real numbers, or points in the plane, or letters in some alphabet. Some objects, such as points on the real line, cannot be counted. These objects form the basis of continuum mathematics. Other objects are countable; they are the basis of discrete mathematics. We use natural numbers in counting, and it is well known that there is no largest natural number. In other words, there is no limit at which counting has to stop. Thus we have a further subdivision of sets of countable objects into infinite sets and finite sets. Since our ultimate interest is in the application of finite discrete devices, digital computers, we shall be primarily concerned with finite sets of objects.

As the concept of a set is basic to mathematics and the theory of sets close to its foundations, we had to refer to sets rather freely in the paragraph above. We should therefore define a set, but this we cannot do. The closest we can come to a definition is to say that a set is a collection of *distinguishable* objects sharing some common feature that qualifies them for membership in the set. The objects comprising a set are called *members* of the set or *elements* of the set. Just as in geometry the terms *point* and *line* cannot be defined, so in set theory the terms *set* and *member* are undefined terms. *Membership* is an undefined relation. A way of specifying a particular set is to enclose its elements in braces: $\{\cdots\}$.

Examples of finite sets

Decimal digits: $\{0, 1, 2, \ldots, 9\}$.
Binary digits: $\{0, 1\}$.
Letters of an alphabet: $\{a, b, c, \ldots, z\}$.
Solutions of $x^2 - x = 0$: $\{0, 1\}$.

Examples of infinite sets

Natural numbers: $\{1, 2, 3, \ldots\}$.
Even natural numbers: $\{2, 4, 6, \ldots\}$.
Integers: $\{0, -1, 1, -2, 2, \ldots\}$.

DEFINITION 1.1 Two sets are equal if and only if they have the same members.

Example

$\{1, 2, 3\} = \{3, 1, 2\} = \{1, 2, 3, 3\} \neq \{1, 3, 4\}$. The first equality holds because elements of a set may be written down in any order. The second equality holds because the two 3s in $\{1, 2, 3, 3\}$ are indistinguishable. Since set $\{1, 2, 3, 3\}$ *is* the set $\{1, 2, 3\}$, one would not normally list an element more than once. Sets $\{1, 2, 3\}$ and $\{1, 3, 4\}$ have the same number of elements, but the elements are not the same.

An alternative notation for specifying the elements of a set uses a formula $P(x)$. If the formular notation is used, a set is defined by $\{x \mid P(x)\}$, read as "the set of *all* objects x such that $P(x)$ is true." Either notation can be used to represent the same set, e.g.,

$$\{0, 1, 2, \ldots, 9\} = \{i \mid i \text{ is a decimal digit}\},$$
$$\{1, 2, 3, \ldots\} = \{n \mid n \text{ is a natural number}\},$$
$$\{0, 1\} = \{x \mid x^2 - x = 0\}.$$

Formular notation specifies the feature common to elements of a set. Therefore, there is more information in $\{x \mid x^2 - x = 0\}$, say, than in $\{0, 1\}$. The increase in the information content is due to the fact that a set of elements can be specified by more than one formula. Thus $\{0, 1\}$ is also equal to $\{b \mid b \text{ is a binary digit}\}$ and to $\{c \mid c = 0 \text{ or } c = 1\}$. Note, however, that sets $\{x \mid x^2 - x = 0\}$, $\{b \mid b \text{ is a binary digit}\}$, and $\{c \mid c = 0 \text{ or } c = 1\}$ are exactly equal (by D.1.1). The increase of information relates entirely to the context in which the sets are being studied.

We shall use lower case Latin letters for elements of sets and Latin capitals for sets, e.g., $I = \{i \mid i \text{ is a decimal digit}\}$, $E = \{x \mid x < 1 \text{ and } x > 2\}$, $L = \{a, b, \ldots, z\}$. Note that E contains no elements. Membership will be indicated by the symbol $\in$, e.g., $k \in L$, $9 \in I$. If an object does not belong to a set, we shall use the symbol $\notin$, e.g., *coffee* $\notin L$, $5 \notin L$, $5 \notin E$, $15 \notin I$, $-5 \notin I$.

DEFINITION 1.2 A *null* (*empty*, *zero*) set has no elements. It is denoted by $\{\ \}$, or by $\varnothing$.

Examples

1. $\{x \mid x < 1 \text{ and } x > 2\} = \varnothing$.
2. $\{x \mid x \neq x\} = \varnothing$.
3. $\{0\} \neq \varnothing$, since the set $\{0\}$ contains an element, namely the digit 0.
4. $\varnothing \neq 0$, since $\varnothing$ is a set and 0 is not a set.
5. $\{\varnothing\} \neq \varnothing$, since set $\{\varnothing\}$ contains an element, the null set $\varnothing$.

DEFINITION 1.3 A *family* (*class*, *collection*) of sets is a set whose elements are themselves sets. Families of sets will be denoted by script letters.

Examples

1. $\mathscr{A} = \{\{0\}, \{1, 2\}, \{3, 4, 5\}, \{6, 7, 8, 9\}\}$.
2. $\mathscr{B} = \{\{0, 9\}, \{1, 8\}, \{2, 7\}, \{3, 6\}, \{4, 5\}\}$.
3. $\mathscr{C} = \{\{1, 2, 3\}, \{1, 2\}, \{2, 3\}, \{3, 1\}, \{1\}, \{2\}, \{3\}, \varnothing\}$.

4. $\mathscr{D} = \{A \mid A = \{x \mid x \text{ is a letter in a particular word}\}\} = \{\{g, u, m\}, \{a, c, t\}, \{a, c, t, i, o, n\}, \ldots\}$. Note that $\{x, y, w\} \notin \mathscr{D}$, that the words *mug* and *gum* both give rise to the same set $\{g, u, m\}$, and that *act*, *cat*, *tact* give rise to the set $\{a, c, t\}$.

DEFINITION 1.4 If A and B are sets, then A is *included* in B, written $A \subseteq B$, if and only if each member of A is a member of B. Note that B *includes* A, written $B \supseteq A$, is synonymous with $A \subseteq B$. If $A \subseteq B$, then A is a *subset* of B, and B is a *superset* of A. If $A \subseteq B$, and there exists an object x such that $x \in B$, $x \notin A$, then A is a *proper* subset of B, written $A \subset B$. If $A \subseteq B$ does not hold, we write $A \nsubseteq B$.

Examples

1. Let $I_1 = \{d \mid d \text{ is a digit}\}$, $I_2 = \{i \mid i \text{ is an integer}\}$. Then $I_1 \subseteq I_2$ and $I_1 \subset I_2$.

2. The family of sets $\mathscr{C}$ in Example 3 of D.1.3 is the family of all subsets of set $\{1, 2, 3\}$. In particular, $\varnothing \subseteq \{1, 2, 3\}$. The subset relation may not be obvious here because we have difficulty associating the phrase *each member* of D.1.4 with something that has no members; see, therefore, Th.1.1 below.

3. $\{1, 2, 3\} \subseteq \{1, 2, 3, 3\}$, and also $\{1, 2, 3\} \supseteq \{1, 2, 3, 3\}$,

THEOREM 1.1 For any set A, $\varnothing \subseteq A$.

Proof. Assume $\varnothing \nsubseteq A$. Then there exists at least one object x such that $x \in \varnothing$, $x \notin A$. But $\varnothing$ has no members (D.1.2). Hence $x \notin \varnothing$. Since the assumption leads to a contradiction, the assumption must be wrong.

THEOREM 1.2 Sets A and B are equal if and only if $A \subseteq B$ and $B \subseteq A$.

Proof. We note first that the *if* part of the theorem tells that conditions $A \subseteq B$ and $B \subseteq A$ are sufficient for $A = B$. The *only if* part tells that the conditions are also necessary. We shall prove sufficiency and necessity separately.

Sufficiency. Let $A \subseteq B$ and $B \subseteq A$. Assume $A \neq B$. Then by D.1.1 there is at least one member in one of the sets but not in the other. Let this member be x, and assume $x \in A$. Then $x \notin B$. But, if $x \in A$ and $A \subseteq B$, then $x \in B$ by D.1.4. This is a contradiction. Similarly, assuming $x \in B$ leads to the contradiction $x \notin A$ and $x \in A$. Hence $A = B$ if $A \subseteq B$ and $B \subseteq A$.

Necessity. Assume that one or other of conditions $A \subseteq B$, $B \subseteq A$ does not hold. Then there is an element in one of the sets and not in the other. But then $A \neq B$ (by D.1.1). Hence $A = B$ only if $A \subseteq B$ and $B \subseteq A$.

DEFINITION 1.5 The family of all subsets of a set A is the *power set* of A, symbolized $\mathscr{P}(A)$ or 2^A.

Example

Let $C = \{1, 2, 3\}$. Then $2^C = \{C, \{1, 2\}, \{1, 3\}, \{2, 3\}, \{1\}, \{2\}, \{3\}, \varnothing\}$.

THEOREM 1.3 If a finite set A has n members, then $\mathscr{P}(A)$ has 2^n members.

Proof. (i) $n = 0$. Then, by Th.1.1, $\varnothing \subseteq A$. Since A has no members, this is the only subset of A, i.e., $\mathscr{P}(A)$ has 1 member. But $1 = 2^0$, as required.

(ii) $n > 0$. Write the n elements of A as the sequence $a_1 a_2 \cdots a_n$. Then describe a subset B of A by a sequence of binary digits $d_1 d_2 \cdots d_n$ in which $d_i = 1$ if $a_i \in B$ and $d_i = 0$ if $a_i \notin B$. To each subset there corresponds just one sequence of digits, and each such sequence uniquely identifies a particular subset. The sequences range from $00 \cdots 0$ (for the null set) to $11 \cdots 1$ (for A itself). We interpret the sequences as binary numbers. The decimal equivalent of a binary number composed of n 1s is $2^n - 1$, and the total number of sequences is 2^n. Consequently A has 2^n distinct subsets.

Example

Let $A = \{1, 2\}$. Then $\mathscr{P}(A)$ has four members. The binary notation associates four binary sequences with these four subsets of A:

{ }	{2}	{1}	{1, 2}
00	01	10	11

If the sequences are interpreted as binary numbers, the leading zeros may be removed.

DEFINITION 1.6 The *union* (*set sum*) of sets A and B is defined by

$$A \cup B = \{x \mid x \in A \quad \text{or} \quad x \in B\},$$

where *or* has the inclusive meaning, i.e., $x \in A$ or $x \in B$ means that one of the following three statements holds: $x \in A$ and $x \notin B$; $x \notin A$ and $x \in B$; $x \in A$ and $x \in B$. The *intersection* (*set product*) of sets A and B is defined by

$$A \cap B = \{x \mid x \in A \quad \text{and} \quad x \in B\}.$$

If $A \cap B = \varnothing$, sets A and B are said to be *disjoint*.

Examples

1. $A = \{1, 3, 5, \ldots\}$, $B = \{2, 4, 6, \ldots\}$. $A \cup B = \{1, 2, 3, 4, \ldots\}$, $A \cap B = \varnothing$.
2. $X = \{1, 2, 3\}$, $Y = \{2, 3, 4, 5\}$. $X \cup Y = \{1, 2, 3, 4, 5\}$, $X \cap Y = \{2, 3\}$.

1b. Indexed Sets

Here we shall study families of sets. Consider the power set of some set A. In formular notation the family can be defined $\mathscr{P}(A) = \{B \mid B \subseteq A\}$. Alternatively, if the elements of A are known, the power set can be written out in full, e.g., $\mathscr{P}(A) = \{A, \{1\}, \{2\}, \varnothing\}$ when $A = \{1, 2\}$. Assume now that we are primarily interested in the number of sets in a family. Formular notation, as we have it, does not give this number. Explicit listing, while enabling us to count the number of sets in the family, is too detailed for our purposes here. The superfluous detail makes the notation cumbersome; imagine listing the 256 elements of the power set of a set having eight elements. (Counting commas, opening braces, and closing braces, there are at least 2,545 symbols in the list.)

A notation that gives the number of elements without requiring the elements to be written out in full is suggested by the binary sequences of Th.1.3. Write the four subsets of $A = \{1, 2\}$ as B_{00}, B_{01}, B_{10}, B_{11}, and collect the subscripts into a set I; we have $I = \{00, 01, 10, 11\}$. We can then write

$$\mathscr{P}(A) = \{B_i \mid i \in I\}, \tag{1.1}$$

where

$$I = \{i \mid i \text{ is a binary integer} \quad \text{and} \quad 00 \leqq i \leqq 11\}. \tag{1.2}$$

If next we want to describe the power set of a set consisting of eight elements, (1.1) does not have to be changed. We simply redefine I:

$$I = \{i \mid i \text{ is a binary integer} \quad \text{and} \quad 00000000 \leqq i \leqq 11111111\}. \tag{1.3}$$

In (1.2) and (1.3) the binary sequences are interpreted as binary numbers. We can even change to decimal numbers and express (1.3), say, as

$$I = \{n \mid n \text{ is a decimal integer} \quad \text{and} \quad 0 \leqq n \leqq 255\}.$$

Sometimes it is more convenient *not* to interpret the binary sequences as numbers. Let A be a set of eight elements. Suppose that we have to define a family $\mathscr{A}$, comprised of subsets of A having less than three elements. For an arbitrary binary sequence s we let $\lambda(s)$ be the length of the sequence, and $\kappa(s)$ be the number of 1s in it. Then

$$\mathscr{A} = \{B_i \mid B_i \in \mathscr{P}(A) \quad \text{and} \quad \kappa(i) < 3\},$$

where $\mathscr{P}(A)$ is defined by (1.1) and (1.3). Alternatively, without direct reference to the power set:

$$\mathscr{A} = \{B_i \mid i \in I\},$$

$$I = \{\, i \mid i \text{ is a binary sequence with } \lambda(i) = 8 \quad \text{and} \quad \kappa(i) < 3\}.$$

DEFINITION 1.7 Let $\mathscr{A}$ be the family of sets $\{A_{s_1}, A_{s_2}, A_{s_3}, \ldots\}$, where $A_{s_i} = A_{s_j}$ if $s_i = s_j$. Then the elements of $\mathscr{A}$ are identified by elements of the set

$$I = \{s_1, s_2, s_3, \ \ldots\}.$$

We can therefore write

$$\mathscr{A} = \{A_i \mid i \in I\}.$$

An element of I is called an *index*, set I itself is the *index set*, and $\mathscr{A}$ is an *indexed set*.

Examples

1. The power set of a set of n elements A is the family $\mathscr{P}(A) = \{A_i \mid i \in I\}$, where the index set I is defined by $I = \{\, i \mid i$ is a binary sequence and $\lambda(i) = n\}$.

2. $I = \{a, b, c\}$. $\mathscr{B} = \{A_i \mid i \in I\} = \{A_a, A_b, A_c\}$.

3. $I = \{2, 4, 4, 6, 6\}$. $\mathscr{X} = \{X_i \mid i \in I\} = \{X_2, X_4, X_6\}$.

4. Consider set $A = \{1, 2, 3, 4\}$ and the family of sets $\mathscr{A} = \{A_i \mid i \in I\}$, where each A_i is a set of two elements, selected from A at random. Let ten selections be made, i.e., let $I = \{1, 2, 3, \ldots, 10\}$. But only six *distinct* selections are possible: $\{1, 2\}$, $\{1, 3\}$, $\{1, 4\}$, $\{2, 3\}$, $\{2, 4\}$, $\{3, 4\}$. This means that some sets in $\mathscr{A}$ will have the same elements. By D.1.1 such sets are indistinguishable. Yet we may need to distinguish among them. Indexing gets around the difficulty in a rather subtle way. Let m and n be members of I. Normally $m \neq n$ means that A_m and A_n do not have the same elements, but note that there is no such *requirement* in D.1.7. This means that A_m and A_n may have the same elements when $m \neq n$. The symbols A_m and A_n are certainly not equal, and this suggests two interpretations of A_i. When appropriate, A_i is interpreted as standing for a set of elements. An alternative interpretation

takes A_i as a symbol, and the indexed set as a set of symbols. Under the first interpretation $\mathscr{A}$ can have at most six elements; if the second interpretation is taken, having ten elements is in order.

DEFINITION 1.8 Let I be an index set. We generalize the operations of union and intersection:

$$\bigcup_{i \in I} A_i = \{a \mid a \in A_i \quad \text{for at least one } i \in I\},$$
$$\bigcap_{i \in I} A_i = \{a \mid a \in A_i \quad \text{for all } i \in I\}.$$

Examples

1. Let $I = \{1, 2, 3, 4\}$, and $A_1 = \{1, 2\}$, $A_2 = \{1, 3\}$, $A_3 = \{1, 4\}$, $A_4 = \{1, 2, 4, 5\}$. Then $\bigcup_{i \in I} A_i = \{1, 2, 3, 4, 5\}$, and $\bigcap_{i \in I} A_i = \{1\}$. Note that $\bigcup_{i \in I} A_i$ can be written also as $\bigcup\{A_i \mid i \in I\}$, or as $\bigcup_i A_i$ (if the index set need not be emphasized). In this example we can also have $\bigcup_{i=1}^{i=4} A_i$ for the union. Similar notational variants can be used for intersection, e.g., $\bigcap_i A_i$.

2. $\bigcup_{i \in \varnothing} A_i = \varnothing$. Since there is no A_i, the set of objects belonging to at least one A_i is empty.

3. $\bigcap_{i \in \varnothing} A_i$ is more difficult to interpret. Let us consider objects that do *not* satisfy the formula of D.1.8. If, for some object a, it is not true that $a \in A_i$ *for all* $i \in I$, then there must exist at least one A_i, with $i \in I$, such that $a \notin A_i$. But if $I = \varnothing$, no A_i exists. This is a contradiction. Hence there are no objects that do not satisfy the formula, i.e., all objects satisfy the formula. We shall discuss the meaning of *all objects* in Section 1c.

4. $\bigcup_{i \in \{1\}} A_i = \bigcap_{i \in \{1\}} A_i = A_1$.

DEFINITION 1.9 A *partition* $\mathscr{A}$ of a set A, $\mathscr{A} = \{A_i \mid i \in I\}$, is a family of nonempty and distinct subsets of A such that $\bigcup_{i \in I} A_i = A$, and $A_i \cap A_j = \varnothing$ for all $i, j \in I$ $(i \neq j)$. Sets A_i are *blocks* of the partition.

Examples

1. Some partitions of $A = \{1, 2, 3, 4\}$: $\mathscr{A}_1 = \{\{1, 2\}, \{3, 4\}\}$, $\mathscr{A}_2 = \{\{1\}, \{2, 4\}, \{3\}\}$, $\mathscr{A}_3 = \{\{1, 2, 3, 4\}\}$.

2. Families $\mathscr{A}$ and $\mathscr{B}$ in Examples 1 and 2 of D.1.3 are partitions of $\{d \mid d \text{ is a decimal digit}\}$.

1c. Complement of a Set

DEFINITION 1.10 If all sets under consideration in a certain discussion are subsets of a set U, then U is the *universal set* or the *universe of discourse* for that discussion.

Examples

1. In elementary number theory U is the set of integers.
2. In plane analytic geometry U is the set of coordinate pairs.
3. If we consider sets of students taking particular courses at a university, U is the set of all students of that university.
4. $\bigcap_{i \in \varnothing} A_i = U$ identifies the *set of all objects*, which arose in Example 3 of D.1.8, with the universal set (or, more precisely, *a* universal set).

The definition of the universal set is rather vague. If I is the set of integers, and S is the set of all students of Example 3, we could use the set $\{x \mid x \in I \text{ or } x \in S\}$ as the universal set for both Examples 1 and 3. Indeed, we might consider the set of *all* objects in the universe (everything we can conceive) as the universal set for any discussion. Let us see what consequences this has. Define the set $R = \{x \mid x \notin x\}$. The defining formula seems reasonable. Consider, for example, the integer 5. We know that 5 is not a set. Hence 5 cannot have any members, $5 \notin 5$ is true, and $5 \in R$. For a further example, we know that $\mathscr{P}(\{1, 2\})$ is not a member of itself (Example of Th.1.3), and hence $\mathscr{P}(\{1, 2\}) \in R$. But if the universal set is the set of all objects, then R itself belongs to it, and the defining formula has to be applied to R. If $R \notin R$ is true, then R qualifies for membership of R, i.e., $R \in R$. On the other hand, if $R \notin R$ is false, which is equivalent to $R \in R$ being true (a possibility that cannot be dismissed, however unlikely it may appear), then R does not qualify for membership of R, i.e., $R \notin R$. Thus we have the contradiction $R \in R$ *if and only if* $R \notin R$. Unrestricted application of the defining formula $x \notin x$ is the basis of this contradiction, which is one of several possible formulations of what is known as the Russell paradox.

The Russell paradox is just one of a number of contradictions that arose in early formulations of set theory. These formulations permitted one to talk about the set of all sets, which means that one could consider a set as a member of itself. The contradictions can be traced back to this particular membership. They disappear if the manner of introducing the objects that set theory can talk about is such that a set can no longer be considered as a member of itself. This is done by making sure that the axioms of set theory can produce the theorem

For any set a, $a \notin a$.

One axiomatic theory in which this is a theorem is known as Zermelo–Fraenkel set theory.

We still have to find out how one should select the universal set for a particular discussion. D.1.10 is very permissive in this respect. We have found, however, that the permissiveness lets us make a selection that leads to contradictions. The proper way is to take an axiomatic set theory that is free of contradictions and to select from the objects this theory can talk about the smallest set that contains as subsets all sets used in the discussion. In practice we let the context in which the sets are used determine the universal set, and we will find that a universal set so selected is an acceptable set in terms of the axiomatic theory. Unless stated otherwise, the universal sets that we shall select will not have sets as members.

DEFINITION 1.11 The *relative complement* of a set A with respect to a set X, written $X - A$ (and sometimes called *set difference*), is the set

$$X - A = \{x \mid x \in X \quad \text{and} \quad x \notin A\}.$$

The *absolute complement* of a set A, written $\bar{A}$, is the set $U - A$. In terms of the absolute complement we have $X - A = X \cap \bar{A}$.

Example

Let $A = \{1, 2, 3, 4\}$ and $B = \{3, 4, 5\}$. Then $A - B = \{1, 2\}$, $B - A = \{5\}$. In this discussion we can take the set of natural numbers for U. This choice gives $\bar{A} = \{5, 6, 7, \ldots\}$, and $B \cap \bar{A} = \{3, 4, 5\} \cap \{5, 6, 7, \ldots\} = \{5\}$. Note that $\bar{A}$ depends on the choice for the universal set, but that $B \cap \bar{A}$ is independent of the choice. In the context here we can equally well select the set of decimal digits for the universal set. Then $\bar{A} = \{0, 5, 6, 7, 8, 9\}$, but $B \cap \bar{A} = \{5\}$ again.

THEOREM 1.4 Let A be a set and let U be the universal set. Then $A \cup \bar{A} = U$ and $A \cap \bar{A} = \varnothing$.

Proof. By D.1.6 $A \cup \bar{A} = \{x \mid x \in A \text{ or } x \in \bar{A}\}$ and $A \cap \bar{A} = \{x \mid x \in A \text{ and } x \in \bar{A}\}$. But (by D.1.11) $\bar{A} = U - A = \{x \mid x \in U \text{ and } x \notin A\} = \{x \mid x \notin A\}$, since it is understood that all elements belong to the universal set. Then we can put $A \cup \bar{A} = \{x \mid x \in A \text{ or } x \notin A\}$ and $A \cap \bar{A} = \{x \mid x \in A \text{ and } x \notin A\}$. The defining formulas are satisfied by all elements and by no elements, respectively. Hence $A \cup \bar{A} = U$, $A \cap \bar{A} = \varnothing$.

COROLLARY If $A \neq \varnothing$ and $A \neq U$, i.e., if $\varnothing \subset A \subset U$, then the family $\{A, \bar{A}\}$ is a partition of U.

Examples

1. Prove that $\bar{A} \subseteq \bar{B}$ if and only if $A \supseteq B$. Assume $A \not\supseteq B$. Then there exists an element $x \in B$ such that $x \notin A$, and (by Th. 1.4) $x \in \bar{A}$. But if $x \in B$, then $x \notin \bar{B}$ (Th.1.4). Thus there exists an element x such that $x \in \bar{A}$ and $x \notin \bar{B}$, i.e., $\bar{A} \not\subseteq \bar{B}$. Hence $\bar{A} \subseteq \bar{B}$ only if $A \supseteq B$. The sufficiency proof is similar.

2. Prove $\overline{A \cap B} = \bar{A} \cup \bar{B}$. We have to show (i) $\overline{A \cap B} \subseteq \bar{A} \cup \bar{B}$, and (ii) $\bar{A} \cup \bar{B} \subseteq \overline{A \cap B}$ (Th.1.2). To show that (i) holds, let $x \in \overline{A \cap B}$. Then $x \notin A \cap B$, i.e., $x \in A$ and $x \in B$ cannot both be true. Consequently $x \in \bar{A}$ or $x \in \bar{B}$, and, by the defining formula of D.1.6, $x \in \bar{A} \cup \bar{B}$. But x is *any* element of $\overline{A \cap B}$. This means that every element of $\overline{A \cap B}$ is an element of $\bar{A} \cup \bar{B}$, or $\overline{A \cap B} \subseteq \bar{A} \cup \bar{B}$. To prove inclusion (ii) *assert* that $\bar{A} \subseteq \overline{A \cap B}$ and $\bar{B} \subseteq \overline{A \cap B}$, and let $x \in \bar{A} \cup \bar{B}$. Then $x \in \bar{A}$ or $x \in \bar{B}$. If $x \in \bar{A}$, then $x \in \overline{A \cap B}$; and also if $x \in \bar{B}$, then $x \in \overline{A \cap B}$ (on the basis of the assertion). Hence $\bar{A} \cup \bar{B} \subseteq \overline{A \cap B}$. It remains to prove the assertion. We show that $\bar{A} \subseteq \overline{A \cap B}$. Let $y \in A \cap B$. Then $y \in A$ and $y \in B$, in particular $y \in A$. Hence $A \supseteq A \cap B$, and consequently, by Example 1, $\bar{A} \subseteq \overline{A \cap B}$. The proof of $\bar{B} \subseteq \overline{A \cap B}$ is similar.

DEFINITION 1.12 The *symmetric difference* of sets A and B, written $A + B$, is the set defined by

$$A + B = (A - B) \cup (B - A).$$

Example

In terms of defining formulas $A - B = \{x \mid x \in A \text{ and } x \notin B\}$, and $B - A = \{x \mid x \in B \text{ and } x \notin A\}$. Then $A + B = \{x \mid x \in A \text{ or } x \in B\}$, where *or* has the exclusive meaning, i.e., $x \in A$ *or* $x \in B$ here means $x \in A$ *or* $x \in B$ *but not* $x \in A$ and $x \in B$ (cf. definition of $A \cup B$ in D.1.6). Let $A = \{1, 2, 3, 4\}$ and $B = \{2, 3, 4, 5, 6\}$. Then $A \cup B = \{1, 2, 3, 4, 5, 6\}$, but $A + B = \{1, 5, 6\}$. Defining formulas of $A + B$ and $A \cup B$ are identical except for the interpretation of the word *or*, and there is nothing in the word itself to indicate the intended meaning. All *natural* languages (e.g., English, French) contain ambiguities. But ambiguities have no place in mathematics. This has led to the use of unambiguous symbolic languages in formal developments. In such languages there are different symbols for *or* (inclusive) and *or* (exclusive).

1d. Algebra of Sets

The properties of the complement of a set discussed in Th.1.4 and in the examples following the theorem are results in an algebra of sets. The results were obtained without a plan: The complement has been discussed, but we still have no knowledge of the basic properties of union and intersection. A consequence of the unplanned approach was the need for separate proofs of $\bar{A} \subseteq \overline{A \cap B}$ and $\bar{B} \subseteq \overline{A \cap B}$ in Example 2 of Th.1.4. We know, of course, that $\bar{B} \subseteq \overline{B \cap A}$ follows from $\bar{A} \subseteq \overline{A \cap B}$ (simply change every A to B, and every

B to A), but we have as yet no theorem that would enable us to deduce $\bar{B} \subseteq \overline{A \cap B}$ from $\bar{B} \subseteq \overline{B \cap A}$; our intuitive belief that $A \cap B$ should equal $B \cap A$ is, of course, no proof. There is a need then for a systematic development of an algebra of sets.

Operators $\cup$, $\cap$, $\bar{\ }$, $-$, $+$ have been defined. Expressions in $-$ or $+$ can be changed to equivalent expressions in which the only operators to appear are $\cup$, $\cap$, and $\bar{\ }$, i.e., any expression in $-$ or $+$ can be interpreted as a shorthand version of the equivalent expression in $\cup$, $\cap$, and $\bar{\ }$, e.g., $A + B$ as an abbreviation for $(A \cap \bar{B}) \cup (B \cap \bar{A})$. Properties of the relative complement and symmetric difference are, therefore, only of secondary interest. Operators $\cup$ and $\cap$ take two operands—union and intersection are binary operations; $\bar{\ }$ takes a single operand—complementation is a unary operation.

We need commutative laws for the binary operations. Interpretation of generalizations $\bigcup_i A_i$ and $\bigcap_i A_i$ of D.1.8 as repeated applications of binary operators $\cup$ and $\cap$, respectively, points out the need for associative laws. Since $\cup$ and $\cap$ can both occur in the one expression, e.g., $(A \cap \bar{B}) \cup (B \cap \bar{A})$, we also need distributive laws. Finally, a set needs to be related to the two special sets U and $\varnothing$, and to the complemented set. The laws are listed in the form of equalities in Th.1.5.

THEOREM 1.5 Let A, B, C be any subsets of the universal set U. Then the following equalities hold.

1A. $A \cup B = B \cup A$.
1B. $A \cap B = B \cap A$.
2A. $A \cup (B \cup C) = (A \cup B) \cup C$.
2B. $A \cap (B \cap C) = (A \cap B) \cap C$.
3A. $A \cup (B \cap C) = (A \cup B) \cap (A \cup C)$.
3B. $A \cap (B \cup C) = (A \cap B) \cup (A \cap C)$.
4A. $A \cup \varnothing = A$.
4B. $A \cap U = A$.
5A. $A \cup \bar{A} = U$.
5B. $A \cap \bar{A} = \varnothing$.

Proof. Parts 5A and 5B are Th.1.4. Proofs of the other equalities can be based on interpretation of defining formulas or on the two-sided inclusion procedure suggested by Th.1.2 and followed in the proof of Example 2 of Th. 1.4.

Union, intersection, and complementation are operations, and they define new sets. Set inclusion is not an operation. It describes a condition. While $A \cap B$ is the set whose elements belong to both A and B, $A \subseteq B$ is not a set—it states that every member of set A belongs also to set B. We know that conditions $A \subseteq B$ and $B \subseteq A$ are equivalent to the equality $A = B$ (Th.1.2). We shall now show that $A \subseteq B$ on its own is equivalent to certain equalities.

THEOREM 1.6 Let A and B be any subsets of the universal set U. The following statements about A and B are equivalent to each other:

(a) $A \subseteq B$.
(b) $A \cup B = B$.
(c) $A \cap B = A$.

Proof. We have to show that statement (b) is true if (a) is true, and that (c) is true if (b) is true. Then it has to be shown that (a) is true if (c) is true. The last part enables us to say for any two of the statements that one is true if and only if the other is true, and this is what is meant by equivalence of the statements.

(a) implies (b). Assume $A \subseteq B$. The defining formula for set union gives $A \cup B = \{x \mid x \in A \text{ or } x \in B\}$. But every member of A is a member of B by assumption. Hence the $x \in A$ part in the defining formula is superfluous, and the definition reduces to $A \cup B = \{x \mid x \in B\}$ here. Also $B = \{x \mid x \in B\}$. Hence $A \cup B = B$.

(b) implies (c). Assume $A \cup B = B$. We shall use parts of Th.1.5 in the proof.

$$
\begin{aligned}
A \cap B &= A \cap (A \cup B) && \text{(Assumption)}\\
&= (A \cup B) \cap A && \text{(1B)}\\
&= (A \cup B) \cap (A \cup \varnothing) && \text{(4A)}\\
&= A \cup (B \cap \varnothing) && \text{(3A)}\\
&= A \cup (B \cap \varnothing) \cup \varnothing && \text{(4A)}\\
&= A \cup (B \cap \varnothing) \cup (B \cap \bar{B}) && \text{(5B)}\\
&= A \cup (B \cap (\varnothing \cup \bar{B})) && \text{(3B)}\\
&= A \cup (B \cap (\bar{B} \cup \varnothing)) && \text{(1A)}\\
&= A \cup (B \cap \bar{B}) && \text{(4A)}\\
&= A \cup \varnothing && \text{(5B)}\\
&= A. && \text{(4A)}
\end{aligned}
$$

(c) implies (a). Assume $A \cap B = A$. Let $y \in A \cap B$. Then $y \in A$ and $y \in B$, in particular $y \in B$. Hence $A \cap B \subseteq B$, and, by the assumption, $A \subseteq B$.

It must be understood that the equalities of Th.1.6 are essentially different from those of Th.1.5, although we have used the same symbol (=) in both theorems. The statements of Th.1.5 hold for all subsets of a universal set. They are therefore identities. In Th.1.6 particular sets have been singled out from the family of all subsets of a universal set, and the equalities hold only for those sets. Thus in $A \cap B = B \cap A$ the symbols A and B stand for any subsets of the universal set, but in $A \cap B = A$ the equality holds only if A is a subset of B. Some writers emphasize the difference by using a special symbol ($\equiv$) for identity.

We have been talking about algebra of sets without making clear what is meant by it. Consider a nonempty set U with power set $\mathscr{P}(U)$. For any sets $A, B \in \mathscr{P}(U)$ we have $A \cup B \in \mathscr{P}(U)$, $A \cap B \in \mathscr{P}(U)$, $\bar{A} \in \mathscr{P}(U)$. If an operation on any members of a set produces an object that is also a member of the set, then the set is said to be *closed* under this operation. The power set $\mathscr{P}(U)$ is closed under union, intersection, and complementation. For some U one can find proper nonempty subsets of $\mathscr{P}(U)$ that are also closed under the three operations. Since $A \cup \bar{A} = U$ and $A \cap \bar{A} = \varnothing$ for any set A, these subsets of $\mathscr{P}(U)$, in order to be closed, necessarily contain $\varnothing$ and U. A nonempty subset of $\mathscr{P}(U)$, closed under union, intersection, and complementation, in which elements $\varnothing$ and U are distinguished, is a closed system known as an algebra of sets. We summarize: The power set $\mathscr{P}(U)$ of a nonempty set U is an algebra of sets based on U (but the $\mathscr{P}(U)$ are not the only algebras of sets).

The use of the plural, algebras of sets, may be somewhat confusing. Set U can be any nonempty set (within the limits discussed in Section 1c) and we have a different algebra for each U. But the form of the theorems is the same in all algebras of sets. Therefore, it is customary to call the theorems, rather loosely, the algebra of sets.

The concept of membership of an element in a set is essential for proof of Th.1.5. This concept belongs to the general theory of sets, and the proof of Th.1.5 places the algebra of sets within the framework of that theory. We assert now that every theorem in the algebra of sets can be deduced from Th.1.5. Note that the statement of Th.1.5 makes no mention of membership and, if we consider the algebra of sets by itself, the question of membership does not arise in it. But then we cannot prove Th.1.5 and have to consider the algebra of sets as an axiomatic theory based on the statement of Th.1.5 as a set of axioms.

1e. Algebra of Sets as an Axiomatic Theory

Most students approach *abstract theories* with trepidation. The adjective *abstract* has come to acquire a forbidding sound; commonly it denotes something that the mind finds hard to grasp. Contrary to popular belief *abstract* does not have this meaning in mathematics. Abstraction in mathematics is the process of eliminating anything that is inessential to a particular discussion;

the aim is clarity and generality. Abstraction is so important that one might even say that mathematics *is* abstraction: A child first experiences mathematics in realizing that putting two blue blocks and three blue blocks together is an operation that in its essence has nothing to do with the color of the blocks or the fact that the toys are blocks rather than beads. Despite appearances, the process that will take us from the algebra of sets to an abstract theory does not differ in kind from that through which a child goes in arriving at an understanding of the nature of addition.

The essentials of an algebra of sets are in the first place symbols representing sets and operations. The interpretation given to the symbols is necessary for proof of Th.1.5. New theorems, however, can be derived from Th.1.5 with no knowledge of the meaning of the symbols. Thus, in proving that $A \cup B = B$ implies $A \cap B = A$ (Th.1.6), symbols are manipulated according to rules supplied by Th.1.5 in a purely mechanical manner. The set of relations given in Th.1.5 is, therefore, another essential feature of the theory, but the meaning the symbols have in Th.1.5 is not. Here then the abstraction process consists of stripping symbols of their meaning. The result is a base for a generalized theory, comprising a system of undefined symbols and Th.1.5 as a set of axioms, expressing relations between symbols in this abstract system.

The idea that a mathematician builds an abstract theory out of nothing by taking a system of symbols, declaring axioms arbitrarily, and hoping for the best is a misconception. Axiomatic theories are abstractions of something already existing. For the purpose of introducing the terminology of axiomatic theories we shall, however, assume that there is nothing to start with. Then the first task is to create a base consisting of certain symbols and properties of the symbols. The symbols in the base cannot be defined; a definition gives meaning to a symbol in terms of other symbols, and there are no other symbols at the time the basic symbols are introduced. The basic symbols are called *primitives*. All other symbols, called *defined symbols*, must be defined in terms of primitives or other defined symbols. If nonprimitives are used in the definition of a new symbol, one must make sure that circular definitions are avoided, i.e., that all nonprimitives in the definition have in fact been defined.

Statements expressing relations between symbols are admitted to a theory only if they are *provable*. A *proof* is a sequence of statements in which each statement follows from one or more preceding statements by rules of logic. All statements in the sequence that are so derived by the system of logic employed are called *theorems*. A small set of statements expressing basic relations between primitives (and thus establishing their basic properties) is accorded special status. These statements, called *axioms*, are admitted to the theory without proof. (Alternatively, to be consistent with the principle that only provable statements are admitted to a theory, axioms can be considered as theorems that prove themselves.) It must be emphasized that *every* statement provable in a theory is a theorem in that theory. Since most provable statements in a theory are uninteresting, we tend to reserve the term *theorem* for those provable statements that we find particularly significant. For example, in Th.1.6 (assuming for the time being that we are dealing with an axiomatic theory), the fact that $A \cap B = A$ follows from $A \cup B = B$ is significant, and we would call "*if* $A \cup B = B$, *then* $A \cap B = A$" a theorem under any interpretation of the term. Looking at the proof of this theorem we see that the statement "*if* $A \cup B = B$, *then* $A \cap B = A \cup (B \cap (\varnothing \cup \bar{B}))$" is also a theorem in the technical sense. But we are not impressed by this theorem.

Most axiomatic theories are *informal*, by which is meant that the formulation presupposes and draws on rules of inference and general set theory. If rules of inference are assumed, one can, for example, infer equality of two things from the observation that the two things are equal to a third. A *formal* theory makes no presuppositions; consequently this rule of inference must be established as a theorem of the formal theory before it may be used in a proof. A formal theory will also contain a general set theory (e.g., the Zermelo–Fraenkel axiomatic set theory). Two clarifying remarks should be made. First, axiomatic set theory is not to be confused with axiomatized algebra of sets. While set theory investigates the relation of a member to a set, the algebra is concerned with relations between sets. Second, an informal theory is informal only in comparison with a formal theory. Even in an informal theory great attention is paid to the form of presentation.

In selecting a set of primitives and constructing the axioms of our theory we shall be guided by Th.1.5. An abstract algebra that results from an axiomatization based on Th.1.5 is known as a *Boolean algebra*. Subsequently the primitives of the general theory can be interpreted in set-theoretical terms, in which case the theory gives an algebra of sets, or they can be given some other interpretation.

Boolean algebra B

Primitives

Set B.

Binary operations $\oplus$ and $*$ under which B is closed.

Unary operation $'$ under which B is closed.

Distinct elements 0 and 1 of B. (By *distinct* we mean here that while symbols a, b, c, and the like stand for any elements of B, symbols 0 and 1 represent themselves.)

Axioms

For all $a, b, c \in B$:

1A. $a \oplus b = b \oplus a$.

1B. $a * b = b * a$.

2A. $a \oplus (b \oplus c) = (a \oplus b) \oplus c$.

2B. $a * (b * c) = (a * b) * c$.

3A. $a \oplus (b * c) = (a \oplus b) * (a \oplus c)$.

3B. $a * (b \oplus c) = (a * b) \oplus (a * c)$.

4A. $a \oplus 0 = a$.

4B. $a * 1 = a$.

5A. $a \oplus a' = 1$.

5B. $a * a' = 0$.

Some symbols appear in the set of axioms that are not listed as primitives. Certain symbols, such as (,), $\in$, *for each*, *there exists*, are symbols in the theories presupposed by the informal Boolean algebra (general set theory and the theory of inference), and as such can be admitted to the theory. General set theory also gives meaning to the terms *binary operation* and *unary operation*. The symbol $=$ is somewhat special in that different meanings are assigned to it in different interpretations of the abstract theory. We shall discuss this symbol in Section 2h.

A set of axioms is *independent* if no axioms can be removed without loss of a theorem in the theory; otherwise the set contains at least one *redundant* axiom. In our set of axioms 2A and 2B are redundant. They can be deduced from the other axioms (Exercise 1.21). As long as the set of axioms remains reasonably small, there is no harm in having redundant axioms. They may in fact help one get a better understanding of the structure of a theory.

Axioms of our algebra come in pairs and axioms in the one pair are similar. This similarity can be precisely defined, and a powerful result can then be deduced from it.

Definition 1.13 Let S be a statement in Boolean algebra B, and replace in S symbols $\oplus$, $*$, 1, 0 according to the following scheme:

$$\oplus \text{ becomes } *, \qquad * \text{ becomes } \oplus;$$
$$1 \text{ becomes } 0, \qquad 0 \text{ becomes } 1.$$

The resulting statement is the *dual* of S.

Example

The dual of $a \oplus b = b \oplus a$ is $a * b = b * a$, and the dual of $a * b = b * a$ is $a \oplus b = b \oplus a$. Each axiom in each of the five pairs of axioms of the Boolean algebra is in fact the dual of the other.

THEOREM 1.7 (*Principle of duality*) If S is a theorem in Boolean algebra B, then its dual is also a theorem in B.

Statements of Boolean algebra express relations between objects *in* the algebra. They certainly do not belong to the algebra as objects that the algebra relates. Therefore, statements *about* theorems are in a different system from that of the theorems. Theorems about theorems are known as *metatheorems*; Th.1.7 is an example of a metatheorem. It functions in the Boolean algebra as a rule of inference. A proper proof of Th.1.7 is difficult, but one might be convinced of its truth by the following argument. A statement S is established as a theorem by means of a proof. Since the dual of every axiom is also an axiom, the sequence of duals of statements in the proof of S should be a proof of the dual of S.

THEOREM 1.8 In Boolean algebra B:

1. If, for all a, $a \oplus b = a$, then $b = 0$.
2. If, for all a, $a * b = a$, then $b = 1$.

Proof. 1. Assuming $a \oplus b = a$, take in particular

$$\text{(i)} \quad a \oplus b_1 = a,$$
$$\text{(ii)} \quad a \oplus b_2 = a.$$

Then, putting b_2 for a in (i) and b_1 for a in (ii) (which we can do, since, by assumption, the relations hold for all a),

$$\text{(iii)} \quad b_2 \oplus b_1 = b_2,$$
$$\text{(iv)} \quad b_1 \oplus b_2 = b_1.$$

From (iv),

$$\text{(v)} \quad b_2 \oplus b_1 = b_1 \qquad \text{(by Axiom 1A).}$$

From (iii) and (v), $b_2 = b_1$ (two things equal to third). Thus the element b in $a \oplus b = a$ is unique. But $a \oplus 0 = a$ (Axiom 4A). Hence the unique element is 0.

2. This follows from Part 1 by principle of duality.

Axioms 4A and 4B give properties of 0 and 1, but they do not say that 0 and 1 are the only elements having these properties. Thus 4A states that $a \oplus 0 = a$ holds, but the question of whether $a \oplus b = a$ can hold for *some* b other than 0 is left open. Part 1 of Th.1.8 excludes this possibility. Th.1.8 in fact states that 0 and 1 are unique elements of B in the sense that in forms $a \oplus b = a$ and $a * c = a$ element b can be only 0 and c can be only 1. A much shorter proof of Th.1.8 can be found (Exercise 1.19), but it does not point out the uniqueness of the distinct elements 0 and 1 as clearly. The next theorem is similar to Th.1.8 in that it establishes the uniqueness of a' for a given element a (an a' exists for every a since B is closed under $'$).

THEOREM 1.9 In Boolean algebra B: For every a, if $a \oplus b = 1$ and $a * b = 0$, then $b = a'$.

Proof. Exercise 1.20.

In defining Boolean algebra B we defined $\oplus$ and $*$ as binary operations and $'$ as a unary operation in this algebra. Use of the term *operation* implies that the application of $\oplus$, $*$, or $'$ gives unique elements (see D.2.1), e.g., if $a \oplus b = c_1$ and $a \oplus b = c_2$, then $c_1 = c_2$ by definition. Our ability to *prove* the uniqueness of a' for a given a shows that $'$ may be defined as something more general than operation (namely, as a relation—see Chapter 2).

THEOREM 1.10 In Boolean algebra B:

1. For all a, $a \oplus a = a$.
2. For all a, $a \oplus 1 = 1$.
3. For all a and b, $a \oplus (a * b) = a$.

Proof. 1.

$$
\begin{aligned}
a \oplus a &= (a \oplus a) * 1 && \text{(4B)}\\
&= (a \oplus a) * (a \oplus a') && \text{(5A)}\\
&= a \oplus (a * a') && \text{(3A)}\\
&= a \oplus 0 && \text{(5B)}\\
&= a. && \text{(4A)}
\end{aligned}
$$

2.

$$
\begin{aligned}
a \oplus 1 &= a \oplus (a \oplus a') && \text{(5A)}\\
&= (a \oplus a) \oplus a' && \text{(2A)}\\
&= a \oplus a' && \text{(Part 1)}\\
&= 1. && \text{(5A)}
\end{aligned}
$$

3. Exercise 1.20.

THEOREM 1.11 In Boolean algebra B:

1. For all a, $a * a = a$.
2. For all a, $a * 0 = 0$.
3. For all a and b, $a * (a \oplus b) = a$.

Proof. From Th.1.10 by principle of duality.

THEOREM 1.12 In Boolean algebra B:

1. For all a, $(a')' = a$.
2. $0' = 1$ and $1' = 0$.
3. For all a and b, $(a \oplus b)' = a' * b'$ and $(a * b)' = a' \oplus b'$.

Proof. 1. Axioms 5A and 5B give the following pairs of theorems:

(i) $a' \oplus (a')' = 1, \quad a' * (a')' = 0;$

(ii) $a \oplus a' = 1, \quad a * a' = 0.$

From (ii), by Axioms 1A and 1B,

(iii) $a' \oplus a = 1, \qquad a' * a = 0.$

Then, by Theorem 1.9, from (i) and (iii), $(a')' = a$.

2. Exercise 1.20.

3. If it can be shown that $(a \oplus b) \oplus (a' * b') = 1$ and $(a \oplus b) * (a' * b') = 0$, then, by Th.1.9, $(a \oplus b)' = a' * b'$.

$$\begin{aligned}
(a \oplus b) \oplus (a' * b') &= ((a \oplus b) \oplus a') * ((a \oplus b) \oplus b') && \text{(3A)}\\
&= ((b \oplus a) \oplus a') * ((a \oplus b) \oplus b') && \text{(1A)}\\
&= (b \oplus (a \oplus a')) * (a \oplus (b \oplus b')) && \text{(2A)}\\
&= (b \oplus 1) * (a \oplus 1) && \text{(5A)}\\
&= 1 * 1 && \text{(Th.1.10, Part 2)}\\
&= 1. && \text{(Th.1.11, Part 1)}
\end{aligned}$$

Similarly one shows that $(a \oplus b) * (a' * b') = 0$, and $(a \oplus b)' = a' * b'$ then follows. Then $(a * b)' = a' \oplus b'$ by principle of duality. These two theorems are known as *de Morgan's laws*.

The theory has as yet no equivalent of the subset relation for sets. We introduce such a relation by definition, taking Th.1.6 for a guide.

DEFINITION 1.14 In Boolean algebra B, for two elements a and b, the relation $a \leqq b$ holds if and only if $a \oplus b = b$. Relation $a \geqq b$ is synonymous with $b \leqq a$.

Examples

1. Show that $a * b = a$ if and only if $a \oplus b = b$. Then the statements $a \leqq b$, $a \oplus b = b$, and $a * b = a$ are all equivalent to each other. First assume $a \oplus b = b$, and show that $a * b = a$ follows:

$$\begin{aligned}
a &= a * (a \oplus b) && \text{(Th.1.11, Part 3)}\\
&= a * b. && \text{(Assumption)}
\end{aligned}$$

Next assume $a * b = a$. Then

$$\begin{aligned}
b &= b \oplus (b * a) && \text{(Th.1.10, Part 3)}\\
&= b \oplus (a * b) && \text{(B1)}\\
&= b \oplus a && \text{(Assumption)}\\
&= a \oplus b. && \text{(A1)}
\end{aligned}$$

2. Statement $a \leqq b$ is equivalent to $a \oplus b = b$ and to $a * b = a$, but they are not duals of each other. Hence, if the dual of a statement containing $\leqq$ or $\geqq$ has to be found, one must replace relations in these symbols by their

equivalents prior to making the replacements. Alternatively, since $a \geqq b$ is equivalent to $a * b = b$, which is the dual of $a \oplus b = b$, one can extend the replacement scheme of D.1.13 by adding to it

$$\leqq \text{ becomes } \geqq, \qquad \geqq \text{ becomes } \leqq.$$

3. For all a in B, $1 \geqq a \geqq 0$.

Nothing has been said about the truth of axioms. In order to tell whether a statement is true one must know what it means. The meaning of a statement is determined by the meaning of the symbols in it. In an abstract system the only meaning that a symbol has is its relation to other symbols in the system, expressed in the first place by axioms. Theorems other than axioms are of course true within the abstract theory, in the sense that they are logical consequences of the axioms, with the axioms having assigned meaning to the symbols in them. But the axioms themselves do not follow from anything. Hence axioms have no meaning, and one cannot talk about their truth. The lack of preconceived notions about the symbols is in fact the strength of an abstract theory. On the other hand, proving theorems in a theory that exists in complete isolation is a sterile game. We emphasize again that in setting up an abstract theory, interpretation of the theory in terms of objects from a world external to the theory is anticipated.

The process of *interpretation* consists in taking objects that have definite meaning in some theory external to the abstract theory and of interpreting the abstract system as these concrete objects. Meaning is thus assigned to the primitives, and the axioms become meaningful statements. One can then ask whether they are true for the concrete objects. If they are, the system of the concrete objects is a *model* of the abstract theory. All theorems of an abstract theory are true statements about objects in a model of the theory,

In using Th.1.5 as a guide for setting up the abstract Boolean algebra B we provided ourselves with a ready-made model—the algebraic system $\mathscr{P}(U)$, $\cup$, $\cap$, $^{-}$, $\varnothing$, U. All objects in this system have been defined earlier. Hence they have a meaning. Th.1.5 establishes the system as a model of algebra B, and all theorems in B are automatically theorems in the Boolean algebra of subsets of U [members of $\mathscr{P}(U)$]. The symbol $=$ is interpreted as set equality. To exemplify this process, we reformulate Th.1.11 and Th.1.12 as theorems about sets.

THEOREM 1.13 For all members of $\mathscr{P}(U)$:

1. $A \cap A = A$.
2. $A \cap \varnothing = \varnothing$.
3. $A \cap (A \cup B) = A$.

THEOREM 1.14 For all members of $\mathscr{P}(U)$:

1. $\bar{\bar{A}} = A$.
2. $\overline{\varnothing} = U$ and $\bar{U} = \varnothing$.
3. $\overline{A \cup B} = \bar{A} \cap \bar{B}$ and $\overline{A \cap B} = \bar{A} \cup \bar{B}$.

Our approach still has the appearance of a game. Obviously Th.1.13 and Th.1.14 could have been deduced directly from Th.1.5. Instead, we axiomatized the algebra of sets, found theorems in the axiomatic theory, switched back to the original system, and restated the theorems in this system. The algebra of subsets of U is not, however, the only model of the abstract theory. This has considerable practical significance. The theorems of an abstract theory are true for all systems that can be shown to be models of the theory, and to establish a system as a model one merely has to show that the axioms of the abstract theory are true for the system. Without an abstract theory every theorem has to be proven separately in each of the systems.

1f. Venn Diagrams

It is much easier to prove something when we know precisely what we have to prove. Thus it is not as easy to "simplify" $A \cup (A \cap B)$ as it is to prove $A \cup (A \cap B) = A$. A graphical device known as a *Venn diagram* (or Venn–Euler diagram) can help one find a precise formulation of a theorem in the algebra of sets before one sets out to prove it. A Venn diagram represents sets by sets of points in plane regions. First, one draws a fairly large rectangle: Points within the rectangle represent the universal set. A subset of the universal set is represented by points in some region, usually the interior of a circle, within the rectangle.

In Figure 1.1 the interiors of the two circles represent sets A and B, the entire shaded region represents $A \cup B$, and the crosshatched region $A \cap B$. Since the entire rectangle and the shaded region represent U and $A \cup B$, respectively, the unshaded region within the rectangle represents $\overline{A \cup B}$. In

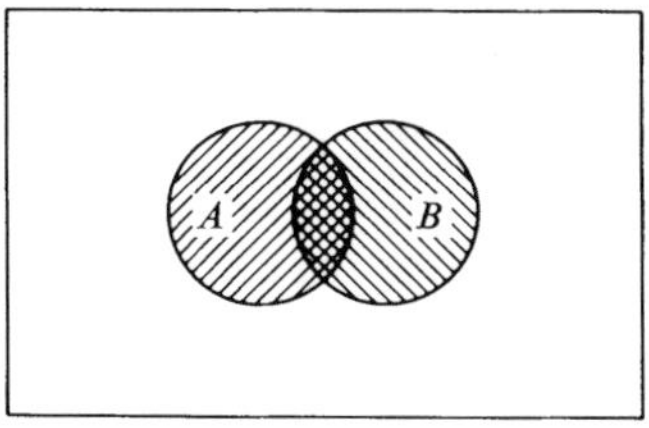

Figure 1.1

Figure 1.2 horizontal shading indicates the region representing $\bar{A}$ and vertical shading the region representing $\bar{B}$. The crosshatched region represents $\bar{A} \cap \bar{B}$, and the unshaded region $\overline{\bar{A} \cup \bar{B}}$. The crosshatched region in Figure 1.2 corresponds to the unshaded region in Figure 1.1, and the unshaded region in Figure 1.2 to the crosshatched region in Figure 1.1. These correspondences suggest $\bar{A} \cap \bar{B} = \overline{A \cup B}$ and $\overline{\bar{A} \cup \bar{B}} = A \cap B$ as possible identities.

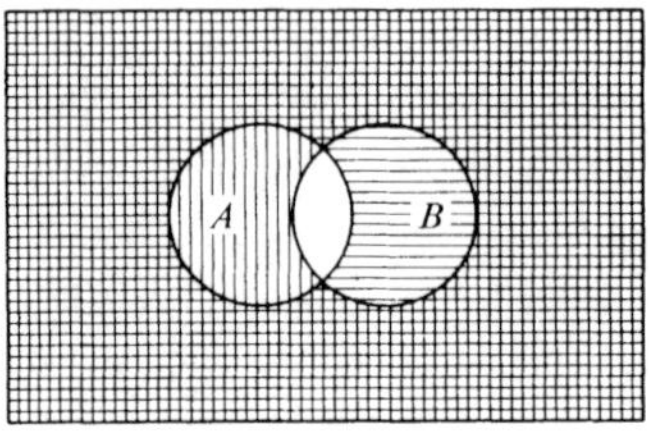

Figure 1.2

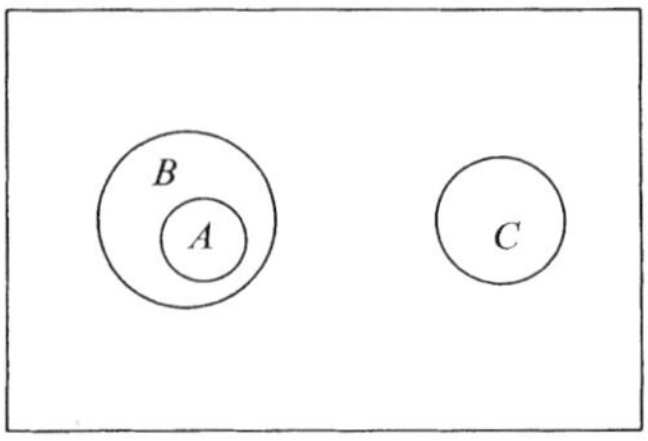

Figure 1.3

Figure 1.3 depicts set inclusion and disjoint sets. The region representing an included set lies entirely within the region representing the including set (circles A and B, for $A \subseteq B$). Disjoint sets, such as B and C, are represented by nonoverlapping regions. Looking again at Figure 1.2, regions shaded by vertical or horizontal lines alone represent the relative complements $A - B$ and $B - A$, respectively. They are two of the four nonoverlapping regions into which the rectangle has been partitioned by circles A and B. Provided each of the four regions contains at least one point, i.e., does not represent a null set, the sets represented by these regions constitute a partition of the universal set, $\{\bar{A} \cap \bar{B}, A - B, B - A, \overline{\bar{A} \cup \bar{B}}\}$.

Although we have shown that arguments based on Venn diagrams are useful for the appreciation of relations between sets, such arguments do not function well as *proofs*. Assume that we have to prove $\bar{A} \cup \bar{B} = \overline{A \cap B}$. In an algebraic proof A and B are variables throughout the proof, including the last line. It is clear that the result is an identity. In a Venn diagram A and B have to be represented by particular regions, and generality is lost at once. We can show that $\bar{A} \cup \bar{B} = \overline{A \cap B}$ is consistent with Figure 1.3, but so too is $\bar{A} \cup \bar{B} = \bar{A}$. There is no way of telling from Figure 1.3 alone that one of the equations is an identity while the other is a consequence of the relation $A \subseteq B$ in which A and B happen to stand as represented by Figure 1.3. To prove that $\bar{A} \cup \bar{B} = \overline{A \cap B}$ is an identity we have to demonstrate consistency with diagrams depicting all the different ways in which A may be related to B. An algebraic proof is much more direct.

1g. The Ordered Pair and Related Concepts

It should be clearly understood that the order in which elements of a set are written down is unimportant. But often we have to deal with composite objects in which the order of components is important. One such object is the coordinate pair of analytic geometry: Points represented by coordinate pairs $\langle 2, 3\rangle$ and $\langle 3, 2\rangle$ are different; sets $\{2, 3\}$ and $\{3, 2\}$ are equal. Another example is a sequence of statements S_1, S_2, S_3, S_4 of a proof. Extending the notation for coordinate pairs, we can write the sequence as $\langle S_1, S_2, S_3, S_4\rangle$. The sequence $\langle S_3, S_2, S_4, S_1\rangle$ is probably not a proof, but, as sets, $\{S_1, S_2, S_3, S_4\} = \{S_3, S_2, S_4, S_1\}$.

We must find a definition of order in terms of sets. For generality we will no longer refer to $\langle a, b\rangle$ as a coordinate pair; we shall call the object an *ordered pair*. The main property of an ordered pair, which must be deducible from its definition, is uniqueness: If $\langle a, b\rangle$ and $\langle x, y\rangle$ are ordered pairs, and $\langle a, b\rangle = \langle x, y\rangle$, then $a = x$ and $b = y$. We shall see that this property is deducible from D.1.15.

DEFINITION 1.15 The *ordered pair* of a and b, written $\langle a, b\rangle$, is the set $\{\{a\}, \{a, b\}\}$. We call a the *first coordinate* and b the *second coordinate* of $\langle a, b\rangle$.

Examples

1. $\{\{a\}, \{a, b\}\} = \{\{b, a\}, \{a\}\}$, but, unless $a = b$, $\{\{a\}, \{a, b\}\} \neq \{\{b\}, \{a, b\}\}$. Indeed $\{\{b\}, \{a, b\}\} = \langle b, a\rangle$.

2. $\langle a, a\rangle = \{\{a\}, \{a, a\}\} = \{\{a\}, \{a\}\} = \{\{a\}\}$. Note that the unordered pair $\{a, b\}$ reduces to $\{a\}$ when $a = b$, and that $\{\{a\}\} \neq \{a\}$.

THEOREM 1.15 If $\langle a, b\rangle = \langle x, y\rangle$, then $a = x$ and $b = y$.

Proof. Write out the ordered pairs in full:

$$\text{(i)}\quad \{\{a\}, \{a, b\}\},$$
$$\text{(ii)}\quad \{\{x\}, \{x, y\}\}.$$

For convenience we shall call sets with two distinct elements pairs and sets with single elements singletons. We shall consider two cases.

(a) $x = y$. Then (ii) reduces to $\{\{x\}\}$. Since $\langle x, y\rangle = \langle a, b\rangle$ by hypothesis, and $\langle x, y\rangle$ is a singleton, $\langle a, b\rangle$ must also be a singleton. This means that $\{a, b\}$ must equal $\{a\}$, giving $a = b$. Hence (i) reduces to $\{\{a\}\}$, and, since (i) and (ii) are equal by hypothesis, $\{\{a\}\} = \{\{x\}\}$, giving $a = x$. It follows that all a, b, x, and y are equal, in particular $a = x$ and $b = y$.

(b) $x \neq y$. Then (ii) contains exactly one singleton and exactly one pair. Since $\langle a, b\rangle = \langle x, y\rangle$, (i) must contain one singleton and one pair. Hence we can equate: $\{a\} = \{x\}$, $\{a, b\} = \{x, y\}$. $\{a\} = \{x\}$ gives $a = x$, and this in turn gives $\{x, b\} = \{x, y\}$. Since $\{x, b\}$ is a pair, $b \neq x$. Hence $b = y$.

DEFINITION 1.16 The *product set* (*Cartesian product*), $A \times B$, of sets A and B is the set of ordered pairs such that the first coordinate of each pair is a member of A, and the second coordinate belongs to B:

$$A \times B = \{\langle a, b\rangle \mid a \in A \quad \text{and} \quad b \in B\}.$$

Examples

1. $A = \{1, 2\}$, $B = \{a, b, c\}$.

$$\begin{aligned} A \times B &= \{\langle 1, a\rangle, \langle 1, b\rangle, \langle 1, c\rangle, \langle 2, a\rangle, \langle 2, b\rangle, \langle 2, c\rangle\}, \\ B \times A &= \{\langle a, 1\rangle, \langle b, 1\rangle, \langle c, 1\rangle, \langle a, 2\rangle, \langle b, 2\rangle, \langle c, 2\rangle\}, \\ A \times A &= \{\langle 1, 1\rangle, \langle 1, 2\rangle, \langle 2, 1\rangle, \langle 2, 2\rangle\}. \end{aligned}$$

If set A has m (distinct) elements and set B has n elements, then the product sets $A \times B$ and $B \times A$ both have mn elements.

2. $X = \{1\}$, $Y = \{1, 2, 3\}$.

$$\begin{aligned} X \times Y &= \{\langle 1, 1\rangle, \langle 1, 2\rangle, \langle 1, 3\rangle\}. \\ (X \times Y) \cap (Y \times X) &= \{\langle 1, 1\rangle\}. \end{aligned}$$

3. $X = \varnothing$, $Y = \{1, 2, 3\}$. $X \times Y = Y \times X = \varnothing$.

4. If A, B, C are sets, then $A \times B$ and $B \times C$ are also sets, and extended product sets such as $A \times (B \times C)$ and $(A \times B) \times C$ have meaning:

$$\begin{aligned} A \times (B \times C) &= \{\langle a, \langle b, c\rangle\rangle \mid a \in A \quad \text{and} \quad \langle b, c\rangle \in B \times C\}, \\ (A \times B) \times C &= \{\langle\langle a, b\rangle, c\rangle \mid \langle a, b\rangle \in A \times B \quad \text{and} \quad c \in C). \end{aligned}$$

5. Note that sets A and B of D.1.16 can be considered as the two members of the family $\{A_i \mid i \in I\}$, with $I = \{1, 2\}$. We can then write the product set $A_1 \times A_2$ as $\mathsf{X}_{i \in I} A_i$.

According to D.1.16 the product set $A \times B$ is constructed by going through all elements of the sets, taking element a from A, element b from B, and writing the elements in the order a, b. We are not at all dependent on the set-theoretical definition of $\langle a, b\rangle$. Similarly we want to look upon the generalized product set $A_1 \times A_2 \times \cdots \times A_n$ as a set of ordered objects $\langle a_1, a_2, \ldots, a_n\rangle$, with $a_1 \in A_1$, $a_2 \in A_2, \ldots, a_n \in A_n$. Intuitively we feel that there should be no essential difference between, say,

$$\langle a_1, \langle a_2, \ldots, a_n\rangle\rangle \quad \text{and} \quad \langle\langle a_1, \ldots, a_{n-1}\rangle, a_n\rangle,$$

i.e., $A_1 \times (A_2 \times \cdots \times A_n)$ should equal $(A_1 \times \cdots \times A_{n-1}) \times A_n$. All that really matters is that a_i comes from A_i and is written down between a_{i-1}

and a_{i+1}. Quite clearly, however, product sets $A \times (B \times C)$ and $(A \times B) \times C$ of Example 4 are not equal; i.e., the operation $\times$ is not associative. This is a consequence of D.1.15, indicating a certain amount of artificiality in the set-theoretical definition of the ordered pair. The reason for persisting with this definition is conceptual economy, which outweighs the possible inconvenience of not being able to consider

$$\langle a_1, \langle a_2, \ldots, a_n\rangle\rangle \quad \text{and} \quad \langle\langle a_1, \ldots, a_{n-1}\rangle, a_n\rangle$$

as identical objects. We have to develop a completely new theory of ordered objects if D.1.15 is rejected. The way out of our difficulty is to define a product set in which the order of parentheses is unambiguously prescribed, to adhere to the definition consistently, and to interpret $A_1 \times A_2 \times \cdots \times A_n$ as the product set so defined. The mechanism for this approach is suggested by Example 5.

Definition 1.17 Let I_n be the index set $\{1, 2, \ldots, n\}$. Define:

(i) $\mathsf{X}_{i \in I_1} A_i = A_1$,
(ii) $\mathsf{X}_{i \in I_k} A_i = (\mathsf{X}_{i \in I_{k-1}} A_i) \times A_k \qquad (k > 1)$.

Examples

1. D.1.17 is a *recursive* definition. Another example of a recursive definition is the well known definition of *n factorial*:

$$0! = 1,$$
$$n! = (n-1)!\,n \qquad (n > 0).$$

2. $A_1 \times A_2 \times \cdots \times A_n$ is interpreted as $\mathsf{X}_{i \in I_n} A_i$. Under this interpretation $A_1 \times A_2 \times A_3 = ((A_1) \times A_2) \times A_3 = (A_1 \times A_2) \times A_3$, but $A_1 \times A_2 \times A_3$ cannot equal $A_1 \times (A_2 \times A_3)$ since D.1.17 does not assign a meaning to $A_2 \times A_3$.

3. Generalization of the ordered pair derives from D.1.17 and D.1.16. The general product set $\mathsf{X}_{i \in I_n} A_i$ is defined as the product set of two sets $\mathsf{X}_{i \in I_{n-1}} A_i$ and A_n, and D.1.16 defines the product set of two sets as a set of ordered pairs. Hence, provided $n > 1$, an element of $\mathsf{X}_{i \in I_n} A_i$ is an ordered pair. The first coordinate of this ordered pair is in turn an ordered pair (provided $n > 2$), and so on. An element of $\mathsf{X}_{i \in I_n} A_i$ then has the form $\langle \cdots \langle\langle a_1, a_2\rangle, a_3\rangle, \ldots, a_n\rangle$, which we shall write simply as $\langle a_1, a_2, \ldots, a_n\rangle$. The element $\langle a_1, a_2, \ldots, a_n\rangle$ is called an *ordered n-tuple*, Some n-tuples have special names: A 3-tuple is called a *triple*, a 4-tuple a *quadruple*, and so forth.

4. $\langle\langle a_1, a_2\rangle, a_3\rangle = \{\{\langle a_1, a_2\rangle\}, \{\langle a_1, a_2\rangle, a_3\}\} = \{\{\{\{a_1\}, \{a_1, a_2\}\}\}, \{\{\{a_1\}, \{a_1, a_2\}\}, a_3\}\}$. By interpretation $\langle\langle a_1, a_2\rangle, a_3\rangle \in A_1 \times A_2 \times A_3$, but $\langle a_1, \langle a_2, a_3\rangle\rangle \notin A_1 \times A_2 \times A_3$.

5. The Boolean algebra of Section 1e is the 6-tuple $\langle B, \oplus, *, ', 0, 1\rangle$. The 6-tuple $\langle \mathscr{P}(U), \cup, \cap, \bar{}, \varnothing, U\rangle$ is a model of the abstract algebra. Writing the systems as 6-tuples makes the interpretation unambiguous; here it is quite clear what the correspondence between the objects is.

6. We write A^n for the product set $A \times A \times \cdots \times A$ taken to n terms, e.g., $A \times A \times A = A^3$.

1h. Permutations and Combinations

In counting permutations and combinations we shall frequently refer to the *number* of elements in a set (the *size* of the set). Therefore, strictly for the sake of convenience, we first define a symbol designating the number of elements in a finite set. In what then follows we shall deal exclusively with *finite nonempty* sets.

DEFINITION 1.18 Let A be a finite set. Then $|A|$ denotes the number of elements in A.

Example

$A = \{1, 2, 2, 5\}$. Here $|A| = 3$.

DEFINITION 1.19 Let A be a set with $|A| = n$. If $\langle a_1, a_2, \ldots, a_m\rangle \in A^m$, then this ordered m-tuple is an *m-sample* of A. If all a_i in an m-sample are distinct, the m-sample is an *m-permutation* of A. In particular, an n-permutation is called simply a *permutation* of A. We shall denote the set of all m-samples of a set A by $S_m(A)$, the set of all m-permutations by $P_m(A)$, and the set of all permutations by $P(A)$.

Examples

1. $A = \{a, b, c\}$. Then $S_2(A) = \{\langle a, a\rangle, \langle a, b\rangle, \langle a, c\rangle, \langle b, a\rangle, \langle b, b\rangle, \langle b, c\rangle, \langle c, a\rangle, \langle c, b\rangle, \langle c, c\rangle\}$, $P_2(A) = \{\langle a, b\rangle, \langle a, c\rangle, \langle b, a\rangle, \langle b, c\rangle, \langle c, a\rangle, \langle c, b\rangle\}$, and $P(A) = \{\langle a, b, c\rangle, \langle a, c, b\rangle, \langle b, a, c\rangle, \langle b, c, a\rangle, \langle c, a, b\rangle, \langle c, b, a\rangle\}$. Note that $S_k(A) = A^k$. In keeping with normal practice we shall abbreviate our notation, writing $\{ab, ac, ba, bc, ca, cb\}$ for $P_2(A)$, $\{abc, acb, bac, bca, cab, cba\}$ for $P(A)$, etc.

2. For all k, $P_k(A) \subseteq S_k(A)$. Specifically, $P_1(A) = S_1(A)$, and $P_k(A) \subset S_k(A)$ for $k > 1$. If k exceeds the size of A, then $P_k(A) = \varnothing$. Samples whose size exceeds $|A|$ do not contradict the definition. Thus, for $A = \{a, b, c\}$,

baabba $\in S_6(A)$. An m-sample is sometimes called an m-permutation with repetition, the terminology being suggested by the subset relation in which permutations stand to samples.

Set A of D.1.19 can be given various interpretations. Consider, for example, the set $\{1, 2, \ldots, 6\}$, interpreted as the set of possible outcomes for the throw of a die. If one throws the die repeatedly and scores the outcomes, the score might be 5 after the first throw, 56 after the second, 561 after the third, and so on to 5612263123 after the tenth throw. The abbreviated notation of Example 1 is very suggestive: In this notation $S_{10}(A)$ is in fact the list of all possible scores for ten throws. Alternatively, set A might represent a deck of 52 cards from which one draws five cards, one after the other. Repetition is not possible; all cards in the hand are distinct. Here $P_5(A)$ lists all possible hands of five cards that may be drawn from the deck. But a straight flush is a straight flush, irrespective of the order in which the cards have been drawn. Therefore we might not want to distinguish between hands that differ only in the order of draw. Similarly we might not be interested in the order of throws of the die, and consider scores 5612263123 and 1122233566 equivalent. There is thus a need to define unordered counterparts of samples and permutations.

Since elements of a permutation are all distinct, the natural unordered counterpart of the ordered m-tuple defining an m-permutation would seem to be the set of its elements. Unfortunately we cannot define the unordered object corresponding to a sample quite this way: the set of elements in *abab*, *ab*, *aaab* is the same, namely $\{a, b\}$. The unordered samples should, however, be all different. One possibility is to take a set in which elements are paired with counts of their occurrences. Then *abab* and *aabb* both correspond to $\{\langle a, 2\rangle, \langle b, 2\rangle\}$, while *ab* corresponds to $\{\langle a, 1\rangle, \langle b, 1\rangle\}$, *aaab* to $\{\langle a, 3\rangle, \langle b, 1\rangle\}$, and *aabbbc* to $\{\langle a, 2\rangle, \langle b, 3\rangle, \langle c, 1\rangle\}$. We do in fact use this scheme in our definition.

DEFINITION 1.20 Let A be a set. Then an m-selection of A is the set $\{\langle a_1, k_1\rangle, \langle a_2, k_2\rangle, \ldots, \langle a_r, k_r\rangle\}$ such that all $a_i \in A$, all k_i are integers $(k_i > 0)$, and $\sum_{i=1}^{r} k_i = m$. In particular, if all $k_i = 1$, the m-selection is called an m-combination of A. The k_i in $\langle a_i, k_i\rangle$ is the *multiplicity* of a_i. We denote the set of all m-selections of a set A by $Q_m(A)$, and the set of all m-combinations by $C_m(A)$.

Examples

1. $A = \{a, b, c\}$. Then $Q_2(A) = \{\{\langle a, 2\rangle\}, \{\langle a, 1\rangle, \langle b, 1\rangle\}, \{\langle a, 1\rangle, \langle c, 1\rangle\}, \{\langle b, 2\rangle\}, \{\langle b, 1\rangle, \langle c, 1\rangle\}, \{\langle c, 2\rangle\}\}$, and $C_2(A) = \{\{\langle a, 1\rangle, \langle b, 1\rangle\}, \{\langle a, 1\rangle, \langle c, 1\rangle\}, \{\langle b, 1\rangle, \langle c, 1\rangle\}\}$. Again we shall abbreviate our notation,

writing $\{(ab), (ac), (bc)\}$ for $C_2(A)$, $\{(aaa), (aab), (aac), (abb), (abc), (acc), (bbb), (bbc), (bcc), (ccc)\}$ for $Q_3(A)$, and so forth.

2. In a combination all multiplicities are equal to 1. Explicit specification of the multiplicities is, therefore, superfluous, and a definition of combinations of A as subsets of A would seem more natural than the definition we have given. Then, however, combinations would differ in *type* from selections. Since $P_k(A) \subseteq S_k(A)$ we also want $C_k(A) \subseteq Q_k(A)$. With our definition $C_k(A) \subseteq Q_k(A)$ for all k. For example, with $A = \{a, b, c\}$, $C_2(A) \cap Q_2(A) = \{\{\langle a, 1\rangle, \langle b, 1\rangle\}, \{\langle a, 1\rangle, \langle c, 1\rangle\}, \{\langle b, 1\rangle, \langle c, 1\rangle\}\} \cap Q_2(A) = C_2(A)$. Hence, by Th.1.6, $C_2(A) \subseteq Q_2(A)$. By contrast, because of the difference in type, $\{\{a, b\}, \{a, c\}, \{b, c\}\} \cap Q_2(A) = \varnothing$. An m-selection is sometimes called an m-combination with repetition.

It may be essential, e.g., for determination of probabilities, to know the total *number* of different m-combinations of a given set (with or without repetition). As preliminaries we define binomial coefficients and find the number of elements in a subset of a product set.

DEFINITION 1.21 Let n and r be nonnegative integers. Define

$$P(n, r) = n(n-1) \cdots (n-r+1),$$
$$C(n, r) = P(n, r)/r!.$$

Numbers $C(n, r)$ are called *binomial coefficients*. (The symbol $\binom{n}{r}$ is often used to denote a binomial coefficient. The reflex-like action, which may automatically "complete" $\binom{n}{r}$ to $\left(\frac{n}{r}\right)$ makes it a dangerous symbol for the occasional user.)

Examples

1. Note that $0! = 1$, $P(n, n) = n!$, and $P(n, r) = 0$ if $n < r$. Hence $C(n, 0) = 1$ and $C(n, n) = 1$ $(n \geqq 0)$, and $C(n, r) = 0$ if $n < r$. If $n \geqq r$ we can write

$$C(n, r) = \frac{n!}{r!\,(n-r)!}$$

but not if $n < r$, since then $(n-r)!$ is undefined.

2. $C(n, r) = C(n, n-r)$.
$C(n, r) = C(n-1, r) + C(n-1, r-1), \qquad (n, r \geqq 1)$.

3. Computation of $n!$ is difficult for large n, on account of the size of the numbers involved. Thus $10!$ is already 3,628,800, and $50! \approx 3{\cdot}04 \times 10^{64}$. An approximation (*Stirling's formula*), $n! \approx \sqrt{2\pi n}\,(n/e)^n$, where e is the base

of the natural logarithms, is widely used in computations involving factorials. The relative error introduced by the approximation is about $1/(12n)$.

ALGORITHM 1.1 Recurrence relation

$$C(n, r) = C(n-1, r) + C(n-1, r-1)$$

is the basis of the following algorithm for the calculation of binomial coefficients for all n up to some maximum value n'.

1. Set $C(0, 0) = 1$. Set $n = 0$.
2. Set $n = n + 1$. If $n > n'$, stop.
3. Set $C(n, 0) = 1$. Set $r = 0$.
4. Set $r = r + 1$. If $r = n$, set $C(n, r) = 1$ and go to Step 2.
5. Set $C(n, r) = C(n-1, r) + C(n-1, r-1)$. Go to Step 4.

TABLE 1.1

BINOMIAL COEFFICIENTS $C(n, r)$

n \ r	0	1	2	3	4	5	...
0	1						
1	1	1					
2	1	2	1				
3	1	3	3	1			
4	1	4	6	4	1		
5	1	5	10	10	5	1	
⋮	⋮	⋮	⋮	⋮	⋮	⋮	⋮

Table 1.1 illustrates Algorithm 1.1. The algorithm generates a triangular array of numbers, known as *Pascal's triangle*. All entries in the array, except the 1s that border the array, are generated by the recurrence relation. With row $n-1$ generated, an element in row n is found by adding two successive elements of row $n-1$. The sum is inserted in row n immediately below the second (rightmost) of the two successive elements.

Our next theorem is difficult to understand. One should not worry about the meaning of this theorem until one has seen how the theorem is put to use in the proof of Th.1.17.

THEOREM 1.16 (*Rule of product*) Consider finite families of sets $\{A_i \mid i \in I\}$ and $\{M_i \mid i \in I\}$. With each M_i associate the number n_i ($n_i \leqq |A_i|$). Sets M_i are defined as follows:

(i) $i = 1$—$M_1 \subseteq A_1$, with $|M_1| = n_1$;

(ii) $i > 1$—$M_i \subseteq M_{i-1} \times A_i$, where M_i is constructed by pairing each element of M_{i-1} with exactly n_i elements of A_i (not necessarily the same n_i elements of A_i).

Then $|M_r| = \prod_{i=1}^{r} n_i$.

Proof. The proof is by induction on r.

(i) By definition, $|M_1| = n_1 = \prod_{i=1}^{1} n_i$.

(ii) Assume that the rule of product holds for $r - 1$ and prove it for r on basis of the assumption. The elements of $M_r (r > 1)$ are formed by pairing each element of M_{r-1} with exactly n_r elements of A_r, i.e., $|M_r| = |M_{r-1}| \times n_r$. By assumption, $|M_{r-1}| = \prod_{i=1}^{r-1} n_i$. Hence $M_r = \prod_{i=1}^{r} n_i$.

Example

If $n_i = |A_i|$ for all $i \leq r$, then $M_r = \mathsf{X}_{i \in I_r} A_i$ and $|M_r| = \prod_{i=1}^{r} |A_i|$.

THEOREM 1.17 Let A be a set with $|A| = n$. Then

(i) $|S_r(A)| = n^r$;
(ii) $|P_r(A)| = P(n, r)$;
(iii) $|C_r(A)| = C(n, r)$;
(iv) $|Q_r(A)| = C(n + r - 1, r)$.

Proof. (i) In Th.1.16 make

$$A_1 = A_2 = \cdots = A_r = A \quad \text{and} \quad n_1 = n_2 = \cdots = n_r = n.$$

Then $|M_r| = n^r$. But $M_r = A^r = S_r(A)$.

(ii) In Th.1.16 make $A_1 = A_2 = \cdots = A_r$ and $n_1 = n$. Since all elements of a permutation must be distinct, only $n - 1$ elements of A can be paired with an element of M_1 in the construction of M_2, only $n - 2$ elements of A remain to be paired with an element of M_2, and so on. Hence $n_2 = n - 1$, $n_3 = n - 2, \ldots, n_r = n - r + 1$. Thus $|P_r(A)| = |M_r| = P(n, r)$.

(iii) The number of r-combinations of A is equal to the number of subsets of A with r elements (see Example 2 of D.1.20). We shall find the number of subsets by investigating their r-permutations. Each subset gives rise to a different set of r-permutations, these sets are disjoint, and each member of $P_r(A)$ belongs to some set. Therefore the family of these sets is a partition of $P_r(A)$. By Part (ii), $P_r(A)$ has $P(n, r)$ elements, and the number of r-permutations of a subset of r elements, i.e., the number of elements in each set belonging to the partition, is $P(r, r)$. Hence the number of sets in the partition is $P(n, r)/P(r, r) = P(n, r)/r! = C(n, r)$. This is also the number of subsets of A with r elements, and hence the number of r-combinations of A.

(iv) Consider sets $S = \{1, 2, \ldots, n\}$ and $T = \{1, 2, \ldots, n + r - 1\}$. Clearly $|Q_r(S)| = |Q_r(A)|$. Also, by Part (iii), $|C_r(T)| = C(n + r - 1, r)$. Therefore, if $Q_r(S)$ and $C_r(T)$ can be shown to have the same number of elements, the theorem is proven. Every member of $Q_r(S)$ can be written in the form $(s_1 s_2 \cdots s_r)$, with $s_1 \leqq s_2 \leqq \cdots \leqq s_r$. Construct the r-combination $\{\langle s_1 + 0, 1\rangle, \langle s_2 + 1, 1\rangle, \ldots, \langle s_r + r - 1, 1\rangle\}$, which is a member of $C_r(T)$. It is now easy to see that $Q_r(S)$ and $C_r(T)$ have the same number of elements: Elements of $Q_r(S)$ run through the sequence $(1\cdots11)$, $(1\cdots12)$, $(1\cdots22)$, $\ldots$, $(n\cdots nn)$, with a different element of $C_r(T)$ corresponding to each element in the sequence, and every element of $C_r(T)$ corresponds to some element in the sequence.

Examples

1. The English alphabet has 26 letters. The number of 5-letter "words" that can be constructed from letters of this alphabet is $26^5 = 11{,}881{,}376$. If repetition of symbols is not permitted, the total reduces to $P(26, 5) = 7{,}893{,}600$. Of course, most of the "words," such as *caabb* or *abxzy*, will not be found in any dictionary.

2. A poker hand is a 5-combination of a deck of 52 cards. The number of different hands is $C(52, 5) = 2{,}598{,}960$. The number of different bridge hands is $C(52, 13) = 635{,}013{,}559{,}600$. In the language of probability theory, each hand is an *event*, and there is a probability associated with the event. By convention the sum of the probabilities for all possible events is 1. Here each event can be assumed to have equal probability of occurrence. Hence the probability of receiving a particular poker hand is $1/2{,}598{,}960$, and that of receiving a particular bridge hand $1{\cdot}57 \times 10^{-12}$.

3. The score for 10 throws of a die, with order disregarded, is a 10-selection of a set of six elements. Hence the total number of unordered scores is $C(15, 10) = 3003$. Here we cannot assume that all selections have an equal probability of occurrence; the events with equal probability are the ordered samples. There is only one sample that results in score 6666666666, but 10 samples result in score 5666666666.

We shall now consider some computational algorithms for generation of combinations and permutations. In some applications it is necessary to generate successive members of a sequence of combinations or permutations in some well-defined order. Normally this order is *lexicographic* order. It is therefore necessary to have at least an intuitive understanding of what is meant by lexicographic order. With m-combinations $(a_1 a_2 \cdots a_m)$ and m-permutations $a_1 a_2 \cdots a_m$ of the alphabet $\{a, b, \ldots, z\}$ lexicographic order is the normal alphabetic order. If the combinations (or permutations) are of a set of digits, a sequence of them is in lexicographic order if and only if, for any two members of the sequence, $(a_1 a_2 \ldots a_m)$ and $(b_1 b_2 \cdots b_m)$, or $a_1 a_2 \cdots a_m$ and $b_1 b_2 \cdots b_m$, $(a_1 a_2 \cdots a_m)$ precedes $(b_1 b_2 \cdots b_m)$ in the sequence, or $a_1a_2 \cdots a_m$ precedes $b_1b_2 \cdots b_m$, if and only if the number $a_1a_2 \cdots a_m$ is smaller than the number $b_1b_2 \cdots b_m$. A rigorous definition of lexicographic order is given in Section 2f.

ALGORITHM 1.2 Two algorithms for finding all m-samples of the set $\{0, 1\}$.

(a) The procedure represented by the flowchart of Figure 1.4 generates the next sample from a given sample $\langle d_1, d_2, \ldots, d_m \rangle$. The sequence of samples is in no special order. Therefore, the first entry to the procedure can be with any m-sample of $\{0, 1\}$. The first 2^m entries generate all 2^m samples. The sequence of samples then starts repeating itself.

(b) The procedure of Figure 1.4 is modified in subroutine NEXTS, which generates m-samples in lexicographic order. First entry must be with $\langle 1, 1, \ldots, 1 \rangle$ in array SAMPLE.

```
      SUBROUTINE NEXTS (SAMPLE, M)
      INTEGER SAMPLE(M)
      K = M
    1 IF (SAMPLE(K).EQ.0) GO TO 2
      SAMPLE(K) = 0
      K = K-1
      IF (K) 1,3,1
    2 SAMPLE(K) = 1
    3 RETURN
      END
```

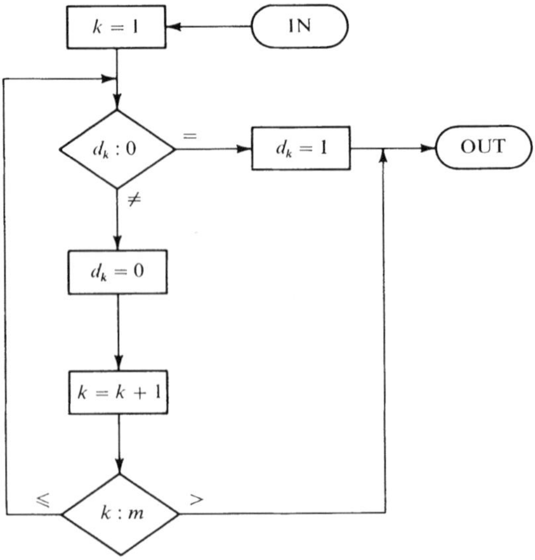

Figure 1.4

ALGORITHM 1.3 Subroutine NEWSAM finds all m-samples corresponding to a particular m-selection of $\{1, 2\}$, i.e., it finds all orderings of this selection. If an m-selection of $\{0, 1\}$ contains n_0 zeros, there are $C(m, n_0)$ such samples. Prior to the first call to NEWSAM array SAM must be set to $\langle 0, 0, \ldots, 0\rangle$. The m-selection is specified by NZ, the value of which determines the number of zeros in the selection. With M and NZ having the values 5 and 3, respectively, successive calls to NEWSAM generate samples 00011, 00101, 00110, 01001, 01010, 01100, 10001, 10010, 10100, 11000. The samples are in lexicographic order. The next call generates 00011 again, i.e., the sequence starts to repeat itself. A given sample can be represented as the pattern: $m - n - z - 1$ digits, zero, n ones, z zeros. The next sample is obtained by rearranging this pattern to $m - n - z - 1$ digits, one, $z + 1$ zeros, $n - 1$ ones. In the case of a 7-sample 1011100 the rearrangement gives 1100011. Exceptions arise in the first call (the given sample is $00\cdots0$) and after the generation of the final sample (when $11\cdots10\cdots0$ has to be changed to $0\cdots011\cdots1$).

```
      SUBROUTINE NEWSAM (SAM, M, NZ)
      INTEGER SAM(M)
      J = M
C COUNT ZEROS (JNIL = NUMBER OF ZEROS + 1)
    1 IF (SAM(J).EQ.1) GO TO 2
      J = J-1
C TEST FOR SAM = 0,0,...,0
      IF (J) 1,102,1
    2 JNIL = M-J+1
C COUNT ONES (JONE = NUMBER OF ONES - 1)
      JONE = 0
    3 J = J-1
C TEST FOR FINAL SAMPLE (1,1,...,1,0,...,0)
      IF (J.EQ.0) GO TO 100
      IF (SAM(J).EQ.0) GO TO 4
      JONE = JONE+1
      GO TO 3
C INSERT A ONE
    4 SAM(J) = 1
C INSERT ZEROS (AT LEAST ONE ZERO)
      DO 5 K = 1,JNIL
      J = J+1
    5 SAM(J) = 0
```

```
C INSERT ONES (PERHAPS NONE)
      IF (JONE.EQ.0) RETURN
      DO 6 K = 1,JONE
      J = J+1
    6 SAM(J) = 1
      RETURN
C INITIATE OR REINITIATE
  100 DO 101 K = 1,NZ
  101 SAM(K) = 0
  102 KA = NZ+1
      DO 103 K = KA,M
  103 SAM(K) = 1
      RETURN
      END
```

ALGORITHM 1.4 Subroutine NEWCOM generates m-combinations of n integers $\{1, 2, \ldots, n\}$ in array COM. Initially COM must be set to $\langle n, n, \ldots, n\rangle$. $C(n, m)$ successive calls generate the sequence of combinations in lexicographic order. Further calls cause the sequence to be repeated.

```
      SUBROUTINE NEWCOM (COM,M,N)
      INTEGER COM(M)
C LOCATE REGION TO BE CHANGED
      DO 1 K = 1,M
      KA = M-K+1
      KB = N-K+1
    1 IF (COM(KA).LT.KB) GO TO 3
C INITIATE -- GENERATE FIRST COMBINATION
      DO 2 K = 1,M
    2 COM(K) = K
      RETURN
C MAKE CHANGES
    3 KB = COM(KA)
      DO 4 K = KA,M
      KB = KB+1
    4 COM(K) = KB
      RETURN
      END
```

There are many algorithms for the generation of permutations. If $|A| = m$, then $|P(A)| = m!$. Hence $m!$ is the minimum number of interchanges of elements necessary for the generation of all permutations of the elements of A. Algorithm 1.5 generates all permutations in just this number of interchanges. It is based on the following principle. Given the set

$$A = \{a_1, a_2, \ldots, a_m\}.$$

For $n = 1, 2, \ldots, m - 1$, if P_n is the set of permutations of elements of $A_n = \{a_1, a_2, \ldots, a_n\}$, then P_{n+1} can be generated from P_n by inserting element a_{n+1} in every position of every permutation belonging to P_n. Finally $P_m = P(A)$. For example, with $A = \{a, b, c\}$, we let $A_1 = \{c\}$, $A_2 = \{b, c\}$, $A_3 = \{a, b, c\}$. Then $P_1 = \{c\}$ and $P_2 = \{bc, cb\}$. Elements of P_3 are generated from those of P_2 : From bc of P_2, by sending a forward, we obtain permutations abc, bac, bca; from cb, by sending a backward, we obtain cba, cab, acb. Given the initial permutation abc, $3! - 1$ successive calls to subroutine PERMER would in fact generate bac, bca, cba, cab, and acb, with the (3!)th call generating abc. The conceptual simplicity of the principle is obscured by programming detail. A program based on the flowchart of A.1.6 would be much simpler, but it would be rather inefficient if written in a higher-level language such as Fortran. An assembler language program for a computer that has an instruction for moving blocks of storage could, however, compare favorably with A.1.5. The storage movements would take more time than the interchanges of A.1.5, but the extreme simplicity of organization of A.1.6 might make total execution times comparable.

ALGORITHM 1.5 Subroutine PERMER generates the $m!$ permutations of m objects in sequence. The initial call is made with FIRST having the value .FALSE. and array PERM containing some permutation of M objects. FIRST is made .TRUE. and remains .TRUE. until the (M!)th call. Then PERM returns the original permutation and FIRST returns the value .FALSE.. Subsequent calls would cause the sequence to repeat itself. The value of M may not exceed 10.

```
      SUBROUTINE PERMER (PERM,M,FIRST)
      DIMENSION PERM(M), NP(10), ND(10)
      LOGICAL FIRST
      IF (FIRST) GO TO 2
C INITIATE REFERENCE ARRAYS FOR SUBSCRIPTING
      DO 1 K = 2,M
      NP(K) = 0
```

```
    1 ND(K) = 1
      FIRST = .TRUE.
    2 N = M
      K = 0
    3 NP(N) = NP(N) + ND(N)
      NQ = NP(N)
      IF (NQ.NE.N) GO TO 4
      ND(N) = -1
      GO TO 5
    4 IF (NQ.NE.0) GO TO 7
      ND(N) = 1
      K = K+1
    5 IF (N.LE.2) GO TO 6
      N = N-1
      GO TO 3
C ALL PERMUTATIONS HAVE BEEN FOUND
    6 NQ = 1
      FIRST = .FALSE.
C INTERCHANGE ELEMENTS
    7 NQ = NQ+K
      TEMP = PERM(NQ)
      PERM(NQ) = PERM(NQ+1)
      PERM(NQ+1) = TEMP
      RETURN
      END
```

(Note that locations `NP(1)` and `ND(1)` are never in use.)

ALGORITHM 1.6 Figure 1.5 represents a procedure for generating all permutations of the set of numbers $\{1, 2, \ldots, m\}$. A new permutation results when all or part of a given permutation is rotated. The meaning of, *rotate the n lowest numbers*, is best explained by means of an example. Suppose we have the permutation 12345. Rotation of the three lowest numbers produces 23145. Schematically, one first transforms

| | 1 | 2 | 3 | 4 | 5 | to | 1 | 2 | 3 | | 4 | 5 |

and then to

| | 2 | 3 | 1 | 4 | 5 |.

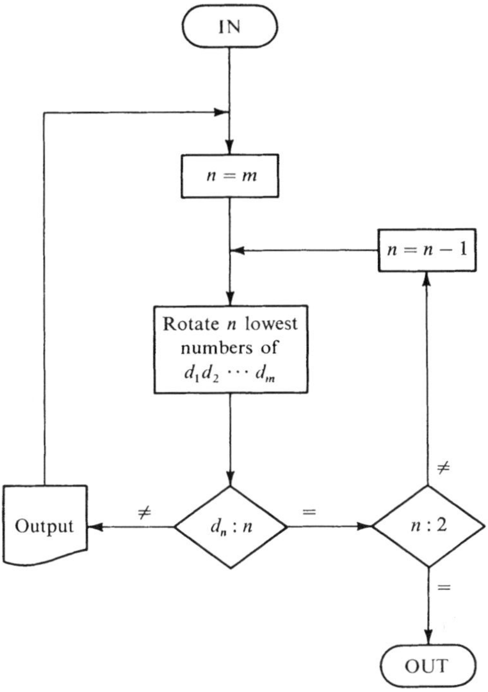

Figure 1.5

Notes

The best way of finding a text to one's liking is to go into a library and browse. Books are like acquaintances: some become friends, others don't. Since the author's style, a subjective manifestation judged according to one's own subjective likes and dislikes, determines to a great extent the regard one has for a book, it is somewhat presumptuous to make recommendations. Nevertheless, I cannot let this opportunity pass without acknowledging my high regard for [St61], which I found a most enjoyable intermediate text covering approximately the same ground as our Chapters 1 and 2. The larger [St63] does not expand on the material covered by [St61]; the increase in size is due to the greater number of topics dealt with. In particular, it derives the basic properties of arithmetic in the sets of natural numbers and real numbers. [Ha60] is another good intermediate text; however, it may be too concise for collateral reading. A huge collection of worked examples and exercises can be found in [Li64]. This is a beginner's book, but Chapters 9 and 11–13 deal with material, which, while of little relevance to practical computing, is an important part of set theory when it is studied for its own sake. [Ko66] is specifically addressed to students of computer science; the first four chapters correspond to our Chapters 1 and 2. The exercises are easy without being trivial, but the bibliography refers to some rather advanced texts. For really solid axiomatic developments of set theory one can consult [Fr66, Su60, Fr58]. Boolean algebras are studied in a highly theoretical manner in [Si64].

Set theory is a well-established discipline, and nearly all the current research is of a very advanced nature. The study of fuzzy sets is an exception. Normally an object either is a member of a set or is not a member. There are, however, sets in which the question of membership cannot be precisely established, i.e., the range of the characteristic function associated with such a set (see Example 4 of D.2.2) is the interval $[0, 1]$ rather than $\{0, 1\}$, and such sets are called fuzzy. Fuzzy sets and extensions of the concept of fuzziness are studied in [Za65a, Za65b, Be66, Go67a, Za68].

For general information on combinatorial mathematics one can consult [Ri58, Ry63, Ha67a]. For properties of binomial coefficients see [Ab64] and [Kn68]. The Algorithms section of CACM contains many algorithms written in Algol for generating permutations and combinations, and most of our algorithms derive from this source: A.1.3 from [Ho63a], A.1.4 from [Ku62], A.1.5 from [Tr62]. (If one intends to use an ACM algorithm, it is advisable to consult a recent subject index to the Algorithms section—generally published in the December issue of CACM—for references to remarks on the algorithm contributed after its publication.) The rotation algorithm A.1.6 derives from [La67a].

Exercises

The letter in parentheses identifies the section to which the exercise has greatest relevance.

1.1 (a) In which instances are the sets A and B equal?

(i) $A = \{1, 2, 3\}$, $B = \{i \mid i \text{ is an integer}\}$.

(ii) $A = \{1, 2, 3, 4, 5\}$, $B = \{1, 10, 11, 100, 101\}$.

(iii) $A = \{\alpha, \beta, \gamma\}$, $B = \{a, b, c\}$.

(iv) $A = \{\alpha\}$, $B = \{x \mid x \text{ is the first letter of the Greek alphabet}\}$.

(v) $A = \{a, b, ab, ba, b, a\}$, $B = \{a, b, ab\}$.

1.2 (a) Prove Th.1.3 by induction. (Consider a statement $P(n)$ about some nonnegative integer n. The *principle of induction* states: If $P(k)$ is true and $P(m)$ implies $P(m+1)$, then $P(n)$ is true for all n such that $k \leqq n \leqq m + 1$. Proving $P(k)$ provides a basis for the proof; normally one takes $k = 0$ or $k = 1$. Showing that $P(m)$ implies $P(m + 1)$ is the induction step.)

1.3 (a) Find $A \cup B$ and $A \cap B$ in the following cases:

(i) $A = \{1, 2, 3\}$, $B = \{i \mid i \text{ is an integer}\}$;
(ii) $A = \{1, 2, 3\}$, $B = \{4, 5, 6, 7, \ldots\}$;
(iii) $A = \{a, b, ab, ba, b, a\}$, $B = \{a, b, ab\}$;
(iv) $A = \{\varnothing\}$, $B = \varnothing$;
(v) Any A, $B = \{x \mid x \notin A\}$.

1.4 (a) Show that if $A \subseteq B$ and $C \subseteq D$, then $A \cap C \subseteq B \cap D$.

1.5 (a) Find $\mathscr{P}(A)$ and $\mathscr{P}(B)$, where $A = \{\varnothing\}$ and $B = \{a, b, c, d\}$. How many members does the power set of $C = \{a, b, c, \ldots, z\}$ have?

1.6 (a) Let M be a set of J integers and N a set of K integers ($J, K \leqq 1{,}000$). Assume that elements of M and N are stored in ascending order in locations `M(1), M(2), ..., M(J)`, and `N(1), N(2), ..., N(K)`, respectively. Write a subroutine for finding $M \cup N$ and $M \cap N$.

1.7 (b) For $\{A_i \mid i \in I\}$, where $I = \{1, 2, \ldots, n\}$, show $B \cap (\bigcup_i A_i) = \bigcup_i (B \cap A_i)$ and $B \cup (\bigcap_i A_i) = \bigcap_i (B \cup A_i)$.

1.8 (b) Let $\mathscr{X} = \{X_1, X_2, \ldots, X_m\}$ and $\mathscr{Y} = \{Y_1, Y_2, \ldots, Y_n\}$ be partitions of a set A. Show that $\mathscr{Z} = \{X_i \cap Y_j \mid X_i \in \mathscr{X},\ Y_j \in \mathscr{Y},\ X_i \cap Y_j \neq \varnothing\}$ is a partition of A (known as the *cross-partition* of $\mathscr{X}$ and $\mathscr{Y}$).

1.9 (b) Let U be the set of students at a university and define the following subsets of U: A—males, B—females, C—undergraduates, D—graduates. $\mathscr{A} = \{A, B\}$ and $\mathscr{B} = \{C, D\}$ are partitions of U. Use formular notation to specify the elements of the sets comprising the cross-partition of $\mathscr{A}$ and $\mathscr{B}$.

1.10 (c) For $\{A_i \mid i \in I\}$, where $I = \{1, 2, \ldots, n\}$, show $\overline{\bigcap_i A_i} = \bigcup_i \bar{A}_i$ and $\overline{\bigcup_i A_i} = \bigcap_i \bar{A}_i$.

1.11 (c) For $\{X_i \mid i \in I\}$, where $I = \{1, 2, \ldots, n\}$, show $U - (\bigcup_i X_i) = \bigcap_i (U - X_i)$.

1.12 (c) Is every family $\{A, \bar{A}\}$ a partition of the universal set?

1.13 (c) Prove: (i) $A + B = (A \cup B) \cap (\bar{A} \cup \bar{B})$,
(ii) $A + B = (A \cup B) - (A \cap B)$.

1.14 (c) Prove: (i) $A + A = \varnothing$, (ii) $A + \varnothing = A$,
(iii) $A + B = \bar{A} + \bar{B}$, (iv) $(A + B) + B = A$.

1.15 (d) Show that $X \cap A = Y \cap A$ and $X \cap \bar{A} = Y \cap \bar{A}$ imply $X = Y$.

1.16 (d) Prove again that $A \subseteq B$ and $C \subseteq D$ imply $A \cap C \subseteq B \cap D$.

1.17 (d) Prove: (i) $A \cap (B + C) = (A \cap B) + (A \cap C)$,
(ii) $A = B$ if and only if $A + B = \varnothing$.

1.18 (d) Modify your subroutine of Exercise 1.6 so that it finds $M - N$ as well. If, in addition, $M + N$ is required, is there a need to design a special routine or to modify the present subroutine for this purpose?

1.19 (e) Give a shorter proof of Th.1.8.

1.20 (e) Prove: (i) Th.1.9, (ii) Part 3 of Th.1.10, (iii) Part 2 of Th.1.12.

1.21 (e) In Boolean algebra B show that $a \oplus (b \oplus c) = (a \oplus b) \oplus c$ and $a * (b * c) = (a * b) * c$ can be deduced from the other axioms. (Hint: Prove first that $x * z = y * z$ and $x * z' = y * z'$ imply $x = y$.)

1.22 (f) Is it true that $A \cup (B + C) = (A \cup B) + (A \cup C)$?

1.23 (f) Use Venn diagrams to find the relation between

$$(W \cup X) + (Y \cup Z) \quad \text{and} \quad (W + Y) \cup (X + Z).$$

Then prove the relation (the proof is tedious and lengthy).

1.24 (g) Find the sets equal to $\langle a, \langle a, a \rangle \rangle$ and $\langle \langle a, a \rangle, a \rangle$.

1.25 (g) It may appear that the definition $\langle a_1, a_2 \rangle = \{\{a_1\}, \{a_1, a_2\}\}$ could be generalized to $\langle a_1, a_2, a_3 \rangle = \{\{a_1\}, \{a_1, a_2\}, \{a_1, a_2, a_3\}\}$. Why is this definition of an ordered triple inadequate?

1.26 (g) Referring to Example 5 of D.1.17, show that $\langle \{\emptyset, U\}, \cup, \cap, \bar{\ }, \emptyset, U \rangle$ is a Boolean algebra.

1.27 (g) If $\langle \mathscr{P}(U), \cup, \cap, \bar{\ }, \emptyset, U \rangle$ is a Boolean algebra, is $\langle \mathscr{P}(U), \cap, \cup, \bar{\ }, U, \emptyset \rangle$ also a Boolean algebra?

1.28 (h) In Fortran the name of a variable may be 1–6 characters long, where the first character must be a letter from the set {A, B, ..., Z}, and the other characters (if any) may be taken from the set of letters or from the set of digits $\{0, 1, \ldots, 9\}$. How many different names are possible?

1.29 (h) Show that $\sum_{k=0}^{h} C(n, k) = 2^n$. (Hint: Use Th.1.3.)

1.30 (h) Given a set of integers $\{a_1, a_2, \ldots, a_n\}$. Write a subroutine that generates m-samples of this set of integers (cf. A.1.2).

1.31 (h) Modify A.1.4 so that it finds an m-combination of n arbitrary distinct integers $\{a_1, a_2, \ldots, a_n\}$.

1.32 (h) Subroutine `NEWSAM` (A.1.3) is not completely "safe." What happens when `NZ` = 0 and when `M` = `NZ`? Make appropriate improvements in the subroutine.

1.33 (h) In the paragraph preceding A.1.5 we state that the algorithm is based on an insertion principle. Convince yourself of the essential correctness of this observation by a study of the pattern of interchanges giving the next permutation in the printout of all permutations of $\{1, 2, 3, 4, 5\}$.

CHAPTER 2

Functions and Relations

2a. Functions

Throughout Chapter 1 we were relating objects: The proof of Th.1.3 depends on a correspondence between sequences of elements of a set and sequences of binary digits; D.1.20 pairs elements of a general set with elements of the set of natural numbers. Intuitive understanding of the correspondences was adequate, but only barely so when the correspondences became complicated, as in the proof of Part (iv) of Th.1.17. We shall now make our intuitive notions precise.

DEFINITION 2.1 The set f is a *function* from set A into set B if and only if it is a subset of a set of ordered pairs $A \times B$, and $\langle a, b\rangle \in f$ and $\langle a, c\rangle \in f$ imply $b = c$. (There are many synonyms for *function*, such as *transformation*, or *operation*, or *mapping* from set A into set B.) If $\langle a, b\rangle \in f$, then b is an *image* of the *argument* a under f, denoted $f(a)$. We also call $f(a)$ the *value* of f for a. When the function is from a set of n-tuples we shall abbreviate $f(\langle a_1, a_2, \ldots, a_n\rangle)$ to $f(a_1, a_2, \ldots, a_n)$.

Examples

1. The set $f = \{\langle$*beer*, 4$\rangle$, $\langle$*is*, 2$\rangle$, $\langle$*food*, 4$\rangle$, $\langle$*for*, 3$\rangle$, $\langle$*some*, 4$\rangle\}$ is a function; the set $\{\langle$4, *five*$\rangle$, $\langle$4, *aces*$\rangle$, $\langle$5, *spell*$\rangle$, $\langle$7, *tragedy*$\rangle\}$ is not. The first coordinates must be distinct. Second coordinates need not be distinct.

We have the images $f(is) = 2, f(for) = 3, f(beer) = f(food) = f(some) = 4$. The reason why first coordinates must be distinct is that normally we identify a function with a computation process: Given an argument, we compute its image. The process has to be deterministic, which it would not be if an argument could have more than one image.

2. Consider sets $f = \{\langle x, x^2\rangle \mid x \in R\}$ and $g = \{\langle x^2, x\rangle \mid x \in R\}$, with R the set of real numbers. Although f is a function, g is not a function (since, for example, both $\langle 1, -1\rangle \in g$ and $\langle 1, 1\rangle \in g$). Here $f(1) = 1, f(1.1) = 1.21$, and, in general, $f(x) = x^2$.

3. Selections and combinations are functions. An m-selection of A is a mapping from A into the set of natural numbers, and an m-combination is a mapping from A into the set $\{1\}$.

4. The binary set operations of union and intersection are mappings from $\mathscr{P}(U) \times \mathscr{P}(U)$ into $\mathscr{P}(U)$, and the unary operation of complementation is a mapping from $\mathscr{P}(U)$ into itself. Consider sets A, B, C. If $A \cup B = C$, the ordered triple $\langle\langle A, B\rangle, C\rangle$ (or $\langle A, B, C\rangle$) is a member of the operation of set union. Unary operations are sets of ordered pairs; binary operations are sets of ordered triples.

5. 2-combinations $\{\langle a, 1\rangle, \langle b, 1\rangle\}$ and $\{\langle b, 1\rangle, \langle a, 1\rangle\}$ are, of course, equal. Since functions are sets, equality of functions is defined by D.1.1.

DEFINITION 2.2 Let $f \subseteq A \times B$ be a function. The *domain* D_f and *range* R_f of the function are defined as follows:

$$D_f = \{a \mid \text{for some } b, \quad \langle a, b\rangle \in f\},$$
$$R_f = \{b \mid \text{for some } a, \quad \langle a, b\rangle \in f\}.$$

If $D_f = A$, function f is said to be *on* the set A or a *total* function with respect to this set. If $R_f = B$, the function is *onto* the set B. A total function from A into B is denoted by the symbol $f: A \to B$, or by the symbol $A \xrightarrow{f} B$.

Examples

1. The domain of $\{\langle beer, 4\rangle, \langle is, 2\rangle, \langle food, 4\rangle, \langle for, 3\rangle, \langle some, 4\rangle\}$ is the set $A = \{beer, is, food, for, some\}$. The range is $\{2, 3, 4\}$. The function is on A, but not on the set of all words in the English language, i.e., it is total with respect to A, but not with respect to the set of all words. The function is both *into* and *onto* set $\{2, 3, 4\}$, but only *into* the set of natural numbers.

2. Let function f be an m-selection of set A, with $m < |A|$. Then $D_f \subset A$ and $R_f \subset \{1, 2, \ldots, m\}$. In general, with $f \subseteq A \times B$, $D_f \subseteq A$ and $R_f \subseteq B$. The notation $f: A \to B$ is used only if $D_f = A$, i.e., if f is a total function with respect to A.

3. Consider $f: A \to B$, with $|A| = m$ and $|B| = n$. Since f is on A, the function has m elements, and no other function from any subset of A can

have a greater number of elements. Hence at most $1/n$ of the mn elements of $A \times B$ can belong to a function from A.

4. The function $f: U \to \{0, 1\}$, where U is a universal set, is known as the *characteristic function* of a set A if $f(a) = 0$ when $a \in A$ and $f(a) = 1$ when $a \notin A$.

DEFINITION 2.3 If $f: A \to B$ and $C \subseteq A$, then the function $f \cap (C \times B)$ is the *restriction* of f to C, written $f \mid C$. A function f is an *extension* of a function g if and only if $g \subseteq f$. The *composite* of functions g and h, symbolized $h \circ g$, is the set $\{\langle x, z\rangle \mid$ there exists a y such that $\langle x, y\rangle \in g$ and $\langle y, z\rangle \in h\}$.

Examples

1. Consider the English alphabet $A = \{a, b, \ldots, z\}$. Let W be the set of all five-letter "words" constructed out of the letters of A (in the sense of Example 1 of Th.1.17), W' the set of five-letter English words, and $S_5(A)$ the set of five-samples of A. Then $f: W \to S_5(A)$ with $f(a_1a_2a_3a_4a_5) = \langle a_1, a_2, a_3, a_4, a_5\rangle$ and $g: W' \to S_5(A)$ with $g(a_1a_2a_3a_4a_5) = \langle a_1, a_2, a_3, a_4, a_5\rangle$ are functions, and $g = f \mid W'$ is the restriction of f to W'.

2. Let N be the set of nonnegative integers $\{0, 1, 2, \ldots\}$. Subtraction as a function into N is defined on the set $\{\langle a, b\rangle \mid a, b \in N$ and $a \geqq b\}$. The function is $S = \{\langle a, b, a - b\rangle \mid a, b \in N$ and $a \geqq b\}$. We define proper subtraction $P: N \times N \to N$ as $\{\langle a, b, a \dot{-} b\rangle \mid a, b \in N\}$. Then

$$\begin{aligned} S(a, b) &= a - b \qquad (a \geqq b); \\ P(a, b) &= a - b \qquad (a \geqq b), \\ &= 0 \qquad\quad\;\; (a < b). \end{aligned}$$

Since $P \supseteq S$, proper subtraction is an extension of subtraction. Whereas P is a total function, S is not total (with respect to $N \times N$).

3. Let R be the set of real numbers, I the set of integers, and N the set of natural numbers $\{1, 2, 3, \ldots\}$. Consider the function $f: N \to N$ with $f(n) = (n - 1)!$. A well-known extension of this factorial function is the gamma function $\Gamma: R - (I - N) \to R$ with $\Gamma(n) = (n - 1)!$ for $n \in N$ (in fact Γ can be defined on $C - (I - N)$, where C is the set of complex numbers). The set $R - (I - N)$ is the set of all real numbers with the set $\{0, -1, -2, \ldots\}$ excluded.

4. The composite $h \circ g$ of functions g and h is also a function. If $D_h \cap R_g \neq D_h$, then $h \circ g$ is a restriction of h to $D_h \cap R_g$. If $g: X \to Y$ and $h: Y \to Z$, then $h \circ g: X \to Z$.

DEFINITION 2.4 A total function $f: A \to B$ is *one to one* if it maps distinct elements of A onto distinct elements of B; i.e., the function is one to one if and only if $f(a_1) = f(a_2)$ implies $a_1 = a_2$ [alternatively, the function is one to one if and only if $a_1 \neq a_2$ implies $f(a_1) \neq f(a_2)$]. A total function that is not one to one is a *many-to-one* function.

Examples

1. Function f of Example 1 of D.2.3 is one to one. Consider $Q_5(A)$, the set of 5-selections of $A = \{a, b, \ldots, z\}$. The function h: $W' \to Q_5(A)$, with W' the set of five-letter English words and $h(a_1a_2a_3a_4a_5) = (a_1a_2a_3a_4a_5)$, is many to one, since $h(heaps) = h(phase) = h(shape) = (aehps)$.

2. Let I be the set of integers and N the set of natural numbers. Define $f(n) = n^2$. Then f: $I \to N$ is many to one, but the restriction $f \mid N$ is one to one.

3. Consider the alphabet $A = \{a, b, c, \ldots, z\}$ and the set of odd numbers $B = \{3, 5, 7, \ldots, 53\}$. Let f: $A \to B$ be a one-to-one function that assigns the number $2n + 1$ to the nth letter of the alphabet, e.g., $\langle a, 3\rangle \in f$, $\langle g, 15\rangle \in f$. Let W be the set of all "words" constructed out of the letters of A and let $p_1, p_2, p_3, \ldots$ be the primes in ascending order. Products of powers of the primes belong to N, the set of natural numbers. A number-theoretical argument shows that the function g: $W \to N$ with $g(a_1a_2 \cdots a_k) = p_1^{n_1}p_2^{n_2} \cdots p_k^{n_k}$, where $n_i = f(a_i)$ is one to one. Some examples: $g(aabca) = 2^3 \cdot 3^3 \cdot 5^5 \cdot 7^7 \cdot 11^3 =$ 739,891,619,775,000, $g(sapphire) = 2^{39} \cdot 3^3 \cdot 5^{33} \cdot 7^{33} \cdot 11^{17} \cdot 13^{19} \cdot 17^{37} \cdot 19^{11}$. This technique of mapping sequences of symbols into the set of natural numbers is known as Gödel numbering. It is very important in the study of the foundations of mathematics, but the size of the numbers makes the technique useless for representation of symbols in a computer.

THEOREM 2.1 Let f: $A \to B$ be a function and consider the *onto* function $g: f \to f'$ with $g(a, b) = \langle b, a\rangle$. The set f' is a function if and only if f is one to one. Moreover, f' is on B if and only if f is onto B.

Proof. Exercise 2.3.

DEFINITION 2.5 If f: $A \to B$ is a one-to-one function, then the *inverse function* of f, symbolized f^{-1}, is the range of the function $g: f \to f'$ with $g(a, b) = \langle b, a\rangle$. (If f is not one to one, f^{-1} does not exist.)

Examples

1. Consider functions f: $W \to S_5(A)$ and g: $W' \to S_5(A)$ of Example 1 of D.2.3. Both functions are one to one. Hence f^{-1} and g^{-1} exist, and f^{-1} is the function f^{-1}: $S_5(A) \to W$ with $f^{-1}(a_1, a_2, a_3, a_4, a_5) = a_1a_2a_3a_4a_5$. The function g^{-1} from $S_5(A)$ onto W' is not total. The abbreviation for a sample, which we introduced in Example 1 of D.1.19, is actually the image of the sample under f^{-1}.

2. Consider function f: $I \to N$ of Example 2 of D.2.4. Here f^{-1} does not exist. The restriction $f \mid N$ has an inverse, but the inverse is not total with respect to N (the range of $f \mid N$ consists of squared numbers alone).

THEOREM 2.2 Consider sets A and B with $|A| = a$ and $|B| = b$. The number of functions on A into B is b^a.

Proof. Exercise 2.4.

2b. Boolean Functions and Forms

In this section we shall deal in general terms with functions of a special kind, the Boolean functions. The importance of some models of Boolean algebras is more a consequence of the part played in them by Boolean functions than of anything else. These interesting models will be discussed in Section 2c.

Although our definitions will be quite general, examples will be based on the simple system $\langle\{0, 1\}, \oplus, *, ', 0, 1\rangle$. As a preliminary we will have to show that this system is a Boolean algebra.

THEOREM 2.3 The system $\langle\{0, 1\}, \oplus, *, ', 0, 1\rangle$ is a Boolean algebra.

Proof. The given system is precisely the Boolean algebra $\langle B, \oplus, *, ', 0, 1\rangle$ of Section 1e, with set B having just two elements, the distinct elements 0 and 1. To prove that the system is a Boolean algebra it suffices to show that $\{0, 1\}$ is closed under $\oplus$, $*$, and $'$. By Axioms 4A and 4B, Parts 1 and 2 of Th.1.10, Parts 1 and 2 of Th.1.11, and Part 2 of Th.1.12 (all of Section 1e) we have the following definitions of operations $\oplus$, $*$, $'$ for members of $\{0, 1\}$:

$$0 \oplus 0 = 0, \qquad 0 \oplus 1 = 1 \oplus 0 = 1 \oplus 1 = 1;$$
$$0 * 0 = 0 * 1 = 1 * 0 = 0, \qquad 1 * 1 = 1;$$
$$0' = 1, \qquad 1' = 0.$$

Operations are functions ($\oplus$: $\{0, 1\}^2 \to \{0, 1\}$ is the function $\{\langle 0, 0, 0\rangle, \langle 0, 1, 1\rangle, \langle 1, 0, 1\rangle, \langle 1, 1, 1\rangle\}$), and the three functions are all into (and onto) the set $\{0, 1\}$. The set is therefore closed under the operations.

DEFINITION 2.6 Consider the Boolean algebra $\langle B, \oplus, *, ', 0, 1\rangle$. A function $f(x_1, x_2, \ldots, x_n)$ on B^n into B, f: $B^n \to B$, is a *Boolean function* of n variables.

Example

Let $B = \{0, 1\}$. There are 2 elements in B and 2^n elements in B^n (Th.1.16). By Th.2.2, the number of functions on B^n into B is, therefore, 2^{2^n}. In the case of $n = 1$ the four functions are $\{\langle 0, 0\rangle, \langle 1, 0\rangle\}$, $\{\langle 0, 0\rangle, \langle 1, 1\rangle\}$, $\{\langle 0, 1\rangle, \langle 1, 0\rangle\}$, $\{\langle 0, 1\rangle, \langle 1, 1\rangle\}$.

DEFINITION 2.7 Let $\langle B, \oplus, *, ', 0, 1\rangle$ be a Boolean algebra and let f, g, h denote functions on B^n into B, i.e., Boolean functions of n variables. We define the following *functional* equalities:

(i) $f \oplus g = h$ if and only if, for *every* element of the set of samples $S_n(B)$, the result of combining the images under f and g in accordance with the interpretation of $\oplus$ in B is equal to the image of the element under h;

(ii) $f * g = h$ is defined analogously;

(iii) $f' = h$ if and only if, for every element of $S_n(B)$, the complement (image under ′) of the image of this element under f is equal to the image of the element under h.

Example

Let $B = \{0, 1\}$ and consider functions from B^2 into B. For this system operations $\oplus$, $*$, and ′ are defined in the proof of Th.2.3. Take functions f and g such that $f(0,0) = 1, f(0, 1) = f(1,0) = f(1, 1) = 0; g(0, 0) = g(0, 1) = g(1, 0) = 1$, $g(1, 1) = 0$. Take also functions $s(x_1, x_2) = x_1 \oplus x_2$ and $p(x_1, x_2) = x_1 * x_2$. Table 2.1 represents f, g, s, p, $f \oplus g$, $f * g$, f', and g' in tabular form, one row for each member of $S_2(B)$. Since $|S_2(B)| = 4$, we have four rows. Consider the first row. The entries for f and g are given by the definitions above. The other entries are generated using definitions of $\oplus$, $*$, and ′ for the set $\{0, 1\}$ given in the proof of Th.2.3. From the table it is seen that $f \oplus g = g$, $f * g = f$, $f' = s$, $g' = p$.

TABLE 2.1

FUNCTIONAL EQUALITIES

$\langle x_1, x_2\rangle$	f	g	s	p	$f \oplus g$	$f * g$	f'	g'
$\langle 0, 0\rangle$	1	1	0	0	1	1	0	0
$\langle 0, 1\rangle$	0	1	1	0	1	0	1	0
$\langle 1, 0\rangle$	0	1	1	0	1	0	1	0
$\langle 1, 1\rangle$	0	0	1	1	0	0	1	1

In D.2.7 we have defined functional equality. We now have to show that $f \oplus g$, $f * g$, and f' as defined in D.2.7 are in fact functions. For example, we have to show that if $f \oplus g = h_1$ and $f \oplus g = h_2$, then $h_1 = h_2$.

THEOREM 2.4 Let $\langle B, \oplus, *, ', 0, 1\rangle$ be a Boolean algebra. Let f and g denote Boolean functions on B^n into B. Then

(i) $f \oplus g$,
(ii) $f * g$,
(iii) f'

are also functions.

Proof. (i) Consider an n-tuple $a \in S_n(B)$, and assume that $(f \oplus g)(a) = c_1$ and $(f \oplus g)(a) = c_2$. By D.2.7, $(f \oplus g)(a) = f(a) \oplus g(a)$. Hence $f(a) \oplus g(a) = c_1$ and $f(a) \oplus g(a) = c_2$. But, by definition of a function, $f(a)$ and $g(a)$ are unique elements of B. Also, by definition of the Boolean algebra, $\oplus$ is an operation in B (i.e., a function from B^2). Hence $c_1 = c_2$, implying that $f \oplus g$ is a function.

(ii) The proof is similar.
(iii) By Th.1.9 (uniqueness of the complement).

Our next theorem asserts that the set of all functions on B^n into B is a Boolean algebra as well.

THEOREM 2.5 Let $\langle B, \oplus, *, ', 0, 1 \rangle$ be a Boolean algebra. Then $\langle F_n, \oplus, *, '$ $f_0, f_1 \rangle$, where F_n is the set of all functions on B^n into B and the operations are defined by D.2.7, is a Boolean algebra.

Proof. We shall not prove the theorem because the proof is rather lengthy. It consists of showing that F_n is closed under the operations and that its elements satisfy the axioms of Section 1e.

Example

Let $B = \{0, 1\}$ and consider F_2. We have $|F_2| = 16$, with each member of F_2 consisting of four triples. We take f_0 and f_1 as the functions with ranges $\{0\}$ and $\{1\}$, respectively, e.g., $f_0 = \{\langle 0, 0, 0 \rangle, \langle 0, 1, 0 \rangle, \langle 1, 0, 0 \rangle, \langle 1, 1, 0 \rangle\}$. Here it is easy to convince oneself that F_2 is closed under the operations and that its elements satisfy the axioms.

Consider the expression $(x_1 \oplus x_2)' \oplus (x_1' * x_2)$, where x_1 and x_2 are variables denoting elements of B. Although we tend to call such an expression a function, it is certainly not a set of ordered pairs and hence not a function. The expression is a *formula* or *form*. The reason why we sometimes improperly describe forms as functions is that a form of n variables has a function on B^n into B *associated* with it. We shall investigate this association further on.

DEFINITION 2.8 Let $\langle B, \oplus, *, ', 0, 1 \rangle$ be a Boolean algebra, and let variables $x_1, x_2, \ldots, x_n$ denote elements of B. A *Boolean form* (*Boolean formula*) is defined recursively as follows:

1. 0 and 1 are Boolean forms.
2. A variable x_i $(i = 1, 2, \ldots, n)$ is a Boolean form.
3. If α is a Boolean form, then so is (α).
4. If α is a Boolean form, then so is α'.
5. If α and β are Boolean forms, then so is $\alpha \oplus \beta$.
6. If α and β are Boolean forms, then so is $\alpha * \beta$.
7. Only expressions given by Statements 1–6 are Boolean forms.

Example

Consider the expression $(x_1 \oplus x_2)' \oplus (x_1' * x_2)$. Let us determine whether or not it is a Boolean form. This is a recognition problem. We find successively that x_1, x_2, $x_1 \oplus x_2$, $(x_1 \oplus x_2)$, $(x_1 \oplus x_2)'$, x_1', $x_1' * x_2$, $(x_1' * x_2)$ are Boolean forms. Finally, by Statement 5, we recognize the entire expression as a Boolean form. It is easy to see that $(x_1 \oplus x_2)' \oplus * (x_1{}' * x_2)$, say, is not a Boolean form. In the model $\langle \mathscr{P}(U), \cup, \cap, \bar{\ }, \varnothing, U \rangle$, with variables A and B, expression $\overline{A \cup B \cup (\bar{A} \cap B)}$ is a Boolean form, but not $\overline{A \cup B \cup \cap (\bar{A} \cap B)}$.

D.2.8 is not a workable algorithm for recognizing Boolean forms. Although every reader should be able to recognize a Boolean form, most readers will be hard put to give a precise formal description of the recognition procedure they used. A detailed algorithm is, however, necessary if recognition is to be by computer. The algorithm will be an exercise in Section 10b. For the time being we shall assume a recognition algorithm and investigate equivalence of Boolean forms.

Let $\langle B, \oplus, *, ', 0, 1 \rangle$ be a Boolean algebra, and consider a Boolean form in which n variables $x_1, x_2, \ldots, x_n$ denote elements of B. In this context we shall give a special name to elements of the set of n-samples of B; we shall call the elements of $S_n(B)$ *value assignments*. For example, with $B = \{0, 1\}$ and three variables x_1, x_2, and x_3, we have $S_n(B) = \{\langle 0, 0, 0\rangle, \langle 0, 0, 1\rangle, \langle 0, 1, 0\rangle, \langle 0, 1, 1\rangle, \langle 1, 0, 0\rangle, \langle 1, 0, 1\rangle, \langle 1, 1, 0\rangle, \langle 1, 1, 1\rangle\}$, a set of eight assignments. We shall associate the ith coordinate of an assignment with the variable x_i. Take, for example, the Boolean form $x_1 * (x_1 \oplus (x_2 * x_3))$ and the assignment $\langle 1, 0, 1\rangle$. With this assignment the form becomes $1 * (1 \oplus (0 * 1))$, which can be evaluated using the definitions of the operations for members of $\{0, 1\}$ as given in the proof of Th.2.3: $1 * (1 \oplus (0 * 1)) = 1$. The association between coordinates of an assignment and the variables thus enables us to reduce the form to a value. The reduction, which we call a *valuation*, gives a link between forms and functions. Here one member of the function on B^3 into B associated with the form $x_1 * (x_1 \oplus (x_2 * x_3))$ is seen to be $\langle\langle 1, 0, 1\rangle, 1\rangle$ or $\langle 1, 0, 1, 1\rangle$. Valuations for all assignments give all elements of the functions. Table 2.2 displays valuations of $x_1 * (x_1 \oplus (x_2 * x_3))$. The principles of construction are as for Table 2.1.

We call two Boolean forms *equivalent* if their valuations are equal for every value assignment. Table 2.2 shows that forms x_1 and $x_1 * (x_1 \oplus (x_2 * x_3))$ are equivalent in Boolean algebra $\langle \{0, 1\}, \oplus, *, ', 0, 1\rangle$. The equivalence lets us write $x_1 * (x_1 \oplus (x_2 * x_3)) = x_1$. Note, however, that this is no new result; the equivalence is only a special case of Part 3 of Th.1.11. Since the forms are equivalent, the functions associated with them are equal. Although x_1 is a form in a single variable, the function associated with it must be taken as

TABLE 2.2

VALUATION OF BOOLEAN FORM $x_1 * (x_1 \oplus (x_2 * x_3))$

$\langle x_1, x_2, x_3 \rangle$	$\alpha = x_2 * x_3$	$\beta = x_1 \oplus \alpha$	$x_1 * \beta$
$\langle 0, 0, 0 \rangle$	0	0	0
$\langle 0, 0, 1 \rangle$	0	0	0
$\langle 0, 1, 0 \rangle$	0	0	0
$\langle 0, 1, 1 \rangle$	1	1	0
$\langle 1, 0, 0 \rangle$	0	1	1
$\langle 1, 0, 1 \rangle$	0	1	1
$\langle 1, 1, 0 \rangle$	0	1	1
$\langle 1, 1, 1 \rangle$	1	1	1

being on $\{0, 1\}^3$ if we are to talk of equality of this function and that associated with $x_1 * (x_1 \oplus (x_2 * x_3))$.

It is feasible to determine equality of Boolean functions by means of valuations only if there are few elements in set B. Consider, for example, the model $\langle \mathscr{P}(U), \cup, \cap, \bar{}, \varnothing, U \rangle$, with $|U| = 3$. As algebras of sets go, the number of elements in U is very small. But, since $|\mathscr{P}(U)| = 8$, the number of 2-samples of $\mathscr{P}(U)$ is 64 and the number of 3-samples 512. Clearly valuations are impracticable here. Thus it would seem that valuations are in general a poor substitute for the more formal techniques of Section 1e. The study of canonical forms, to which we now turn, does, however, provide us with a powerful decision procedure for establishing equality of functions despite its dependence on a particular type of valuations.

DEFINITION 2.9 Let $\langle B, \oplus, *, ', 0, 1 \rangle$ be a Boolean algebra and $\alpha(x_1, \ldots, x_n)$ a Boolean form in the variables $x_1, \ldots, x_n$, which stand for elements of B. Denote elements of the set of n-samples $S_n(\{0, 1\})$, which we call *binary value assignments*, by t_m^n ($m = 0, 1, \ldots, 2^n - 1$). The sample $\langle m_1, \ldots, m_n \rangle$ is denoted by t_m^n when the binary number $m_1 \cdots m_n$ is equal to the (decimal) number m. Let F be the set of forms $\alpha(m_1, \ldots, m_n)$, and define a function $q: S_n \to F$ such that $q(t_m^n) = \alpha(m_1, \ldots, m_n)$ if and only if $t_m^n = \langle m_1, \ldots, m_n \rangle$. Next define the *binary valuation function* $v: F \to \{0, 1\}$ such that $v(\alpha(m_1, \ldots, m_n)) = 0$ if and only if $\alpha(m_1, \ldots, m_n)$ has the value 0 when evaluated according to the definitions of operations for members of $\{0, 1\}$ derived in the proof of Th.2.3.

Examples

1. Consider $\alpha(x_1, x_2, x_3) = x_1 * (x_1 \oplus (x_2 * x_3))$. Members of S_3 are given by the first column of Table 2.2, e.g., $t_5^3 = \langle 1, 0, 1 \rangle$. Then $q(t_5^3) = \alpha(1, 0, 1) = 1 * (1 \oplus (0 * 1))$, and $v(q(t_5^3)) = 1$. Images of $q(t_m^3)$ under v are given by the

last column of the table. If $B = \{0, 1\}$, then the set $\{\langle t^3_m, v(q(t^3_m))\rangle \mid m = 0, \ldots, 7\}$ is the function associated with the form $x_1 * (x_1 \oplus (x_2 * x_3))$.

2. The form $(x_1 \oplus x_2) * (x_1 \oplus x_3')$ is processed in Table 2.3.

TABLE 2.3

BINARY VALUATION OF $(x_1 \oplus x_2) * (x_1 \oplus x_3')$

m	t^3_m	$q(t^3_m)$	$v(q(t^3_m))$
0	$\langle 0, 0, 0\rangle$	$(0 \oplus 0) * (0 \oplus 0')$	0
1	$\langle 0, 0, 1\rangle$	$(0 \oplus 0) * (0 \oplus 1')$	0
2	$\langle 0, 1, 0\rangle$	$(0 \oplus 1) * (0 \oplus 0')$	1
3	$\langle 0, 1, 1\rangle$	$(0 \oplus 1) * (0 \oplus 1')$	0
4	$\langle 1, 0, 0\rangle$	$(1 \oplus 0) * (1 \oplus 0')$	1
5	$\langle 1, 0, 1\rangle$	$(1 \oplus 0) * (1 \oplus 1')$	1
6	$\langle 1, 1, 0\rangle$	$(1 \oplus 1) * (1 \oplus 0')$	1
7	$\langle 1, 1, 1\rangle$	$(1 \oplus 1) * (1 \oplus 1')$	1

DEFINITION 2.10 Let $\langle B, \oplus, *, ', 0, 1\rangle$ be a Boolean algebra, let variables $x_1, x_2, \ldots, x_n$ stand for elements of B, and let a_i denote either x_i or x_i'. The form $a_1 * a_2 * \cdots * a_n$ is called a *minterm* (*minimal polynomial*, *complete product*, *fundamental product*), and the form $a_1 \oplus a_2 \oplus \cdots \oplus a_n$ is called a *maxterm* (*maximal polynomial*, *complete sum*, *fundamental sum*) of the n variables. We shall write x_i^1 for x_i and x_i^0 for x_i' in these forms. Under this convention we shall denote the minterm $x_1^{m_1} * x_2^{m_2} * \cdots * x_n^{m_n}$ by min^n_m and the maxterm $x_1^{m_1'} \oplus x_2^{m_2'} \oplus \cdots \oplus x_n^{m_n'}$ by max^n_m when the binary number $m_1 m_2 \cdots m_n$ is equal to m.

Example

With n variables there are 2^n minterms and 2^n maxterms. Thus, if $n = 3$, there are 8 minterms and 8 maxterms. The minterms are $x_1^0 * x_2^0 * x_3^0$, $x_1^0 * x_2^0 * x_3^1$, $x_1^0 * x_2^1 * x_3^0$, $x_1^0 * x_2^1 * x_3^1$, $x_1^1 * x_2^0 * x_3^0$, $x_1^1 * x_2^0 * x_3^1$, $x_1^1 * x_2^1 * x_3^0$, $x_1^1 * x_2^1 * x_3^1$, denoted by min^3_0, min^3_1, $\ldots$, min^3_7, respectively; $max^3_7 = x_1^0 \oplus x_2^0 \oplus x_3^0$.

THEOREM 2.6 Let $\langle B, \oplus, *, ', 0, 1\rangle$ be a Boolean algebra, and let variables $x_1, x_2, \ldots, x_n$ denote elements of B. Every Boolean form $\alpha(x_1, x_2, \ldots, x_n)$ of n variables is equivalent to an expansion in minterms:

$$\alpha(x_1, x_2, \ldots, x_n) = \bigoplus_{m=0}^{m=2^n-1} v(q(t^n_m)) * min^n_m$$

(where we have used an obvious abbreviation for representing a repeated $\oplus$ operation).

Proof. We shall prove the theorem by induction on n.
(i) $n = 1$. Since

$$\begin{aligned} x_1 \oplus 1 &= x_1' \oplus 1 = x_1' \oplus x_1 = 1, \\ x_1 * 0 &= x_1' * 0 = x_1' * x_1 = 0, \\ x_1 * 1 &= x_1 \oplus 0 = x_1 * x_1 = x_1 \oplus x_1 = x_1, \\ x_1' * 1 &= x_1' \oplus 0 = x_1' * x_1' = x_1' \oplus x_1' = x_1', \end{aligned}$$

every $\alpha(x_1)$ ultimately reduces to 1, 0, x_1, or x_1'. The images of the two binary value assignments $t_0^1 = 0$ and $t_1^1 = 1$ under q are $q(t_0^1) = \alpha(0)$ and $q(t_1^1) = \alpha(1)$. We have to consider the images $v(\alpha(0))$ and $v(\alpha(1))$. Clearly $v(\alpha(0)) = v(\alpha(1)) = 1$ when $\alpha(x_1) = 1$, and $v(\alpha(0)) = v(\alpha(1)) = 0$ when $\alpha(x_1) = 0$. When $\alpha(x_1) = x_1$ we have $\alpha(0) = 0$ and $\alpha(1) = 1$. Consequently $v(\alpha(0)) = 0$ and $v(\alpha(1)) = 1$. The images are reversed when $\alpha(x_1) = x_1'$: $\alpha(0) = 1$ and $\alpha(1) = 0$, giving $v(\alpha(0)) = 1$ and $v(\alpha(1)) = 0$. Evaluation of $(v(\alpha(0)) * x_1') \oplus (v(\alpha(1)) * x_1)$ completes the proof for $n = 1$:

$$\begin{aligned} \alpha(x_1) &= 1, & (1 * x_1') \oplus (1 * x_1) &= 1, \\ \alpha(x_1) &= 0, & (0 * x_1') \oplus (0 * x_1) &= 0, \\ \alpha(x_1) &= x_1, & (0 * x_1') \oplus (1 * x_1) &= x_1, \\ \alpha(x_1) &= x_1', & (1 * x_1') \oplus (0 * x_1) &= x_1'. \end{aligned}$$

(ii) Assume the theorem true for n. This enables us to write

$$\alpha(x_1, \ldots, x_{n+1}) = \bigoplus_{m=0}^{m=2^n-1} \alpha(m_1, \ldots, m_n, x_{n+1}) * min_m^n. \tag{A}$$

Also

$$\begin{aligned} \alpha(m_1, \ldots, m_n, x_{n+1}) = (&v(\alpha(m_1, \ldots, m_n, 0)) * x_{n+1}') \\ &\oplus (v(\alpha(m_1, \ldots, m_n, 1)) * x_{n+1}), \end{aligned}$$

and it should be reasonably easy to see that use of this expression in (A) gives the theorem for $n + 1$:

$$\alpha(x_1, \ldots, x_{n+1}) = \bigoplus_{m=0}^{m=2^n-1} v(\alpha(x_1, \ldots, x_{n+1})) * min_m^{n+1}.$$

THEOREM 2.7 For a given Boolean form $\alpha(x_1, \ldots, x_n)$ the equivalent expansion in minterms defined by Th. 2.6 is unique.

Proof. Assume that two different expanded forms are equivalent to $\alpha(x_1, \ldots, x_n)$. Since the forms are different, there exists at least one $t_m^n = \langle m_1, \ldots, m_n \rangle$ such that $v(q(t_m^n))$ is 0 in one of the forms and 1 in the other, but then, by D.2.9, we have the contradiction $\alpha(m_1, \ldots, m_n) = 0$ and $\alpha(m_1, \ldots, m_n) = 1$.

THEOREM 2.8 Let $\langle B, \oplus, *, ', 0, 1\rangle$ be a Boolean algebra, and let variables $x_1, x_2, \ldots, x_n$ denote elements of B. Every Boolean form $\alpha(x_1, x_2, \ldots, x_n)$ of n variables has a unique equivalent expansion in maxterms:

$$\alpha(x_1, x_2, \ldots, x_n) = \mathop{*}_{m=0}^{m=2^n-1} v(q(t_m^n)) \oplus max_m^n$$

(where an obvious abbreviation is used for a repeated $*$ operation).

Proof. By principle of duality from Th.2.6 and Th.2.7.

Since the expanded forms are unique, we call them *canonical forms* or *normal forms*. The canonical expansion of Th.2.6 is known as the *disjunctive normal form*, and that of Th.2.8 as the *conjunctive normal form*. An obvious consequence of the uniqueness of normal forms is that two forms are equivalent if they have the same normal form. Then the associated functions are, of course, equal. Inspection of the canonical expansions shows that two Boolean forms $\alpha(x_1, \ldots, x_n)$ and $\beta(x_1, \ldots, x_n)$ have the same normal form, i.e., have equal associated functions, if $v(\alpha(m_1, \ldots, m_n)) = v(\beta(m_1, \ldots, m_n))$ for all n-samples $\langle m_1, \ldots, m_n\rangle$ of set $\{0, 1\}$.

Examples

1. Clearly a term min_m^n for which the corresponding $v(q(t_m^n))$ is 0 does not have to be included in a disjunctive normal form. Similarly, one does not have to include a term max_m^n for which $v(q(t_m^n))$ is 1 in a conjunctive normal form. From Table 2.3, the disjunctive and conjunctive normal forms equivalent to $(x_1 \oplus x_2) * (x_1 \oplus x_3')$ are, respectively, $(x_1' * x_2 * x_3') \oplus (x_1 * x_2' * x_3') \oplus (x_1 * x_2' * x_3) \oplus (x_1 * x_2 * x_3') \oplus (x_1 * x_2 * x_3)$ and $(x_1 \oplus x_2 \oplus x_3') * (x_1 \oplus x_2 \oplus x_3) * (x_1 \oplus x_2' \oplus x_3')$.

2. Sometimes we may be given a function and asked to find a formula describing it. Let function f: $\{0, 1\}^2 \to \{0, 1\}$ be the set $\{\langle 0, 0, 0\rangle, \langle 0, 1, 1\rangle, \langle 1, 0, 1\rangle, \langle 1, 1, 0\rangle\}$. The disjunctive normal form of the function is $(x_1' * x_2) \oplus (x_1 * x_2')$. The conjunctive normal form is $(x_1 \oplus x_2) * (x_1' \oplus x_2')$. Since the normal forms represent the same function, they are equivalent. Hence, in model $\langle \mathscr{P}(U), \cup, \cap, \bar{}, \varnothing, U\rangle$,

$$(\bar{A} \cap B) \cup (A \cap \bar{B}) = (A \cup B) \cap (\bar{A} \cup \bar{B}).$$

Since the left-hand side of the equation defines the symmetric difference, we can write $A + B = (A \cup B) \cap (\bar{A} \cup \bar{B})$.

3. Tables 2.4 and 2.5 show that the two forms $\alpha = ((x_1 * x_3) \oplus (x_2 * x_3'))'$ and $\beta = (x_1' * x_3) \oplus (x_2' * x_3')$ are equivalent. The disjunctive normal form of α and β is $(x_1' * x_2' * x_3') \oplus (x_1' * x_2' * x_3) \oplus (x_1' * x_2 * x_3) \oplus (x_1 * x_2' * x_3')$. The conjunctive form is $(x_1 \oplus x_2' \oplus x_3) * (x_1' \oplus x_2 \oplus x_3') * (x_1' \oplus x_2' \oplus x_3) * (x_1' \oplus x_2' \oplus x_3')$.

TABLE 2.4

$$\alpha = ((x_1 * x_3) \oplus (x_2 * x_3'))' = (\alpha_1 \oplus \alpha_2)'$$

m	t_m^3	$v(\alpha_1)$	$v(\alpha_2)$	$v(\alpha_1 \oplus \alpha_2)$	$v(\alpha)$
0	⟨0, 0, 0⟩	0	0	0	1
1	⟨0, 0, 1⟩	0	0	0	1
2	⟨0, 1, 0⟩	0	1	1	0
3	⟨0, 1, 1⟩	0	0	0	1
4	⟨1, 0, 0⟩	0	0	0	1
5	⟨1, 0, 1⟩	1	0	1	0
6	⟨1, 1, 0⟩	0	1	1	0
7	⟨1, 1, 1⟩	1	0	1	0

TABLE 2.5

$$\beta = (x_1' * x_3) \oplus (x_2' * x_3')$$

m	t_m^3	$v(x_1' * x_3)$	$v(x_2' * x_3')$	$v(\beta)$
0	⟨0, 0, 0⟩	0	1	1
1	⟨0, 0, 1⟩	1	0	1
2	⟨0, 1, 0⟩	0	0	0
3	⟨0, 1, 1⟩	1	0	1
4	⟨1, 0, 0⟩	0	1	1
5	⟨1, 0, 1⟩	0	0	0
6	⟨1, 1, 0⟩	0	0	0
7	⟨1, 1, 1⟩	0	0	0

4. There is a fundamental difference between the valuations of Table 2.2 and the binary valuations here. In Table 2.2 variables x_1, x_2, x_3 stand for elements of the very special algebra $\langle\{0, 1\}, \oplus, *, ', 0, 1\rangle$, and the valuations define the elements of a function. In Tables 2.4 and 2.5 the x_1, x_2, x_3 represent elements of the general Boolean algebra $\langle B, \oplus, *, ', 0, 1\rangle$. The binary valuations of these tables let us make decisions about the equivalence of forms (and thus about equality of functions), but they do not define the elements of functions. Consider the model $\langle \mathscr{P}(U), \cup, \cap, \bar{\ }, \varnothing, U\rangle$ and functions $g\colon (\mathscr{P}(U))^3 \to \mathscr{P}(U)$, $h\colon (\mathscr{P}(U))^3 \to \mathscr{P}(U)$, defined by the forms

$$g(A, B, C) = (A \cup B) \cap (B \cup C) \cap (C \cup A),$$
$$h(A, B, C) = (A \cap B) \cup (B \cap C) \cup (C \cap A),$$

where A, B, C represent elements of $\mathscr{P}(U)$. The equality of g and h can be demonstrated by standard procedures of the algebra of sets applied to the formulas $(A \cup B) \cap (B \cup C) \cap (C \cup A)$ and $(A \cap B) \cup (B \cap C) \cup (C \cap A)$.

Alternatively, the forms can be shown to be equivalent by showing that they have the same normal forms. This we do in Table 2.6. The disjunctive normal form of $(A \cup B) \cap (B \cup C) \cap (C \cup A)$ is $(\bar{A} \cap B \cap C) \cup (A \cap \bar{B} \cap C) \cup (A \cap B \cap \bar{C}) \cup (A \cap B \cap C)$. (Clearly, we have not been defining elements of the function here. Let $U = \{1, 2, 3, 4, 5\}$, say. Then $|\mathscr{P}(U)| = 32$, and the number of 3-samples of $\mathscr{P}(U)$ is 32,768. Since this is the number of elements in the function g—and in any other function on $(\mathscr{P}(U))^3$ when U has five elements—listing the elements would be a rather meaningless exercise.)

TABLE 2.6

EQUALITY OF FUNCTIONS g AND h

m	t_m^3	$v(g(A, B, C))$	$v(h(A, B, C))$
0	$\langle \varnothing, \varnothing, \varnothing \rangle$	$\varnothing$	$\varnothing$
1	$\langle \varnothing, \varnothing, U \rangle$	$\varnothing$	$\varnothing$
2	$\langle \varnothing, U, \varnothing \rangle$	$\varnothing$	$\varnothing$
3	$\langle \varnothing, U, U \rangle$	U	U
4	$\langle U, \varnothing, \varnothing \rangle$	$\varnothing$	$\varnothing$
5	$\langle U, \varnothing, U \rangle$	U	U
6	$\langle U, U, \varnothing \rangle$	U	U
7	$\langle U, U, U \rangle$	U	U

5. Consider $\alpha = (x_1 * x_2 * x_3) \oplus (x_1' * x_2 * x_3) \oplus x_2' \oplus x_3'$. Here $v(\alpha) = 1$ for all $\langle m_1, m_2, m_3 \rangle$. Hence the disjunctive normal form has eight terms. In the conjunctive normal form each term appears as $1 \oplus max_m^n$. Since, by convention, such terms are not included in the normal form, the conjunctive normal form would be written as 1 here.

2c. Applications of Boolean Functions

The system $\langle \{0, 1\}, \oplus, *, ', 0, 1 \rangle$, which by Th.2.3 is a Boolean algebra, is sometimes called *switching algebra*. Functions s: $\{0, 1\}^n \rightarrow \{0, 1\}$ are known as *switching functions*. The terminology reflects an important application of the functions s. Wherever information passes along wires, as in a computer, telephone switching system, or some other such system, the simplest information processing is performed by switches. A *switch* is a two-state device. It is either *closed* and passes information through, or it is *open* and stops the information flow. Switching functions have been found useful for describing circuits that contain switches and in the design of such circuits.

A circuit that links two terminal points t_1 and t_2, and consists of interconnected switches is known as a *combinatorial two-terminal switching circuit*. Depending on the states of the switches, information passes or does not pass from one terminal to the other. If information does pass, the circuit is *closed*. Otherwise it is *open*. Figure 2.1 shows two very simple circuits containing two switches x_1 and x_2. In the *series* connection the circuit is closed only if both switches are closed. In the *parallel* connection it is sufficient to have just one switch closed for the circuit to be closed.

(a) Series connection of two switches.

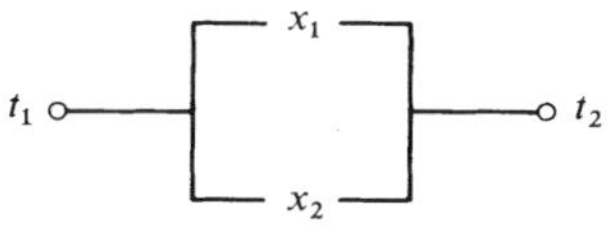

(b) Parallel connection of two switches.

Figure 2.1

We associate the element 0 of {0, 1} with an open switch and the element 1 with a closed switch. The n switches in a circuit are associated with *circuit variables* $x_1, x_2, \ldots, x_n$, and the circuit itself with a function of these variables, $s(x_1, x_2, \ldots, x_n)$. We associate elements of {0, 1} with the circuit as well: 0 with an open circuit and 1 with a closed circuit. States of the circuits of Figure 2.1 are investigated in Table 2.7. The four rows of the table correspond to the four possible states of the two switches (both open, x_1 alone open, x_2

TABLE 2.7

STATES OF SERIES AND PARALLEL CONNECTIONS OF TWO SWITCHES

$\langle x_1, x_2 \rangle$	Series connection	Parallel connection	$x_1 * x_2$	$x_1 \oplus x_2$
$\langle 0, 0 \rangle$	0	0	0	0
$\langle 0, 1 \rangle$	0	1	0	1
$\langle 1, 0 \rangle$	0	1	0	1
$\langle 1, 1 \rangle$	1	1	1	1

alone open, both closed). The table shows the states of the circuits corresponding to the states of the switches, and $x_1 * x_2$ and $x_1 \oplus x_2$ (functions p and s of Table 2.1). We see that a series connection of x_1 and x_2 can be represented by $x_1 * x_2$, and a parallel connection by $x_1 \oplus x_2$.

Consider now the circuit of Figure 2.2. It can be represented by the switching function $s(x_1, x_2, x_3, x_4, x_5) = x_1 * (x_2 \oplus (x_3 * (x_4 \oplus x_5)))$. This form is equivalent to the form $(x_1 * x_2) \oplus (x_1 * x_3 * x_4) \oplus (x_1 * x_3 * x_5)$. The question arises: Can one represent a form containing more than one occurrence of a variable by a switching circuit? The answer is affirmative, with an understanding that in the physical realization of the circuit there is a mechanism to ensure that switches bearing the same label are simultaneously all open or all closed. In practice, however, one is not interested in designing an equivalent circuit more complicated than the original one. Rather, the practical problem is one of simplification.

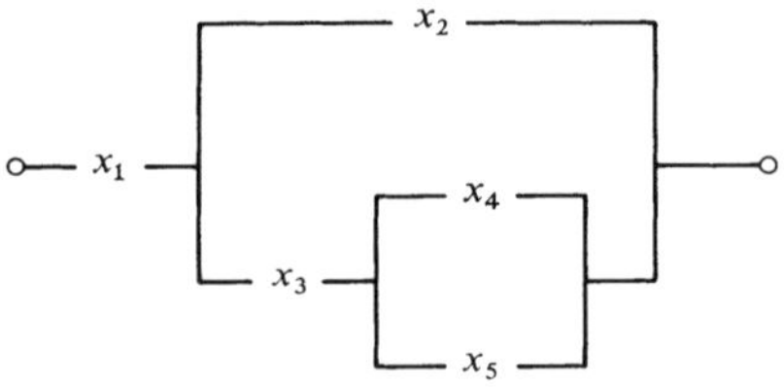

Figure 2.2

The important application of switching functions is in the design of circuits. Assume that we are given some switching function and are required to design a switching circuit corresponding to this function. The circuit is to be open for those combinations of switches $x_1, x_2, \ldots, x_n$ for which $s(x_1, x_2, \ldots, x_n) = 0$ and closed for combinations giving $s(x_1, x_2, \ldots, x_n) = 1$. The theory of normal forms of Section 2b permits us to write down formulas of the function. If combinations with $s(x_1, x_2, \ldots, x_n) = 0$ predominate in the specification of the function, we choose the disjunctive normal form. Otherwise we select the conjunctive normal form. Probably there will be terms in x_i' as well as in x_i in the normal form. The state of switch x_i' is completely determined by that of x_i: If x_i is closed, x_i' is open; if x_i is open, x_i' is closed. In what follows we shall abbreviate $x_1 * x_2 * \cdots * x_n$ to $x_1 x_2 \cdots x_n$.

Example

Suppose we require a combinatorial two-terminal switching circuit of four switches to be closed whenever exactly two of the switches are closed and to be open otherwise. First, in Table 2.8, we specify the switching function.

TABLE 2.8

A SWITCHING FUNCTION

$\langle x_1, x_2, x_3, x_4\rangle$	$s(x_1, x_2, x_3, x_4)$
$\langle 0, 0, 0, 0\rangle$	0
$\langle 0, 0, 0, 1\rangle$	0
$\langle 0, 0, 1, 0\rangle$	0
$\langle 0, 0, 1, 1\rangle$	1
$\langle 0, 1, 0, 0\rangle$	0
$\langle 0, 1, 0, 1\rangle$	1
$\langle 0, 1, 1, 0\rangle$	1
$\langle 0, 1, 1, 1\rangle$	0
$\langle 1, 0, 0, 0\rangle$	0
$\langle 1, 0, 0, 1\rangle$	1
$\langle 1, 0, 1, 0\rangle$	1
$\langle 1, 0, 1, 1\rangle$	0
$\langle 1, 1, 0, 0\rangle$	1
$\langle 1, 1, 0, 1\rangle$	0
$\langle 1, 1, 1, 0\rangle$	0
$\langle 1, 1, 1, 1\rangle$	0

There are 6 nonzero images against 10 zero images. Therefore, the disjunctive normal form of the function is the simpler form:

$$s(x_1, x_2, x_3, x_4) = x_1'x_2'x_3x_4 \oplus x_1'x_2x_3'x_4 \oplus x_1'x_2x_3x_4' \oplus x_1x_2'x_3'x_4 \oplus x_1x_2'x_3x_4' \oplus x_1x_2x_3'x_4'.$$

Figure 2.3 shows the circuit corresponding to this function.

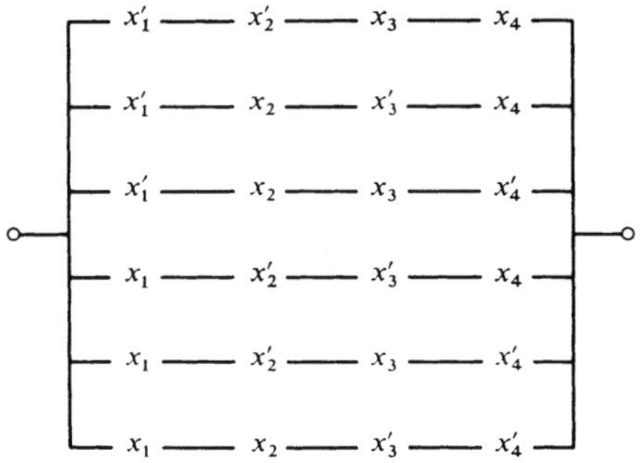

Figure 2.3

The circuit of Figure 2.3 contains 24 switches. This number seems excessive. Let us therefore try to simplify the formula. Using the theorems of Section 1e, we find

$$s(x_1, x_2, x_3, x_4) = x_1'x_2'x_3x_4 \oplus x_1x_2x_3'x_4' \oplus (x_1'x_2 \oplus x_1x_2')(x_3'x_4 \oplus x_3x_4').$$

The simpler circuit (16 switches) corresponding to this form is shown in Figure 2.4.

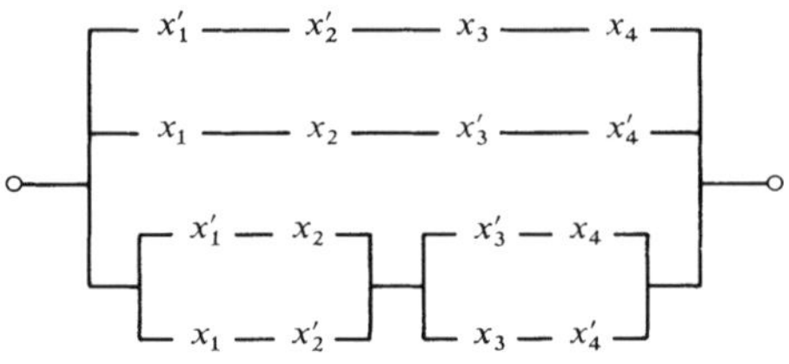

Figure 2.4

Next we consider *gating circuits*. In a gating circuit there are *n inputs* $x_1, x_2, \ldots, x_n$, and each input is either 0 (signal absent) or 1 (signal present). There is a single *output*, also 0 or 1. The simplest gating circuits consist of just two inputs, x_1 and x_2, and of a *gate*, either an OR-*gate* or an AND-*gate*. If the gate is an OR-gate, the output signal is 0 when $x_1 = x_2 = 0$ and 1 otherwise. If the gate is an AND-gate, the signal is 1 when $x_1 = x_2 = 1$ and 0 otherwise. In addition to gates we may have an *inverter*, which has a single input x. The output of the inverter is 1 when $x = 0$ and 0 when $x = 1$.

A gating circuit with n inputs corresponds to a *gating function* $g: \{0, 1\}^n \to \{0, 1\}$. The OR-gate corresponds to the $\oplus$-operation in $\{0, 1\}$, the AND-gate to the $*$-operation, and the inverter to complementation. There is a great similarity between gating circuits and switching circuits. In fact, there is a one-to-one correspondence between gating circuits with n inputs and switching circuits with n switches. Switches x_i and x_j in parallel correspond to an OR-gate with inputs x_i and x_j, and switches in series correspond to an AND-gate. Wherever there is an x_i' in the switching circuit, signal x_i is passed through an inverter in the gating circuit. To avoid confusion between switching and gating problems, and to adhere to generally accepted notation we shall use the symbol $\vee$ for the $\oplus$-operation and the symbol $\wedge$ for the $*$-operation; i.e., here our Boolean algebra, sometimes called *gating algebra*, is $\langle\{0, 1\}, \vee, \wedge, ', 0, 1\rangle$. Again we shall abbreviate $x_i \wedge x_j$ to x_ix_j.

Figure 2.5 shows the three fundamental gating circuits. In Figure 2.6 we show the gating circuit corresponding to the switching circuit of Figure 2.2,

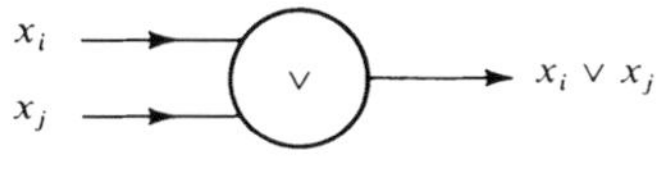

(a) OR-gate

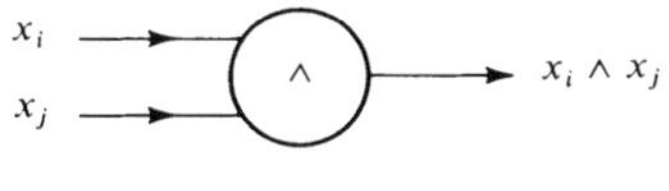

(b) AND-gate

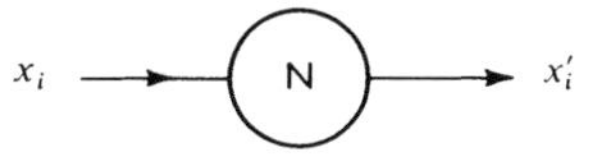

(c) Inverter

Figure 2.5

and in Figure 2.7 the gating circuit corresponding to the switching circuit of Figure 2.4.

The complexity of Figure 2.7 is best evidence that it would be difficult to construct a gating circuit corresponding to the function of Table 2.8 if we did not have the techniques based on canonical forms. These techniques enable us to find a formula involving functions $\oplus$, $*$, and $'$ for any Boolean function $f: B^n \to B$. The question arises whether it is still possible to find formulas for all functions if the basic set of operations is other than $\{\oplus, *, '\}$. This is the problem of functional completeness.

DEFINITION 2.11 A set of operations $\{p_1, \ldots, p_k\}$ in set B is *functionally complete* if and only if *every* Boolean function $f: B^n \to B$ can be represented by a form $\alpha(x_1, \ldots, x_n)$ in variables $x_1, x_2, \ldots, x_n$ and operations $p_1, \ldots, p_k$.

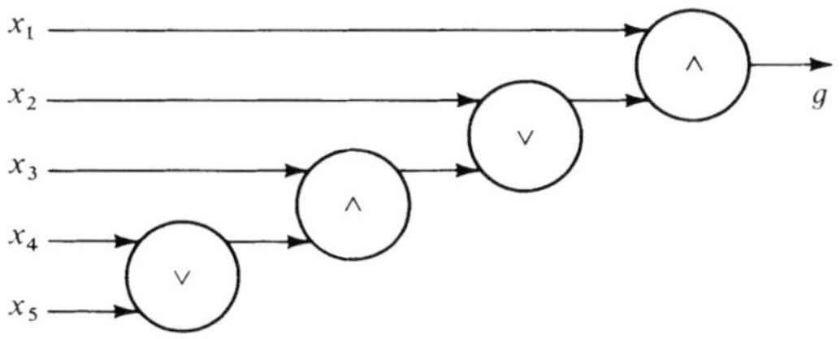

Figure 2.6

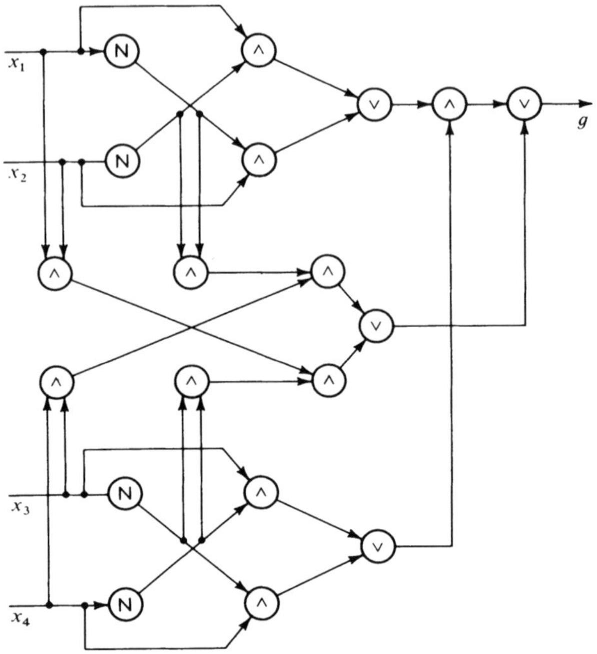

$$g = x_1 x_2 x_3' x_4' \vee x_1' x_2' x_3 x_4 \vee (x_1 x_2' \vee x_1' x_2)(x_3 x_4' \vee x_3' x_4)$$

Figure 2.7

Examples

1. The set of functions $\{\oplus, *\}$ on $\{0, 1\}^2$ is not functionally complete. Since $0 \oplus 0 = 0 * 0 = 0$, there is no way of constructing a formula corresponding to a function that contains $\langle 0, 0, 1\rangle$. Also, since $1 \oplus 1 = 1 * 1 = 1$, there is no way of constructing a formula of a function with $\langle 1, 1, 0\rangle$ as a member. Hence only 4 of the 16 functions on $\{0, 1\}^2$ into $\{0, 1\}$ can be represented by forms in $\oplus$ and $*$ alone. These forms are x_1, x_2, $x_1 \oplus x_2$, $x_1 * x_2$ (or forms equivalent to them). In the context of gating circuits this means that very few functions can be represented by gating circuits containing no inverters.

2. In Boolean algebra $\langle B, \oplus, *, ', 0, 1\rangle$ we have, for any elements a, $b \in B$,

$$a \oplus b = (a' * b')'; \qquad a * b = (a' \oplus b')'.$$

Hence sets of functions $\{*, '\}$ and $\{\oplus, '\}$ are functionally complete in B.

3. In Table 2.1 we define function g on set $\{0, 1\}$. Since $g' = p$, where p is

defined by the form $x_1 * x_2$, we have $g(x_1, x_2) = (x_1 * x_2)'$. Function $g(x_1, x_2)$ is written $x_1 \mid x_2$ and called *stroke function*. We have

$$\begin{aligned} x_1' &= x_1 \mid x_1, \\ x_1 \oplus x_2 &= x_1' \mid x_2' = (x_1 \mid x_1) \mid (x_2 \mid x_2), \\ x_1 * x_2 &= (x_1 \mid x_2)' = (x_1 \mid x_2) \mid (x_1 \mid x_2). \end{aligned}$$

Hence $\{\ |\ \}$ is functionally complete. The gating circuit corresponding to the stroke function is known as a NAND-*gate* (NAND for Not AND). Any Boolean function can be represented by a circuit consisting of nothing but NAND-gates.

Our final application deals with statements. By a *statement* or *proposition* we mean a sentence that is either true or false. It is possible to determine the truth or falsity of the sentences

There are 23 lines of print on this page (p)

$10 + 10 = 100$ (q)

Jack and Jill went up the hill (r)

They are therefore statements. The sentence

What time is it?

is not a statement.

One can modify a statement by the word *not* or connect statements with words *and*, *or*, *if–then*, *if and only if*. The five words or combinations of words (or their synonyms, e.g., *implies* for *if–then*) are called *connectives*. Statements built up from simpler statements by means of the connectives are called *composite statements*. The different types of composite statements have been given special names. A statement modified by the connective *not* is the *negation* of the original statement. The statement that results when two statements are joined by *and* is the *conjunction* of the two statements. When two statements are joined by *or* the resulting statement is the *disjunction* of the two statements. The statement *if p then q* is a *conditional* statement having statement p as its *antecedent* and statement q as its *consequent*. The connective *if and only if* generates a *biconditional* statement.

Statements p and q are not composite. Such statements are called *prime statements* or *atomic statements*. Statement r, on the other hand, is composite if it is regarded as the conjunction of statements

Jack went up the hill (s)

Jill went up the hill (t)

(The presence of *and* in a statement does not necessarily make the statement composite. If Jill's actions always match Jack's actions, then r can be regarded as an atomic statement.) The following statements are composite:

$$10 + 10 \neq 100 \quad (u)$$

$$10 + 10 = 100 \quad \text{or} \quad 10 + 10 \neq 100 \quad (v)$$

$$10 + 10 = 100 \quad \text{if and only if} \quad 10 + 10 \neq 100 \quad (w)$$

$$\text{If} \quad 3 < 5 \quad \text{and} \quad 4 < 5 \quad \text{then} \quad 4 < 3 \quad \text{or} \quad 3 < 5 \quad (x)$$

Statement u is the negation of q. It is true if q is false, and false if q is true. Hence v is true irrespective of the truth or falsity of q. Indeed, the disjunction of any statement and its negation is necessarily true. The biconditional of a statement and its negation, of which w is an example, is always false. The fact that the truth or falsity of certain statements is a consequence of their form alone suggests that statements can be studied in the framework of an abstract theory.

We shall denote statements by letters $p, q, r, \ldots$, and use the following symbols for the connectives:

$\vee$	for	*or,*
$\wedge$	for	*and,*
$\neg$	for	*not,*
$\rightarrow$	for	*if–then,*
$\leftrightarrow$	for	*if and only if.*

With this symbolism we can write statement r as $s \wedge t$, statement u as $\neg q$, statement v as $q \vee \neg q$ and statement w as $q \leftrightarrow \neg q$. We shall indicate the truth or falsity of a statement by assigning a *truth value* to it: T if the statement is true, F if it is false.

Ordinary (natural) language is ambiguous. Consider the phrase *visitors or guests*. Here the *or* may indicate *guest* as a synonym of *visitor*; i.e., *visitors or guests* may refer to just the one class of people. Under a different interpretation visitors and guests may be considered to belong to disjoint classes. Then *or* has the *exclusive* meaning. Under yet another interpretation a visitor may, but need not, be a guest, and *or* has the *inclusive* meaning. Usually the context enables one to decide which meaning is intended. But a process of abstraction isolates statements from context. Hence there is no way of determining the intended meaning of *or* in the statement $p \vee q$ of the abstract theory unless the connective is precisely defined. In Table 2.9 we define connectives by giving the truth value of the composite statement resulting from prime statements p and q for all assignments of truth values to p and q.

If every F in Table 2.9 is replaced by 0 and every T by 1, the table becomes familiar. Then the columns for $\neg p$, $p \vee q$, $p \wedge q$ represent binary valuations of

TABLE 2.9

DEFINITIONS OF CONNECTIVES

$\langle p, q\rangle$	$\neg p$	$p \vee q$	$p \wedge q$	$p \to q$	$p \leftrightarrow q$
⟨F, F⟩	T	F	F	T	T
⟨F, T⟩	T	T	F	T	F
⟨T, F⟩	F	T	F	F	F
⟨T, T⟩	F	T	T	T	T

Boolean forms p', $p \oplus q$, $p * q$, respectively. Hence the system $\langle\{F, T\}, \vee, \wedge, \neg, F, T\rangle$ is a Boolean algebra. It is called the *algebra of truth values*. Just as a switch may be open or closed, a statement may be false or true, and just as a switching circuit may be open or closed depending on the states of the switches, a composite statement may be false or true depending on the assignment of truth values to its prime statements. In analogy with switching theory we denote the prime statements of a composite statement by variables $p_1, p_2, \ldots, p_n$, and associate the composite statement with a function $f(p_1, p_2, \ldots, p_n)$ of these variables, f: $\{F, T\}^n \to \{F, T\}$.

Clearly, the set of operations $\{\vee, \wedge, \neg\}$ is functionally complete on $\{F, T\}^n$. Although $\{\vee, \neg\}$ and $\{\wedge, \neg\}$ are also functionally complete (see Example 2 of D.2.11), by tradition the set of connectives $\{\vee, \wedge, \neg\}$ is considered basic to the theory, and its members are called *primary connectives*. Connectives $\to$ and $\leftrightarrow$ are called *secondary connectives*. Functional equality (in the sense of D.2.7) of $p \to q$ and $\neg p \vee q$ is demonstrated in Table 2.10. The table shows also that $p \leftrightarrow q$ is equivalent to $(p \to q) \wedge (q \to p)$, i.e., to $(\neg p \vee q) \wedge (\neg q \vee p)$. (A valuation table, such as Table 2.10, has a special name in the statement calculus; it is called *truth table*.)

We can show also that $p \to q$ is functionally equal to $p \vee q \leftrightarrow q$ (Exercise 2.19). Comparing this result with D.1.14 we see that the $p \to q$ of our model represents $p \leqq q$ of the abstract Boolean algebra $\langle B, \oplus, *, ', 0, 1\rangle$. Note further that the statement $(p \to q) \leftrightarrow (p \vee q \leftrightarrow q)$ is true for every value

TABLE 2.10

FUNCTIONAL EQUALITIES

$\langle p, q\rangle$	$\neg p$	$\neg p \vee q$	$(p \to q)$	$\wedge$	$(q \to p)$
⟨F, F⟩	T	T	T	T	T
⟨F, T⟩	T	T	T	F	F
⟨T, F⟩	F	F	F	F	T
⟨T, T⟩	F	T	T	T	T

assignment $\langle p, q\rangle$. Statements that are true for all value assignments are called *tautologies*. The symbol $\leftrightarrow$ has special significance in the context of a biconditional tautology. In the *tautology* $p \leftrightarrow q$ it symbolizes equivalence of formulas p and q. In the model $\langle\{F, T\}, \vee, \wedge, \neg, F, T\rangle$ it replaces the symbol $=$ of the general Boolean algebra of Section 1e.

Examples

1. Statement *If* 3 < 5 *and* 4 < 5 *then* 4 < 3 *or* 3 < 5 has the form *if p and q then r or p*, i.e., $(p \wedge q) \rightarrow (r \vee p)$. This form represents a function on $\{F, T\}^3$ into $\{F, T\}$. Table 2.11 is the truth table of this function. It shows that the statement $(p \wedge q) \rightarrow (r \vee p)$ is a tautology. Some readers may find the definition of the conditional of Table 2.9 hard to accept. Somehow it seems wrong that a conditional having a false antecedent and a true consequent should be true. The definition may be easier to accept if we rephrase *if p then q* to *p is a sufficient condition for q*. Then we see that a false antecedent is immaterial to the truth of $p \rightarrow q$ and that $p \rightarrow q$ can only be false if a false consequent results when a sufficient condition is satisfied, i.e., when the antecedent is true.

TABLE 2.11

A TAUTOLOGY

$\langle p, q, r\rangle$	$(p \wedge q)$	$\rightarrow$	$(r \vee p)$
⟨F, F, F⟩	F	T	F
⟨F, F, T⟩	F	T	T
⟨F, T, F⟩	F	T	F
⟨F, T, T⟩	F	T	T
⟨T, F, F⟩	F	T	T
⟨T, F, T⟩	F	T	T
⟨T, T, F⟩	T	T	T
⟨T, T, T⟩	T	T	T

2. Table 2.12 shows that the statement $((p \rightarrow q) \wedge (q \rightarrow r)) \rightarrow (p \rightarrow r)$ is a tautology. This tautology, known as the *law of syllogism*, is very important in logic. (The statement calculus is part of *symbolic logic*.)

Most higher level programming languages, such as Fortran and Algol, provide facilities for representation and evaluation of logical expressions. In Fortran operations `.OR.`, `.AND.`, `.NOT.` represent, respectively, $\vee$, $\wedge$, and $\neg$. Truth values, or logical constants, T and F are written `.TRUE.` and `.FALSE.`. Secondary connectives are not provided (Algol does provide them). For example, $p \rightarrow p \vee q$ has to be reformulated to $\neg p \vee (p \vee q)$ before it can be written as a Fortran expression, namely `.NOT.P.OR.(P.OR.Q)`.

TABLE 2.12

THE LAW OF SYLLOGISM

$\langle p, q, r\rangle$	$((p \to q) \wedge (q \to r)) \to (p \to r)$				
⟨F, F, F⟩	T	T	T	T	T
⟨F, F, T⟩	T	T	T	T	T
⟨F, T, F⟩	T	F	F	T	T
⟨F, T, T⟩	T	T	T	T	T
⟨T, F, F⟩	F	F	T	T	F
⟨T, F, T⟩	F	F	T	T	T
⟨T, T, F⟩	T	F	F	T	F
⟨T, T, T⟩	T	T	T	T	T

ALGORITHM 2.1 This is an algorithm, expressed as a Fortran function, for testing whether a given statement is a tautology. Logical function `TAUT` returns the value `.TRUE.` if the statement specified by the function `FORM` is a tautology and `.FALSE.` if it is not. Subroutine `NEXT` must be provided. Given an n-sample of {`.FALSE.`, `.TRUE.`} in array `SAMPLE`, `NEXT` generates in `SAMPLE` the next n-sample (in lexicographic order, say). This subroutine would be very similar to that of A.1.2.

```
      LOGICAL FUNCTION TAUT (FORM, SAMPLE, N)
      LOGICAL SAMPLE(N), FORM
      DO 1 K = 1,N
    1 SAMPLE(K) = .FALSE.
      NN = 2**N
      TAUT = .FALSE.
      DO 2 K = 1,NN
      IF (.NOT.FORM(SAMPLE,N)) RETURN
    2 CALL NEXT (SAMPLE,N)
      TAUT = .TRUE.
      RETURN
      END
```

Example

Let us test whether $(p \to q) \to ((q \to r) \to (p \to r))$ is a tautology. We rewrite the statement as $\neg(\neg p \vee q) \vee (\neg(\neg q \vee r) \vee (\neg p \vee r))$, which can be simplified to $(p \wedge \neg q) \vee (q \wedge \neg r) \vee (\neg p \vee r)$ by means of the theorems of Section 1e. (The advisability of making the simplification is debatable. It may be argued that it defeats the purpose of mechanization.) Function `FORM` is now written as follows:

```
      LOGICAL FUNCTION FORM (S, N)
      LOGICAL S(N)
      FORM = (S(1).AND..NOT.S(2)).OR.(S(2).AND..NOT.
     X       S(3)).OR.(.NOT.S(1).OR.S(3))
      RETURN
      END
```

Symbolic logic reduces the determination of the validity of an argument to a mechanical procedure, which may employ the following laws.

Law of detachment: If p does imply q, and if p is true, then q is true.

Law of substitution: If $p \leftrightarrow q$, then the substitution of p for q, or of q for p, in an argument does not affect the validity of the argument.

Law of excluded middle: $p \wedge \neg p$ is always false, i.e., a statement may not be simultaneously true and false.

Law of syllogism: $((p \rightarrow q) \wedge (q \rightarrow r)) \rightarrow (p \rightarrow r)$ is a tautology.

From the conditional $p \rightarrow q$ we can construct *derived conditionals* by negating or interchanging statements p and q in various combinations. The more important of the derived conditionals are the *converse* $(q \rightarrow p)$, the *inverse* $(\neg p \rightarrow \neg q)$, and the *contrapositive* $(\neg q \rightarrow \neg p)$. It can be shown that $(\neg q \rightarrow \neg p) \leftrightarrow (p \rightarrow q)$ (Exercise 2.20).

An argument is expressed as follows: Given that statements $p_1, p_2, \ldots, p_n$ are true, statement p_{n+1} is also true. Statements $p_1, \ldots, p_n$ are called *premises* and statement p_{n+1} *conclusion* of the argument. The proof of the validity of an argument consists in showing that the truth of the conclusion does in fact derive from the premises by the laws of logic. In what follows we shall abbreviate the phrase *p is true* to p; i.e., we shall assume that the act of writing down p is in itself an assertion of the truth of p.

Examples

1. Given that p, $p \rightarrow q$, and $\neg r \rightarrow \neg q$, prove r. The forms $\neg r \rightarrow \neg q$, $\neg(\neg q) \rightarrow \neg(\neg r)$ (its contrapositive), and $q \rightarrow r$ (the contrapositive simplified) are equivalent. Hence the argument may be rewritten as p, $p \rightarrow q$, $q \rightarrow r$, r (by the law of substitution). Since $p \rightarrow q$ and $q \rightarrow r$, then also $p \rightarrow r$ (by the laws of detachment and syllogism). But if p and $p \rightarrow r$, then r (by the law of detachment). The following statements provide a specific example of the general argument.

Premises:	Prices are high.
	If prices are high, one should sell stocks.
	If productivity is not low, one should not sell stocks.
Conclusion:	Productivity is low.

2. In Example 1 a *direct proof* is employed. In a direct proof one derives the conclusion from the premises. Now we shall set up an *indirect proof* (also called *proof by contradiction*, or *reductio ad absurdum proof*) in which the premises and the *negation* of the conclusion are assumed, and an attempt is made to derive a contradiction $p \wedge \neg p$ in which p is any statement. If a contradiction can be derived, one of the assumed statements must be false. Since the premises are true, the false statement is the negated conclusion. Hence the conclusion is true. Consider the following argument.

Premises:	There is unemployment, or productivity is high.
	Prices are high if and only if there is an inflation.
	If there is an inflation, there is no unemployment.
	If prices are low, productivity is high.
Conclusion:	Productivity is high.

We write the proof as a column of statements. The first five statements are the four premises and the negation of the conclusion. We interpret *low* as *not high.*

(1)	$u \vee p$	
(2)	$h \leftrightarrow i$	
(3)	$i \rightarrow \neg u$	
(4)	$\neg h \rightarrow p$	
(5)	$\neg p$	
(6)	$\neg p \rightarrow h$	(contrapositive of 4)
(7)	h	(from 5 and 6, law of detachment)
(8)	i	(from 7 and 2, substitution)
(9)	$\neg u$	(from 8 and 3, detachment)
(10)	p	(from 1 and 9, see Exercise 2.20)
(11)	$\neg p \wedge p$	(from 5 and 10)

Let $p_1, \ldots, p_n, p_{n+1}$ be an argument in which $p_1, \ldots, p_n$ are premises and p_{n+1} is the conclusion. The argument is valid if $(p_1 \wedge \cdots \wedge p_n) \rightarrow p_{n+1}$ is a tautology. This observation provides an alternative method for proving the validity of an argument. The proof is an examination of the Boolean function $f\colon B^{n+1} \rightarrow B$, where $B = \{\mathrm{F}, \mathrm{T}\}$, corresponding to the form $(p_1 \wedge \cdots \wedge p_n) \rightarrow p_{n+1}$. The argument is valid if the range of f is $\{\mathrm{T}\}$.

Example

The forms corresponding to the arguments in Examples 1 and 2 above are

$$(p \wedge (p \rightarrow q) \wedge (\neg r \rightarrow \neg q)) \rightarrow r,$$
$$((u \vee p) \wedge (h \leftrightarrow i) \wedge (i \rightarrow \neg u) \wedge (\neg h \rightarrow p)) \rightarrow p.$$

2d. Relations

Correspondences between elements of sets A and B are defined by subsets of the set of ordered pairs $A \times B$. Correspondences may be one–one, many–one, one–many, and many–many. Only one–one and many–one correspondences have distinct first coordinates; they alone are functions from A into B. By a subterfuge we may also represent one–many correspondences by functions. While the subsets of $A \times B$ that define such correspondences are not functions, interchange of coordinates produces sets of ordered pairs in which first coordinates are all distinct. These sets are therefore functions from B into A. In Example 1 of D.2.1 we had the nonfunction $\{\langle 4, \textit{five}\rangle, \langle 4, \textit{aces}\rangle$ $\langle 5, \textit{spell}\rangle, \langle 7, \textit{tragedy}\rangle\}$. The set $\{\langle \textit{five}, 4\rangle, \langle \textit{aces}, 4\rangle, \langle \textit{spell}, 5\rangle, \langle \textit{tragedy}, 7\rangle\}$, however, is a function. This expedient clearly does not work for many–many correspondences. Therefore, we need a theory more general than that of functions.

Let us assume that the four children of some family are called Thomas, Mary, Richard, and Henry. Abbreviating their names to initials, we have the set $C = \{T, M, R, H\}$. Consider now a subset of $C \times C$, the set of ordered pairs $\{\langle T, M\rangle, \langle T, R\rangle, \langle T, H\rangle, \langle R, T\rangle, \langle R, M\rangle, \langle R, H\rangle, \langle H, T\rangle,$ $\langle H, M\rangle, \langle H, R\rangle\}$. The correspondence defined by this set is expressed by the formula x *is the brother of* y. Here x and y are variables, and *is the brother of* is a relation. If, on substituting constants for the variables, the formula becomes a true statement, the ordered pair of the constants belongs to the relation. Order is important: T *is the brother of* M is true, but M *is the brother of* T is not. Consequently $\langle T, M\rangle$ is a member of the relation, but not $\langle M, T\rangle$. All members of the set of ordered pairs given above belong to the relation *is the brother of*, and no other member of $C \times C$ belongs to this relation. The relation with formula x *and* y *are brothers* is the set $\{\langle T, R\rangle,$ $\langle T, H\rangle, \langle R, T\rangle, \langle R, H\rangle, \langle H, T\rangle, \langle H, R\rangle\}$.

DEFINITION 2.12 A *binary relation* from set A to set B is a subset of $A \times B$. If R is a relation, we write $\langle x, y\rangle \in R$ and xRy interchangeably. If $\langle x, y\rangle \in R$, we say that x is *R-related* to y. (If $\langle x, y\rangle \notin R$, we may write $x\not{R}y$.)

Examples

1. Let $A = \{a, b, c\}$, and let $C_n(A)$ be the set of n-combinations and $P_n(A)$ the set of n-permutations of A. Let R_n be a relation from C_n to P_n such that qR_ns if and only if the n elements of A in s are the same as in q. Here $C_2 = \{(ab), (ac), (bc)\}$ and $C_3 = \{(abc)\}$. Relation R_2 is $\{\langle (ab), ab\rangle,$ $\langle (ab), ba\rangle, \langle (ac), ac\rangle, \langle (ac), ca\rangle, \langle (bc), bc\rangle, \langle (bc), cb\rangle\}$. All six members of R_3 have the same first coordinate. Some of them are $\langle (abc), abc\rangle$, $\langle (abc),$ $bac\rangle$, $\langle (abc), cab\rangle$. Note that $R_3 = C_3 \times P_3$, but that $R_2 \subset C_2 \times P_2$ $(\langle (ab), ca\rangle \notin R_2)$.

2. Every function is, of course, a relation. Recall our note following Th.1.9 to the effect that ′ does not have to be introduced in Boolean algebra B as a function. It is sufficient to introduce it more generally as a relation; the fact that it is also a function is derived in the theory as Th.1.9.

3. Let I be the set of integers. The formula $a < b$, where variables a and b stand for members of I and $<$ has the conventional meaning, defines a binary relation from I to I. Thus $2 < 5$ or $\langle 2, 5\rangle \in <$, but $\langle 5, 2\rangle \notin <$. We can, however, have a relation $\nless$ as well: then $\langle 5, 2\rangle \in \nless$, and $\langle 2, 5\rangle \notin \nless$.

Definition 2.13 A subset of $A \times A$ is a binary relation *in* the set A. In particular, the set $A \times A$ is the *universal* relation in A.

Examples

1. Any relation in a set is a subset of the universal relation in this set. The relations $<$ and $\nless$ of Example 3 of D.2.12 are relations *in* I. Their union, $< \cup \nless$, is the universal relation in I, and $< \cap \nless = \varnothing$.

2. The set of 2-samples $S_2(A)$ is a universal relation in A.

3. Let R be the set of real numbers. The square root relation can be defined by $\{\langle x^{1/2}, x\rangle \mid x \in R\}$ or by $\{\langle x, x^{1/2}\rangle \mid x \in R\}$. Both sets are relations in R. Only one of the sets is a function (cf. Example 2 of D.2.1).

4. We shall see that the theory of relations is almost exclusively concerned with relations in a set. This has led some writers to define relations in a set as the only relations. Since $A \times B$ is a subset of $(A \cup B) \times (A \cup B)$, a relation from A to B is also a relation in $A \cup B$; i.e., every relation is in fact a relation in some set. The alternative definition is, therefore, not unduly restrictive.

Definition 2.14 Let R be a relation and let A be a set. Then

$$R[A] = \{y \mid \text{for some } x \text{ in } A,\ xRy\}$$

is called the set of *R-relatives* of the elements of A.

Examples

1. For relation R_2 of Example 1 of D.2.12, $R_2[\{(ab)\}] = \{ab, ba\}$ and $R_2[C_2(A)] = \{ab, ba, ac, ca, bc, cb\}$. $R_2[C_3(A)] = \varnothing$.

2. Let U be a universal relation in a set A. Then $U[A] = A$.

3. Let $I_n = \{1, 2, \ldots, n\}$. For relation $>$ in I_{100}, $>[I_1] = \varnothing$ and $>[I_n] = I_{n-1}$ $(n = 2, \ldots, 100)$.

Definition 2.15 The *domain* D_ρ and *range* R_ρ of a relation ρ are defined as follows:

$$D_\rho = \{x \mid \text{for some } y,\quad \langle x, y\rangle \in \rho\},$$
$$R_\rho = \{y \mid \text{for some } x.\quad \langle x, y\rangle \in \rho\}.$$

Examples

1. The domain of a relation is the set of all first coordinates and the range is the set of all second coordinates of the relation. Let ρ be a relation. Then $\rho[D_\rho] = R_\rho$ and, for any set A, $\rho[A] \subseteq R_\rho$.

2. $D_{A \times B} = A$ provided $B \neq \varnothing$, and $R_{A \times B} = B$ provided $A \neq \varnothing$. If $B = \varnothing$ or $A = \varnothing$, then $A \times B = \varnothing$. Clearly a null relation cannot have a nonnull domain or range.

3. Consider the set $C = \{T, M, R, H\}$ defined at the beginning of this section. The domain of the relation *x is the brother of y* in C is $\{T, R, H\}$. The range is $\{T, M, R, H\}$. The relation *x and y are brothers* has the same domain, but the range contracts to $\{T, R, H\}$, i.e., the relation has the same set for its domain and range.

DEFINITION 2.16 If R is a relation, the *converse* (*reversed*) *relation* of R, written R^{-1}, is a relation such that $yR^{-1}x$ if and only if xRy.

Examples

1. Let R be the relation $\{\langle n^{1/2}, n\rangle \mid n \in I\}$ in the set of integers. The converse R^{-1} is the relation $\{\langle n, n^{1/2}\rangle \mid n \in I\}$.

2. The converse of relation R_2 of Example 1 of D.2.12 is $\{\langle ab, (ab)\rangle, \langle ba, (ab)\rangle, \langle ac, (ac)\rangle, \langle ca, (ac)\rangle, \langle bc, (bc)\rangle, \langle cb, (bc)\rangle\}$.

3. The set $\{x \mid \text{for some } y \text{ in set } B,\ x\rho y\}$ is the set of ρ^{-1}-relatives of B. Thus, $\rho^{-1}[R_\rho] = D_\rho$, and $\rho^{-1}[\{b\}]$ is the set of first coordinates of all ordered pairs belonging to ρ that have b as the second coordinate; e.g., for R_2 of Example 1 of D.2.12, $R_2^{-1}[(ac)] = \{ac, ca\}$.

4. The relation *x and y are brothers* is its own converse.

DEFINITION 2.17 A relation R in a set A is

(i) *reflexive* if xRx for all $x \in A$;
(ii) *irreflexive* if xRx for no $x \in A$;
(iii) *symmetric* if xRy implies yRx for all $x, y \in A$;
(iv) *antisymmetric* if xRy and yRx imply $x = y$ for all $x, y \in A$;
(v) *transitive* if xRy and yRz imply xRz for all $x, y, z \in A$.

Examples

1. The relation *x is the brother of y* in a set of males is irreflexive (no one is his own brother) and symmetric. At first sight the relation appears to be transitive: *Henry is the brother of Thomas* and *Thomas is the brother of Richard* do imply *Henry is the brother of Richard.* But *Henry is the brother of Thomas* implies *Thomas is the brother of Henry*, and, if the relation were transitive, the unacceptable *Henry is the brother of Henry* would be implied.

2. Let R be the relation *course a is a prerequisite for course b* in the set of courses offered by a university. Relation R is irreflexive and transitive. It is not symmetric.

3. The relation $\leqq$ in a set of numbers is reflexive, antisymmetric, and transitive. The relation $<$ is irreflexive, antisymmetric, and transitive.

If a proper substitution is made in the ordered pair $\langle\langle$*father*, *mother*$\rangle$, *their child*$\rangle$ for the variables, the resulting ordered pair is a member of the parenthood relation. Since here the ordered pair can be considered also an ordered triple, the parenthood relation may be called a *ternary relation.* Generalizing, a binary relation from a set of $(n - 1)$-tuples to a set of simple elements can be considered a set of n-tuples. It may be called, therefore, an *n-ary relation.* This is simply a matter of terminology: An n-ary relation is not a new concept; it is still essentially a set of ordered pairs of ordered $(n - 1)$-tuples and simple elements.

2e. The Equivalence Relation

DEFINITION 2.18 A relation in a set is an *equivalence relation* if it is reflexive, symmetric, and transitive.

Examples

1. Equality in a set of numbers is an equivalence relation. In the set of integers the relation $=$ is an identity relation, and this relation is a rather trivial equivalence relation (it makes little sense to speak of symmetry and transitivity when $x = y$ if and only if x and y are identical; in fact the relation $=$ is both symmetric and antisymmetric in a trivial sense here). In the set of rational numbers we can have, for example, $2/3 = 4/6$ and $4/6 = 200/300$. Therefore, equality in the set of rational numbers is a nontrivial equivalence relation. Equality of moduli in a set of complex numbers is another nontrivial equivalence relation: $\langle 2 + 5i, 5 + 2i\rangle$ belongs to the relation.

2. Let $\langle B, \oplus, *, ', 0, 1\rangle$ be a Boolean algebra. The relation $=$ in the set of forms in variables standing for elements of B, the distinct elements 0 and 1, and the operations $\oplus$, $*$, and $'$ is an equivalence relation.

3. The relation $\leftrightarrow$ in a set of statements (prime or composite) is an equivalence relation. Note that the symbol $\leftrightarrow$ has two interpretations in the statement calculus: $p \leftrightarrow q$ can denote a statement, which may be false for certain truth values of p and q, or it can represent a biconditional tautology. We have the second interpretation in mind when we speak of the *relation* $\leftrightarrow$. (The tautology $p \leftrightarrow q$ is sometimes denoted $\vdash p \leftrightarrow q$.)

4. The relation x *has the same image as* y in the domain of a many–one function is an equivalence relation. The domain of the function is both domain and range of the relation. It is clear that the domain of any equivalence relation is the set in which the relation is defined. One can therefore speak of an equivalence relation *on* a set. The equivalence relation x *has the same image as* y is on the domain of a function.

DEFINITION 2.19 Let R be an equivalence relation on a set A. Consider an element a of A. The set of R-relatives of a in A, $R[\{a\}]$, is called the *R-equivalence class generated* by a. Where there is no danger of confusion, the symbol $R[\{a\}]$ can be abbreviated to $[a]$.

THEOREM 2.9 Let R be an equivalence relation on A and let $a, b \in A$. Then

(i) $a \in [a]$,
(ii) if aRb, then $[a] = [b]$.

Proof. The first part is a direct consequence of reflexivity of an equivalence relation. To prove the second part, assume aRb and let x be any element such that $x \in [b]$. Then bRx, and, by transitivity, aRx. Hence $x \in [b]$ implies $x \in [a]$, i.e., $[b] \subseteq [a]$. By symmetry, bRa. A similar argument gives $[a] \subseteq [b]$.

Part (i) of Th.2.9 implies that every element of a set on which an equivalence relation R is defined belongs to some R-equivalence class, i.e., that the union of all R-equivalence classes generated by elements of this set is the set itself. Part (ii) means that any element of an equivalence class can be used to represent the equivalence class. Looking at the examples of D.2.18 in the light of these results one begins to suspect that the R-equivalence classes constitute a partition of the set on which R is defined. In our next theorem we turn suspicion into fact.

THEOREM 2.10 Let X be the set of equivalence relations on a set A and Y the set of partitions of A. Let ρ be any member of X. There exists a one-to-one onto function f: $X \to Y$ such that $f(\rho)$ is the set of ρ-equivalence classes generated by elements of A.

Proof. The proof can be broken down into four parts.

(i) The set of ρ-equivalence classes is a partition of A. Part (i) of Th.2.9 implies that the union of the ρ-equivalence classes is A. Let $a, b \in A$. These elements generate $[a]$ and $[b]$. Now let $x \in [a]$ and $x \in [b]$. By part (ii) of Th.2.9 we have $[x] = [a]$ and $[x] = [b]$. Hence $[a] = [b]$, i.e., equivalence classes generated by members of A are either disjoint or equal.

(ii) The set of ρ-equivalence classes is unique, i.e., f exists. Assume that ρ defines two different partitions. Then there exists some $a \in A$ such that the equivalence classes generated by a are different in the two partitions. This, in turn, means that there exists some element b that belongs to one of the classes but not to the other, i.e., $\langle a, b\rangle \in \rho$ and $\langle a, b\rangle \notin \rho$. This is a contradiction.

(iii) Function f is one to one. Let ρ_1 and ρ_2 be equivalence relations on A. By D.2.4, function f is one to one if and only if $\rho_1 \neq \rho_2$ implies $f(\rho_1) \neq f(\rho_2)$. Assume $\rho_1 \neq \rho_2$. Then we may assume that there exist elements a and b such that $\langle a, b\rangle$ belongs to one of the relations but not to the other. Let the equivalence classes generated by a be denoted by $[a]_1$ in $f(\rho_1)$ and by $[a]_2$ in $f(\rho_2)$. Then $b \in [a]_1$ and $b \notin [a]_2$, or $b \notin [a]_1$ and $b \in [a]_2$. In either case $[a_1] \neq [a]_2$. But $[a]_1$ and $[a]_2$ are the only sets in the respective partitions containing element a. Hence $f(\rho_1) \neq f(\rho_2)$.

(iv) Function f is onto. Consider any partition $\mathscr{A} = \{A_1, A_2, \ldots, A_n\}$ of A and a relation α such that $\langle a, b\rangle \in \alpha$ if and only if both a and b belong to one and the same A_i for some $i \in \{1, 2, \ldots, n\}$. If α is an equivalence relation, then members of $\mathscr{A}$ are the α-equivalence classes, and—since $\mathscr{A}$ is any partition—f is onto. We show that α is an equivalence relation. Let $a \in A$. Since $\mathscr{A}$ is a partition, $a \in A_i$ for some $i \in \{1, 2, \ldots, n\}$ and $a \notin A_j$ when $j \neq i$. Hence $\langle a, a\rangle \in \alpha$; i.e., α is reflexive. Relation α is symmetric: If $\langle a, b\rangle \in \alpha$, then $a, b \in A_i$ for some i; hence $\langle b, a\rangle \in \alpha$. If $\langle a, b\rangle \in \alpha$ and $\langle b, c\rangle \in \alpha$, then $a, b \in A_i$ and $b, c \in A_j$ for some i and j. But if $b \in A_i$, then $b \in A_j$ only if $i = j$. Hence $a, c \in A_i$, giving $\langle a, c\rangle \in \alpha$; i.e., the relation is transitive.

Examples

1. It is easy to make errors in defining the equivalence relation that corresponds to a given partition. A population of males can be partitioned into sets of brothers. The relation of brotherhood, however, is not reflexive. Hence it is not an equivalence relation. The relation that corresponds to this partition might be formulated as *x is y or the brother of y* or as *x has the same parents as y*. The set $\{1, 2, \ldots, 100\}$ can be partitioned as $\{\{1, 100\}, \{2, 99\}, \ldots, \{50, 51\}\}$. The relation $101 - a = b$ has nothing to do with the partition.

2. Let a, b, m be integers ($m \neq 0$). The relation *congruence modulo m* in the set of integers is defined as follows: a is congruent to b, modulo m, if and only if $a - b$ is divisible by m. We write $a = b(\text{mod } m)$, e.g., $15 = 5(\text{mod } 5)$, but $15 \neq 5(\text{mod } 4)$. Congruence modulo m is an equivalence relation with the number of equivalence classes in the partition defined by this relation being equal to m. Thus, when $m = 4$, the set of all integers is partitioned into the four equivalence classes $[0] = \{\ldots, -4, 0, 4, 8, \ldots\}, [1] = \{\ldots, -3, 1, 5, \ldots\}$, $[2] = \{\ldots, -2, 2, 6, \ldots\}$, $[3] = \{\ldots, -1, 3, 7, \ldots\}$.

2f. Ordering Relations

Definition 2.20 A reflexive, antisymmetric, and transitive relation in a set is a *partial order relation* or a *partial ordering* in that set. If R is a partial ordering in A, the ordered pair $\langle A, R\rangle$ is a *partially ordered set.*

Examples

1. The relation *less than or equal*, symbolized $\leqq$, in any subset of the set of real numbers is a partial ordering. The relation *less than* is irreflexive, and, hence, not a partial ordering.

2. The subset relation in a collection of sets is a partial ordering, but the proper subset relation is not.

3. The relations *is an integral multiple of* and *divides* in the natural numbers are partial orderings. Note that the relations are converses of each other. Generalizing, if a relation is a partial ordering, then so is its converse (Exercise 2.38).

4. Let R and R^* denote the relation *divides* in the set of natural numbers and in the set $\{3, 5, 15\}$, respectively. Then $R^* = \{\langle 3, 3\rangle, \langle 3, 15\rangle, \langle 5, 5\rangle, \langle 5, 15\rangle, \langle 15,15\rangle\}$. This is a partial ordering. Now let $A = \{5, 15\}$ and $B = \{3, 30\}$. Then $R^* \cap (A \times A) = \{\langle 5, 5\rangle, \langle 5, 15\rangle, \langle 15, 15\rangle\}$, a partial ordering in A. But $R^* \cap (B \times B)$ is not a partial ordering in B; since $\langle 30, 30\rangle \notin R^* \cap (B \times B)$, the relation is not reflexive. Relation $R \cap (B \times B)$, however, is a partial ordering. If A and B are any sets, and R is a partial ordering in A, we say that R *partially orders* B if and only if $R \cap (B \times B)$ is a partial ordering in B. In particular, R partially orders B if $B \subseteq A$.

It is customary to use the symbols $\leqq$ and $\geqq$ for the designation of partial order relations. Although the symbols derive from the natural ordering of numbers, here they need not have anything to do with the comparison of numerical values. Thus, one may denote the partial ordering *is an integral multiple of* by $\leqq$. We have then, for example, $\langle 30, 5\rangle \in\ \leqq$ or $30 \leqq 5$. (It would be more suggestive, however, to use the symbol $\geqq$ here.)

Definition 2.21 Define relation $<$ in a set A as follows: For $a, b \in A$, $a < b$ if and only if $a \leqq b$ and $a \neq b$. If $a < b$ we say that *a precedes* (is *less than*) b, or that *b follows* (is *greater than*) a.

Example

Consider the set R^* of Example 4 of D.2.20. It is a partial ordering, which we denote by $\leqq$. Here relation $<$ is $\{\langle 3, 15\rangle, \langle 5, 15\rangle\}$. The set $R^* - <$ is the identity relation $\{\langle 3, 3\rangle, \langle 5, 5\rangle, \langle 15, 15\rangle\}$.

Definition 2.22 A partial order relation $\leqq$ in a set A is a *simple* (or *linear*) *ordering* if and only if $a \leqq b$ or $b \leqq a$ for all $a, b \in A$. If $\leqq$ is a simple ordering in A, the ordered pair $\langle A, \leqq \rangle$ is a *simply ordered set* or *chain*. (If, for some $a, b \in A$, neither $a \leqq b$ nor $b \leqq a$, then a and b are said to be *incomparable*.)

Examples

1. The partial ordering *less than or equal* in the set of natural numbers $\{1, 2, 3, \ldots\}$ is a simple ordering. For any two numbers in the set, one is less than or equal to the other.

2. Consider $A = \{1, 2, 3, 5, 6, 10, 15, 30\}$, the set of numbers that divide 30. Set A is partially ordered by relation *less than or equal*, which we denote by $\leqq$, or by the relation *divides*, denoted by $\leqq'$. Then $\langle A, \leqq \rangle$ and $(A, \leqq' \rangle$ are different partially ordered sets. Only $\langle A, \leqq \rangle$ is a chain.

3. The subset relation in any family of sets is a partial ordering. Only rarely is it a simple ordering. Let $A = \{1, 2, 3\}$. Then $\mathscr{P}(A) = \{\varnothing, \{1\}, \{2\}, \{3\}, \{1, 2\}, \{1, 3\}, \{2, 3\}, A\}$. The pair $\langle \mathscr{P}(A), \subseteq \rangle$ is not a simply ordered set: Since, for example, $\{1, 2\} \nsubseteq \{2, 3\}$ and $\{2, 3\} \nsubseteq \{1, 2\}$, sets $\{1, 2\}$ and $\{2, 3\}$ are incomparable. The pair $\langle \mathscr{A}, \subseteq \rangle$, where $\mathscr{A} = \{\varnothing, \{1\}, \{1, 3\}, A\}$, is a simply ordered set.

DEFINITION 2.23 Let ρ be a simple ordering in a set A.

We define a *lexicographic ordering* ρ' in set B, where B is A^n or $\bigcup_n A^n (n = 1, 2, 3, \ldots)$. Let $\langle a_1, \ldots, a_p \rangle$, $\langle b_1, \ldots, b_q \rangle \in B$ and assume $p \leqq q$. Then $\langle a_1, \ldots, a_p \rangle \rho' \langle b_1, \ldots, b_q \rangle$ if one of the following holds:

(i) $\langle a_1, \ldots, a_p \rangle = \langle b_1, \ldots, b_p \rangle$;

(ii) $a_1 \neq b_1$ and $a_1 \rho b_1$ in A;

(iii) $a_i = b_i$, $i = 1, 2, \ldots, k$ $(k < p)$; and $a_{k+1} \neq b_{k+1}$ and $a_{k+1} \rho b_{k+1}$ in A.

Otherwise $\langle b_1, \ldots, b_q \rangle \rho' \langle a_1, \ldots, a_p \rangle$.

Examples

1. Let $A = \{0, 1, \ldots, 9\}$ and $B = A^5$. Then, for example, using our abbreviated notation for samples, 12304, 00012, 12302 $\in B$. Let ρ be the relation $\leqq$ (less than or equal) in A. By Condition (i), $12304 \rho' 12304$. Consider 12304 and 00012. Since the first digits differ, and $0 \leqq 1$, we have $00012 \rho' 12304$. Similarly, also by Condition (ii), $00012 \rho' 12302$. With 12304 and 12302, since the first four digits agree, and the fifth digits are different, we consider Condition (iii). Since $2 \leqq 4$, $12302 \rho' 12304$. A natural interpretation of the samples is to consider them as numbers. The relation ρ' is then the numerical relation *less than or equal*.

2. Let $A = \{a, b, c, \ldots, z\}$ and $B = \bigcup_{i=1}^{i=6} A^i$; B is the set of all "words" of six letters or less on the alphabet A. We let ρ be a relation $\leqq$ in A such that

$a \leqq b \leqq c \leqq \cdots \leqq z$. By Condition (i), *tablep'tablet*. By Condition (ii) *sablep'table* and *sablep'tablet*. By Condition (iii), *tabletp'taboo*. Relation ρ' defines the normal alphabetic order of words.

DEFINITION 2.24 A *least* member of a set A relative to a partial ordering $\leqq$ in A is an element $b \in A$ such that, for all $a \in A$, $b \leqq a$. Similarly, a *greatest* member of A relative to $\leqq$ is a $b \in A$ such that, for all $a \in A$, $a \leqq b$.

Examples

1. The least and greatest members of the set B of Example 1 of D.2.23 are 00000 and 99999, respectively. For set B of Example 2 of D.2.23 they are, respectively, *a* and *zzzzzz*.

2. Consider the family of sets $\mathscr{A} = \{\{1\}, \{2\}, \{3\}, \{1, 2\}, \{1, 3\}, \{2, 3\}\}$. Relative to partial ordering $\subseteq$ in $\mathscr{A}$ there is neither a least nor a greatest member of $\mathscr{A}$. Now consider $\mathscr{A} \cup \{\varnothing\}$. There is still no greatest member, but $\varnothing$ is the least member. With $\mathscr{A} \cup \{\varnothing\} \cup \{\{1, 2, 3\}\}$ we have both a greatest and a least member relative to $\subseteq$. The greatest member is $\{1, 2, 3\}$. Relative to some partial ordering, a set can have at most one least member and one greatest member (Exercise 2.41).

DEFINITION 2.25 A *minimal* member of set A relative to a partial ordering $\leqq$ in A is an element $b \in A$ such that, for no $a \in A$, $a < b$ (where $<$ is defined by D.2.21). Similarly, a *maximal* member is an element b such that $b < a$ for no $a \in A$.

Examples

1. If a set has a least (greatest) element, then the least (greatest) element is also a minimal (maximal) element (Exercise 2.41). If A is finite and $\langle A, \leqq \rangle$ is a chain, then A has a least and a greatest element. $\langle B, \rho' \rangle$, where ρ' is a lexicographic ordering in B, is a chain, and the least (greatest) elements of sets B in the examples of D.2.24 are unique minimal (maximal) elements of the respective sets. The infinite set of integers $I = \{\ldots, -2, -1, 0, 1, 2, \ldots\}$, together with the *less than or equal* relation $\leqq$, constitutes the chain $\langle I, \leqq \rangle$. Set I has neither a least nor a greatest element, and there are no minimal or maximal elements.

2. The fact that a set has a least (greatest) element implies that the set has exactly one minimal (maximal) element. The converse need not be true: The union of the infinite set of odd integers and the set $\{2, 4\}$ has exactly one maximal element, namely 4, with respect to the partial ordering *divides*. There is no greatest element.

3. Consider family $\mathscr{A}$ of Example 2 of D.2.24. Relative to partial ordering $\subseteq$ the family has three minimal elements $\{1\}, \{2\}, \{3\}$, and three maximal

elements $\{1, 2\}$, $\{1, 3\}$, $\{2, 3\}$. The family $\mathscr{A} \cup \{\varnothing\} \cup \{\{1, 2, 3\}\}$ has unique minimal and maximal elements.

DEFINITION 2.26 Let $\leqq$ be a partial ordering in set A, and let $<$ be the relation defined by D.2.21. Then $b \in A$ is a *cover* of $a \in A$ if and only if $a < b$ and there exists no $u \in A$ such that $a < u < b$.

Examples

1. The family $\{\{1\}, \{2\}, \{1, 2\}, \{1, 2, 3\}\}$ is partially ordered by $\subseteq$. Here $\{1, 2, 3\}$ is the cover of $\{1, 2\}$, and $\{1, 2\}$ covers both $\{1\}$ and $\{2\}$. But $\{1, 2, 3\}$ covers neither $\{1\}$ nor $\{2\}$ ($\{1\} \subset \{1, 2\} \subset \{1, 2, 3\}$ and $\{2\} \subset \{1, 2\} \subset \{1, 2, 3\}$). Members $\{1\}$ and $\{2\}$ are incomparable. Hence neither can cover the other.

2. Consider the partially ordered sets $\langle A, \leqq \rangle$ and $\langle A, \leqq' \rangle$ of Example 2 of D.2.22. We have 6 covering 5 with respect to $\leqq$ but not with respect to $\leqq'$ (6 covers 2 and 3 with respect to $\leqq'$).

DEFINITION 2.27 Let set A be partially ordered by $\leqq$ and let $\varnothing \subset B \subseteq A$. Then an element $a \in A$ is an *upper bound* (*lower bound*) of B if and only if, for all $b \in B$, $b \leqq a$ ($a \leqq b$). The least (greatest) element of the set of all upper (lower) bounds of B is the *least upper bound* (*greatest lower bound*) or *supremum* (*infimum*) of B, symbolized lub B (glb B) or sup B (inf B).

Examples

1. Let the power set $\mathscr{P}(\{1, 2, 3\})$ be partially ordered by $\subseteq$. Let $B_1 = \{\{1\}, \{2\}\}$ and $B_2 = \{\{1\}, \{1, 2\}\}$. The set of upper bounds of B_1 is $\{\{1, 2\}, \{1, 2, 3\}\}$. There is only one lower bound, the null set $\varnothing$. Then $\sup B_1 = \{1, 2\}$ and $\inf B_1 = \varnothing$. Neither belongs to B_1. The set of upper bounds of B_2 is also $\{\{1, 2\}, \{1, 2, 3\}\}$, but the set of lower bounds is $\{\varnothing, \{1\}\}$. Then $\sup B_2 = \{1, 2\}$ and $\inf B_2 = \{1\}$. Both supremum and infimum belong to B_2.

2. Let set $A = \{a, b, c\}$ be partially ordered by the identity relation $=$. The only subsets of A with upper or lower bounds are $\{a\}$, $\{b\}$, $\{c\}$. In each case the upper and lower bound is the single element belonging to the set. Each upper bound is a supremum and each lower bound an infimum, e.g., $\sup \{a\} = \inf \{a\} = a$.

Algorithm 2.2 is a procedure for the construction of diagrams that represent partially ordered sets. The diagrams help identify least and greatest elements, minimal and maximal elements, and the various bounds of subsets. The basis of the algorithm is the observation that if $\langle A, \leqq \rangle$ is a finite partially ordered set, then $a_1 < a_n$ in A if and only if there exists a chain $a_1 < a_2 < \cdots < a_n$ in which a_{i+1} covers a_i for $i = 1, 2, \ldots, n - 1$.

Algorithm 2.2 Consider the partially ordered set $\langle A, \leqq \rangle$. Represent each element $a_i \in A$ by a node α_i in the plane, and consider all ordered pairs $\langle a_i, a_j \rangle$ belonging to $A \times A$. Draw node α_i above node α_j if and only if $a_j < a_i$, and join nodes α_i and α_j by a line if a_i covers a_j. The result is a diagram for $\langle A, \leqq \rangle$ in which there is a sequence of joined lines ascending from node α_n to node α_m if $a_n < a_m$.

Example

Diagrams (a) and (b) of Figure 2.8 represent, respectively, the partially ordered sets $\langle A, \leqq \rangle$ and $\langle A, \leqq' \rangle$ of Example 2 of D.2.22. Diagram (c) represents the partially ordered set $\langle \mathscr{P}(\{1, 2, 3\}), \subseteq \rangle$. Diagram (d) represents $\langle \mathscr{A}, \subseteq \rangle$ of Example 2 of D.2.24, and Diagram (e) the partially ordered set $\langle A, = \rangle$ of Example 2 of D.2.27. Diagrams (b) and (c) are equal.

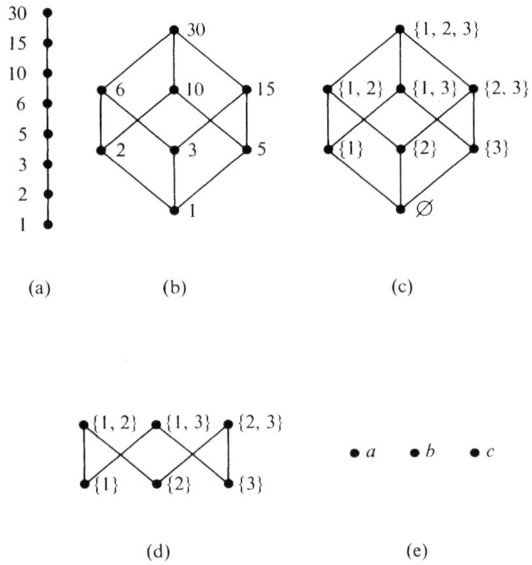

Figure 2.8

Any node from which there is no ascending line represents a maximal element, and, if there is only one such node, it also represents the greatest element. Nodes from which there are no descending lines represent minimal elements. If there is only one node with no descending lines, this node represents also the least element. We can have $A = B$ in the context of D.2.27. Therefore, the node representing the greatest (least) element of a partially ordered set also represents the supremum (infimum) of this set.

THEOREM 2.11 Every nonempty finite partially ordered set can be represented by a diagram.

Proof (by induction on n, the number of elements in the set).

(i) Basis. The only ordering of a set $\{a_1\}$ can be $a_1 \leqq a_1$. This is represented by a single node.

(ii) Induction step. Assume the theorem for a partially ordered set on $n-1$ elements. Let A be a set of n elements, partially ordered by $\leqq$. Since A is finite, it has at least one maximal element, say a_m (see Exercise 2.42). If a_m is omitted, the set $A - \{a_m\}$ is still partially ordered by $\leqq$, and it has $n-1$ elements. But this set has a diagram (induction hypothesis). Therefore, take this diagram and place a node α_m above it. Then, for all a_i that are covered by a_m, join α_i and α_m in the diagram. The result is a diagram for A.

DEFINITION 2.28 A function $f\colon A \to A'$ is *order preserving* relative to an ordering $\leqq$ for A and an ordering $\leqq'$ for A' if and only if $a \leqq b$ in A implies $f(a) \leqq' f(b)$ in A'. If there exist order preserving functions $f\colon A \to A'$ and $f^{-1}\colon A' \to A$, then the partially ordered sets $\langle A, \leqq\rangle$ and $\langle A', \leqq'\rangle$ are said to be *order-isomorphic*.

Examples

1. Consider the sets $\{1, 2, 3, 5, 6, 10, 15, 30\}$, partially ordered by *divides*, and $\mathscr{P}(\{1, 2, 3\})$, partially ordered by the subset relation. Let f be the function $\{\langle 1, \varnothing\rangle, \langle 2, \{1\}\rangle, \langle 3, \{2\}\rangle, \langle 5, \{3\}\rangle, \langle 6, \{1, 2\}\rangle, \langle 10, \{1, 3\}\rangle, \langle 15, \{2, 3\}\rangle, \langle 30, \{1, 2, 3\}\rangle\}$. It is easy to see that both f and f^{-1} are order preserving. This is particularly easy to see by reference to the diagrams of the partially ordered sets, Diagrams (b) and (c) of Figure 2.8.

2. Let $A = \{a_1, a_2, a_3, a_4\}$ and $B = \{b_1, b_2, b_3, b_4\}$. Define a partial ordering in A: $a_1 \leqq a_2$, $a_1 \leqq a_3$, $a_1 \leqq a_4$, $a_3 \leqq a_4$. Define also a partial ordering in B: $b_1 \leqq' b_2 \leqq' b_3 \leqq' b_4$. Then $f\colon A \to B$, defined by $\{\langle a_i, b_i\rangle \mid 1 \leqq i \leqq 4\}$, is order preserving. Clearly f is one to one. Hence $f^{-1}\colon B \to A$ exists. But f^{-1} is not order preserving; we have $b_2 \leqq' b_3$, but $f^{-1}(b_2) = a_2$ and $f^{-1}(b_3) = a_3$ are incomparable. Figure 2.9 represents $\langle A, \leqq\rangle$ and $\langle B, \leqq'\rangle$.

THEOREM 2.12 Two partially ordered sets can be represented by the same diagram if and only if they are order-isomorphic.

Proof. Denote two partially ordered sets by $\langle A, \leqq\rangle$ and $\langle A', \leqq'\rangle$, and assume that both sets are represented by the one diagram. In the coincident diagram each node α_i represents an $a_i \in A$ and an $a_i' \in A'$. Therefore, there exist functions $f\colon A \to A'$ and $f^{-1}\colon A' \to A$ such that $\langle a_i, a_i'\rangle \in f$ and $\langle a_i', a_i\rangle \in f^{-1}$. Since we have the same diagram, clearly $a_i \leqq a_j$ if and only if

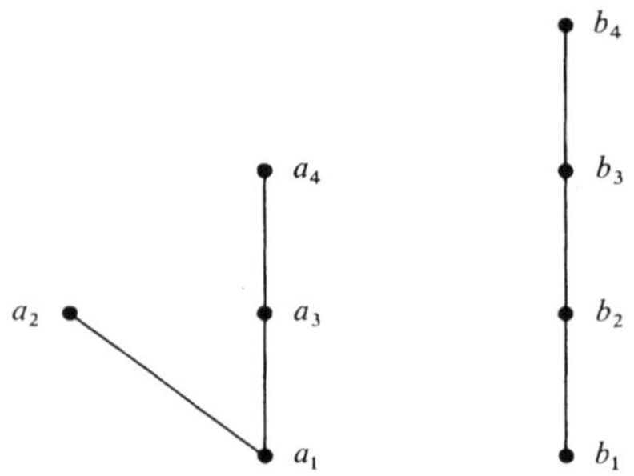

Figure 2.9

$a'_i \leqq a'_j$. Hence having the same diagram implies order-isomorphism. Conversely, assume that $\langle A, \leqq \rangle$ and $\langle A', \leqq' \rangle$ are order-isomorphic, and construct a diagram of A. There exists an order preserving function $f\colon A \to A'$, and, since f^{-1} also exists, f is one to one. Hence we can let node α_i, which represents $a_i \in A$, also represent the image of a_i under f. Clearly, since order-isomorphism means that $f(a_i) \leqq' f(a_j)$ if and only if $a_i \leqq a_j$, the diagram represents A' as well.

2g. Lattices

DEFINITION 2.29 Let $\langle A, \leqq \rangle$ be a partially ordered set. The system $\langle A, \leqq \rangle$ is a *lattice* if and only if each pair of elements a_i and a_j of A has a supremum and an infimum in A. We denote sup $\{a_i, a_j\}$ by $a_i \oplus a_j$ and inf $\{a_i, a_j\}$ by $a_i * a_j$.

Examples

1. Let $X = \{1, 2, 3, 5, 6, 10, 15, 30\}$ be partially ordered by relation *divides*, which we denote by D here. Every pair of elements of X has a supremum and an infimum. $\langle X, D \rangle$ is therefore a lattice. Since no subset of a set has more than one supremum and one infimum (relative to a given partial ordering $\leqq$), $\oplus$ and $*$ are well-defined operations (see Exercise 2.43). For a finite lattice $\langle A, \leqq \rangle$ we can therefore draw up a table of images for all elements of $A \times A$ under $\oplus$ and $*$. Table 2.13 is such an "addition" and "multiplication" table for $\langle X, D \rangle$. Since $x_i \oplus x_j = x_j \oplus x_i$, $x_i * x_j = x_j * x_i$, and $x_i \oplus x_i = x_i * x_i$, both operations can be defined by a single array in which the diagonal is shared. Entries above the diagonal constitute the rest of the "multiplication" table, and entries below the diagonal complete the "addition" table. We can denote a lattice $\langle A, \leqq \rangle$ by $\langle A, \oplus, * \rangle$. This notation emphasizes the operations in the lattice rather than its structure as a partially ordered set.

TABLE 2.13

OPERATIONS $\oplus$ AND $*$ IN $\langle X, D\rangle$

x_j \ x_i	1	2	3	5	6	10	15	30
1	1	1	1	1	1	1	1	1
2	2	2	1	1	2	2	1	2
3	3	6	3	1	3	1	3	3
5	5	10	15	5	1	5	5	5
6	6	6	6	30	6	2	3	6
10	10	10	30	10	30	10	5	10
15	15	30	15	15	30	30	15	15
30	30	30	30	30	30	30	30	30

Upper part (right of the diagonal): $x_i * x_j$; lower part (diagonal and below): $x_i \oplus x_j$

2. Any partially ordered set $\langle A', \leqq'\rangle$ that is order-isomorphic to a lattice $\langle A, \leqq\rangle$ is also a lattice. Let $f\colon A \to A'$ be the order preserving function associated with the isomorphism. Then, if a_k is the supremum (infimum) of $\{a_i, a_j\}$ relative to $\leqq$, $f(a_k)$ is the supremum (infimum) of $\{f(a_i), f(a_j)\}$ relative to $\leqq'$. The partially ordered set $\langle\mathscr{P}(\{1, 2, 3\}), \subseteq\rangle$ is order-isomorphic to the lattice $\langle X, D\rangle$ of Example 1; it is therefore also a lattice; i.e., the lattice may be denoted by $\langle\mathscr{P}(\{1, 2, 3\}), \cup, \cap\rangle$. The two lattices are represented by Diagrams (b) and (c) in Figure 2.8.

3. A partially ordered set need not be a lattice. Let $A = \{a, b, c\}$ be partially ordered by the identity relation $=$. Then $\langle A, =\rangle$ is a partially ordered set, but subsets $\{a, b\}$, $\{a, c\}$, $\{b, c\}$ of A have neither suprema nor infima (see Diagram (e) of Figure 2.8).

4. Consider lattice $\langle A, \oplus_A, *_A\rangle$. A lattice $\langle B, \oplus_B, *_B\rangle$, where $B \subseteq A$, is a *sublattice* of $\langle A, \oplus_A, *_A\rangle$ if $\oplus_B$ and $*_B$ are restrictions of $\oplus_A$ and $*_A$ to $B \times B$. Let $\langle X, D\rangle$ be the lattice of Example 1. Define the following subsets of X: $X_1 = \{2\}$, $X_2 = \{1, 2, 3\}$, $X_3 = \{2, 3, 6\}$, $X_4 = \{1, 2, 3, 6\}$, $X_5 = \{1, 2, 5, 6, 15, 30\}$. Now consider partially ordered sets $\langle X_i, D \cap (X_i \times X_i)\rangle$. They are lattices when $i = 1, 4, 5$, but not when $i = 2, 3$. The lattices are sublattices of $\langle X, D\rangle$ when $i = 1, 4$. But $2 \oplus 5 = 10$ in $\langle X, D\rangle$ and $2 \oplus 5 = 30$ in $\langle X_5, D \cap (X_5 \times X_5)\rangle$. Hence $\oplus$ is not a restriction in this instance (neither is $*$), and the lattice of X_5 is not a sublattice of $\langle X, D\rangle$.

5. Define a relation R in the family $\mathscr{P} = \{\mathscr{A}_i \mid \mathscr{A}_i$ is a partition of set $A\}$ as follows: $\mathscr{A}_m R \mathscr{A}_n$ if and only if every block of $\mathscr{A}_m$ is the subset of some block of $\mathscr{A}_n$. Relation R is expressed in words as *is a refinement of* (since $\mathscr{A}_m R \mathscr{A}_n$ implies that $\mathscr{A}_m$ has more blocks than $\mathscr{A}_n$ when $m \neq n$). Relation R is a partial ordering in $\mathscr{P}$ and the system $\langle\mathscr{P}, R\rangle$ is a lattice.

DEFINITION 2.30 A lattice $\langle A, \oplus, *\rangle$ is *distributive* if, for all $a, b, c \in A$,

$$a \oplus (b * c) = (a \oplus b) * (a \oplus c),$$
$$a * (b \oplus c) = (a * b) \oplus (a * c).$$

Examples

1. Diagrams (a), (b), and (c) of Figure 2.8 represent distributive lattices.
2. The lattice represented by Diagram (a) of Figure 2.10 is not distributive. We have $x_2 \oplus (x_3 * x_4) = x_2 \oplus x_5 = x_2$, but $(x_2 \oplus x_3) * (x_2 \oplus x_4) = x_1 * x_1 = x_1$. Diagram (b) of Figure 2.10 represents the lattice

$$\langle\{\varnothing, \{1\}, \{2\}, \{3\}, \{1, 2, 3\}\}, \subseteq\rangle.$$

Lattices represented by these diagrams are clearly order-isomorphic. Note here that lattice operations $\oplus$ and $*$ cannot be interpreted as set operations $\cup$ and $\cap$ in the lattice of the sets. We have, for example, $\{1\} \oplus \{2\} = \{1, 2, 3\} \neq \{1\} \cup \{2\}$.

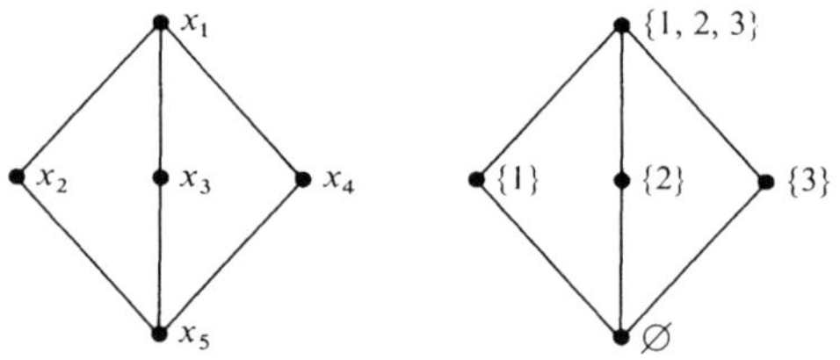

Figure 2.10

THEOREM 2.13 Let $\langle A, \leqq\rangle$ be a lattice. Then A has exactly one minimal (maximal) element relative to $\leqq$ if and only if it has a least (greatest) element relative to $\leqq$.

Proof. For any partially ordered set $\langle A, \leqq\rangle$, if A has a least element, then A has a unique minimal element (Exercise 2.41). The converse is true only for some partially ordered sets. To prove it for a lattice, let a and b be arbitrary elements of A. Since $a * b = \inf\{a, b\}$, it follows from D.2.27 that $a * b \leqq a$. Now, in particular, if a is a minimal element of A, then $a * b < a$ does not hold, and we have $a * b = a$. Since $a * b = a$ implies $a \leqq b$ (Exercise 2.52), and b is completely arbitrary, a is a least element. Analogously we can prove identity of maximal and greatest elements.

Example

Since finiteness of A guarantees existence of minimal and maximal elements (Exercise 2.42), we have a corollary: If $\langle A, \leqq\rangle$ is a finite lattice, A has unique minimal and maximal elements. In the lattices of Figure 2.10 elements x_5 and $\varnothing$ are minimal, and elements x_1 and $\{1, 2, 3\}$ maximal.

DEFINITION 2.31 If $\langle A, \leqq \rangle$ is a lattice, and A has maximal and minimal elements relative to $\leqq$, the (unique) maximal and minimal elements are called *bounds* of the lattice and the lattice is called a *bounded* lattice. The bounds are distinguished by special names: The maximal and minimal elements are given names 1 and 0, respectively.

DEFINITION 2.32 Let $\langle A, \leqq \rangle$ be a bounded lattice. Then $\langle A, \leqq \rangle$ (or $\langle A, \oplus, * \rangle$) is a *complemented* lattice if, for every $a \in A$, there exists an element $a' \in A$ such that $a \oplus a' = 1$ and $a * a' = 0$. An element a' satisfying these conditions is a *complement* of a.

Examples

1. If a lattice is not bounded, then it cannot be complemented. Even a bounded lattice need not be complemented. Lattice $\langle \mathscr{A}, \subseteq \rangle$ of Figure 2.11 has bounds $\varnothing$ and $\{1, 2, 3\}$. The only elements b that satisfy the equation $\{1, 3\} \oplus b = \{1, 2, 3\}$ are $\{1, 2\}$, $\{2, 3\}$, and $\{1, 2, 3\}$. But for these elements $\{1, 3\} * b \neq \varnothing$. On the other hand, in the lattice of Diagram (c) of Figure 2.8, $\{1, 3\} \oplus \{2\} = \{1, 2, 3\}$ and $\{1, 3\} * \{2\} = \varnothing$.

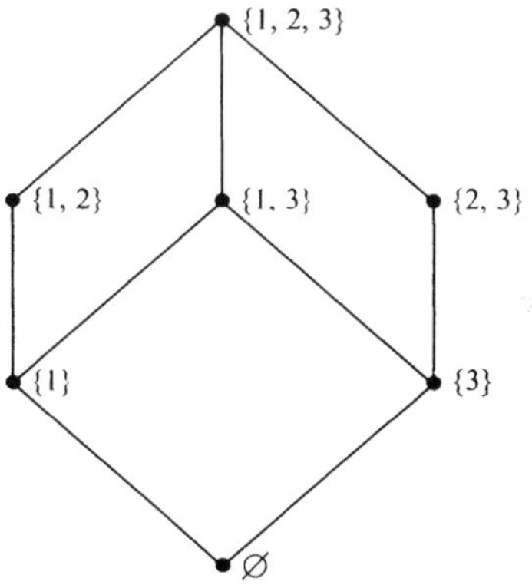

Figure 2.11

2. The lattices of Figure 2.10 are complemented, but only elements 1 and 0 have unique complements: $1' = 0$ and $0' = 1$. In the lattice of Diagram (a) we have that x_3 is a complement of x_2 ($x_2 \oplus x_3 = 1$ and $x_2 * x_3 = 0$), and also that x_4 is a complement of x_2 ($x_2 \oplus x_4 = 1$ and $x_2 * x_4 = 0$). Similarly, x_3 has complements x_2 and x_4, and x_4 has complements x_2 and x_3.

DEFINITION 2.33 Let $\langle A, \leqq \rangle$ be a complemented lattice. If each $a \in A$ has a unique complement in A, then $\langle A, \leqq \rangle$ is *uniquely complemented*.

Example

Lattice $\langle \mathscr{P}(\{1, 2, 3\}), \subseteq \rangle$ is uniquely complemented. We have already seen that the complemented lattices of Example 2 of D.2.32 are not uniquely complemented.

Let us now summarize our knowledge about lattices. Let $\langle A, \leqq \rangle$ be a lattice and let a, b, c be any elements of A. The following results hold for all lattices (see Exercise 2.53):

$$a \oplus b = b \oplus a, \qquad a * b = b * a;$$
$$a \oplus (b \oplus c) = (a \oplus b) \oplus c, \qquad a * (b * c) = (a * b) * c.$$

If $\langle A, \leqq \rangle$ is distributive, we have, in addition,

$$a \oplus (b * c) = (a \oplus b) * (a \oplus c),$$
$$a * (b \oplus c) = (a * b) \oplus (a * c).$$

If $\langle A, \leqq \rangle$ is bounded, then $0 \leqq a$ and $a \leqq 1$ for all $a \in A$. But this implies (see Exercise 2.52)

$$a \oplus 0 = a, \qquad a * 1 = a.$$

Finally, if $\langle A, \leqq \rangle$ is complemented, then, for each $a \in A$, there exists some a' such that

$$a \oplus a' = 1, \qquad a * a' = 0.$$

The properties of distributive complemented lattices are precisely the axioms of Boolean algebra. Boolean algebras are therefore lattices of a special type. Since lattices are more general algebraic systems than Boolean algebras, all theorems that hold for lattices hold for Boolean algebras. Since Boolean algebras are in fact distributive complemented lattices, every theorem that has been proven for Boolean algebras holds for such lattices, but not necessarily for lattices in general.

THEOREM 2.14 A complemented distributive lattice is uniquely complemented.

Proof. Th. 2.14 is Th. 1.9.

2h. Abstract Algebras

We have seen (in Section 2c) that some algebraic systems deriving from different application areas, such as switching algebra and the algebra of truth values, are so similar in most respects that they can be taken to be essentially the same system. The notion of similarity and sameness were, however, left vague. A formal discussion of these concepts has been deliberately postponed because their proper understanding requires reasonable familiarity with the abstract approach. This, we hope, has now been gained.

In quite general terms an algebra consists of a set A with some distinguished elements and functions from A^n into A. Since the partial ordering $\leqq$ is not a function, a lattice written as $\langle L, \leqq\rangle$ is not an algebra; written in the form $\langle L, \oplus, *\rangle$ is it an algebra.

DEFINITION 2.34 Let A be a nonempty set. Every function $f\colon A^n \to A$ ($n = 0, 1, 2, \ldots$) is an *n-argument operation* in A. By a 0-argument operation x in A we understand a *constant* (*distinguished*) *element* $x \in A$.

Example

In the complemented distributive lattice or Boolean algebra $\langle L, \oplus, *, ', 0, 1\rangle$ we have the set of operations $\{\oplus, *, ', 0, 1\}$. Operations $\oplus$ and $*$ are functions from L^2 into L, operation $'$ is a function from L into L, and 0 and 1 are zero-argument operations.

DEFINITION 2.35 Let f be an n-argument operation in A. Then a set B, such that $B \subseteq A$, is said to be *closed* under f if $f(x_1, x_2, \ldots, x_n) \in B$ for all $\langle x_1, x_2, \ldots, x_n\rangle \in B^n$.

Example

Consider algebra $\langle X, \oplus, *\rangle$ of Example 1 of D.2.29 and subsets X_i of X as defined in Example 4 of D.2.29. Inspection of Table 2.13 shows that sets X_1 and X_4 are closed under both operations, set X_2 under $*$, set X_3 under $\oplus$, but that set X_5 is not closed under either operation.

DEFINITION 2.36 Let A be a nonempty set. The system $\langle A, f_1, f_2, \ldots, f_m\rangle$, where each f_i is an n_i-argument operation in A, is an *abstract algebra*. (The set of operations of an abstract algebra does not have to be finite. We have made it finite purely for convenience.)

Example

$\langle L, \oplus, *\rangle$, where $\oplus$ and $*$ are two-argument operations in L, and $\langle B, \oplus, *, ', 0, 1\rangle$, where $\oplus$ and $*$ are two-argument operations in B, $'$ is a one-argument operation, and 0 and 1 are zero-argument operations, are abstract algebras. It is customary to denote an algebra by its set; here we may speak of the abstract algebras L and B. We cannot call L a lattice or B a a Boolean algebra unless we also specify the appropriate sets of axioms that the systems have to satisfy. But if we do so, then L and B cease to be *abstract* algebras.

DEFINITION 2.37 Let A be the abstract algebra $\langle A, f_1, f_2, \ldots, f_m \rangle$ and consider a subset B of A. An abstract algebra $\langle B, h_1, h_2, \ldots, h_m \rangle$, where each h_i is a restriction $f_i \mid B^{n_i}$ and B is closed under f_i, is a *subalgebra* of A.

Examples

1. It is customary to denote an operation and its restriction by the same symbol. The algebra $\langle \{0, 1\}, \oplus, *, ', 0, 1 \rangle$ is a subalgebra of $\langle B, \oplus, *, ', 0, 1 \rangle$. If $B \supset \{0, 1\}$, the $\oplus$ in algebra $\{0, 1\}$ is not the same as the $\oplus$ in algebra B. We use the same symbol because $\oplus$ in $\{0, 1\}$ is a restriction of the $\oplus$ in B to $\{0, 1\}^2$.

2. Consider algebra $\langle X, \oplus, * \rangle$ of Example 1 of D.2.29 and subsets X_i defined in Example 4 of D.2.29. The systems $\langle X_1, \oplus, * \rangle$ and $\langle X_4, \oplus, * \rangle$ are subalgebras of $\langle X, \oplus, * \rangle$. Although $\langle X_5, \oplus, * \rangle$ is an abstract algebra, it is not a subalgebra of $\langle X, \oplus, * \rangle$. Systems $\langle X_2, \oplus, * \rangle$ and $\langle X_3, \oplus, * \rangle$ are not even algebras ($2 \oplus 3$ is not defined in X_2 and $2 * 3$ is not defined in X_3). Note that $\langle X, \oplus, *, ', 1, 30 \rangle$ is also an abstract algebra. Then $\langle X_i, \oplus, *, ', 1, 30 \rangle$ can be an algebra only if 1 and 30 are members of X_i. Element 30 is not a member of X_1, X_2, X_3, or X_4. Both 1 and 30 are members of X_5, but $'$ is not an operation in X_5 (for example, 6 has both 5 and 15 for complements). The system $\langle X_4, \oplus, *, ', 1, 6 \rangle$ is an algebra, but, since an algebra and its subalgebras must have the same constant elements, this algebra cannot be a subalgebra of X.

DEFINITION 2.38 Abstract algebras $\langle A, f_1, f_2, \ldots, f_m \rangle$ and $\langle B, h_1, h_2, \ldots, h_n \rangle$ are *similar* if $m = n$ and each h_i has as many arguments as the corresponding f_i. Consider similar algebras A and B. A function $\phi\colon A \to B$ is a *homomorphism* of A into B (or onto B if ϕ is an onto function) if and only if $\phi(f_i(x_1, x_2, \ldots, x_{n_i})) = h_i(\phi(x_1), \phi(x_2), \ldots, \phi(x_{n_i}))$ for all $i = 1, 2, \ldots, m$ and for every n_i-sample of A. A homomorphism that is one to one and onto is an *isomorphism*. In particular, if sets A and B are equal, a homomorphism is called an *endomorphism* and an isomorphism is called an *automorphism*.

Examples

1. An algebra of sets $\langle A, \cup, \cap, \bar{}, \varnothing, U \rangle$ and the algebra of truth values $\langle \{\mathrm{F}, \mathrm{T}\}, \vee, \wedge, \neg, \mathrm{F}, \mathrm{T} \rangle$ are similar. Let $A = \mathscr{P}(\{a, b\})$ and define functions $\alpha\colon \{\mathrm{F}, \mathrm{T}\} \to A$, $\beta\colon \{\mathrm{F}, \mathrm{T}\} \to A$, $\gamma\colon A \to \{\mathrm{F}, \mathrm{T}\}$ such that $\alpha(\mathrm{F}) = \{a, b\}$ and $\alpha(\mathrm{T}) = \varnothing$, $\beta(\mathrm{F}) = \varnothing$ and $\beta(\mathrm{T}) = \{a, b\}$, and $\gamma(\{a, b\}) = \gamma(\{a\}) = \mathrm{T}$ and $\gamma(\{b\}) = \gamma(\varnothing) = \mathrm{F}$. Since, for example, $\alpha(\mathrm{F} \wedge \mathrm{T}) = \{a, b\}$ and $\alpha(\mathrm{F}) \cap \alpha(\mathrm{T}) = \varnothing$, α is not a homomorphism. Function β is a homomorphism into A (but not onto A). Function γ is a homomorphism onto $\{\mathrm{F}, \mathrm{T}\}$. Since, however, it is many to one, the function is not an isomorphism.

2. Let $A = \{\varnothing, U\}$ and $B = \{F, T\}$, and consider the algebras $\langle A, \cup, \cap, ^{-}, \varnothing, U\rangle$ and $\langle B, \vee, \wedge, \neg, F, T\rangle$. The function $\phi: A \to B$ such that $\phi(\varnothing) = F$, $\phi(U) = T$ is an isomorphism. If $\phi: A \to B$ is an isomorphism, then $\phi^{-1}: B \to A$ is also an isomorphism. We say then that algebras A and B are *isomorphic* to each other and that each algebra is an *isomorphic image* of the other.

3. Consider algebras of sets

$$\langle \mathscr{P}(A_1), \cup, \cap, ^{-}, \varnothing, U\rangle$$

and

$$\langle \mathscr{P}(A_2), \cap, \cup, ^{-}, U, \varnothing\rangle,$$

and let $A_1 = A_2$. Then the function $\phi: \mathscr{P}(A_1) \to \mathscr{P}(A_2)$ such that $\phi(X) = \overline{X}$ for all $X \in \mathscr{P}(A_1)$ is an automorphism.

THEOREM 2.15 Two lattices, considered as algebras, are isomorphic if and only if they are order-isomorphic when considered as partially ordered sets.

Proof. The proof is not particularly difficult. It does, however, depend on several theorems in the theory of lattices that we consider irrelevant to the purposes of this book.

Example

The partially ordered sets represented by Diagrams (b) and (c) of Figure 2.8 are order-isomorphic. They are lattices. The lattice of Diagram (b) can be considered as the algebra $\langle L_1, \mu, \delta\rangle$, where the two-argument operations μ and δ have as their respective images the least common multiple and the greatest common divisor of the arguments. The lattice of Diagram (c) can be considered as the algebra $\langle L_2, \cup, \cap\rangle$, where $\cup$ and $\cap$ have their usual set-algebraic meaning. The function $\{\langle 1, \varnothing\rangle, \langle 2, \{1\}\rangle, \langle 3, \{2\}\rangle, \langle 5, \{3\}\rangle, \langle 6, \{1, 2\}\rangle, \langle 10, \{1, 3\}\rangle, \langle 15, \{2, 3\}\rangle, \langle 30, \{1, 2, 3\}\rangle\}$ is an isomorphism. Hence L_1 and L_2 are isomorphic.

With the introduction of the concept of isomorphism we hope to have clarified the notion of essential sameness. If two abstract algebras are isomorphic, then, with respect to their structure, they can be regarded as one and the same algebra. We shall now link the concepts of isomorphism and equivalence classes and find that a number of results of great practical importance arise. Isomorphism is of particular importance in the study of Boolean algebras. Since a distributive complemented lattice is a Boolean algebra, we know already that finite Boolean algebras are isomorphic if and only if they have the same diagram when considered as partially ordered sets

(Th.2.12 and Th.2.15). But we are interested in infinite (not necessarily Boolean) algebras as well. For example, turning to statement calculi, we find that no bound can be put on a sequence of statements in just p and $\vee$ alone: $p,\ p \vee p,\ p \vee p \vee p,\ \ldots$. We shall, therefore, discuss isomorphism in rather general terms.

DEFINITION 2.39 If R is an equivalence relation on a set A, the partition of A induced by R is the *quotient set* of A by R, denoted A/R (and read *A modulo R*).

DEFINITION 2.40 An equivalence relation R on A is a *congruence* with respect to an operation $f\colon A^m \to A$ if, for all m-samples $\langle x_1, x_2, \ldots, x_m \rangle$ and $\langle y_1, y_2, \ldots, y_m \rangle$ of A, the conditions $x_1 R y_1, x_2 R y_2, \ldots, x_m R y_m$ imply $\langle f(x_1, x_2, \ldots, x_m), f(y_1, y_2, \ldots, y_m) \rangle \in R$. If relation R is a congruence with respect to every operation in an abstract algebra A, we call it a *congruence in A*.

Examples

1. The universal relation is a congruence. Then A/R has exactly one member. This member is the set A itself. If a congruence R is not a universal relation on A (i.e., if the quotient set has more than one member), relation R is called a *proper* congruence.

2. Consider a set of statements S and operations $\vee, \wedge, \neg$ in S. The biconditional $\leftrightarrow$, interpreted as a relation, is clearly an equivalence. We can show that $p \leftrightarrow s$ and $q \leftrightarrow t$ for $p, s, q, t \in S$ imply $(p \vee q) \leftrightarrow (s \vee t)$, $(p \wedge q) \leftrightarrow (s \wedge t)$, and $\neg p \leftrightarrow \neg s$ by means of truth tables (we do so for the operation $\vee$ in Table 2.14, where we need only consider truth value assignments to p, s, q, t for which the conditions $p \leftrightarrow s$ and $q \leftrightarrow t$ hold). Hence $\leftrightarrow$ is a congruence with respect to $\vee$, $\wedge$, and $\neg$. Since, for any $p \in S$, p and $\neg p$ belong to different equivalence classes, it is a proper congruence. Relation $\leftrightarrow$ is a congruence in the algebra $\langle S, \vee, \wedge, \neg \rangle$.

TABLE 2.14

CONGRUENCE OF $\leftrightarrow$ WITH RESPECT TO $\vee$

p	q	s	t	$p \leftrightarrow s$	$q \leftrightarrow t$	$(p \vee q) \leftrightarrow (s \vee t)$
T	T	T	T	T	T	T T T
T	F	T	F	T	T	T T T
F	T	F	T	T	T	T T T
F	F	F	F	T	T	F T F

THEOREM 2.16 Let R be a congruence in an algebra $\langle A, f_1, f_2, \ldots, f_m \rangle$. Then, corresponding to every f_i on A^{n_i}, there exists an operation f_i' on $(A/R)^{n_i}$ such that $f_i'([x_1], [x_2], \ldots, [x_{n_i}]) = [f_i(x_1, x_2, \ldots, x_{n_i})]$, i.e.,

$$\langle A/R, f_1', f_2', \ldots, f_m' \rangle$$

is an algebra similar to algebra A. (A/R is called the *quotient algebra* of A by R.)

Proof. Since $x_1 R y_1$, $x_2 R y_2$, $\ldots$, $x_{n_i} R y_{n_i}$ imply $\langle f_i(x_1, x_2, \ldots, x_{n_i}), f_i(y_1, y_2, \ldots, y_{n_i}) \rangle \in R$, we have that $[f_i(x_1, x_2, \ldots, x_{n_i})]$ is independent of the choice of elements $x_1, x_2, \ldots, x_{n_i}$ from the equivalence classes $[x_1], [x_2], \ldots, [x_{n_i}]$. Therefore, the definition of f_i' does in fact define an operation, $\langle A/R, f_1', f_2', \ldots, f_m' \rangle$ is an algebra, and clearly this algebra is similar to algebra $\langle A, f_1, f_2, \ldots, f_m \rangle$.

THEOREM 2.17 If R is a congruence in an abstract algebra A, then $f\colon A \to A/R$ such that $f(x) = [x]$ is an onto homomorphism (called the *natural* homomorphism from A onto A/R). Conversely, if $g\colon A \to B$ is a homomorphism, there exists a congruence Q in A, and the algebra A/Q is isomorphic to B.

Proof. That f is a homomorphism follows at once from the definition of homomorphism and the definition of the operations in A/R. The homomorphism must be onto: If $[x] \in A/R$, then clearly $x \in A$ and $f(x) = [x]$. Next consider an n-argument operation k in algebra A and the corresponding operation k' in the similar algebra B. Define a relation Q in A such that xQy if and only if $g(x) = g(y)$, and assume that conditions $x_1 Q y_1, \ldots, x_n Q y_n$ hold for some $x_1, \ldots, x_n, y_1, \ldots, y_n \in A$. By definition of homomorphism and the equalities $g(x_1) = g(y_1), \ldots, g(x_n) = g(y_n)$,

$$\begin{aligned} g(k(x_1, \ldots, x_n)) &= k'(g(x_1), \ldots, g(x_n)) \\ &= k'(g(y_1), \ldots, g(y_n)) \\ &= g(k(y_1, \ldots, y_n)). \end{aligned}$$

But this means that $\langle k(x_1, \ldots, x_n), k(y_1, \ldots, y_n) \rangle \in Q$. Hence Q is a congruence in A. Now define relation h as follows: $h = \{\langle [x], g(x) \rangle \mid [x] \in A/Q\}$. Clearly, by the definition of Q, h is a one-to-one function. Therefore, if h is a homomorphism, it is also an isomorphism. By Th.2.16, algebras A and A/Q are similar. Since g is a homomorphism, A and B are also similar, and, since similarity of algebras is clearly a transitive relation, algebras A, B, and A/Q are all similar. We introduce an n-argument operation ϕ in A/Q corresponding to operations k in A and k' in B. Then

$$\begin{aligned} h(\phi([x_1], \ldots, [x_n])) &= h([k(x_1, \ldots, x_n)]) \\ &= g(k(x_1, \ldots, x_n)) \\ &= k'(g(x_1), \ldots, g(x_n)) \\ &= k'(h([x_1]), \ldots, h([x_n])), \end{aligned}$$

and we have shown h to be a homomorphism and hence an isomorphism. (Since ϕ is an operation in a quotient algebra, Line 1 follows from Th.2.16. Lines 2 and 4 are given by the definition of h, and Line 3 is a consequence of the assumption that g is a homomorphism.)

By Th.2.17, the existence of a homomorphism $g: A \to B$ implies existence of a congruence R in A, and of an isomorphism from A/R onto B. Further, if homomorphism $g': A \to C$ also determines congruence R, then A/R, B, and C are all isomorphic, i.e., they are essentially the same algebra, namely A/R. Since the algebras are isomorphic, homomorphisms g and g' are not essentially different from the natural homomorphism $f: A \to A/R$. Discussion of onto homomorphisms of an algebra and the images under the homomorphisms can therefore be reduced to a consideration of natural homomorphisms and quotient algebras. It should be clear that natural homomorphisms on an algebra are in one-to-one correspondence with congruences in the algebra.

In the discussion of Boolean algebra B in Section 1e we could not explain the significance of the relation $=$. We are now in a position to assign a precise meaning to it. From an examination of the axioms and theorems we note that this relation must be an equivalence relation, and that if $\alpha(x_1, \ldots, x_n)$ and $\alpha(y_1, \ldots, y_n)$ are Boolean forms in variables $x_1, \ldots, x_n, y_1, \ldots, y_n$, which stand for elements of B, then $x_1 = y_1, \ldots, x_n = y_n$ should imply $\alpha(x_1, \ldots, x_n) = \alpha(y_1, \ldots, y_n)$. We assert that these conditions constitute an adequate definition of the relation. Comparison of the definition with D.2.40 shows that $=$ is a congruence in B. In a model of an algebra the congruence $=$ should be precisely defined. The congruence is then called the *natural* congruence for the model. In an algebra of sets the natural congruence is equality of sets, and the symbol $=$ denotes it nicely. Sometimes, however, it is more suggestive to use a symbol other than $=$ for a natural congruence. Thus, the symbol for the natural congruence in the algebra of truth values is $\leftrightarrow$.

In Example 2 of D.2.40 we found that relation $\leftrightarrow$ is a congruence in an algebra of statements $\langle S, \vee, \wedge, \neg\rangle$. Perhaps we could find zero-argument operations that would make S a Boolean algebra. The algebra of truth values suggests that we should look for some statement that is always true and some statement that is always false. But $(p \vee \neg p)$ is always true and $(p \wedge \neg p)$ always false for *every* $p \in S$. Consequently there is nothing unique about the constants. Indeed we can find no unique pair of constants that would make S a Boolean algebra. But $S/\leftrightarrow$ is a Boolean algebra. Clearly $(p \vee \neg p) \leftrightarrow (q \vee \neg q)$ and $(p \wedge \neg p) \leftrightarrow (q \wedge \neg q)$ for any $p, q \in S$. We can therefore define equivalence classes $T = [p \vee \neg p]$ and $\mathrm{F} = [p \wedge \neg p]$. Then the quotient algebra $\langle S/\leftrightarrow, \vee, \wedge, \neg, \mathrm{F}, \mathrm{T}\rangle$ is a Boolean algebra. Operations $\vee$, $\wedge$, and $\neg$ are analogs of the operations in S; they are defined by Th.2.16.

Notes

The general references of Chapter 1 cover the material of this chapter as well. To those we add [Ho66], which deals with applications of Boolean algebras, and [Ru65], which is a clear introduction to lattice theory.

Although not considered here at all, minimization of Boolean forms is an important topic of great practical significance. [Ha65a] contains a thorough discussion of minimization techniques. Besides parallel and series arrangements, there are bridge switching circuits (for an example see Exercise 2.14), which may need fewer switches than their equivalents in series and parallel form. They, too, are studied in [Ha65a]. The algebra of two-state switches can be generalized to a system in which switches have more than two states. A ternary switching algebra is studied in [He68].

The CODASYL Development Committee has published a report on an Algebra of Information based on the theory of functions. [Zz62a]—[Cl62] is a shortened version of the report; [Mc63a] contains an example of the use of the algebra of information in the formulation of data processing problems. [Me67] is a more recent paper taking the same approach.

A proof of Th.2.15 is found on p. 41 of [Sz63]. Stirling numbers of the second kind of Exercise 2.35 are tabulated in [Ab64].

Exercises

2.1 (a) Which of the following sets are functions?

(i) $\{\langle 1, 1\rangle, \langle 2, 4\rangle, \langle 3, 9\rangle, \langle 4, 16\rangle, \ldots\}$.
(ii) $\{\langle a, b, c\rangle, \langle a, c, b\rangle, \langle b, a, d\rangle\}$.
(iii) $\{\langle a, \langle b, c\rangle\rangle, \langle a, \langle c, b\rangle\rangle, \langle b, \langle a, d\rangle\rangle\}$.
(iv) $\{\langle\{a, b\}, c\rangle, \langle\{a, c\}, b\rangle, \langle\{b, a\}, d\rangle\}$.

2.2 (a) Consider the Cartesian product $A \times B$, with $|A| = m$ and $|B| = n$.

(i) What is the greatest number of elements that a set $C \subseteq A \times B$ may have if it is to be a function?
(ii) If f: $A \to B$ and $m < n$, can f be onto B?

2.3 (a) Prove Th.2.1.

2.4 (a) Prove Th.2.2.

2.5 (a) A function on a set S into the same set is said to be *in* S. How many one-to-one functions are there in a set of n elements? How many onto functions are there in this set? How many into functions?

2.6 (a) Show that the elements of the Cartesian product $A \times B$ are in one-to-one correspondence with the elements of $B \times A$.

2.7 (b) Let $B = \{0, 1\}$ and consider functions from B^2 into B. Let $f(0, 0) = f(0, 1) = 0$, $f(1, 0) = f(1, 1) = 1$; $g(0, 0) = g(0, 1) = 1$, $g(1, 0) = g(1, 1) = 0$. Define $f \oplus g$, $f * g$, $f * f'$, $g * f'$ by a table similar to Table 2.1.

2.8 (b) With reference to D.2.8, which of the following are Boolean forms?

(i) $x_1 \oplus x_2 * x_3 \oplus x_4 * x_5 \oplus x_6$;
(ii) $(((x_1))) * ((x_2))$;
(iii) $(((x_1) * (x_2 \oplus x_3))) * x_4)$;
(iv) $(((x_1)')')')$.

2.9 (b) What is the Boolean function corresponding to Boolean form $(x_1 \oplus x_2) * (x_1 \oplus x_3')$ of Table 2.3 when $B = \{0, 1\}$?

2.10 (b) Simplify the conjunctive normal form of Tables 2.4 and 2.5.

2.11 (b) Repeat the proof of Exercise 1.23.

2.12 (c) Write down the switching function corresponding to the circuit of Figure 2.12. Simplify the circuit.

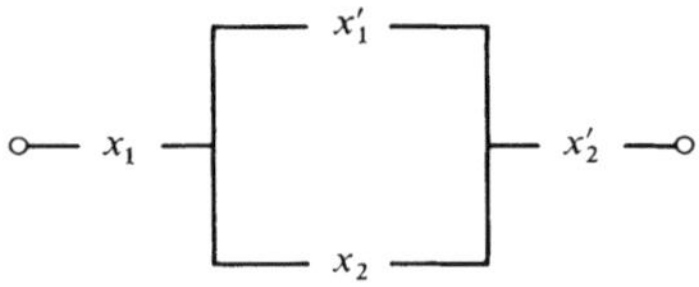

Figure 2.12

2.13 (c) Draw a circuit corresponding to the switching function $((x_1 * x_3) \oplus (x_2 * x_3'))'$.

2.14 (c) In all switching circuits considered in the text connections have been in series or in parallel. It is possible to reduce the number of switches necessary for the implementation of a switching function by means of bridge circuits. Figure 2.13 shows the bridge circuit corresponding to the series–parallel circuit of Figure 2.3. Show that the circuits are in fact equivalent.

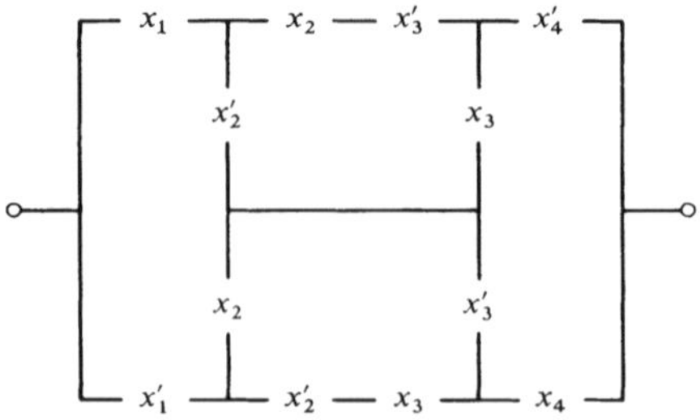

Figure 2.13

2.15 (c) Function f of Table 2.1 defines a NOR-gate. Using NOR-gates alone, design circuits that function as (i) inverter, (ii) OR-gate, (iii) AND-gate, (iv) NAND-gate.

2.16 (c) Design gating circuits with three inputs and a single output that give an output signal when

(i) exactly one of the inputs carries a signal,
(ii) at least two inputs carry signals,
(iii) exactly two inputs carry signals,
(iv) at least one and at most two inputs carry signals.

2.17 (c) A binary half-adder is a circuit having two inputs and two outputs, where the outputs are defined by Table 2.15. A full adder has three inputs (p, q, k) and two outputs (s, c), defined by Table 2.16. Design a half-adder, and, hence, using two half-adders and an OR-gate, the full adder.

TABLE 2.15

Input		Output	
p	q	s	c
0	0	0	0
0	1	1	0
1	0	1	0
1	1	0	1

TABLE 2.16

Input			Output	
p	q	k	s	c
0	0	0	0	0
0	0	1	1	0
0	1	0	1	0
0	1	1	0	1
1	0	0	1	0
1	0	1	0	1
1	1	0	0	1
1	1	1	1	1

2.18 (c) Define a (Boolean) statement form in the algebra of truth values.

2.19 (c) Show that $p \to q$ is functionally equal to $p \vee q \leftrightarrow q$.

2.20 (c) Show that the following composite statements are tautologies.

(i) $\neg q \wedge (q \vee p) \to p$;
(ii) $(\neg q \to \neg p) \leftrightarrow (p \to q)$;
(iii) $((p \to q) \wedge (q \to p)) \leftrightarrow (p \leftrightarrow q)$;
(iv) $((p \to q) \wedge (\neg p \to \neg q)) \leftrightarrow (p \leftrightarrow q)$.

2.21 (c) By means of a computer program show that the following are tautologies:

(i) $(p \wedge (p \to q) \wedge (\neg r \to \neg q)) \to r$;
(ii) $((p \leftrightarrow q) \wedge (q \to r) \wedge (\neg r \vee s) \wedge (\neg p \to s)) \to s$.

2.22 (c) Show that $(p \to q) \wedge (q \to r) \wedge (r \to p)$ implies (i) $p \leftrightarrow r$, (ii) $p \leftrightarrow q$, (iii) $q \leftrightarrow r$. (Note that these implications justify the form of the proof of Th.1.6.)

2.23 (c) Are the operations $\to$ and $\leftrightarrow$ associative?

2.24 (c) Given that $a < b$ if and only if both $a \leqq b$ and $a \neq b$, show that the statement, "if $a \leqq b$ then $a < b$ or $a = b$" is true.

2.25 (c) Examine the validity of the following argument, in which the last sentence is the conclusion.

> If the balance of payments situation deteriorates, imports are high and exports are low. A surtax has been imposed. A surtax causes money to become scarce. If money is scarce, both imports and exports are low. The balance of payments situation does not deteriorate.

2.26 (c) Instead of using truth tables to arrive at the truth value of a composite statement, one may use an arithmetic procedure with the following representations:

$$\begin{aligned} \neg p &\quad - \quad 1 + p, \\ p \wedge q &\quad - \quad p + q + pq, \\ p \vee q &\quad - \quad pq. \end{aligned}$$

When the value T (F) is assigned to a prime statement, the value 0 (1) is assigned to the corresponding variable in the arithmetic expression. Addition is performed modulo 2, i.e., we have

$$0 + 0 = 0, \qquad 0 + 1 = 1 + 0 = 1, \qquad 1 + 1 = 0.$$

A tautology is identically equal to 0. The representation of $\neg s \wedge (s \vee t)$, for example, is $(1 + s) + st + (1 + s)st = 1 + s + st + st + sst = 1 + s + st$ (since $p + p \equiv 0$ and $pp \equiv p$). We have then, corresponding to assignments F to s and T to t, $1 + 1 + 1 \cdot 0 = 0$. Find arithmetic representations of $p \to q$, $p \leftrightarrow q$

and $p \wedge q \wedge r$. Demonstrate the validity of the argument of Exercise 2.25 by arithmetic means.

2.27 (d) If $|A| = m$ and $|B| = n$, how many binary relations are there from set A to set B? How many binary relations are there from A to B if $A \subseteq B$?

2.28 (d) Let $I = \{1, 2, 3, 4, 5\}$. List members of the following relations in I:

(i) $I \times I$;
(ii) $\{\langle x, y \rangle \mid x, y \in I \text{ and } x < y\}$;
(iii) $\{\langle x, y \rangle \mid x, y \in I \text{ and } x = y\}$.

Which of the relations is a function?

2.29 (d) Can $\{\langle a, b \rangle, \langle a, b, c \rangle, \langle b, c \rangle, \langle b, c, d \rangle\}$ be a relation in a set?

2.30 (d) Give examples of relations ρ in $I = \{1, 2, 3, 4, 5\}$ such that $R_\rho = D_\rho$ and (i) $\rho = \rho^{-1}$, (ii) $\rho \neq \rho^{-1}$.

2.31 (d) Characterize the following relations according to D.2.17:

(i) $A \times A$ considered as a relation in A;
(ii) $x > y$ in a set of integers;
(iii) x and y are brothers;
(iv) identity relation;
(v) square root relation in the set of integers.

2.32 (d) Show that a nonempty transitive and symmetric relation cannot be irreflexive but need not be reflexive.

2.33 (e) Which of the following are equivalence relations?

(i) *x and y take the same course* in the set of students of a university.
(ii) $\{\langle\langle x_1, x_2 \rangle, \langle x_3, x_4 \rangle\rangle \mid x_1 + x_4 = x_2 + x_3\}$.
(iii) $\{\langle\langle x_1, x_2 \rangle, \langle x_3, x_4 \rangle\rangle \mid x_1 + x_3 = x_2 + x_4\}$.

2.34 (e) Find the equivalence classes corresponding to the following equivalence relations:

(i) $A \times A$;
(ii) identity relation in a set $\{x_1, x_2, \ldots, x_n\}$;
(iii) the relation of Part (ii) of Exercise 2.33 in the set $N = \{\langle 1, 2 \rangle, \langle 3, 4 \rangle, \langle 5, 6 \rangle, \ldots\}$.
(iv) the relation of part (ii) of Exercise 2.33 in the set $\{\langle 0, 2 \rangle, \langle 1, 2 \rangle, \langle 2, 4 \rangle, \langle 3, 4 \rangle, \langle 4, 6 \rangle, \langle 5, 6 \rangle, \ldots\}$.

2.35 (e) Define the numbers $S(n, m)$:

$$S(0, n) = S(n, 0) = 0 \qquad (n > 0),$$
$$S(n, n) = 1 \qquad (n \geqq 0),$$
$$S(n, m) = mS(n-1, m) + S(n-1, m-1) \qquad (n > 0).$$

These numbers are called Stirling numbers of the second kind. Show that $\sum_{m=1}^{m=n} S(n, m)$ is the number of equivalence relations in a set of n elements.

2.36 (f) Let X be the set $\{1, 4, 9, 16, \ldots\}$. Suggest one or more partial ordering relations on X.

2.37 (f) Show that the identity relation in a set is the only relation in this set that is both an equivalence relation and a partial ordering.

2.38 (f) Show that if relation R is a partial ordering, then R^{-1} is also a partial ordering.

2.39 (f) A reflexive and transitive relation is a preordering. Every partial ordering is, of course, a preordering. Give an example of a preordering that is not a partial ordering.

2.40 (f) Consider set $A = \{2, 6, 10\}$, partially ordered by the relation *divides*. Find all subsets of A that are simply ordered by *divides*.

2.41 (f) Show that, relative to a partial ordering, a set can have at most one least member and one greatest member, and that existence of a least (greatest) member implies existence of a unique minimal (maximal) member, which coincides with the least (greatest) member.

2.42 (f) Show that, relative to a partial ordering, a finite set has at least one minimal (maximal) element.

2.43 (f) Let $\langle A, \leqq \rangle$ be a partially ordered set. Show that each subset of A has at most one supremum (infimum).

2.44 (f)=Represent the following sets, partially ordered by the relation *divides*, by diagrams:

(i) $\{2, 3, 5, 7, 210\}$, (ii) $\{2, 3, 4, 5, 9, 1080\}$, (iii) $\{2, 3, 5, 10, 15, 300\}$.

2.45 (f) Find least and greatest elements, and minimal and maximal elements of the partially ordered sets of Exercise 2.44.

2.46 (f) Find subsets of the partially ordered sets of Exercise 2.44 that have three elements and are linearly ordered by the relation *divides*.

2.47 (f) Find all upper bounds and the supremum for subset $\{2,5\}$ of each of the partially ordered sets of Exercise 2.44.

2.48 (f) Represent the partially ordered set $\langle \mathscr{P}(\{a, b, c, d\}), \subseteq \rangle$ by a diagram and find a subset of the natural numbers, which, when partially ordered by the relation *divides*, is order-isomorphic to this partially ordered set. What does the diagram turned through 180° represent? What does the diagram turned through 90° represent?

2.49 (f) Define sets $X = \{2, 4, 8, 16, 32\}$, $Y = \{2, 4, 8, 16, 96\}$, and $Z = \{2, 4, 8, 12, 96\}$, and denote relation *divides* by D, relation *less than or equal* by $\leqq$ and relation *greater than or equal* by $\geqq$. Which of the partially ordered sets of $\{X, Y, Z\} \times \{D, \leqq, \geqq\}$ are order-isomorphic to each other?

2.50 (g) Are any of the partially ordered sets of Exercise 2.44 lattices? Is any of the partially ordered sets of Exercise 2.49 not a lattice?

2.51 (g) Draw the operation table for lattice $\langle Z, D\rangle$ of Exercise 2.49.

2.52 (g) Let $\langle A, \leqq\rangle$ be a lattice. Show that statements $a * b = a$, $a \leqq b$, $a \oplus b = b$ are equivalent for any $a, b \in A$.

2.53 (g) Let $\langle A, \leqq\rangle$ be a lattice. Show that, for any $a, b, c \in A$,

$$a \oplus b = b \oplus a, \qquad a * b = b * a;$$
$$a \oplus (b \oplus c) = (a \oplus b) \oplus c, \qquad a * (b * c) = (a * b) * c.$$

2.54 (g) Let $\langle A, \leqq\rangle$ be a lattice. Show that, for any $a, b \in A$, $a \oplus (a * b) = a = a * (a \oplus b)$.

2.55 (g) Show that either of the distributive laws of D.2.30 implies the other.

2.56 (g) (i) Let $\langle A, \leqq\rangle$ be a lattice and let $x, y, z \in A$. Show that $x \leqq y$ implies $x \oplus z \leqq y \oplus z$ and $x * z \leqq y * z$, but that $x \oplus z \leqq y \oplus z$ or $x * z \leqq y * z$ need not imply $x \leqq y$.

(ii) Let $\langle A, \leqq\rangle$ be a distributive lattice with $x, y, z \in A$. Show that $x \oplus z = y \oplus z$ and $x * z = y * z$ imply $x = y$.

2.57 (g) Show that lattice $\langle A, \leqq\rangle$ is distributive if and only if, for any $a, b, c \in A$, $(a \oplus b) * c \leqq a \oplus (b * c)$.

2.58 (g) Let $\leqq$ be a partial ordering and let $\geqq$ be the converse of $\leqq$. Show that if $\langle A, \leqq\rangle$ is a lattice, then there exists a lattice $\langle A, \geqq\rangle$, and that if $\langle A, \leqq\rangle$ is distributive, then so is $\langle A, \geqq\rangle$.

2.59 (g) Develop the theory of lattices as an axiomatic theory.

2.60 (g) Given a lattice $\langle X, \leqq\rangle$, where $X = \{x_1, x_2, \ldots, x_n\}$ is a set of integers. A convenient representation of the lattice in a computer is by means of its addition–multiplication table. A program for computing this table might consist of a logical function `TEST`, which returns `.TRUE.` if $x_i \leqq x_j$ and `.FALSE.` otherwise, subroutine `SETUP`, which generates an `N`-by-`N` array `LESS` in which `LESS(I,J)` $= 1$ if $x_i \leqq x_j$ and `LESS(I,J)` $= 0$ otherwise, and subroutine `TABLE`, which computes the addition–multiplication table (`SETUP` and `TABLE` are independent of the particular partial ordering).

(i) Write the program and test it on the lattice of Example 1 of D.2.29. For ease of programming assume that for elements $x_1, x_2, \ldots, x_n$, which are stored as `LAT(1)`, `LAT(2)`, ..., `LAT(N)`, $x_i \leqq x_j$ does not hold if i is greater than j (this means that all `LESS(I,J)` with `I` greater than `J` are zero).

(ii) Indicate how the program could be used to determine whether or not the lattice is a chain and to compute an element such as $(x_1 \oplus x_2) * (x_2 \oplus (x_4 * x_6))$, say.

2.61 (g) Rewrite the program of Exercise 2.60 for the general case in which $x_i \leqq x_j$ may hold if i is greater than j. Test the modified program on the lattice of Example 1 of D.2.29 with LAT containing the elements of X in the order 2, 5, 15, 30, 10, 6, 1, 3.

2.62 (h) (i) Consider algebra $\langle B, \oplus, *, ', 0, 1\rangle$ and define operation $+$ by $x + y = (x * y') \oplus (x' * y)$. Is B closed under $+$?

(ii) Let $N = \{0, 1, 2, 3, \ldots\}$. Give examples of operations under which N is closed.

2.63 (h) Let $\oplus$ and $*$ be two-argument operations, let $'$ be a one-argument operation, and let 0 and 1 be zero-argument operations in $B = \{0, 1, 2\}$. Which of the following algebras are similar?

(i) $\langle B, \oplus, *, ', 0, 1\rangle$.
(ii) $\langle B, \oplus, *, 0, 1\rangle$.
(iii) $\langle B, *, \oplus, ', 0, 1\rangle$.
(iv) $\langle B, ', *, \oplus, 0, 1\rangle$.
(v) $\langle \{0, 1\}, \oplus, *, ', 0, 1\rangle$.

2.64 (h) Show that every sublattice and every homomorphic image of a distributive lattice is a distributive lattice. (Sublattices are subalgebras of lattices.)

2.65 (h) Consider an algebra $\langle X, *\rangle$, where $X = \{x_1, x_2, x_3, x_4\}$ and the operation is defined by Table 2.17. How many automorphisms f_k: $X \to X$ are there such that the image algebra is also $\langle X, *\rangle$?

TABLE 2.17

$x_i * x_j$

x_i \ x_j	x_1	x_2	x_3	x_4
x_1	x_1	x_2	x_3	x_4
x_2	x_2	x_1	x_4	x_3
x_3	x_3	x_4	x_1	x_2
x_4	x_4	x_3	x_2	x_1

2.66 (h) Show that, for Boolean algebras B and C, if f: $B \to C$ is an isomorphism, then f^{-1}: $C \to B$ is also an isomorphism.

2.67 (h) Show that a sublattice of a Boolean algebra need not be a Boolean algebra.

2.68 (h) Develop an algorithm for the construction of the diagram of a lattice from its addition table. Use the algorithm to show that lattices $\langle A, R_1\rangle$ and $\langle A, R_2\rangle$, where $A = \{a, b, c, d, e\}$ and Table 2.18 defines addition in the two lattices, are isomorphic.

TABLE 2.18

	a	b	c	d	e
a	a				
b	a	b			
c	a	a	c		
d	a	b	c	d	
e	a	b	c	d	e

$\langle A, R_1\rangle$

	a	b	c	d	e
a	a				
b	b	b			
c	a	b	c		
d	a	b	c	d	
e	b	b	e	e	e

$\langle A, R_2\rangle$

CHAPTER 3

Graph Theory

3a. Diagrams and Graphs

Consider again the diagrams of Figure 2.8. Assume that we have to store enough information in a computer for reconstruction of a diagram at some later date. First, we should store a list of the nodes. Secondly, we will need to know which pairs of nodes are joined by lines, and, for each line, which node covers the other. For example, we should store Diagram (b) as

Nodes	$\{1, 2, 3, 5, 6, 10, 15, 30\}$;
Lines	$\{\langle 1, 2\rangle, \langle 1, 3\rangle, \langle 1, 5\rangle, \langle 2, 6\rangle, \langle 2, 10\rangle, \langle 3, 6\rangle, \langle 3, 15\rangle, \langle 5, 10\rangle, \langle 5, 15\rangle, \langle 6, 30\rangle, \langle 10, 30\rangle, \langle 15, 30\rangle\}$,

where it is understood that the node represented by the second coordinate of an ordered pair covers the node represented by the first coordinate. This pair, a set of nodes and a set of lines, is precisely a directed graph.

DEFINITION 3.1 A *directed graph* (*digraph*, *oriented graph*) is the ordered pair $D = \langle A, R\rangle$, where A is a set of *nodes* (*points*, *vertices*) and R is a relation in A, i.e., R is a set of ordered pairs, which are called *arcs* (*lines*, *pointers*).

Set A may be infinite. We, however, shall deal only with finite digraphs, and the unqualified term *digraph* will always mean a *finite* digraph. Elements of A can then be represented by drawing points in the plane. An arc $\langle a, b\rangle$

is represented by joining a and b with a line, and providing the line with an arrowhead pointing from a to b. Note that if $a \in A$, and there exists no $b \in A$ such that $\langle a, b \rangle \in R$, then a point is still drawn for a in the plane. The notational device of arrowheads is much more explicit than the positional convention we adopted for diagrams of partially ordered sets. If our familiar diagram for $\{1, 2, 3, 5, 6, 10, 15, 30\}$, partially ordered by *divides*, is turned upside down, then the diagram represents partial ordering by the relation, *is a multiple of*. No such misunderstanding can arise with the arrowhead notation; every arrowhead has to be reversed before a picture of digraph $\langle A, R \rangle$ becomes a picture of digraph $\langle A, R^{-1} \rangle$.

A picture of a digraph can give very useful insight into the structure of the relation associated with the digraph, and we sometimes refer to pictures of digraphs as digraphs or even as graphs. We must clearly understand, however, that a digraph is the object defined by D.3.1; it is not a picture. Drawings of the same digraph can, in fact, be greatly different. The three drawings of Figure 3.1 represent the same digraph; it is Diagram (b) of Figure 2.8 interpreted as the digraph $\langle A, R \rangle$, where $A = \{1, 2, 3, 5, 6, 10, 15, 30\}$ and R is the relation *is covered by*. The picture of Figure 3.2, on the other hand, represents a different digraph, despite its agreement in appearance with one of the

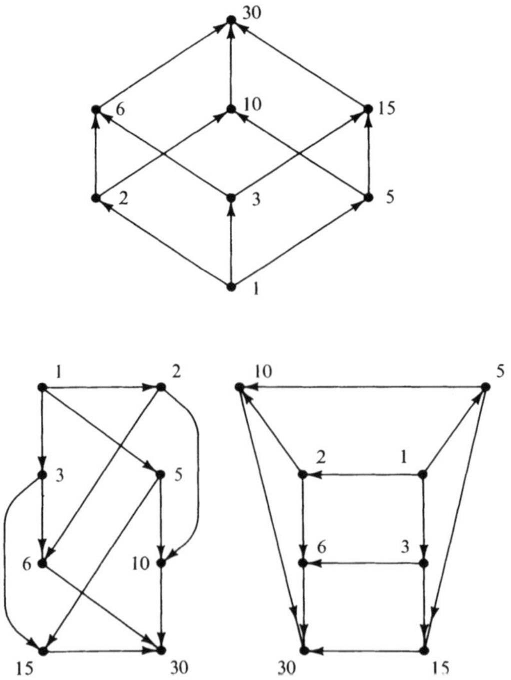

Figure 3.1

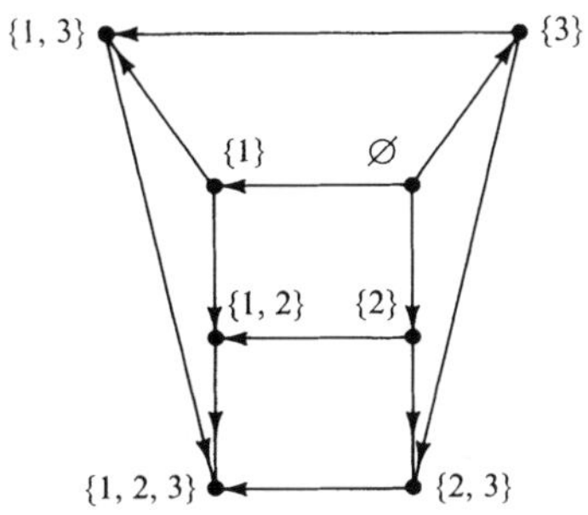

Figure 3.2

drawings of Figure 3.1. Figure 3.2 represents the digraph $\langle \mathscr{P}(\{1, 2, 3\}), R\rangle$, where R is again the relation *is covered by*.

What has come to be known as the theory of graphs really comprises two distinct theories: the theory of digraphs and the theory of (nonoriented) graphs. D.3.1 makes *digraph* a synonym for a relation and the set in which the relation is defined. The theory of digraphs can, therefore, be considered an extension of the theory of relations. One should not, however, put too great an emphasis on the relations as relations. An empirical structure, which may not easily submit itself to characterization by a single meaningful relation, can still be profitably studied as a digraph. One such structure is a system of routes along which something or other moves, be it information, or control, or automobiles, or some commodity such as water or electricity. We might be interested in the accessibility of certain points in the system, or in the elimination of circular routes. The theory of digraphs provides algorithms for the solution of such problems for a general digraph $\langle A, R\rangle$, which are independent of any external interpretation given to relation R.

The theory of graphs proper has more of a combinatorial flavor. For example, the theory of graphs might address itself to the following problem: Given a crystal lattice, how many figures of some type can be formed from points of the lattice and lines between the points? Direction of lines has no importance here.

DEFINITION 3.2 A *graph* is the ordered pair $G = \langle A, P\rangle$, where A is a set of nodes (*points*, *vertices*) and P is a set of (unordered) pairs of elements of A. Elements of P are called *edges*.

One can convert any graph to a digraph by replacing every $\{a, b\} \in P$ by two ordered pairs $\langle a, b\rangle$ and $\langle b, a\rangle$. It would seem, then, that the theory of digraphs is more general and includes the theory of graphs. This is not so. The two theories deal with different classes of problems and employ different techniques. One is as important as the other. We shall be considering the theory of digraphs almost exclusively here for no other reason but that it has more links with computer science.

It should be noted that there is no standard terminology in graph theory. Not only do synonyms abound, but the same word is sometimes used for different concepts. We shall try not to be too original in our choice of terminology. Nevertheless, one somewhat unorthodox term will be used. Normally an arc $\langle a, a\rangle$ is called a *loop*. To avoid confusion with the meaning of *loop* in programming, we shall call arcs of the form $\langle a, a\rangle$ *slings*. Moreover, we have included the term *pointer* in D.3.1 to emphasize the fact that certain data structures in which this term has become standard are in fact digraphs.

3b. Basic Definitions in the Theory of Digraphs

DEFINITION 3.3 If $G = \langle A, R\rangle$ is a digraph and Y is a subset of A, then digraph $G' = \langle Y, (Y \times Y) \cap R\rangle$ is a *subdigraph* of G; it is a *proper subdigraph* if $G \neq G'$. If Q is a subset of R, a digraph $G'' = \langle A, Q\rangle$ is a *partial digraph* of G. The concept of a *partial subdigraph* is an obvious extension.

Example

Consider a digraph $\langle A, R\rangle$ representing a chart of the system of streets in a city. Intersections and ends of streets are represented by nodes. If a street is two way, the segment between two adjacent intersections a and b is represented by arcs $\langle a, b\rangle$ and $\langle b, a\rangle$. If a street is one way in the direction from intersection c to an adjacent intersection d, we represent this segment by $\langle c, d\rangle$. The charts of all two-way streets and of all one-way streets are partial digraphs of $\langle A, R\rangle$. A chart showing only intersections provided with traffic lights and the arcs between such intersections when they adjoin is a subdigraph of $\langle A, R\rangle$. Other subdigraphs are the charts for sections of the city. A chart showing only the one-way streets in some section of the city is a partial subdigraph. Figure 3.3 shows digraph $\langle A, R\rangle$ and a subdigraph representing the system of streets in the eastern end of the city. Figure 3.4 shows the digraph of all one-way streets for the entire city (a partial digraph of $\langle A, R\rangle$) and for the eastern end (a partial subdigraph). Figure 3.5 shows the subdigraph of intersections with traffic lights.

DEFINITION 3.4 A digraph in which some or all nodes have labels associated with them (in addition to the identifying names of the nodes) is called a *labeled digraph*. A digraph in which arcs have weights associated with them is called a *weighted digraph* or *network*. A digraph that is both labeled and weighted is a *labeled network*.

Examples

1. In the digraph of Figure 3.3 let nodes representing intersections provided with traffic lights receive the label T. The resulting object is a labeled digraph (with nodes 11, 14, 16, 17, 18, 23, 27, 29 labeled). Further, let the arcs

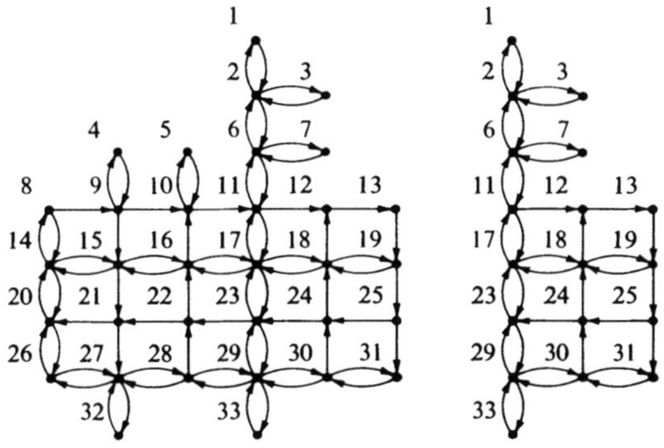

(a) (b)

Figure 3.3

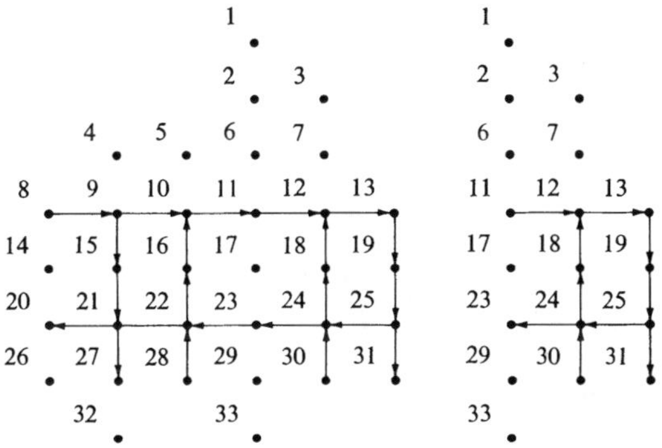

(a) (b)

Figure 3.4

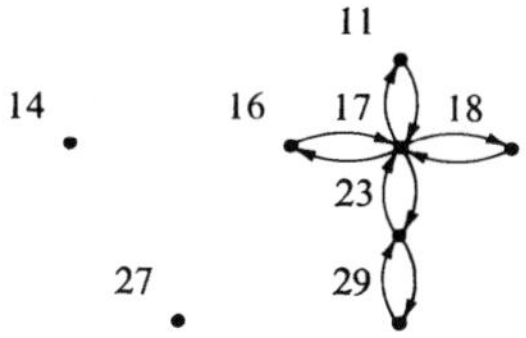

Figure 3.5

have street names attached to them: the name *South Lane* to arcs $\langle 25, 24\rangle$, $\langle 24, 23\rangle$, etc.; the name *Main Street* to arcs $\langle 14, 15\rangle$, $\langle 15, 14\rangle$, $\langle 15, 16\rangle$, $\langle 16, 15\rangle$, etc.; and so on. These street names are weights and the digraph is now a labeled network.

2. The term *weight* is given a more general meaning here than is generally the case. Originally only numbers were used as weights, perhaps expressing the capacity of the arcs for the flow of some commodity along them. As Example 1 shows, we permit a weight to be any piece of information one wishes to associate with an arc. In Figure 3.6 we show a network in which the weights are capacities for flow. The network might represent a system of pipelines.

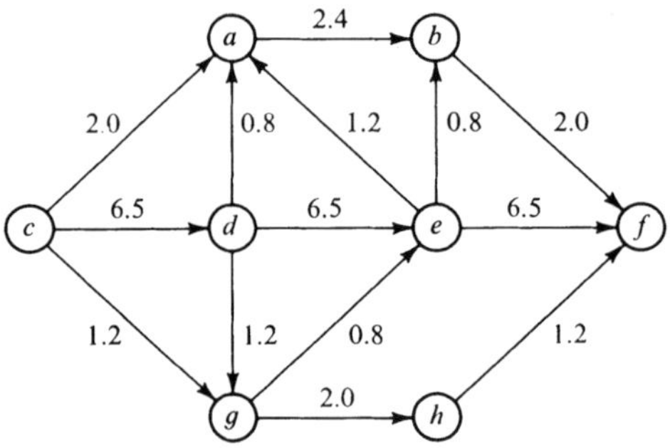

Figure 3.6

3. An $n \times n$ matrix X can be represented by a weighted digraph $\langle A, R\rangle$ in which $\langle i, j\rangle \in R$ if and only if element x_{ij} of matrix X is nonzero. The weight of arc $\langle i, j\rangle$ is the value of x_{ij}. Figure 3.7 shows a matrix and the network that represents it. The weighted digraph can be represented by an ordered pair $\langle A, Q\rangle$, where Q is a set of ordered triples $\langle i, j, x_{ij}\rangle$. $Q = \{\langle 1, 1, 4\rangle, \langle 1, 2, 3\rangle, \langle 2, 1, 2\rangle, \langle 2, 3, 4\rangle, \langle 3, 5, -1\rangle, \langle 4, 1, 8\rangle, \langle 4, 4, -2\rangle, \langle 5, 3, 1\rangle, \langle 5, 5, 2\rangle\}$.

DEFINITION 3.5 Let $\langle A, R\rangle$ be a digraph with arc $\langle a, b\rangle \in R$. Then a is the *initial node* and b the *terminal node* of $\langle a, b\rangle$. Arc $\langle a, b\rangle$ is said to *originate from* node a and to *terminate at* node b. Extending these concepts, consider $X \subseteq A$. Then $\langle a, b\rangle$ originates from set X if $a \in X$ and $b \notin X$, and the arc terminates in X if $a \notin X$ and $b \in X$. Arcs that originate from or terminate in a set are said to be *incident* with the set.

DEFINITION 3.6 Let $\langle A, R\rangle$ be a digraph and let $X \subseteq A$. We define the following subsets of R:

$$R_X^+ = \{\langle a, b\rangle \mid a \in X, \quad b \notin X\},$$
$$R_X^- = \{\langle a, b\rangle \mid a \notin X, \quad b \in X\}.$$

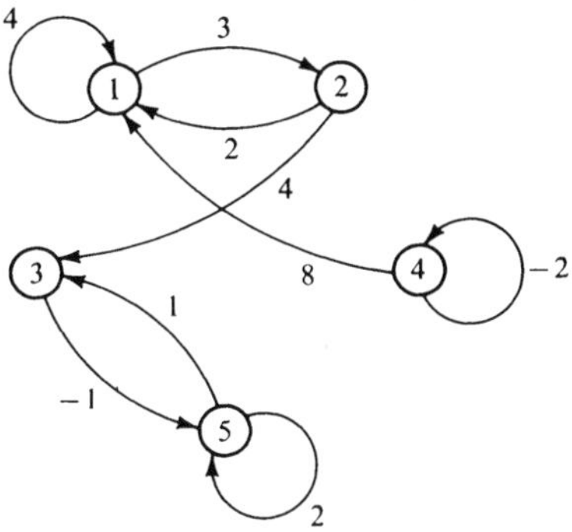

$$\begin{bmatrix} 4 & 3 & 0 & 0 & 0 \\ 2 & 0 & 4 & 0 & 0 \\ 0 & 0 & 0 & 0 & -1 \\ 8 & 0 & 0 & -2 & 0 \\ 0 & 0 & 1 & 0 & 2 \end{bmatrix}$$

Figure 3.7

Then $|R_X^+|$ is the *outdegree* of X, written $od(X)$, and $|R_X^-|$ is the *indegree* of X, written $id(X)$. The *total degree* of X, written $td(X)$, is defined by $td(X) = od(X) + id(X)$. When X consists of a single node, say a, we write $od(a)$, $id(a)$, $td(a)$, and speak of the outdegree, indegree, and total degree of node a.

Example

In the digraph of Figure 3.7 let $X = \{1, 3, 5\}$. Then $R_X^+ = \{\langle 1, 2\rangle\}$, $R_X^- = \{\langle 2, 1\rangle, \langle 4, 1\rangle, \langle 2, 3\rangle\}$, and $od(X) = 1$, $id(X) = 3$, $td(X) = 4$. For node 1 we have $od(1) = 1$, $id(1) = 2$. Sling $\langle 1, 1\rangle$ cannot contribute to the indegree or outdegree of node 1. The sum of the indegrees over all nodes of a digraph (or the sum of the outdegrees over all nodes) is the number of those arcs in the digraph that are not slings. The result would be aesthetically more satisfactory if we did not have the qualifier concerning slings. There are many other results in the theory of digraphs that would be simpler to express were it not for slings. This has led some authors to go as far as to completely exclude digraphs containing slings from their developments of the theory of digraphs.

DEFINITION 3.7 Let $D = \langle A, R\rangle$ be a digraph. If certain members of R can be placed in a sequence of the form $\langle a_1, a_2\rangle, \langle a_2, a_3\rangle, \ldots, \langle a_{n-1}, a_n\rangle$, then the set $P = \{\langle a_1, a_2\rangle, \langle a_2, a_3\rangle, \ldots, \langle a_{n-1}, a_n\rangle\}$ is a *path* from a_1 to a_n in D. The path is a *cycle* if $a_n = a_1$. A sequence of $n - 1$ arcs can be replaced by an equivalent sequence of n nodes $(a_1, a_2, a_3, \ldots, a_n)$. If all nodes in this sequence are distinct, the corresponding path is *simple* (*elementary*). If the node sequence corresponds to a cycle, and nodes $a_1, a_2, \ldots, a_{n-1}$ are distinct, then the cycle is *simple* (*elementary*). If the node sequence corresponds to a simple path (cycle), and it contains every node of A, then the path (cycle) is *Hamiltonian*. The *length* of a simple path (cycle) is $|P|$.

Example

In the digraph of Figure 3.7 we have an arc sequence $\langle 4, 1\rangle, \langle 1, 2\rangle, \langle 2, 3\rangle, \langle 3, 5\rangle$, and the equivalent node sequence (4, 1, 2, 3, 5). Hence there exists a path $\{\langle 4, 1\rangle, \langle 1, 2\rangle, \langle 2, 3\rangle, \langle 3, 5\rangle\}$. The path is simple and Hamiltonian. Its length is 4. We also have a sequence $\langle 4, 1\rangle, \langle 1, 2\rangle, \langle 2, 1\rangle, \langle 1, 2\rangle, \langle 2, 1\rangle, \langle 1, 2\rangle, \langle 2, 3\rangle, \langle 3, 5\rangle$, and the path corresponding to this sequence is $\{\langle 4, 1\rangle, \langle 1, 2\rangle, \langle 2, 1\rangle, \langle 2, 3\rangle, \langle 3, 5\rangle\}$. This path is not simple. Let us add the arc $\langle 5, 4\rangle$ to the digraph, and to the two arc sequences. The equivalent node sequences are then (4, 1, 2, 3, 5, 4) and (4, 1, 2, 1, 2, 1, 2, 3, 5, 4). The former corresponds to a cycle that is simple and Hamiltonian; the cycle corresponding to the latter is not simple and, therefore, cannot be Hamiltonian. The digraph of Figure 3.7 contains simple cycles corresponding to node sequences (1, 1), (1, 2, 1), (3, 5, 3), (4, 4), and (5, 5). None of these is Hamiltonian.

D.3.7 warrants several remarks. First, cycles are also paths, but a simple cycle cannot be a simple path. Second, the length of a path or cycle that is not simple is not defined. Sequences of different lengths can correspond to a single path. This prevents us from using the length of a sequence as a measure of the length of the corresponding path. Consider the path $\{\langle 4, 1\rangle, \langle 1, 2\rangle, \langle 2, 1\rangle, \langle 2, 3\rangle\}$ in the digraph of Figure 3.7. The path has only four members, but tracing of this path in the drawing involves traversal of at least five lines. Hence the number of arcs in a nonsimple path is not a satisfactory measure of its length either. Third, there is no fundamental reason why sequences should not define paths and cycles. Thus, in the digraph of Figure 3.7, sequences (3, 5, 3) and (5, 3, 5) could define two distinct cycles. But our intuitive feeling is that there should be only one cycle here. This intuitive feeling is recognized in our definition of paths and cycles in terms of sets rather than sequences. For reasons of economy we shall at times speak of a "cycle" (3, 5, 3), or of a "cycle" (5, 3, 5), say, but it will be understood that the two sequences are equivalent, both standing for the set $\{\langle 3, 5\rangle, \langle 5, 3\rangle\}$.

DEFINITION 3.8 A digraph is *cyclic* if it contains at least one cycle; otherwise it is *acyclic*.

THEOREM 3.1 A path in a digraph D is not simple if and only if some subset of the path defines a cycle.

Proof. Assume that P is a nonsimple path in D. Then any node sequence corresponding to P contains a subsequence of the form $(a_i, \ldots, a_i)$, and this subsequence corresponds to a cycle. Next assume that subset Q of a path P in D defines a cycle. Any node sequence corresponding to Q has the form $(a_i, \ldots, a_i)$, and the sequence corresponding to any superset of Q will, therefore, contain a subsequence of this form. Consequently neither Q nor any superset of Q can define a simple path.

DEFINITION 3.9 In a digraph node b is *reachable* from node a if there exists a path from a to b. We assume that every node is reachable from itself along a path of zero length.

DEFINITION 3.10 If node a in digraph D is reachable from no other node in D, and no other node is reachable from a, then a is an *isolated* node. (Alternatively, a is isolated if $id(a) = od(a) = 0$.)

Example

In the digraph of Figure 3.8 every node in the set $N = \{a_1, a_2, a_3, a_5, a_6\}$ is reachable from any other node in N. Node a_7 is reachable from any node in N, but no node in N is reachable from it. Node a_7 is reachable also from a_8, but a_8 is reachable only from itself. Node a_4 is isolated, and so is node a_9.

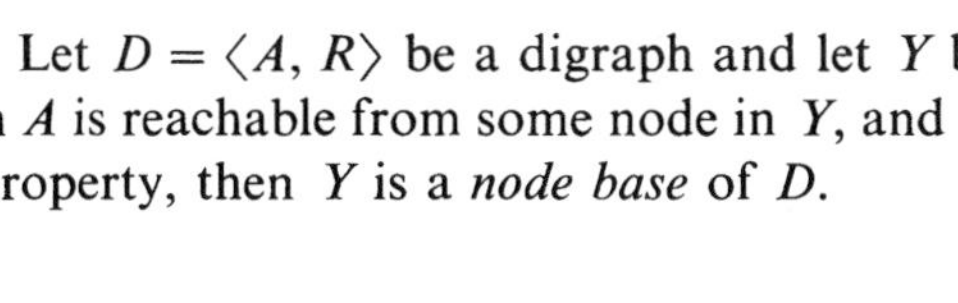

Figure 3.8

DEFINITION 3.11 Let $D = \langle A, R \rangle$ be a digraph and let Y be a subset of A. If every node in A is reachable from some node in Y, and no proper subset of Y has this property, then Y is a *node base* of D.

Example

The node base of the digraph of Figure 3.8 comprises the nodes a_4, a_8, a_9, and any one member of $\{a_1, a_2, a_3, a_5, a_6\}$. This digraph, therefore, has five distinct node bases, one of which is $\{a_3, a_4, a_8, a_9\}$.

THEOREM 3.2 A node belonging to a node base is not reachable from any other node in the base.

Proof. Let nodes a and b belong to a node base, and assume that there exists a path $(a, \ldots, b)$. Then every node reachable from b is reachable from a

as well, i.e., removal of b from the node base produces a proper subset of the node base such that all nodes of the digraph are still reachable from its members. Therefore, a path $(a, \ldots, b)$ cannot exist.

THEOREM 3.3 A node that does not have zero indegree and does not lie on a cycle cannot belong to a node base.

Proof. Let N be a node base of a digraph. Consider a node a that has nonzero indegree and does not lie on a cycle. Then there exists an arc that terminates at a, say $\langle b, a \rangle$. Node b belongs to N or is reachable from some node c that belongs to N. Neither b nor c is identical with a (a does not lie on a cycle). Hence a is reachable from some node other than itself in N, and, by Th.3.2, cannot belong to N.

THEOREM 3.4 The node base of an acyclic digraph (or a digraph in which all cycles are slings) $D = \langle A, R \rangle$ is defined by $N = \{a_i \mid a_i \in A, id(a_i) = 0\}$.

Proof. We note first that the presence of a sling affects the reachability of a node from itself alone. Since a node is reachable from itself in any case, the proof can be confined to acyclic digraphs. In the case of an acyclic digraph Th.3.3 states that a node that does not have zero indegree cannot belong to a node base. Hence all nodes belonging to the node base must have zero indegree. Moreover, every such node must belong to the node base. Otherwise, since a node having zero indegree can only be reached from itself, some node would not be reachable from any node belonging to the node base.

THEOREM 3.5 All node bases of a digraph have the same number of elements.

Proof. Consider node bases N_1 and N_2. We shall prove $|N_1| = |N_2|$ by showing that reachability, considered as a relation from N_1 to N_2, is a total one-to-one function. Clearly any member of N_2 must be reachable from some member of N_1, and vice versa. Assume that nodes a and b in N_2 are both reachable from node t in N_1. But t is in turn reachable from some node c in N_2, and both a and b are then reachable from c. By Th. 3.2 this can be only if $a = b = c$. Similarly we can show that a node in N_2 is reachable from at most one node in N_1. Next assume that there exists a member of N_1 from which no member of N_2 is reachable. Then there exists a proper subset of N_1 from which all nodes of N_2, and hence of the digraph, are reachable; i.e., N_1 is then not a node base.

3c. Digraphs, Matrices, and Relations

A digraph having n nodes $a_1, a_2, \ldots, a_n$ can be completely specified by a square matrix of order n.

DEFINITION 3.12 Let $D = \langle A, R \rangle$ be a digraph, with $A = \{a_1, a_2, \ldots, a_n\}$. The *adjacency matrix* X of D is defined as follows:

$$
\begin{aligned}
x_{ij} &= 1 \quad \text{if} \quad \langle a_i, a_j \rangle \in R, \\
x_{ij} &= 0 \quad \text{if} \quad \langle a_i, a_j \rangle \notin R.
\end{aligned}
$$

If D is weighted and w_{ij} is the weight associated with arc $\langle a_i, a_j \rangle$, then, assuming that w_{ij} is zero if and only if $\langle a_i, a_j \rangle$ is not an arc, the w_{ij} define a matrix W. This is the *variable adjacency matrix* of the weighted digraph.

Examples

1. The adjacency matrix of the digraph of Figure 3.8 is

$$
X = \begin{bmatrix}
0 & 1 & 0 & 0 & 0 & 0 & 0 & 0 & 0 \\
0 & 0 & 0 & 0 & 0 & 1 & 0 & 0 & 0 \\
0 & 1 & 0 & 0 & 0 & 0 & 1 & 0 & 0 \\
0 & 0 & 0 & 0 & 0 & 0 & 0 & 0 & 0 \\
1 & 0 & 0 & 0 & 0 & 0 & 0 & 0 & 0 \\
0 & 1 & 1 & 0 & 1 & 0 & 0 & 0 & 0 \\
0 & 0 & 0 & 0 & 0 & 0 & 0 & 0 & 0 \\
0 & 0 & 0 & 0 & 0 & 0 & 1 & 0 & 0 \\
0 & 0 & 0 & 0 & 0 & 0 & 0 & 0 & 1
\end{bmatrix}.
$$

2. The matrix shown in Figure 3.7 is the variable adjacency matrix associated with the weighted digraph of that figure. Our definition of the variable adjacency matrix is very general, but the main need for the matrix arises in the rather special situation in which arcs are given distinct identifying labels and one wants to derive new results from the pattern of the arcs (see A.6.3).

Another matrix of importance, which does not, however, completely specify a digraph, is the path matrix of the digraph.

DEFINITION 3.13 Let $D = \langle A, R \rangle$ be a digraph, with $A = \{a_1, a_2, \ldots, a_n\}$. The *path matrix* P of D is defined as follows:

$$
\begin{aligned}
p_{ij} &= 1 \quad \text{if there exists a path of nonzero length from } a_i \text{ to } a_j, \\
p_{ij} &= 0 \quad \text{otherwise.}
\end{aligned}
$$

Example

The path matrix of the digraph of Figure 3.8 is

$$P = \begin{bmatrix} 1 & 1 & 1 & 0 & 1 & 1 & 1 & 0 & 0 \\ 1 & 1 & 1 & 0 & 1 & 1 & 1 & 0 & 0 \\ 1 & 1 & 1 & 0 & 1 & 1 & 1 & 0 & 0 \\ 0 & 0 & 0 & 0 & 0 & 0 & 0 & 0 & 0 \\ 1 & 1 & 1 & 0 & 1 & 1 & 1 & 0 & 0 \\ 1 & 1 & 1 & 0 & 1 & 1 & 1 & 0 & 0 \\ 0 & 0 & 0 & 0 & 0 & 0 & 0 & 0 & 0 \\ 0 & 0 & 0 & 0 & 0 & 0 & 1 & 0 & 0 \\ 0 & 0 & 0 & 0 & 0 & 0 & 0 & 0 & 1 \end{bmatrix}.$$

The path matrix remains unchanged when arc $\langle a_6, a_2 \rangle$ is removed from the digraph.

THEOREM 3.6 Let X be the adjacency matrix of digraph D, and let $Y = X^h$. Then y_{ij} is the total number of distinct sequences $\langle a_i, \ldots \rangle, \ldots, \langle \ldots, a_j \rangle$ that (i) have length h, and (ii) correspond to paths in D.

Proof. We prove the theorem by induction. The base is provided by D.3.12: With $h = 1$ the theorem is in fact D.3.12. For the induction step assume that the theorem is true for $h = h'$. Let $P = X^{h'}$. Then, by assumption, p_{ik} is the number of sequences of length h' having the form $\langle a_i, \ldots \rangle, \ldots, \langle \ldots, a_k \rangle$, and this is also the number of sequences of length $h' + 1$ having the form $\langle a_i, \ldots \rangle, \ldots, \langle \ldots, a_k \rangle, \langle a_k, a_j \rangle$, i.e., $p_{ik}x_{kj} = p_{ik}$ if $\langle a_k, a_j \rangle$ is an arc, and $p_{ik}x_{kj} = 0$ if $\langle a_k, a_j \rangle$ is not an arc. The total number of sequences of length $h' + 1$ having the form $\langle a_i, \ldots \rangle, \ldots, \langle \ldots, a_j \rangle$ is therefore equal to the sum $\sum_{k=1}^{k=n} p_{ik}x_{kj}$, where n is the order of X. But this is the (i, j)th element of $X^{h'+1}$.

COROLLARY 1 If $X^h = 0$ for some $h \leq n$, then D is acyclic.

COROLLARY 2 If P is the path matrix of D and $Q = X + X^2 + \cdots + X^n$, then $p_{ij} = 1$ if and only if q_{ij} is nonzero.

By Corollary 2, the path matrix can be found by generating powers of the adjacency matrix, but this brute force approach produces a very slow procedure. A faster method for generating the path matrix will be discussed in Section 6a.

Example

For the digraph of Figure 3.8 we have

$$X^3 = \begin{bmatrix} 0 & 1 & 1 & 0 & 1 & 0 & 0 & 0 & 0 \\ 1 & 1 & 0 & 0 & 0 & 1 & 1 & 0 & 0 \\ 0 & 1 & 1 & 0 & 1 & 0 & 0 & 0 & 0 \\ 0 & 0 & 0 & 0 & 0 & 0 & 0 & 0 & 0 \\ 0 & 0 & 0 & 0 & 0 & 1 & 0 & 0 & 0 \\ 0 & 2 & 1 & 0 & 1 & 1 & 0 & 0 & 0 \\ 0 & 0 & 0 & 0 & 0 & 0 & 0 & 0 & 0 \\ 0 & 0 & 0 & 0 & 0 & 0 & 0 & 0 & 0 \\ 0 & 0 & 0 & 0 & 0 & 0 & 0 & 0 & 1 \end{bmatrix}.$$

There are two sequences of length 3 corresponding to paths from a_6 to a_2: $\langle a_6, a_5\rangle, \langle a_5, a_1\rangle, \langle a_1, a_2\rangle$ and $\langle a_6, a_2\rangle, \langle a_2, a_6\rangle, \langle a_6, a_2\rangle$. The elements of X^9 are quite large:

$$X^9 = \begin{bmatrix} 2 & 8 & 4 & 0 & 4 & 5 & 2 & 0 & 0 \\ 4 & 11 & 5 & 0 & 5 & 8 & 4 & 0 & 0 \\ 2 & 8 & 4 & 0 & 4 & 5 & 2 & 0 & 0 \\ 0 & 0 & 0 & 0 & 0 & 0 & 0 & 0 & 0 \\ 2 & 5 & 2 & 0 & 2 & 4 & 2 & 0 & 0 \\ 5 & 17 & 8 & 0 & 8 & 11 & 5 & 0 & 0 \\ 0 & 0 & 0 & 0 & 0 & 0 & 0 & 0 & 0 \\ 0 & 0 & 0 & 0 & 0 & 0 & 0 & 0 & 0 \\ 0 & 0 & 0 & 0 & 0 & 0 & 0 & 0 & 1 \end{bmatrix}.$$

A digraph is a set and a relation in the set, but the theory of digraphs is not the theory of relations. Although there is no distinct demarcation between the two theories, they do pursue different ends and use different techniques to achieve their ends. The theory of digraphs emphasizes explicit listing of the members of a relation; the theory of relations is more concerned with relations as defined by formulas. Selecting the proper approach to a problem involving a relation is a matter of utility. The pattern of flow of some commodity, say, is best represented by a digraph, and, if we were interested in accessibility of one point from another, or in some related problem, we would not try to express the relation in words or attempt to determine its type. Conversely, given a relation defined by some simple formula (e.g., the *less than* relation in a set of numbers), considering the relation as a digraph, i.e., listing all its elements, will rarely serve a useful purpose. Nevertheless, there are times when it may be profitable to apply the techniques of one theory to the problems of the other. We shall see that operations on the adjacency matrix, which belong to the theory of digraphs rather than the theory of relations, can give useful information about relations.

DEFINITION 3.14 Let $D = \langle A, R\rangle$. Digraph D is *reflexive* (*irreflexive, symmetric, antisymmetric, transitive*) if and only if R is a reflexive (irreflexive, symmetric, antisymmetric, transitive) relation.

In a reflexive digraph there is a sling on every node; in an irreflexive digraph there are no slings. In a symmetric digraph $\langle a, b\rangle \in R$ implies $\langle b, a\rangle \in R$, and in an antisymmetric digraph $\langle a, b\rangle \in R$ implies $\langle b, a\rangle \notin R$ when $a \neq b$. By definition, the existence of a simple path of length 2, $\{\langle i, k\rangle, \langle k, j\rangle\}$, in a transitive digraph implies existence of an arc $\langle i, j\rangle$. Equivalently, if $\langle i, j\rangle$ is not an arc, then there cannot exist a path $\{\langle i, k\rangle, \langle k, j\rangle\}$ for any $k \in A$. The following algorithm makes use of this property of a transitive digraph in determining whether or not a relation is a partial ordering.

ALGORITHM 3.1 Function `ISPORD` returns `.TRUE.` if the relation defined by the adjacency matrix `MX` is a partial ordering.

```
      LOGICAL FUNCTION ISPORD (MX,N)
      DIMENSION MX(N,N)
      ISPORD = .FALSE.
C1    TEST FOR REFLEXIVITY
      DO 10  K = 1,N
   10 IF (MX(K,K).EQ.0) RETURN
C2    TEST FOR ANTISYMMETRY AND TRANSITIVITY
      DO 20  I = 2,N
      JTOP = I - 1
      DO 20  J = 1,JTOP
C3    ANTISYMMETRY TEST
      IF (MX(I,J) * MX(J,I).EQ.1) RETURN
C4    TRANSITIVITY TEST - TEST FOR PATHS (I,K,J) IF
```

```
C4  MX(I,J) = 0  AND FOR PATHS (J,K,I) IF  MX(J,I) = 0
      IF (MX(I,J).EQ.1) GO TO 14
      DO 12  K = 1,N
   12 IF (MX(I,K) * MX(K,J).EQ.1) RETURN
      IF (MX(J,I).EQ.1) GO TO 20
   14 DO 16  K = 1,N
   16 IF (MX(J,K) * MX(K,I).EQ.1) RETURN
   20 CONTINUE
C5  EXIT FROM LOOP IMPLIES THAT THE RELATION IS A
C5  PARTIAL ORDERING
      ISPORD = .TRUE.
      RETURN
      END
```

A relation can be shown to be an equivalence by a procedure similar to A.3.1 (Exercise 3.13), but a faster procedure is one based on the fact that an equivalence relation is uniquely determined by a partition of the nodes in which each element belonging to a block is related to every element in that block, but to no element outside the block. Therefore, a relation is an equivalence if there exists a permutation of rows and corresponding columns of the adjacency matrix that transforms it to a block diagonal form exemplified by Figure 3.9 in which the shaded blocks consist of elements that are all 1, and all elements in the unshaded regions are 0. In our algorithm we shall not interchange the actual rows and columns. Instead, a reference vector will be used to keep track of the interchanges.

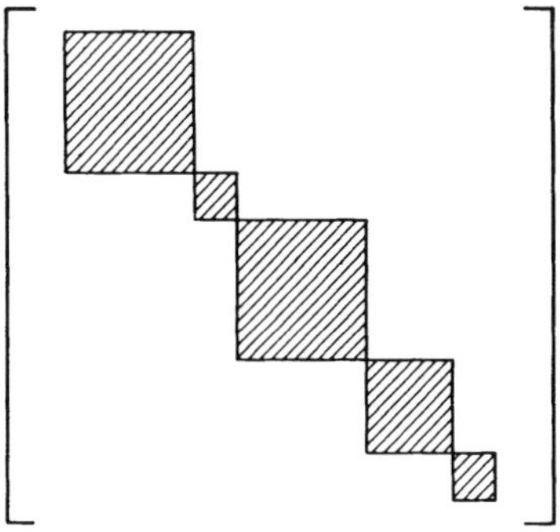

Figure 3.9

ALGORITHM 3.2 Function `ISEQUI` returns `.TRUE.` if the relation defined by the adjacency matrix **MX** is an equivalence. Vector `IREF` returns a permutation that would take **MX** to block diagonal form.

```
      LOGICAL FUNCTION ISEQUI (MX,IREF,N)
      DIMENSION MX(N,N), IREF(N)
C1   INITIALIZE THE REFERENCE VECTOR
      DO 5  I = 1,N
    5 IREF(I) = I
C2   IN (N-M) INTERCHANGES, WHERE M IS THE NUMBER OF
C2   BLOCKS, GENERATE A FORM THAT IS BLOCK DIAGONAL IF
C2   AND ONLY IF THE RELATION IS AN EQUIVALENCE
      IR = 1
   10 II = IREF(IR)
      IR = IR + 1
      IF (IR.GE.N) GO TO 25
      J = IR
   15 JJ = IREF(J)
      IF (MX(II,JJ).EQ.0) GO TO 20
      IREF(J) = IREF(IR)
      IREF(IR) = JJ
      IR = IR + 1
   20 J = J + 1
      IF (N-J) 10,15,15
C3   TEST THAT FORM IS BLOCK DIAGONAL
   25 ISEQUI = .FALSE.
      ILOW = 1
C4   FIND UPPER LIMIT OF BLOCK
   30 II = IREF(ILOW)
      DO 35  J = ILOW,N
      JJ = IREF(J)
   35 IF (MX(II,JJ).EQ.0) GO TO 40
C5   THIS IS THE LAST BLOCK - SKIP ZERO TESTS
      JM = N
      GO TO 50
C6   TEST DIAGONAL ELEMENT
   40 IF (J.EQ.ILOW) RETURN
C7   TEST THAT ALL ELEMENTS TO THE RIGHT AND BENEATH
C7   THE BLOCK ARE ZERO
      JM = J - 1
      DO 45  IROW = ILOW,JM
      II = IREF(IROW)
      DO 45  ICOL = J,N
      JJ = IREF(ICOL)
   45 IF (MX(II,JJ) + MX(JJ,II).GT.0) RETURN
C8   TEST THAT ALL ELEMENTS IN BLOCK ARE ONE (FIRST ROW
```

```
C8  OF BLOCK HAS ALREADY BEEN DONE)
  50   IF (ILOW.EQ.JM) GO TO 60
       ILOWP = ILOW + 1
       DO 55  IROW = ILOWP,JM
       II = IREF(IROW)
       DO 55  ICOL = ILOW,JM
       JJ = IREF(ICOL)
  55   IF (MX(II,JJ).EQ.0) RETURN
C9  TEST NEXT BLOCK
  60   ILOW = JM + 1
       IF (ILOW.LE.N) GO TO 30
       ISEQUI = .TRUE.
       RETURN
       END
```

Example

Given

$$\mathbf{MX} = \begin{bmatrix} 1 & 0 & 0 & 0 & 1 & 1 \\ 0 & 1 & 0 & 1 & 0 & 0 \\ 0 & 0 & 1 & 0 & 0 & 0 \\ 0 & 1 & 0 & 1 & 0 & 0 \\ 1 & 0 & 0 & 0 & 1 & 1 \\ 1 & 0 & 0 & 0 & 1 & 1 \end{bmatrix},$$

`ISEQUI` returns `.TRUE.`, and `IREF` returns (1, 5, 6, 4, 2, 3). To see how `IREF` is used, assume that we have to generate the block diagonal matrix explicitly, and that we are generating row 2 of this matrix. Since `IREF(2)` = 5, we go to the fifth row in **MX**: (1, 0, 0, 0, 1, 1). `IREF` now tells us to take elements of this row in the sequence 1, 5, 6, 4, 2, 3. By doing so we obtain (1, 1, 1, 0, 0, 0). The complete block diagonal matrix is

$$\begin{bmatrix} 1 & 1 & 1 & 0 & 0 & 0 \\ 1 & 1 & 1 & 0 & 0 & 0 \\ 1 & 1 & 1 & 0 & 0 & 0 \\ 0 & 0 & 0 & 1 & 1 & 0 \\ 0 & 0 & 0 & 1 & 1 & 0 \\ 0 & 0 & 0 & 0 & 0 & 1 \end{bmatrix}.$$

DEFINITION 3.15 A digraph $\langle A, R\rangle$ is *complete* if, for every pair of nodes a and b in A, $\langle a, b\rangle \notin R$ implies $\langle b, a\rangle \in R$.

By D.2.22, if R is a partial ordering in A, and digraph $\langle A, R\rangle$ is complete, the relation is a simple ordering.

3d. Connectedness in a Digraph

DEFINITION 3.16 Let $D = \langle A, R\rangle$ be a digraph. If, for every nonempty proper subset X of A one or both of $od(X) \neq 0$ and $id(X) \neq 0$ holds, then D is *connected.* Otherwise D is *disconnected.*

DEFINITION 3.17 A connected subdigraph $D' = \langle X, (X \times X) \cap R\rangle$ of D, such that $id(X) = od(X) = 0$, is a *(connected) component* of D.

Examples

1. Let $D = \langle A, R\rangle$ be a connected digraph, and let $X \subseteq A$. Condition $id(X) = od(X) = 0$ holds if and only if $X = A$. Therefore, a connected digraph has a single connected component, which is the digraph itself.

2. Every block of the partition induced by an equivalence relation defines a component. The only equivalence whose digraph is connected is the universal relation. In the digraph of the identity relation every node is isolated; consequently there are as many components as there are nodes.

In D.3.7 the node sequence $(a_1, a_2, \ldots, a_n)$ stands for the arc sequence $\langle a_1, a_2\rangle, \langle a_2, a_3\rangle, \ldots, \langle a_{n-1}, a_n\rangle$. We require the first coordinate of an arc to be the same as the second coordinate of its predecessor. Let us drop this requirement; i.e., let $(a_1, a_2, \ldots, a_n)$ stand for an arc sequence of which no more is required than that a_1 and a_2, a_2 and $a_3, \ldots, a_{n-1}$ and a_n be coordinates of successive arcs. For example, our node sequence could stand for $\langle a_1, a_2\rangle$, $\langle a_3, a_2\rangle, \ldots, \langle a_n, a_{n-1}\rangle$. Although we can no longer say that a_n is reachable from a_1, the relation between a_1 and a_n is similar to reachability, and we speak of a *semipath* between a_1 and a_n, or say that a_1 and a_n are *connected*, where we assume that a node is always connected with itself. An alternative definition of a connected digraph can now be given: A digraph is connected if all of its nodes lie on one semipath. In Figure 3.10 all five nodes lie on the semipath (a, b, c, e, d). The corresponding sequence of arcs is $\langle b, a\rangle$, $\langle b, c\rangle$, $\langle c, e\rangle$, $\langle d, e\rangle$. In Figure 3.11 the semipath is (a, e, b, e, c, e, d), and the arc sequence is $\langle a, e\rangle$, $\langle b, e\rangle$, $\langle b, e\rangle$, $\langle e, c\rangle$, $\langle e, c\rangle$, $\langle d, e\rangle$.

THEOREM 3.7 Let $\langle C_i, (C_i \times C_i) \cap R\rangle$ be the connected components of $\langle A, R\rangle$. Sets C_i form a partition of A.

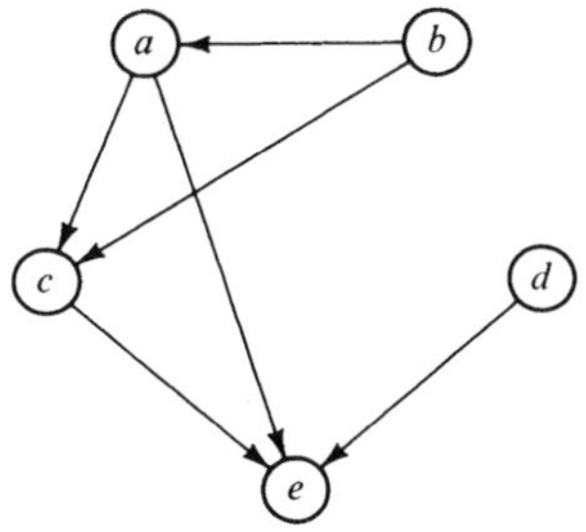

Figure 3.10

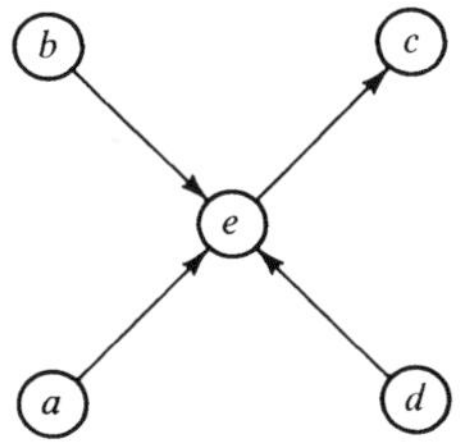

Figure 3.11

Proof. We use the terminology of the paragraph above. By definition, connectedness in a set of nodes is reflexive. It is also symmetric and transitive. Therefore it is an equivalence relation. It is easy to see that the partition induced by this equivalence relation consists of the C_i.

It is important to realize that the equivalences discussed in Example 2 of D.3.17 differ from that of the proof of Th.3.7. The latter is not the relation R. Therefore, although the adjacency matrix of $\langle A, R\rangle$ can be transformed to block diagonal form, elements of a block need no longer be all ones. But, if Q denotes the relation *a and b are connected*, elements of blocks of the block diagonalized adjacency matrix of $\langle A, Q\rangle$ are all ones, and these blocks define the C_i of Th.3.7. Exercise 3.18 asks for the adjacency matrix of $\langle A, Q\rangle$ to be generated from the adjacency matrix of $\langle A, R\rangle$. Application of A.3.2 to $\langle A, Q\rangle$ produces the block diagonal form. It is then an easy matter to find the connected components of $\langle A, R\rangle$.

DEFINITION 3.18 A digraph is *strongly connected* if every node in the digraph is reachable from every other node.

DEFINITION 3.19 Subdigraph $D' = \langle X, (X \times X) \cap R\rangle$ of D is a *strongly connected component* (*strong component*) of D if it is strongly connected and there exists no pair of nodes $a \in X$, $b \notin X$ such that a and b lie on the same cycle in D.

Example

The strong components of the digraph of Figure 3.8 are $\{a_1, a_2, a_3, a_5, a_6\}$, $\{a_4\}$, $\{a_7\}$, $\{a_8\}$, $\{a_9\}$. Those of the digraph of Figure 3.10 are $\{a\}$, $\{b\}$, $\{c\}$, $\{d\}$, $\{e\}$.

DEFINITION 3.20 A partial digraph D'' of D is the *cycle digraph* of D if it contains all arcs belonging to cycles in D and only such arcs.

THEOREM 3.8 Digraph $D = \langle A, R\rangle$ is strongly connected if and only if R is a cycle.

Proof. Exercise 3.19.

COROLLARY The strong components of a digraph are precisely the connected components of its cycle digraph.

ALGORITHM 3.3 Given the adjacency matrix X of digraph D.

1. Find path matrix P of D.
2. Compute matrix C, defined by $c_{ij} = x_{ij} \times p_{ji}$.

Matrix C is the adjacency matrix of the cycle digraph of D.

Example

The adjacency and path matrices of the digraph of Figure 3.8 are given by Example 1 of D.3.12 and the example of D.3.13, respectively. Then

$$C = \begin{bmatrix} 0 & 1 & 0 & 0 & 0 & 0 & 0 & 0 & 0 \\ 0 & 0 & 0 & 0 & 0 & 1 & 0 & 0 & 0 \\ 0 & 1 & 0 & 0 & 0 & 0 & 0 & 0 & 0 \\ 0 & 0 & 0 & 0 & 0 & 0 & 0 & 0 & 0 \\ 1 & 0 & 0 & 0 & 0 & 0 & 0 & 0 & 0 \\ 0 & 1 & 1 & 0 & 1 & 0 & 0 & 0 & 0 \\ 0 & 0 & 0 & 0 & 0 & 0 & 0 & 0 & 0 \\ 0 & 0 & 0 & 0 & 0 & 0 & 0 & 0 & 0 \\ 0 & 0 & 0 & 0 & 0 & 0 & 0 & 0 & 1 \end{bmatrix}.$$

DEFINITION 3.21 Let $D = \langle A, R\rangle$ be a digraph and define the following sets:

$$\mathcal{S} = \{S_i \mid S_i \text{ is a strong component of } D\},$$
$$Q = \{\langle S_i, S_j\rangle \mid s_i \in S_i, \quad s_j \in S_j, \quad \langle s_i, s_j\rangle \in R\}.$$

Digraph $\langle \mathcal{S}, Q\rangle$ is the *condensation* of D.

Example

Figure 3.12 shows a digraph D and its condensation D^*. A condensed digraph is acyclic. This implies that, if $\langle h, i\rangle$ in D were replaced by $\langle i, h\rangle$, the condensation of the new digraph would consist of a single node.

THEOREM 3.9 Let $D = \langle A, R\rangle$ and $D^* = \langle \mathcal{S}, Q\rangle$ be a digraph and its condensation, respectively. Define a subset of $\mathcal{S}: \mathcal{S}' = \{S_i \mid id(S_i) = 0\}$. A node base of D is generated by taking exactly one element from each S_i in $\mathcal{S}'$.

Proof. Exercise 3.21.

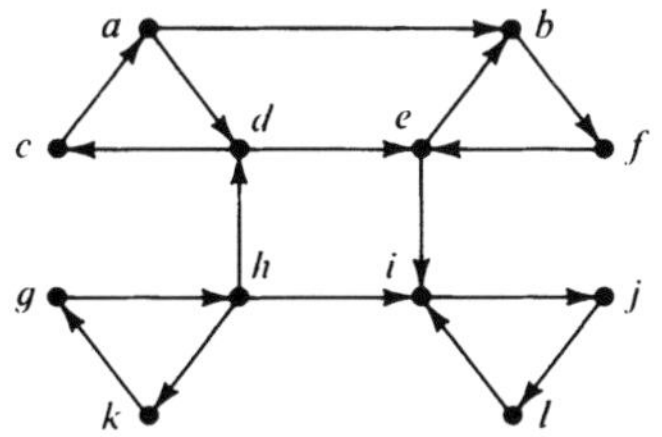

Digraph D

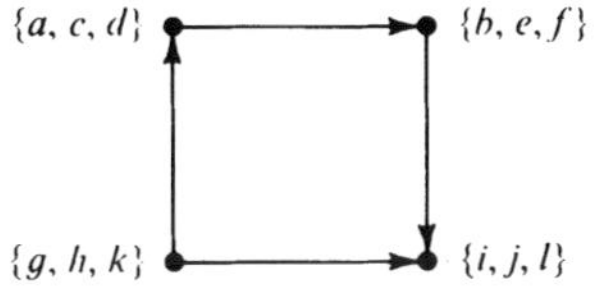

Condensation D^*

Figure 3.12

3e. Linear Formulas of Digraphs

Instead of representing a digraph by its adjacency matrix, we can represent it by a set of linear formulas. Let us first discuss the notation informally. It is based on the representation of an arc by an operator $*$ applied to the node symbols: Arc $\langle a, b\rangle$ is represented by $*ab$. We shall call operator $*$ the K-operator. If more than one arc originates from a node, a formula representing all these arcs consists of as many K-operators as there are arcs, followed by the symbol of the node from which the arcs originate, followed in turn by symbols of the nodes in which the arcs terminate. Referring to Figure 3.13, we have, for arcs originating from nodes a, b, and c, formulas

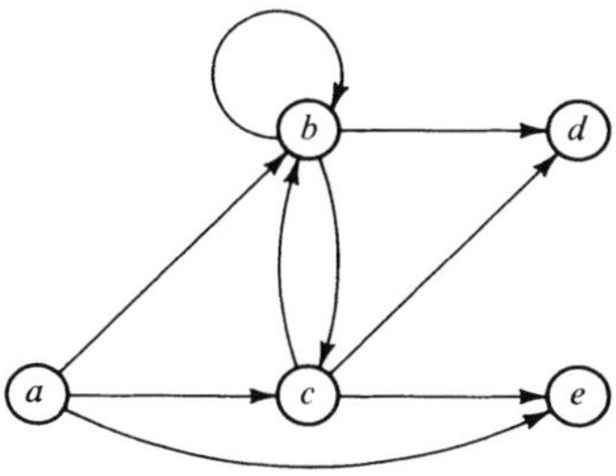

Figure 3.13

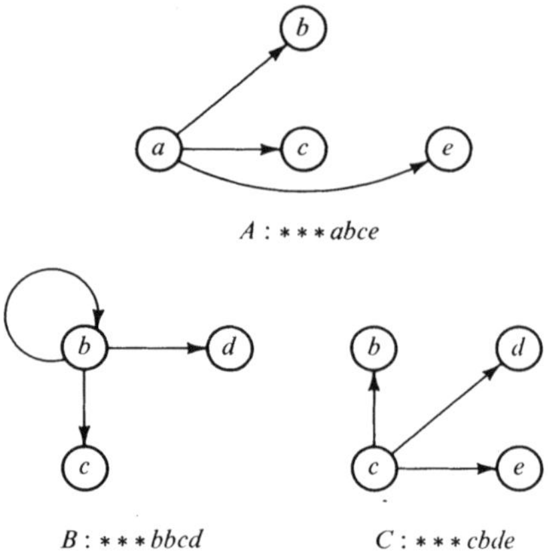

Figure 3.14

***abce, ***bbcd, ***cbde, respectively. The three formulas represent the three digraphs shown in Figure 3.14. Note that the symbols of the terminal nodes can be written in any order. Thus we can rewrite formula ****abce* as ****aebc*, or as ****aecb*, and so forth.

The formulas may be combined. We replace the *b* in the formula for *A* by the formula for *B* to get ****a***bbcdce*, a formula representing the digraph of Figure 3.15. If next we substitute ****cbde* for one of the occurrences of *c* in this formula, we obtain ****a***bb***cbdedce* or ****a***bbcd***cbdee*, depending on which occurrence of *c* we substitute for. Either of these final formulas represents the digraph of Figure 3.13. We call the formulas K-formulas.

DEFINITION 3.22 We define a K-formula recursively:

(a) A node symbol is a K-formula.
(b) If α and β are K-formulas, then $*\alpha\beta$ is a K-formula.

[To be quite explicit, we should add (c): K-formulas are only those entities that are constructed under (a) and (b).]

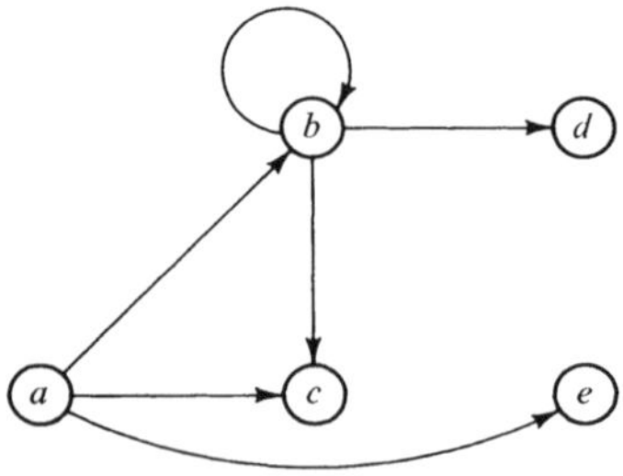

Figure 3.15

DEFINITION 3.23 A K-formula is a K-formula *of the node* whose symbol is the leftmost node symbol in the K-formula, and this node is the *leading node* of the K-formula.

Example

$***c***bbcdde$ is a K-formula of leading node c.

Since D.3.22 is a recursive definition, we can consider subformulas of K-formulas as K-formulas in their own right, provided they are consistent with the definition. The definition enables us to tell whether or not a given formula is "well formed," but it does not relate a K-formula to a particular digraph. Hence we require an algorithm for generating K-formulas of given digraphs. In what follows we shall always assume that K-formulas are considered in the context of particular digraphs rather than as abstract objects.

ALGORITHM 3.4 Let $D = \langle A, R \rangle$ be a digraph.

1. For every isolated node $a \in A$ that has no sling on it write the K-formula a.
2. For every arc $\langle a, b \rangle \in R$ write the K-formula $*ab$.
3. Combine the K-formulas according to the following *substitution rule:* If there exists a K-formula of a node and there exists another K-formula in which a symbol of the node appears, substitute the K-formula of the node for this symbol.

Apply the substitution rule until it can no longer be applied, taking care that the K-formula resulting from the substitution is always the longest that can be generated at the time.

4. (Check step) Denote the K-formulas produced in Step 3 by $k_1, k_2, \ldots, k_n$, and the leading node of a k_i by a_i. If some k_i contains as a subformula a K-formula of node b in which a_i appears, and the b occurs in one of $k_1, \ldots, k_{i-1}, k_{i+1}, \ldots, k_n$, extract the K-formula of b from k_i, inserting b in its place, substitute what now remains of k_i into this K-formula, and return to Step 3.

Examples

1. Consider the digraph of Figure 3.16. There are no isolated nodes, so, in applying A.3.4, we start in Step 2, and obtain the set

0. $\{*ad, *bc, *cd, *db, *de, *ea\}$.

Successive applications of the substitution rule produce

1. $\{*a*de, *bc, *cd, *db, *ea\}$,
2. $\{*a*d*ea, *bc, *cd, *db\}$,

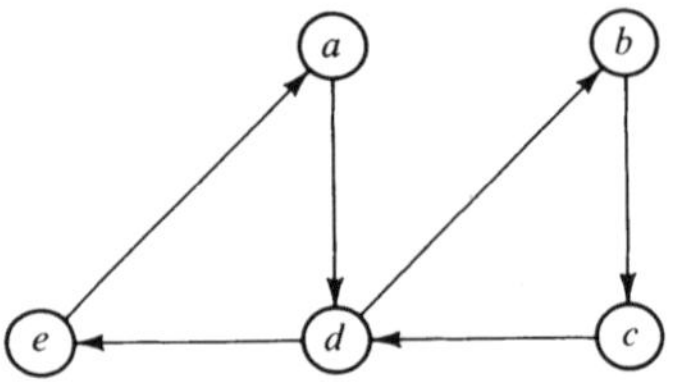

Figure 3.16

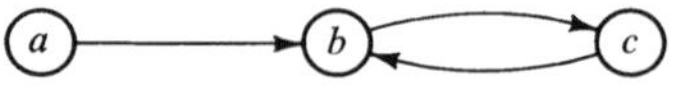

Figure 3.17

3. {*a**db*ea, *bc, *cd},
4. {*a**d*bc*ea, *cd},
5. {*a**d*b*cd*ea}.

2. With the digraph of Figure 3.17 we proceed from {*ab, *bc, *cb} to {*ab, *c*bc}, but find then that the substitution rule cannot be applied any further. Therefore we go to Step 4 of the algorithm and find that subformula *bc of *c*bc contains c, and that *ab contains b. We extract the *bc, and substitute into it what is left after the extraction, namely *cb. The resulting K-formula *b*cb is substituted into *ab to produce *a*b*cb.

DEFINITION 3.24 For a digraph D, a set of K-formulas produced by A.3.4 is a *minimal* set of K-formulas of D.

Although in practice one is mostly interested in minimal sets of K-formulas, for theoretical reasons we must be able to associate more general sets of K-formulas with a particular digraph. D.3.25 enables us to do so.

DEFINITION 3.25 A set of K-formulas *represents* a digraph D if and only if it can be obtained by applying the substitution rule of A.3.4 to the K-formulas generated from D in Steps 1 and 2 of A.3.4, where the number of applications of the rule may vary from zero to as many as are required to take A.3.4 to completion.

In D.3.25 we do not make the stipulation of Step 3 of A.3.4 as to how the substitution rule is to be applied. To see what happens when the stipulation is disregarded let us take the set produced by the second substitution in Example 1 of D.3.23, and proceed as follows:

3a. {*a*d*ea, *b*cd, *db},
4a. {*a*d*ea, *b*c*db}.

Although the substitution rule can be applied no further, the final set is not minimal. Still, the two K-formulas in this set can be combined by Step 4 of A.3.4; the stipulation has been included in Step 3 merely to reduce the number of calls on Step 4.

The following remarks apply to any set of K-formulas representing a digraph D. Since the substitution rule neither creates nor destroys any K-operators, all representative sets of D contain the same number of K-operators, and this number is equal to the number of arcs in D. Furthermore, the number of K-operators preceding occurrences of a particular node symbol a is equal to the number of arcs originating from a. For example, in a representation $\{**a***a*bc**dcbef, **bf**e**bad*ac\}$, symbol a occurs four times, preceded by 2, 3, 0, and 1 K-operators, respectively. Therefore, the total number of arcs originating from a is 6. Let us see where these arcs terminate. If a is preceded by n K-operators, then it is the leading node of a K-formula $** \cdots *a\alpha_1\alpha_2 \cdots \alpha_n$, where the $\alpha_1, \alpha_2, \ldots, \alpha_n$ are again K-formulas. The leading nodes of these K-formulas are terminal nodes of arcs originating from a. Continuing with our example, we have the following relevant K-formulas of a: $**a(***a*bc**dcbe)(f)$, $***a(*bc)(**dcb)(e)$, $*a(c)$, where the α_i have been enclosed in parentheses for clarity. The set of arcs originating from a is, therefore, $\{\langle a, a\rangle, \langle a, f\rangle, \langle a, b\rangle, \langle a, d\rangle, \langle a, e\rangle, \langle a, c\rangle\}$. Applying the procedure to occurrences of b we get the set $\{\langle b, c\rangle, \langle b, f\rangle, \langle b, e\rangle, \langle b, a\rangle, \langle b, d\rangle\}$. There are no arcs originating from nodes c and f, and arcs originating from d and e constitute the sets $\{\langle d, c\rangle, \langle d, b\rangle\}$ and $\{\langle e, b\rangle, \langle e, a\rangle\}$, respectively. The union of these four sets is precisely the set of arcs of the digraph.

The recursive definition of a K-formula is not particularly well suited to identification of the α_i. Therefore, we give an equivalent iterative definition.

DEFINITION 3.22a Consider a formula $s_1s_2 \cdots s_i \cdots s_m$. Let k_i and n_i denote numbers of K-operators and node symbols, respectively, in the subformula $s_1 \cdots s_i$. The formula is a K-formula if and only if the following conditions are satisfied:

$$n_i \leqq k_i, \qquad i = 1, 2, \ldots, m-1;$$
$$n_m = k_m + 1.$$

Example

Consider the formula $**a***a*bc**dcbef$. Set $n = 0$ and $k = 0$. Scan the formula from the left, setting $n = n + 1$ when a scanned symbol represents a node, or $k = k + 1$ when it is a K-operator. Throughout the scan we have $n \leqq k$, except when the symbol f is reached, and then $n = k + 1$. Therefore the formula is a K-formula. Writing the formula as $**a\alpha_1\alpha_2$, we can determine α_1 by repeating the procedure on the subformula $***a*bc**dcbef$. Again $n \leqq k$ while we are scanning $***a*bc**dcb$, and $n = k + 1$ when the e is reached. Therefore α_1 is $***a*bc**dcbe$. The rest of the subformula, namely f, must be α_2. In a similar fashion, expressing $***a*bc**dcbe$ as $***a\alpha_1\alpha_2\alpha_3$, we find that $\alpha_1 = *bc$, $\alpha_2 = **dcb$, $\alpha_3 = e$.

THEOREM 3.10 The leading nodes of K-formulas in a minimal set of a digraph constitute a base of the digraph.

Proof. It is easy to show that every node whose symbol appears in a K-formula of a node is reachable from this node. Also, K-formulas of a minimal set (indeed, of any representative set) contain symbols of all nodes. Hence all nodes of the digraph are reachable from the leading nodes of the K-formulas of a minimal set. We shall show that no proper subset of the leading nodes has this property. Let k_i and k_j be any two K-formulas in the minimal set, with leading nodes a and b, respectively, and assume that a is reachable from b. Then there exists a finite path $(b, n_1, n_2, \ldots, n_t, a)$. The original $*n_t a$ could not have been substituted into any formula other than k_i (otherwise k_i could now be substituted into this formula). Therefore n_t appears in k_i. Moreover, n_t is the leading node of a K-formula that contains a. Therefore, if $*n_{t-1}n_t$ had been substituted into a formula other than k_i, the check step of A.3.4 would be applicable. In a like manner we show that $*n_{t-2}n_{t-1}, \ldots, *n_1n_2, *bn_1$ have been substituted into k_i. But $*bn_1$ could not have been so substituted (otherwise k_j could now be substituted into k_i). This contradiction establishes that the leading nodes are not reachable one from another, i.e., that there exists no proper subset of the leading nodes from which every node of the digraph could be reached.

DEFINITION 3.26 Let F be a set of K-formulas.

(a) If a member of F or a subformula of a member has the form $** \cdots *a\alpha_1\alpha_2 \cdots \alpha_n$, where node symbol a is preceded by n K-operators ($n \geqq 2$), and $\alpha_1, \ldots, \alpha_n$ are K-formulas, a reordering of the $\alpha_1, \ldots, \alpha_n$ is an application of the *switch rule*.

(b) The interchange of two K-formulas of a node a, occurring as members or subformulas of members of F, provided the interchange does not produce a new member of F consisting of a single node symbol, is an application of the *interchange rule*.

D.3.25 does not specify the order in which the substitution rule is to be applied. Application of switch and interchange rules merely converts a given representative set to the set that would have resulted had a different sequence of substitutions been followed. Consider K-formula $***a***bbcd***cbdee$ in which $\alpha_1 = ***bbcd$, $\alpha_2 = ***cbde$, $\alpha_3 = e$. One possible reordering produces $***ae***cbde***bbcd$. Interchange of $**cbd$ and c in this formula produces $***ae*ce***bb**cbdd$. For another example, consider the set $\{***a***bbcdce, ***cbde\}$. Interchange of $***bbcd$ and b to produce $\{***abce, ***c***bbcdde\}$ is valid, and so is the interchange of c and $**cbd$, but we may not interchange the c with $***cbde$. In the last instance $\{***a***bbcd***cbdee, c\}$ would be produced, which is not a representative set according to D.3.25.

The set of leading nodes of a minimal set of K-formulas is invariant under applications of switch and interchange rules. If a digraph has more than one node base, the leading nodes of the minimal set produced by A.3.4 can be made to constitute any one of the bases. Therefore, if a digraph has more than one node base, it is impossible to generate every minimal set of K-formulas of the digraph, starting from a given set, by means of switch and interchange rules alone. In order to generate all minimal sets we have to apply the procedure of Step 4 of A.3.4 as well.

3f. Trees

DEFINITION 3.27 An acyclic digraph in which exactly one node has indegree 0 and every other node has indegree 1 is a *directed tree*. The node with indegree 0 is the *root* of the tree. Nodes with zero outdegree are *terminal nodes*. The length of the path from the root to a node is the *level* of the node.

Examples

1. Usually a directed tree is depicted with the root placed at the top of the drawing, nodes reachable from the root along a path of length 1 immediately below the root, nodes on level 2 immediately below these, etc. Figure 3.18 shows a directed tree with root a and terminal nodes d, e, f, h, i. Node a is on 0 level; nodes b and c are on level 1; nodes d, e, f, and g on level 2; nodes h and i on level 3.

2. A single node is a directed tree.

The left-to-right order in which one draws arcs originating from a node of a directed tree is significant in most applications, so that usually the picture

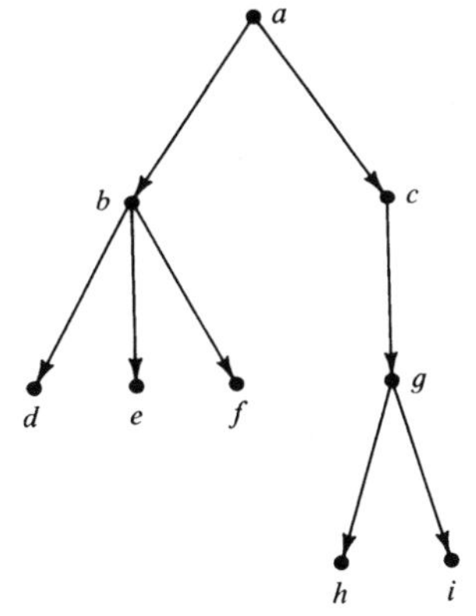

Figure 3.18

of a directed tree does not really correspond to a digraph. For reasons of conceptual economy we do, however, wish to consider directed trees as a species of digraphs. This can be done if the arcs are labeled according to some scheme that expresses the order they would have in a drawing. A canonical labeling scheme results if arcs originating from the one node are simply labeled 0, 1, 2, ..., reading from left to right. An alternative canonical scheme can be devised that labels nodes. This, however, may prove confusing when the nodes already carry information connected with the application, as they do in most applications.

DEFINITION 3.28 A directed tree in which every node has outdegree 0 or 2 is a *binary tree*.

Example

The directed trees of Figure 3.19 are binary.

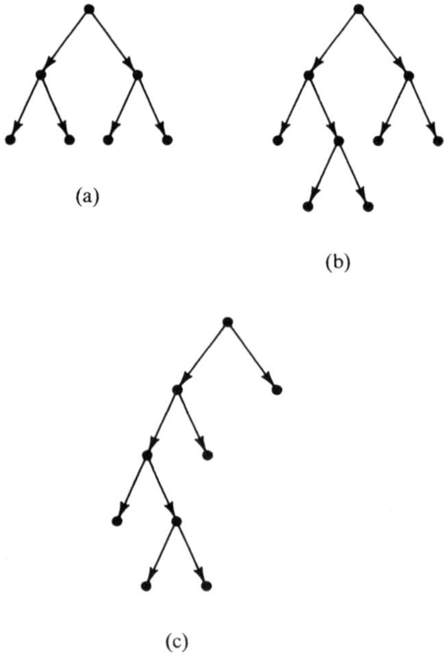

Figure 3.19

THEOREM 3.11 Let n be any positive nonzero integer. A binary tree T_n with n terminal nodes exists.

Proof. Assume that $T_n = \langle A_n, R_n \rangle$ exists. Let $a \in A_n$ be any terminal node in T_n. Introduce nodes $b, c \notin A_n$ and draw arcs $\langle a, b\rangle$, $\langle a, c\rangle$. Now $id(b) =$

$id(c) = 1$, $od(b) = od(c) = 0$, $od(a) = 2$, and indegrees and outdegrees of all other members of A_n remain unchanged. Clearly the new structure is again a binary tree. Node a is no longer terminal, but two new terminal nodes have been created; i.e., the number of terminal nodes has increased to $n + 1$. Hence T_{n+1} exists if T_n exists. But T_1 exists; it is a single isolated node. (If you are not happy with basing the proof on the degenerate tree T_1, construct T_2 as well.)

THEOREM 3.12 Let $T = \langle A, R \rangle$ be a binary tree. Then $r = |R| = 2(n_t - 1)$, where n_t is the number of terminal nodes in T.

Proof. Exercise 3.29.

DEFINITION 3.29 Let n be the number of terminal nodes in a binary tree, and let m be a nonnegative integer; let d be the length of a path from the root to a terminal node. A binary tree is *balanced* if

(a) $n = 2^m$ implies $d = m$,
(b) $2^m < n < 2^{m+1}$ implies $d = m$ or $d = m + 1$.

Example

Binary trees (a) and (b) of Figure 3.19 are balanced. Tree (a) has four terminal nodes, and the length of every path from the root to a terminal node is 2. Tree (b) has five terminal nodes. Three of the paths have length 2, the other two have length 3. Binary tree (c) is not balanced.

3g. Isomorphism of Digraphs

DEFINITION 3.30 Two digraphs are *isomorphic* if some permutation of the rows and corresponding columns in the adjacency matrix of one of the digraphs produces the adjacency matrix of the other; i.e., isomorphic digraphs are identical except for the identifying names carried by their nodes.

Example

The digraphs of Figure 3.20 are isomorphic. Their adjacency matrices are equal when we make elements of $\{1, 2, 3, 4, 5\}$ and $\{a, b, c, d, e\}$ correspond to the $a_1, \ldots, a_5$ of D.3.12 as follows:

a_1	a_2	a_3	a_4	a_5
1	2	3	4	5
c	d	a	e	b

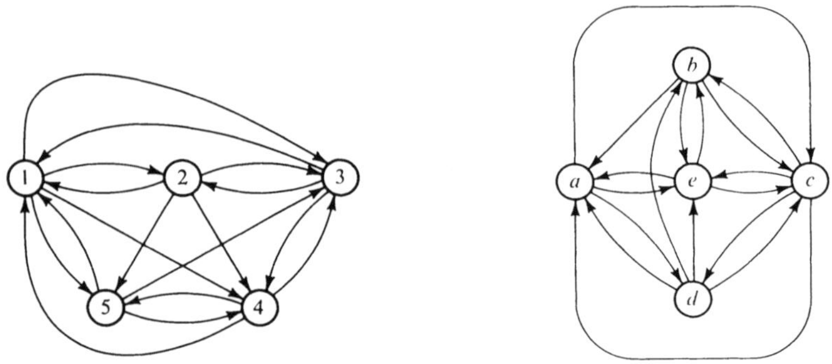

Figure 3.20

The isomorphism problem is very difficult. An algorithm for testing pairs of digraphs for isomorphism certainly exists: Permute the rows and corresponding columns of one of the adjacency matrices until it matches the other, or—in case the digraphs are not isomorphic—stop after $n!$ permutations. The algorithm fails on practical grounds: even with a digraph of only 16 nodes, a hypothetical computer that generates a new permutation and checks the two matrices for equality once every microsecond could take close to 40 years for the job. It is for this reason that heuristic procedures that examine the fine structure of the digraphs have been developed. Let us illustrate two such procedures with reference to the digraphs of Figure 3.20.

First, let us list the indegrees and outdegrees of the nodes.

n	$id(n)$	$od(n)$	n	$id(n)$	$od(n)$
1	4	4	a	4	3
2	2	4	b	3	3
3	4	3	c	4	4
4	4	3	d	2	4
5	3	3	e	4	3

The list shows at once that nodes 1, 2, and 5 can correspond only to nodes c, d, and b, respectively, and that nodes in the set $\{3, 4\}$ can correspond only to nodes in $\{a, e\}$. This is how far the first procedure takes us here.

Next let us list the arcs originating from 3 and 4, and from a and e.

$$\langle 3, 1\rangle, \langle 3, 2\rangle, \langle 3, 4\rangle \qquad \langle a, c\rangle, \langle a, d\rangle, \langle a, e\rangle$$
$$\langle 4, 1\rangle, \langle 4, 3\rangle, \langle 4, 5\rangle \qquad \langle e, a\rangle, \langle e, b\rangle, \langle e, c\rangle$$

Assume the correspondences 1–c, 2–d, 5–b, and assume that node 3 corresponds to node e. Existence of $\langle 3, 1\rangle$ implies existence of $\langle e, c\rangle$, and existence of $\langle 3, 2\rangle$ implies existence of $\langle e, d\rangle$. But $\langle e, d\rangle$ does not exist. Therefore 3 cannot correspond to e, and can correspond only to a. The only node to which 4 can then correspond is e. All that remains is to set up the two adjacency matrices according to the scheme

a_1	a_2	a_3	a_4	a_5
1	2	3	4	5
c	d	a	e	b

The digraphs are isomorphic if and only if the two matrices are equal. Here they are.

DEFINITION 3.31 Let $D = \langle A, R\rangle$ be a digraph. Then $D' = \langle A, R'\rangle$ is the *complement* of D if the following condition is satisfied: For all $a, b \in A$, $\langle a, b\rangle \in R'$ if and only if $\langle a, b\rangle \notin R$.

Example

Figure 3.21 shows the complements of the two digraphs of Figure 3.20.

THEOREM 3.13 Let A and B be two digraphs, and let A' and B' be their complements. A and B are isomorphic if and only if A' and B' are.

Proof. Consider adjacency matrices of A and A'. An element in the adjacency matrix of A is 1 just when the corresponding element in the adjacency matrix of A' is 0, and vice versa. The two matrices are therefore identical in *form*, and the permutation of rows and corresponding columns of the adjacency matrix of B that makes it equal to the adjacency matrix of A is precisely the permutation that has to be applied to the adjacency matrix of B' to make it equal to the adjacency matrix of A'.

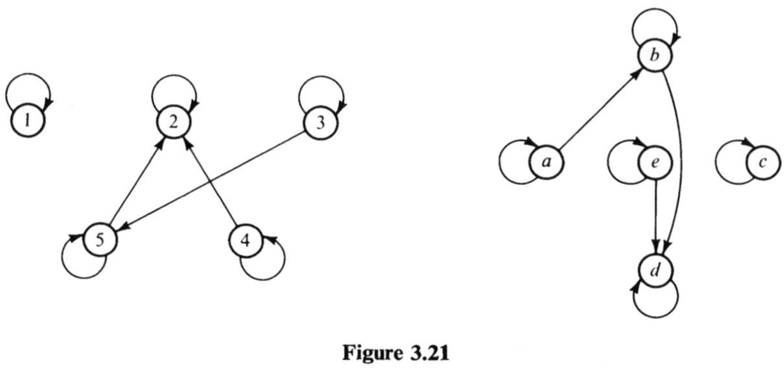

Figure 3.21

Quite often it is easier to establish isomorphism of complements than of the digraphs themselves. The digraphs of Figures 3.20 and 3.21 provide a good example. Let $D' = \langle A, R' \rangle$ be the complement of $D = \langle A, R \rangle$. Then $|R \cup R'| = |A|^2$. Complement D' has fewer arcs than D when $|R| > \frac{1}{2}|A|^2$, and there may then be an advantage in using complements in tests for isomorphism.

Isomorphism of digraphs can be defined also in terms of functions. Consider digraphs $\langle A, R \rangle$ and $\langle B, S \rangle$. They are isomorphic if and only if there exists a one-to-one onto function $f: A \to B$ such that $\langle a, b \rangle \in R$ implies $\langle f(a), f(b) \rangle \in S$ and $\langle c, d \rangle \in S$ implies $\langle f^{-1}(c), f^{-1}(d) \rangle \in R$ for all members of R and S. There may be more than one function on A onto B that satisfies the conditions. Functions $f: A \to A$ satisfying these conditions also exist, e.g., the identity function. In the extreme case, letting A be a set of n nodes, there are $n!$ such functions associated with, for example, the digraph $\langle A, A \times A \rangle$.

The formalism of D.2.38 is not applicable here. Taking a digraph $\langle A, R \rangle$, with $A = \{a_1, \ldots, a_n\}$, we can denote the set of arcs terminating at a_i by f_i. The $(n + 1)$-tuple $\langle A, f_1, \ldots, f_n \rangle$ defines the digraph just as well as $\langle A, R \rangle$, but, although the f_i are functions, they need not be operations.

In the heuristic approach to the isomorphism problem sets of arcs are the data. One develops a number of procedures, each designed to extract from a set of arcs information relating to a particular aspect of the structure of a digraph. The procedures are finally combined and a very complicated program results. Although a set of arcs contains all the information required, the information is not directly accessible; it must be computed. Moreover, the computation takes a different form in each of the procedures. Hence the complexity. For example, in setting up a table of outdegrees, a list of first coordinates of the arcs has to be examined. The indegrees, on the other hand, are computed from a list of second coordinates. The procedure for extracting the arcs that originate from the same node is again different in form, and so on.

The linear formulas introduced in the preceding section contain structural information in a more readily accessible form. As a consequence, instead of having to test digraphs for isomorphism by a sequence of dissimilar tests, one can perform the task by letting a relatively simple unified procedure operate on minimal sets of K-formulas. We shall give only a very sketchy outline of the procedure.

Definition 3.32 Consider a K-formula containing t node symbols (not necessarily distinct), denoted $s_1, s_2, \ldots, s_t$. Let n_i be the *total* number of K-operators to the left of s_i in the K-formula, and denote by $\sum_K$ the sum of the n_i over all t node symbols. A K-formula has *standard structural form* (*ssf*) when $\sum_K$ has been maximized by means of switches and interchanges.

Example

Consider the K-formula $*a*b**da**cae$. Here $\sum_K = 1 + 2 + 4 + 4 + 6 + 6 + 6 = 29$. Let us apply the switch rule to change the distribution of the K-operators: $*a*b**d**caea$ results, and now $\sum_K = 31$, a maximum.

Although an *ssf* of a given K-formula need not be unique, there is generally only a small number of possible *ssf*s. This suggests that digraphs can be tested for isomorphism by comparing the *ssf*s of the K-formulas representing them. There is, however, the difficulty that it is impossible to transform a minimal set of K-formulas having one base of a digraph for their leading nodes to a set of K-formulas whose leading nodes belong to a different base by switches and interchanges alone. The procedure that tests digraphs for isomorphism must therefore have the transformation technique of Step 4 of A.3.4 incorporated in it. In Figure 3.22 each node has two names to imply that the picture represents two digraphs. Any member of $\{a, b, c\}$ constitutes a node base in one digraph; any member of $\{1, 2, 3\}$ does so in the other. Two possible *ssf*s representing the digraphs are $*a*b**cad$ and $*2**3*124$. We have to transform the second K-formula into $\{*12, *2**314\}$, and this set into $*1*2**314$.

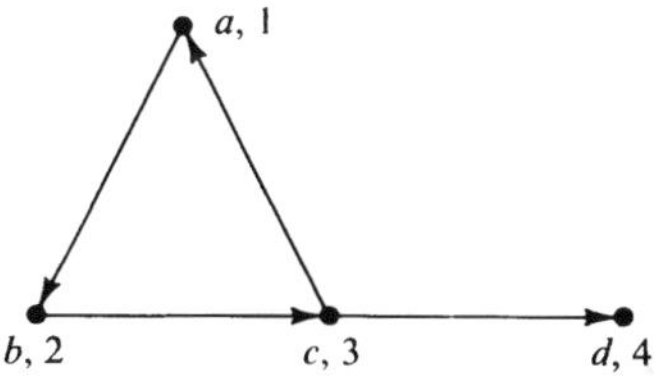

Figure 3.22

3h. Planar Graphs

Planarity is a property studied in the theory of undirected graphs. Our problem: Given a graph, can it be drawn on a sheet of paper in such a way that no two edges cut each other? As usual with problems of some complexity, the solution is found in stages. In advancing to the solution we shall be developing a terminology, and a number of interesting results will be obtained, not always properly relevant to the problem at hand. This section, then, while motivated by the specific problem of planarity, serves as a more general introduction to the theory of graphs.

DEFINITION 3.33 Let $G = \langle A, P \rangle$ be a graph. Then $G' = \langle B, Q \rangle$, where $B \subseteq A$, is a *subgraph* of G if Q contains every edge of P whose elements are both in B. It is a *proper subgraph* if $G' \neq G$. The graph $G'' = \langle A, R \rangle$, with $R \subseteq P$, is a *partial graph* of G.

Definition 3.34 If certain edges of a graph can be placed in a sequence of the form $\{a_1, a_2\}$, $\{a_2, a_3\}$, ..., $\{a_{n-1}, a_n\}$, where all edges are distinct, the set of these edges is a *chain*. It is a *circuit* if, moreover, $a_1 = a_n$. If all nodes in the corresponding sequence of nodes $(a_1, a_2, a_3, \ldots, a_n)$ are distinct, the chain is *simple*. If $a_1, \ldots, a_{n-1}$ are distinct, but $a_1 = a_n$, then the set of edges is a *simple* circuit.

Theorem 3.14 A chain is simple if and only if it has no circuit for a subset.

Proof. Exercise 3.32.

Definition 3.35 A graph G is *connected* if every two distinct nodes of the graph are joined by a chain. Let G' be a connected subgraph of G. G' is a *connected component* of G if there exists no further connected subgraph of which G' is a proper subgraph. A *disconnected* graph is not connected.

In studying a disconnected graph one can treat each of its connected components as a graph in its own right. Therefore, we can limit the discussion that follows to connected graphs with no loss of generality.

Definition 3.36 A *tree* is a connected graph that contains no circuits. Considering the tree $T = \langle B, Q\rangle$ as a partial subgraph of a connected graph $G = \langle A, P\rangle$, members of $P - Q$ are *chords* of T. If $B = A$, then T is a *spanning tree* of G.

Example

In Figure 3.23 trees T_1 and T_2 span G, and they are not the only spanning trees of G. With respect to T_1 we have the set of chords $\{\{a, b\}, \{a, c\}, \{a, d\}, \{b, c\}, \{b, d\}, \{c, d\}\}$. With respect to T_2 the set of chords is $\{\{a, b\}, \{a, c\}, \{b, c\}, \{b, e\}, \{c, e\}, \{d, e\}\}$. One aspect of the theory of graphs is its concern with enumeration. Finding all trees on a given number of nodes would be a typical problem.

Theorem 3.15 Let $\langle A, P\rangle$ be a tree with n nodes. Then $|P| = n - 1$.

Proof. Consider some $a_1 \in A$. Since a tree is connected and has no circuits, there is one and only one chain between a_1 and a member of $\{a_2, a_3, \ldots, a_n\}$. The final edges in the $n - 1$ sequences defining the chains are, of course, distinct. To show that the set of final edges is in fact P assume that there exists a final edge $\{a_i, a_k\}$ and some other edge $\{a_j, a_k\}$. But then there exists a chain $(a_1, \ldots, a_i, a_k, a_j)$; i.e., $\{a_k, a_j\}$ is a final edge.

Theorem 3.16 Let $T = \langle A, Q\rangle$ be a spanning tree of $G = \langle A, P\rangle$, and let edge e be a chord of T. The graph $\langle A, Q \cup \{e\}\rangle$ contains exactly one circuit.

Proof. Exercise 3.35.

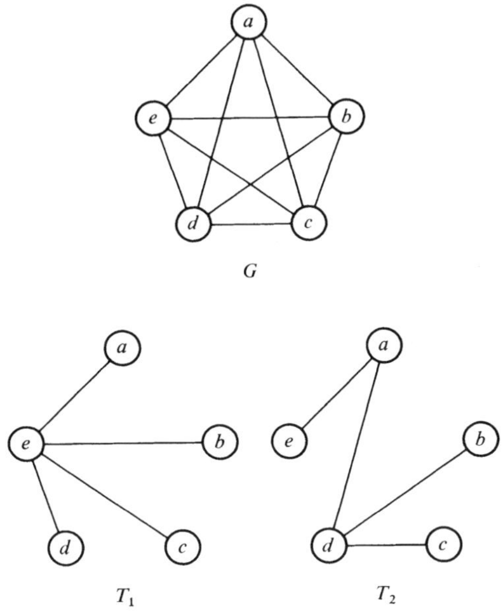

Figure 3.23

DEFINITION 3.37 Let $T = \langle A, Q \rangle$ be a spanning tree of $G = \langle A, P \rangle$, and write $P - Q = \{e_1, e_2, \ldots, e_k\}$. The k circuits contained in the k graphs $\langle A, Q \cup \{e_i\} \rangle$ form a *fundamental set of circuits* (not necessarily unique) of G. The quantity $|P - Q|$ is the *cyclomatic number* of G.

Let $T = \langle A, Q \rangle$ be a spanning tree of $G = \langle A, P \rangle$, and let $|A| = n$ and $|P| = m$. Then, by Th.3.15, $|P - Q| = m - n + 1$. This is the number of fundamental circuits of G. It is obvious that a graph contains no circuits when its cyclomatic number is zero. It contains a single circuit when $m - n + 1 = 1$, i.e., when $m = n$. When $m > n$, a graph may contain simple circuits additional to those in a fundamental set. Algorithms have been found for deriving the fundamental set and, from it, the set of all simple circuits of a graph.

DEFINITION 3.38 Represent a connected graph in the geometric plane by drawing nodes as distinct points, and edges as simple curves. The graph is *planar* if it possesses a geometric representation in which edges intersect only at points representing nodes. In this representation a region of the plane that is bounded by a circuit, and that encloses no other circuit sharing a common edge with it, is a *face*. The unbounded infinite region exterior to the finite faces is also considered a face. (A finite face may enclose another face, provided that the circuits defining the two faces are disjoint.)

Example

Graph G_1 of Figure 3.24 is planar, as shown by the construction. Its faces are z_0, z_1, z_2, z_3 (z_0 is infinite, the others are finite). Graph G_2 is non-planar.

THEOREM 3.17 The circuits defining the finite faces of a planar connected graph form a fundamental set of circuits.

Proof. Consider a planar graph G_{f+1} with $f + 1$ finite faces. The graph is constructed from a planar graph G_f with f finite faces by drawing a simple chain between two nodes a and b in G_f, $(a, n_1, \ldots, n_k, b)$, where $n_1, \ldots, n_k$ do not belong to G_f. (If some n_i belonged to G_f, the resulting structure would have at least $f + 2$ finite faces.) The construction increases the number of nodes by k and the number of edges by $k + 1$; i.e., the cyclomatic number, in going from G_f to G_{f+1}, is increased by $(k + 1) - k = 1$. Therefore, by D.3.37, the theorem is true for G_{f+1} if it is true for G_f. But the theorem is true for G_1, a graph with one finite face—G_1 contains a single circuit, and this circuit must be the fundamental set.

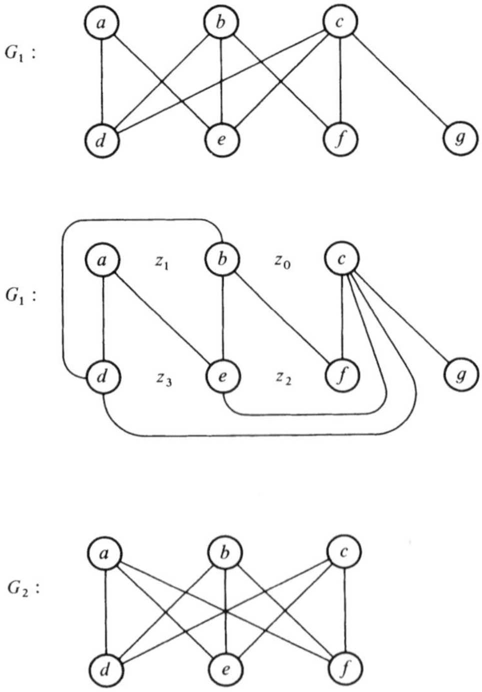

Figure 3.24

We have shown that the cyclomatic number of a planar connected graph is equal to the number of finite faces in the graph. Therefore, putting n for the number of nodes, m for the number of edges, and f for the *total* number of faces (including the infinite face), we can express Th.3.17 as a formula: $n - m + f = 2$. This expression is known as *Euler's formula.* A connected graph is planar if and only if it satisfies Euler's formula.

DEFINITION 3.39 A graph $G = \langle A, P \rangle$ is *complete* if P contains every 2-element subset of A. A complete graph on n nodes is denoted by K_n.

Example

Graph G of Figure 3.23 is a complete graph on 5 nodes. A complete graph $\langle A, P \rangle$ having n nodes contains the greatest possible number of edges on n nodes. This number is equal to the number of 2-combinations of A, and, by Th.1.17, $C_2(A) = C(n, 2) = \frac{1}{2}n(n-1)$. Note that a set $\{a, a\}$ has only one element and cannot be an edge; i.e., we have no counterpart of slings in the theory of graphs.

DEFINITION 3.40 A graph $\langle B \cup C, P \rangle$ is *bipartite* if $B \cap C = \varnothing$ and every member of P has one element in B and the other in C. A bipartite graph in which every member of B is joined by an edge to every member of C is known as a *utility graph.* The utility graph in which the disjoint sets of nodes have s and t members, respectively, is denoted by $K_{s,t}$.

Example

Graphs G_1 and G_2 of Figure 3.24 are bipartite. In the case of G_1 we have $B = \{a, b, c\}$ and $C = \{d, e, f, g\}$. In the case of G_2 the sets are $B = \{a, b, c\}$ and $C = \{d, e, f\}$. G_2 is a utility graph. The term has arisen from the use of this species of graphs to represent situations in which each of s consumers is to be supplied with each of t public utilities (e.g. water, gas, electricity) from their supply stations.

THEOREM 3.18 Graphs K_5 (graph G of Figure 3.23) and $K_{3,3}$ (graph G_2 of Figure 3.24) are nonplanar.

Proof. Assume that K_5 is planar. Then Euler's formula gives $f = 2 - n + m = 2 - 5 + 10 = 7$. A circuit defining a face must contain at least three edges, and each edge lies on a boundary of two faces. Hence $m/3 \geqq f/2$ or $2m \geqq 3f$, leading to the contradiction $20 \geqq 21$. Next assume that $K_{3,3}$ is planar. Circuits in a bipartite graph must contain at least four edges. (If a circuit had only three edges, one edge would have to have both members in one of the disjoint sets.) Hence we must have $m/4 \geqq f/2$ or $m \geqq 2f$. But here $f = 2 - 6 + 9 = 5$, again implying an absurdity, namely $9 \geqq 10$.

DEFINITION 3.41 A simple chain $(a, n_1, \ldots, n_k, b)$ in a graph is *free* if all edges of the graph that contain nodes $n_1, \ldots, n_k$ belong to this chain. Replacement of a free chain $(a, n_1, \ldots, n_k, b)$ by the edge $\{a, b\}$ and removal of nodes $n_1, \ldots, n_k$ is a *reduction* of the free chain.

Example

Graph G_1 of Figure 3.25 contains free chains (a, z, b) and (e, x, y, d). Reduction of these chains produces G_2. Chain (a, z, b) reduces to edge $\{a, b\}$, and, since this edge is already in G_1, reduction does not add a new edge to the graph. The other free chain reduces to the edge $\{e, d\}$. In this instance the free chain is replaced by a new edge.

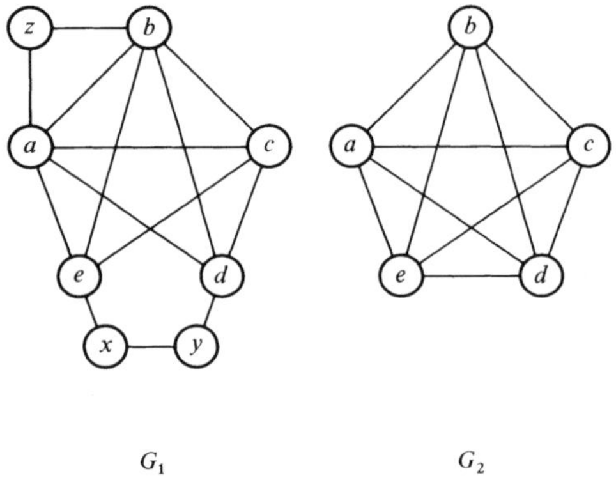

Figure 3.25

The property of planarity or the lack of it is invariant under the operation of reduction. This is recognized in the formulation of the next theorem.

THEOREM 3.19 (Kuratowski's theorem). A graph is planar if and only if it does *not* contain as a subgraph the graph K_5, or the graph $K_{3,3}$, or a graph that can be reduced to K_5 or $K_{3,3}$ by the operation of reduction.

Proof. We have already demonstrated necessity. The sufficiency proof is very difficult. We shall not give it here.

Notes

The most useful general text for this chapter appears to be [Bu65]; this book gives the basic theory of directed and undirected graphs, and contains numerous examples of applications. [Ha65b] is a specific text on digraphs. It is primarily addressed to social scientists, and is rather easy to read. The theory of undirected graphs is covered by [Ha69]. The present interest in applications of graph theory has been to a large extent due to Berge's *Theory of Graphs and its Applications*, [Be58], which puts a strong stress on the use of graphs in operations research. A more recent work, [Be62], continues on the same lines. [Ka64] has an interesting structure; the first part introduces graphs informally as a tool in operations research, the more formal development being held over to the later sections of the book.

A.3.3 has been published in [Be68a]. The K-operator and K-formulas of Section 3e were introduced by Krider in [Kr64] as an aid to automatic flowcharting of computer programs, but most of the material in this section is new. Th.3.10 has been stated without proof in [Be69a]. A different linear notation is described in [Ba67a]. The heuristic procedures for testing pairs of digraphs (or graphs) for isomorphism were independently developed by Sussenguth [Sa64] and Unger [Un64]. A somewhat different heuristic procedure, based on the matrix representation of a graph, is to be found in [Bo64a]. This paper contains an Algol program of the procedure. Some attempts have been made to find a function from the set of graphs under which two graphs would have the same image if and only if they were isomorphic. To date all conjectures in this area have come to nothing—see [Tu68]. A canonical scheme for labeling the nodes of a tree is described in [Go63]. Solutions to the problem of finding circuits in graphs are to be found in [We66a, Go67b, Pa69a]. The algorithm for generating all cycles from the fundamental set in [We66a] has been found faulty and is corrected in [Gi69]. [Mc69] is an algorithm for finding all spanning trees of a graph. A proof of Th.3.19 can be found in [Be58, Bu65]. Because of the difficulty of telling whether or not K_5 or $K_{3,3}$ is the subgraph of a graph, Kuratowski's theorem has little practical use. An algorithm for deriving a planar geometric representation of a planar graph can be found in [De64]; this paper contains also a discussion of the reduction of the number of intersections of edges in a nonplanar graph.

Exercises

3.1 (a) Draw pictures of the following digraphs:

(i) $\langle A, A \times A\rangle$, where $A = \{a, b, c, d\}$;

(ii) $\langle\{1, 2, 3, 4, 5\}$, greater than$\rangle$;

(iii) $\langle A, R\rangle$, where $A = \{a, b, c, d\}$ and R is the identity relation;

(iv) $\langle \mathscr{P}(\{1, 2, 3\}), R\rangle$, where R is the subset relation;

(v) $\langle \mathscr{P}(\{1, 2, 3\}), Q\rangle$, where Q is the relation *covers* with respect to the subset relation in $\mathscr{P}(\{1, 2, 3\})$;

(vi) $\langle S \cup C, R\rangle$, where S is a set of five students of your acquaintance, C is the set of courses attended by these students, and $R = \{\langle a, b\rangle \mid$ student a takes course $b\}$;

(vii) $\langle S \cup L, R\rangle$, where S is the set of students of Part (vi) and L is the set of localities where they live, and $R = \{\langle a, b\rangle \mid \text{student } a \text{ lives in } b\}$;

(viii) the *structure* of Figure 2.7.

3.2 (a) The digraphs of Parts (vi) and (vii) of Exercise 3.1 are *bipartite* digraphs. Try to abstract their distinguishing feature and hence devise a definition of bipartite digraphs.

3.3 (a) Using the information you collected for Part (vi) of Exercise 3.1, draw an undirected graph $\langle S, P\rangle$, where $P = \{\{a, b\} \mid a, b \in S, a \text{ and } b \text{ have at least one course in common}\}$.

3.4 (b) Find indegrees and outdegrees of all nodes in the digraphs of Figures 3.7 and 3.8.

3.5 (b) Find all cycles in the digraph of Figure 3.8.

3.6 (b) Show that if a digraph contains paths $(a, \ldots, b)$ and $(b, \ldots, a)$, then the digraph contains a cycle, but that there need not exist a *simple* cycle $(a, \ldots, b, \ldots, a)$.

3.7 (b) Show that if A and B are cycles, and the node sequences that define them have at least one node in common, then $A \cup B$ is also a cycle.

3.8 (b) Consider a digraph that has a node base of n elements. Show that at least $n - 1$ arcs have to be added to the digraph to convert it to a digraph whose node base consists of a single node.

3.9 (c) Write a Fortran subroutine that computes the path matrix of a digraph from its adjacency matrix. Would it matter if the input to the subroutine were a variable adjacency matrix?

3.10 (c) Write a Fortran subroutine that computes from the adjacency matrix of a digraph $\langle A, R\rangle$ a matrix S such that s_{ij} is the length of the *shortest* path from a_i to a_j and $s_{ij} = 10^{10}$ (or some other very large number) if there is no path from a_i to a_j. Can S be computed from the path matrix of the digraph?

3.11 (c) Characterize the digraphs of Exercise 3.1 according to D.3.14.

3.12 (c) In A.3.1 make `MX` a logical array in which an element is `.TRUE.` if the corresponding element of the adjacency matrix is 1, and `.FALSE.` if it is 0. Modify `ISPORD` accordingly.

3.13 (c) What is to be changed in `ISPORD` to make it into a function subprogram that determines whether or not a relation is an equivalence?

3.14 (c) Write a Fortran logical function that, given the *path* matrix of a digraph, returns `.TRUE.` if the digraph is cyclic, and `.FALSE.` if it is not.

3.15 (c) Attempt to get an understanding of `ISEQUI` of A.3.2 by following through the actions of the procedure on the matrix of the example of A.3.2.

Then modify `ISEQUI` so that it also prints the equivalence classes generated by the equivalence relation defined by the adjacency matrix `MX` (as sets of row numbers).

3.16 (c) Write a Fortran function subprogram that determines whether or not a digraph is complete.

3.17 (d) What is the form of the path matrix of a strongly connected digraph?

3.18 (d) Write a Fortran subroutine that computes from the adjacency matrix of a digraph $\langle A, R\rangle$ a matrix C such that $c_{ij} = 1$ if a_i and a_j are connected and $c_{ij} = 0$ if they are not connected.

3.19 (d) Prove Th.3.8.

3.20 (d) Devise an algorithm for finding the strong components of a digraph.

3.21 (d) Prove Th.3.9, and devise an algorithm for finding all node bases of a digraph.

3.22 (e) Find minimal sets of K-formulas for the digraphs of Figures 3.7 and 3.8 (disregard weights in Figure 3.7).

3.23 (e) Implement A.3.4 as a Fortran program.

3.24 (e) Which of the following are K-formulas?

(i) $***aba*dc$. (ii) $***aba*dcd$.
(iii) $**aba**dc$. (iv) $*a*b*a*dc$.

Draw the digraphs corresponding to the K-formulas.

3.25 (e) Implement a recognition algorithm for K-formulas based on D.3.22a as a Fortran program.

3.26 (e) Show that the interchange rule of D.3.26 does not lose any of its power if it is defined in the following, more restricted, fashion: The interchange of a K-formula of node a, occurring as a (proper) subformula of a member of F, and a node symbol a, occurring in any member of F, is an application of the interchange rule.

3.27 (f) Produce an example of a digraph in which exactly one node has indegree 0 and every other node has indegree 1 that is not a directed tree.

3.28 (f) Show that D.3.27 defines the same objects when the condition that the digraph must be acyclic is replaced by the condition that the digraph must be connected and without slings.

3.29 (f) Prove Th.3.12.

3.30 (g) Let $D_1 = \langle X_1, R_1\rangle$, $D_2 = \langle X_2, R_2\rangle$, where $X_1 = \{a, b, c, d, e, f\}$ and $X_2 = \{1, 2, 3, 4, 5, 6\}$, $R_1 = \{\langle c, a\rangle, \langle b, c\rangle, \langle e, b\rangle, \langle e, f\rangle, \langle f, a\rangle, \langle b, f\rangle, \langle c, f\rangle, \langle b, e\rangle, \langle a, c\rangle, \langle d, a\rangle, \langle d, f\rangle, \langle e, d\rangle, \langle d, b\rangle, \langle f, b\rangle, \langle f, e\rangle, \langle c, e\rangle, \langle b, d\rangle, \langle a, f\rangle, \langle a, d\rangle, \langle d, e\rangle, \langle a, e\rangle, \langle f, c\rangle, \langle f, d\rangle, \langle d, c\rangle\}$, $R_2 = \{\langle 1, 2\rangle, \langle 5, 4\rangle, \langle 6, 4\rangle, \langle 4, 1\rangle, \langle 2, 4\rangle, \langle 1, 3\rangle, \langle 6, 1\rangle, \langle 5, 6\rangle, \langle 3, 2\rangle, \langle 2, 5\rangle, \langle 1, 6\rangle, \langle 3, 4\rangle, \langle 2, 3\rangle, \langle 3, 1\rangle, \langle 4, 2\rangle, \langle 3, 5\rangle, \langle 6, 5\rangle, \langle 2, 6\rangle, \langle 1, 4\rangle, \langle 2, 1\rangle, \langle 4, 3\rangle, \langle 5, 1\rangle, \langle 1, 5\rangle, \langle 6, 2\rangle\}$. Are the digraphs D_1 and D_2 isomorphic?

3.31 (g) Let $D = \langle X, R\rangle$ be a complete symmetric digraph on n nodes $x_1, x_2, \ldots, x_n$. Let $D' = \langle X, R'\rangle$ be a digraph that results when the subscripts of the nodes are permuted. Discuss the isomorphism of D and D'.

3.32 (h) Prove Th.3.14.

3.33 (h) Find all spanning trees for graph G of Figure 3.23.

3.34 (h) Show that the definition of a tree in D.3.36 is consistent with the following definition: A tree is a connected graph that becomes disconnected when *any* one of its edges is removed.

3.35 (h) Prove Th.3.16.

CHAPTER 4

Strings

4a. Algebraic Structures

As we have pointed out before, abstraction is the process of eliminating everything that is inessential to a particular investigation. Thus, when we had to *define* concepts such as similarity, homomorphism, and the like, we found the truly abstract algebras of Section 2h most effective, precisely because they are devoid of structure. On the other hand, when we look for a useful mathematical model of a real system, we want a model with as much structure as possible. The more structure an algebra possesses, the richer its stock of theorems. Consequently, the more features of the system are reflected in the model, the more information about the system is the model capable of providing.

Structure is imposed on an abstract algebra by means of rules that must be satisfied by operations in the algebra. We might require an operation to be associative, or commutative. For a pair of operations, we might have an axiom defining the way one operation distributes over the other. The simplest nontrivial abstract algebra is the system $\langle A, * \rangle$, where $*$ denotes a binary operation in A. D.4.1 is a set of statements, subsets of which define specific algebras A. These statements are then the axioms of A.

DEFINITION 4.1 Let $\langle A, * \rangle$ be an algebra. We say that algebra A satisfies the associative law if statement A_1 is true. Similarly for statements A_2, A_3, and A_4.

A_1. *Associative law*: $a * (b * c) = (a * b) * c$ for all $a, b, c \in A$.
A_2. *Commutative law*: $a * b = b * a$ for all $a, b \in A$.
A_3. *Identity law*: There exists an element $e \in A$ such that $e * a = a * e = a$ for every $a \in A$.
A_4. *Inverse law*: For every $a \in A$ there exists an element $a' \in A$ such that $a' * a = a * a' = e$.

Statement A_3 actually comprises two identity laws: a left identity law, expressed by $e * a = a$, and a right identity law, expressed by $a * e = a$. Similarly A_4 comprises left inverse and right inverse laws.

DEFINITION 4.2 An algebra $\langle A, * \rangle$ that satisfies A_1 of D.4.1 is a *semigroup*. If, in addition, the algebra satisfies A_3, it is a *semigroup with identity* or a *monoid*. (It is easy to show that the identity element of a monoid is unique.)

Examples

1. If A is a nonempty set, then algebras $\langle \mathscr{P}(A), \cup \rangle$ and $\langle \mathscr{P}(A), \cap \rangle$ are semigroups with identity, the identity elements being the null set and the set A, respectively.
2. If N is the set of nonnegative integers, then $\langle N, + \rangle$ and $\langle N, \cdot \rangle$ are semigroups with identity elements 0 and 1, respectively.

In keeping with D.2.34 we should consider the identity element a 0-argument operation, and write the algebras of the examples above as $\langle \mathscr{P}(A), \cup, \varnothing \rangle$, $\langle \mathscr{P}(A), \cap, A \rangle$, $\langle N, +, 0 \rangle$, and $\langle N, \cdot, 1 \rangle$.

Semigroups without identity elements are not particularly interesting. Therefore, it is quite normal to find the term *semigroup with identity* abbreviated to *semigroup*. This usage is based on the assumption that no confusion would arise because semigroups without identity elements are not worth talking about. When one goes as far as to speak of a *semigroup* $\langle A, *, e \rangle$, of course no confusion can arise.

DEFINITION 4.3 A semigroup with identity that satisfies A_4 is a *group*. If, in addition, it satisfies A_2, it is an *Abelian group*.

Examples

1. The set of all integers under addition is a group. Addition is clearly associative. There exists an identity element, namely $e = 0$, and the inverse law is satisfied if we take $a' = -a$. The group is Abelian.
2. The set of all integers is not a group under multiplication (A_4 does not hold), but the set of all rational numbers is an Abelian group. The inverse of an element of this group is defined by $x' = 1/x$.

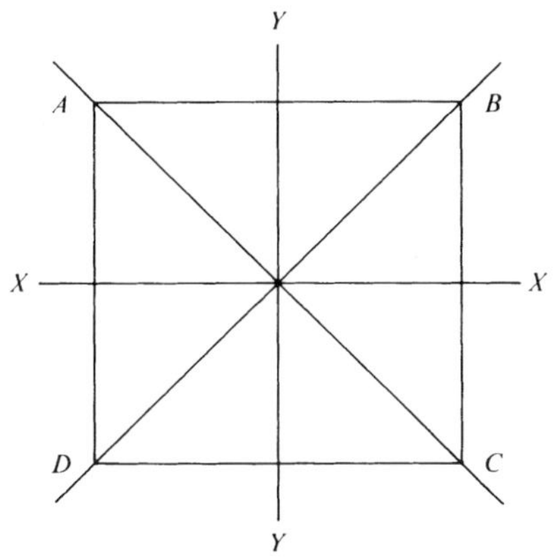

Figure 4.1

3. Consider the square $ABCD$ shown in Figure 4.1. The square can be rotated (clockwise) through 90°, 180°, 270°, and 360° with no change to its appearance. It can be rotated also about axes XX, YY, AC, and BD. We denote these eight rotations by a, b, c, d, e, f, g, and h, respectively, and let $x * y$ stand for rotation x followed by rotation y. It is easy to see that $*$ is an operation in $\{a, \ldots, h\}$. For example, $a * e = h$ and $e * a = g$. Table 4.1 is a "multiplication" table for this operation. The operation is associative, but not commutative. Rotation d is a (unique) identity element, and each element possesses an inverse. Algebra $\langle\{a, \ldots, h\}, *, d\rangle$ is, therefore, a group (known as the dihedral group D_4).

TABLE 4.1

OPERATION $x * y$ IN $\{a, \ldots, h\}$

x \ y	a	b	c	d	e	f	g	h
a	b	c	d	a	h	g	e	f
b	c	d	a	b	f	e	h	g
c	d	a	b	c	g	h	f	e
d	a	b	c	d	e	f	g	h
e	g	f	h	e	d	b	a	c
f	h	e	g	f	b	d	c	a
g	f	h	e	g	c	a	d	b
h	e	g	f	h	a	c	b	d

DEFINITION 4.4 If a group G has n elements, it is called a group of *order* n. If G is infinite, it is called a group of *infinite order* or an *infinite group*.

Example

The dihedral group D_4 is a finite group of order 8. The groups of the other examples of D.4.3 are infinite.

We shall now study permutations in the context of the theory of groups. In Section 1h a permutation of n objects was defined as something static, an n-tuple with distinct elements. Here we take a dynamic approach; we identify a permutation with the *action* taken to transform one n-tuple into another by defining a permutation of a set A having n elements as a one-to-one transformation (function) $p: A \to A$. There are $n!$ such transformations. Since an algorithm generates permutations by a sequence of transformations, and algorithms are of greater relevance in computer science than static definitions, the definition given here should hold a greater appeal for us.

We write a permutation as

$$p = \begin{pmatrix} a_1 & a_2 & a_3 & \cdots & a_n \\ b_1 & b_2 & b_3 & \cdots & b_n \end{pmatrix},$$

where $\langle a_i, b_i \rangle \in p$. The order in which the a_i are written is immaterial, but if $\langle a_i, b_i \rangle \in p$, the b_i must be written below the a_i:

$$p = \begin{pmatrix} a_1 & a_2 & \cdots & a_n \\ b_1 & b_2 & \cdots & b_n \end{pmatrix} = \begin{pmatrix} a_2 & a_n & \cdots & a_1 \\ b_2 & b_n & \cdots & b_1 \end{pmatrix}.$$

The product of permutations p and q, written pq, is obtained by carrying out transformations p and q in sequence. For example, if

$$p = \begin{pmatrix} 1 & 2 & 3 \\ 2 & 1 & 3 \end{pmatrix} \quad \text{and} \quad q = \begin{pmatrix} 1 & 2 & 3 \\ 3 & 1 & 2 \end{pmatrix},$$

then

$$pq = \begin{pmatrix} 1 & 2 & 3 \\ 2 & 1 & 3 \end{pmatrix}\begin{pmatrix} 1 & 2 & 3 \\ 3 & 1 & 2 \end{pmatrix} = \begin{pmatrix} 1 & 2 & 3 \\ 2 & 1 & 3 \end{pmatrix}\begin{pmatrix} 2 & 1 & 3 \\ 1 & 3 & 2 \end{pmatrix} = \begin{pmatrix} 1 & 2 & 3 \\ 1 & 3 & 2 \end{pmatrix}.$$

Note that $pq \neq qp$ here:

$$qp = \begin{pmatrix} 1 & 2 & 3 \\ 3 & 1 & 2 \end{pmatrix}\begin{pmatrix} 1 & 2 & 3 \\ 2 & 1 & 3 \end{pmatrix} = \begin{pmatrix} 1 & 2 & 3 \\ 3 & 1 & 2 \end{pmatrix}\begin{pmatrix} 3 & 1 & 2 \\ 3 & 2 & 1 \end{pmatrix} = \begin{pmatrix} 1 & 2 & 3 \\ 3 & 2 & 1 \end{pmatrix}.$$

Let us consider $P(A)$, the set of all permutations of a set A. Obviously the product function is an operation in $P(A)$. It is easy to prove that this operation obeys the associative law. The identity function provides an identity element; we write

$$i_n = \begin{pmatrix} a_1 & a_2 & \cdots & a_n \\ a_1 & a_2 & \cdots & a_n \end{pmatrix},$$

and each $p \in P(A)$ has a unique inverse $p^{-1} \in P(A)$ such that if

$$p = \begin{pmatrix} a_1 & a_2 & \cdots & a_n \\ b_1 & b_2 & \cdots & b_n \end{pmatrix}, \qquad \text{then} \qquad p^{-1} = \begin{pmatrix} b_1 & b_2 & \cdots & b_n \\ a_1 & a_2 & \cdots & a_n \end{pmatrix}.$$

Clearly $pp^{-1} = p^{-1}p = i_n$. $P(A)$ is therefore a group under the product operation. It is called the *symmetric permutation group* or *symmetric group* of order $n!$ (when A has n elements).

Let $C = \{c_1, c_2, \ldots, c_m\}$ be a subset of $A = \{a_1, a_2, \ldots, a_n\}$. A *cycle* of length m is a permutation p such that

$$\begin{aligned} p(c_i) &= c_{i+1} \qquad (i = 1, 2, \ldots, m-1), \\ p(c_m) &= c_1, \\ p(a_i) &= a_i \qquad (a_i \in A - C). \end{aligned}$$

For example,

$$\begin{pmatrix} 1 & 2 & 3 & 4 & 5 & 6 & 7 \\ 1 & 4 & 3 & 6 & 2 & 5 & 7 \end{pmatrix}$$

is a cycle of length 4 ($c_1 = 2$, $c_2 = 4$, $c_3 = 6$, $c_4 = 5$). It is customary to abbreviate a cycle to just those elements changed in the transformation, e.g., to (2 4 6 5) here. Two or more cycles that have no elements in common in the abbreviated notation are called *disjoint*, and every permutation can be expressed as a product of disjoint cycles. The disjoint cycles are very easily detected when a permutation is represented by a digraph. Figure 4.2 is the digraph representing the permutation

$$p = \begin{pmatrix} 1 & 2 & 3 & 4 & 5 & 6 \\ 5 & 6 & 3 & 1 & 4 & 2 \end{pmatrix}.$$

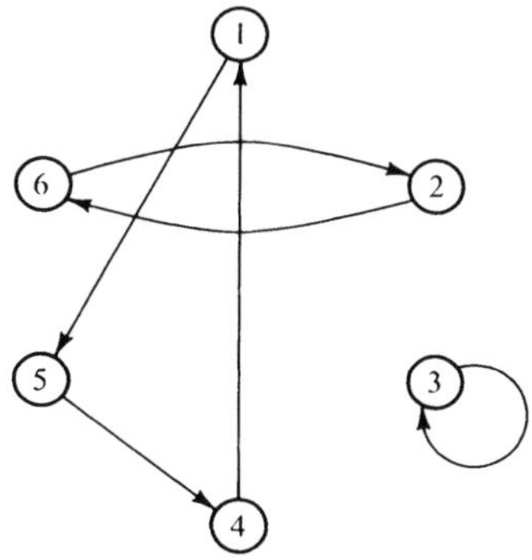

Figure 4.2

We see at once that $p = (1\ 5\ 4)(2\ 6)(3)$. A cycle of length 1 is the identity function. Therefore, it is shorter to write $p = (1\ 5\ 4)(2\ 6)$, but a notation that includes every element in the domain of the permutation is preferable.

A cycle of length 2 is called a *transposition*, and every permutation can be expressed as the product of transpositions. If it is the product of an even number of transpositions, the permutation is called *even*; otherwise it is called *odd*. A permutation cannot be both even and odd. Half the permutations in $P(A)$ are even, half are odd. The permutation $(1\ 5\ 4)(2\ 6)$ becomes, in terms of transpositions, $(1\ 5)(1\ 4)(2\ 6)$; it is odd.

We shall now consider algebras with two binary operations: $\langle R, +, \cdot\rangle$. It is common practice to call operations $+$ and $\cdot$ "addition" and "multiplication," respectively, and to abbreviate expressions of the form $a \cdot b$ to ab.

DEFINITION 4.5 The algebra $\langle R, +, \cdot, 0, 1\rangle$ is a *ring* if $\langle R, +, 0\rangle$ is an Abelian group, $\langle R, \cdot, 1\rangle$ is a semigroup, and the following additive distribution laws hold for all $a, b, c \in R$:

$$a(b + c) = ab + ac, \qquad (b + c)a = ba + ca.$$

Examples

1. The algebra $\langle I, +, \cdot, 0, 1\rangle$, where I is the set of integers, and the operations are the normal arithmetic operations, is a ring.
2. The Boolean algebra $\langle B, \oplus, *, ', 0, 1\rangle$ is not a ring ($a' \oplus a \neq 0$ in B), but if we define a new operation in B, $a + b = (a * b') \oplus (b * a')$, then $\langle B, +, *, 0, 1\rangle$ is a ring.

Strictly speaking the structures defined in D.4.5 are *rings with multiplicative identity*. The more precise definition of a "bare" ring would not be as restrictive as D.4.5 because it would not require the semigroup to have an identity element, but, as we have pointed out before, semigroups without identity elements excite little interest.

DEFINITION 4.6 The algebra $\langle Q, +, \cdot, 0, 1\rangle$ is a *Q-semiring* if $\langle Q, +, 0\rangle$ and $\langle Q, \cdot, 1\rangle$ are semigroups, and the following laws hold for all $a, b, c \in Q$:

$$a + b = b + a;$$
$$a(b + c) = ab + ac, \qquad (b + c)a = ba + ca;$$
$$a + 1 = 1;$$
$$a \cdot 0 = 0 \cdot a = 0.$$

Examples

1. The Boolean algebra $\langle B, \oplus, *, 0, 1\rangle$ is a Q-semiring.
2. $\langle R_+^\infty, min, +, \infty, 0\rangle$, where R_+^∞ is the set of nonnegative real numbers,

together with positive infinity, *min* is an operation that takes the smaller of two numbers, and + is normal addition, is a Q-semiring, and $\langle R_+^\infty$, *max*, *min*, 0, $\infty\rangle$, where *max* is the operation of taking the larger of two numbers, is also a Q-semiring.

Our interest in Q-semirings will be justified in Section 6a.

4b. Algebra of Strings

DEFINITION 4.7 An *alphabet* is a finite set of symbols. An alphabet will always be denoted by the letter V.

Example

The set $\{a, b, c, \ldots, z\}$ can be an alphabet, and so can the set {*apple*, *pear*, *banana*, *carrot*}, but, if the second set is used for an alphabet, every element in the set must be considered a single indivisible object.

DEFINITION 4.8 A *string* over V of length $m\,(m \geqq 0)$ is an m-sample of V. The string of length 0 is distinguished as the *empty string*; we write Λ for it. A string $\langle x_1, x_2, \ldots, x_m\rangle$ will be written $x_1x_2 \cdots x_m$ for convenience, possibly enclosed in quotes, "$x_1x_2 \cdots x_m$". The set of all strings over V is denoted by V^*. Members of V^* will be denoted by Greek letters: α, β, $\gamma, \ldots$.

Example

Let $V = \{a, b\}$. We have $V^* = \bigcup_{i \in N} S_i(V)$,where N is the set of non-negative integers, and $S_0(V) = \{\Lambda\}$, $S_1(V) = \{a, b\}$, $S_2(V) = \{aa, ab, ba, bb\}$, and so on. It is meaningless to talk of strings of infinite length, but obviously no bound can be set on the length of a string.

In linguistics the term *vocabulary* is sometimes used as a synonym for our *alphabet*, and its members are called *words*. In logic, on the other hand, *word* is sometimes used to denote our *string*.

DEFINITION 4.9 Let α be the string $x_1x_2 \cdots x_m$ and β the string $y_1y_2 \cdots y_n$. The *concatenation* (*complex product*) of α and β, written $\alpha \cdot \beta$ or simply $\alpha\beta$, is the string $x_1x_2 \cdots x_my_1y_2 \cdots y_n$.

Example

Let $\alpha = ter$, $\beta = ra$. Then $\alpha\beta = terra$, $\beta\alpha = rater$.

Sometimes one has to consider samples of V^* as strings. Let $V = \{e, g, i, m, n, r, s, t\}$. Then $semi \in V^*$ and $string \in V^*$, and one 2-sample of V^* is $\langle semi, string\rangle$. We shall call the taking of a 2-sample and the 2-sample itself *juxtaposition*, and use the symbol + for it (making sure that $+ \notin V$). Then ⟨*semi*, *string*⟩ becomes the string *semi* + *string*. Juxtaposition must not be confused with concatenation. Since $+ \notin V$, juxtaposition cannot be an operation in V.

THEOREM 4.1 The algebra of strings $\langle V^*, \cdot, \Lambda \rangle$ is a semigroup (with identity).

Proof. Obvious.

DEFINITION 4.10 If $A \subseteq V^*$ and $B \subseteq V^*$, then the *complex product* of A and B is the set of strings $A \cdot B = AB = \{\alpha\beta \mid \alpha \in A, \beta \in B\}$.

DEFINITION 4.11 Let $A \subseteq V^*$. Then the powers of A are defined by $A^0 = \{\Lambda\}$, $A^n = A^{n-1}A$ $(n \geqq 1)$. We put $A^* = \bigcup_{i \in N} A^i$, where N is the set of nonnegative integers, and call A^* the *star* of A. If $A = \{a\}$, where $a \in V$, we write $a^0 = \Lambda$, $a^n = a^{n-1}a$.

Examples

1. $\{a\}^* = \{\Lambda, a, aa, aaa, \ldots\}$.
2. If $A \supseteq V$ holds, then $A^* = V^*$.

If $\langle A, * \rangle$ is a semigroup, then the smallest subset of A from which every element of A can be generated by repeated application of the operation $*$ is called the set of *generators* of A. In the case of $\langle V^*, \cdot, \Lambda \rangle$ the set of generators is V. The algebra $\langle V^*, \cdot, \Lambda \rangle$ is sometimes referred to as a *free* semigroup. It is very difficult to explain the difference between a free algebra and one that is not free. The definition itself is straightforward. Let A be a semigroup with generators B, and let F be the set of functions $\phi: B \to M$, where M denotes any semigroup. Then A is a free semigroup if and only if every member of F possesses an extension $f: A \to M$ that is a homomorphism. Other free algebras are defined analogously. The definition suggests that the elements of a free algebra are as unrestricted as they can possibly be, and still have the structure of this algebra; i.e., only properties that follow from the axioms can be of any consequence in a free algebra. Consider a semigroup V^* with generators $\{a, \ldots, z\}$, say. If V^* is to be free, we can accept only the following definition of equality in V: $a_1a_2 \cdots a_n = b_1b_2 \cdots b_m$ if and only if $n = m$ and $a_i = b_i$ for $1 \leqq i \leqq n$. This is simply the definition of equality of samples. If we were to extend the meaning of equality by rules such as *two* · *tens* = *twenty*, say, which cannot be derived from the axioms, V^* would cease to be a *free* semigroup.

DEFINITION 4.12 String β is a *substring* of string σ if there are strings α and γ (possibly empty) such that $\sigma = \alpha\beta\gamma$. If at least one of α and γ is nonempty, then β is a *proper substring* of σ.

Examples

1. Let $\sigma = ***a***bbcd***cbdee$. Then $***bbcd$ is a substring of σ, and so is $***a*$, and so forth.

2. The string *aabb* has nine substrings. They are Λ, *a*, *b*, *aa*, *ab*, *bb*, *aab*, *abb*, *aabb*.

3. Consider strings *semi* and *tone*. Their concatenation is *semitone*, their juxtaposition *semi* + *tone*. The concatenation has *emit* for a substring, but not the juxtaposition. The string *one* is a substring of both *semitone* and *semi* + *tone*.

Concatenation is the only operation in a semigroup of strings. If two strings are concatenated, the length of the resulting string cannot be smaller than the length of the longer of the two arguments. Obviously concatenation can operate on strings only as indivisible entities. In most practical situations, however, substrings have to be extracted, or deleted, or rearranged into new patterns. We need, therefore, to define a process of substitution.

DEFINITION 4.13 If $\sigma = \alpha\beta\gamma$ and $\sigma' = \alpha\beta'\gamma$ are strings, then σ' is the result of a *substitution* of β' for β in σ. If, moreover, $\alpha\beta$ cannot be expressed as $\alpha'\beta\gamma'$, where γ' is nonempty, the substitution is *canonical*. Substitution is a function in a given V^*. We use the notation $f(\sigma, \alpha, \beta, \beta')$. In the case of canonical substitution only three arguments are needed, and we write $f_c(\sigma, \beta, \beta')$. (The functions are defined only when β is a substring of σ.)

Examples

1. $f(ararat, ar, ara, \Lambda) = art$ and $f(ararat, ar, ara, den) = ardent$, but $f_c(ararat, ara, \Lambda) = rat$.

2. $f_c(tun, \Lambda, s) = stun$. This is an example of a canonical substitution for the null string. We can say that σ' is the result of a canonical substitution for the null string if and only if $\sigma' = \beta'\sigma$. If $\sigma = \Lambda$, this becomes $\sigma' = \beta'$.

4c. Markov Algorithms

Any manipulation of a string can be formulated as a sequence of canonical substitutions. In this section we define a notation for specifying sequences of canonical substitutions for particular tasks, give examples of such sequences, and discuss general properties of the formalism.

DEFINITION 4.14 Let V be an alphabet $\{a, b, c, \ldots\}$ and V' an *auxiliary* alphabet $\{a', b', c', \ldots\}$ such that $V \cap V' = \varnothing$. Let $\rightarrow$ and . be symbols that are not members of $V \cup V'$. If $\beta, \beta_1 \in (V \cup V')^*$, then the string

$\beta \to \beta_1$ is a *simple* (*Markov*) *production* and the string $\beta \to . \beta_1$ is a *terminal* (*Markov*) *production*. In these productions β is the *antecedent* and β_1 the *consequent*.

DEFINITION 4.15 A production with antecedent β and consequent β_1 is *applicable* to a string σ if and only if the canonical substitution $f_c(\sigma, \beta, \beta_1)$ is defined (i.e., if and only if β is a substring of σ). The image of $\langle \sigma, \beta, \beta_1 \rangle$ is then the *result* of the production.

Example

Productions $ara \to \Lambda$ and $ara \to den$ are both applicable to the string *ararat*. The results are *rat* and *denrat*, respectively. Neither production is applicable to either of strings *rat* and *denrat*. Production $\Lambda \to s$ is applicable to the string *tun*. The result is *stun*.

DEFINITION 4.16 A *Markov algorithm* (*normal algorithm*) is a finite sequence of productions $P_1, P_2, \ldots, P_n$ applied to a string $\sigma_0 \in V^*$ or to a string $\sigma_i \in (V \cup V')^*$, $i > 0$, according to the following procedure.

1. Set $i = 0$.
2. Set $j = 1$.
3. If P_j is applicable to σ_i, go to 5.
4. Set $j = j + 1$. If $j \leqq n$, go to 3. Else algorithm is *blocked*.
5. Apply P_j to σ_i and obtain σ_{i+1}. Set $i = i + 1$. If P_j was simple, go to 2. Otherwise algorithm *terminates*.

DEFINITION 4.17 A Markov algorithm is *applicable* to a string σ_0 if and only if it stops (by termination or blocking) after a finite number of steps.

Examples

1. The algorithm $\Lambda \to . \Lambda$ is applicable to every string. It corresponds to the identity function. The algorithm $\Lambda \to \Lambda$, since it never stops, is not applicable to any string.

2. Let $V = \{a, b, c, d\}$. The following Markov algorithm removes the first d and every symbol following it from any string over V.

$$\begin{aligned} da &\to d \\ db &\to d \\ dc &\to d \\ dd &\to d \\ d &\to . \Lambda \end{aligned}$$

Applied to the string *aabdcbdda*, the algorithm produces in turn *aabdcbdd*, *aabdbdd*, *aabddd*, *aabdd*, *aabd*, *aab*. If the first four productions were written in some other order, then the strings produced at intermediate stages could differ from the ones above, but the final string would still be *aab*. We shall abbreviate the algorithm by writing a single expression for the first four productions. In this notation the algorithm becomes

$$\begin{aligned} dx &\rightarrow d \qquad (x \in V) \\ d &\rightarrow .\,\Lambda \end{aligned}$$

Although the composite production $dx \rightarrow d$ represents the four separate productions in some *definite* order, one can, for convenience, give a different interpretation: Take *da* for the antecedent if the first *d* in the string is followed by *a*, take *db* if it is followed by *b*, take *dc* if it is followed by *c*, and take *dd* if it is followed by another *d*. Under this interpretation the abbreviated algorithm changes *aabdcbdda* to *aabdbdda* (instead of to *aabdcbdd*). If the production $d \rightarrow .\Lambda$ were changed to $d \rightarrow \Lambda$ the algorithm would stop by blocking.

Some authors consider a Markov algorithm inapplicable if it stops by blocking. Whether or not one agrees with this view is largely a matter of aesthetics. On the one hand, a formalism that permits only one type of stop has greater simplicity. On the other hand, it is sometimes possible to shorten an algorithm (by one line) if blocking is a legitimate mode of stopping.

We should distinguish two types of applicability. That of D.4.17 is concerned only with stopping. The other type, external to the formalism, is concerned with the proper completion of a specified task, and in this sense a particular Markov algorithm is applicable only if it produces the string that we expect it to produce. Applicability in the sense of D.4.17 is, of course, a pre-condition for applicability in the narrower sense. The distinction is analogous to that between a computer program that is free of syntactic errors, but may still contain "logical bugs," and a completely debugged program.

We have not used an auxiliary alphabet this far. Let us therefore take as our next example an algorithm for duplicating strings. Auxiliary symbols have to be used. In the first stage of the algorithm a copy is made of every symbol in the string; an auxiliary symbol a' moves along the string pointing to the symbol that is to be copied next. In the second stage all copies are moved to one end of the string. To make sure that only the copies are moved, they must be suitably marked; auxiliary symbol b' is the marker.

$$\begin{aligned} a'x &\rightarrow xb'xa' \qquad (x \in V) \\ b'xy &\rightarrow yb'x \qquad (x, y \in V) \\ b' &\rightarrow \Lambda \\ a' &\rightarrow .\,\Lambda \\ \Lambda &\rightarrow a' \end{aligned}$$

The pointer is introduced by the last production to ensure that it will not be produced more than once. Let us apply the algorithm to a string over $V = \{a, b, c\}$, say *acbcc*. We have

$$\begin{aligned} acbcc &\to a'acbcc \\ &\to ab'aa'cbcc \\ &\vdots \\ &\to ab'acb'cbb'bcb'ccb'ca' \\ &\to acb'ab'cbb'bcb'ccb'ca' \\ &\vdots \\ &\to acbccb'ab'cb'bb'cb'ca' \\ &\to acbccab'cb'bb'cb'ca' \\ &\vdots \\ &\to acbccacbcca' \\ &\to acbccacbcc \end{aligned}$$

Our examples of Markov algorithms have corresponded to mappings from V^* into V^*, but this is not an essential feature of Markov algorithms. The string produced by an applicable algorithm may belong to $(V \cup V')^*$ without belonging to V^*. The following algorithm, which counts occurrences of a specified substring, is a case in point. The symbol $+$ $(+ \in V')$ is appended to the string for every occurrence of the substring. We shall write the specified substring α as $\beta\gamma$, where β stands for the first symbol in α. The general form of the algorithm is

$$\begin{aligned} a'\beta\gamma &\to \beta + a'\gamma \\ a'x &\to xa' \qquad (x \in V) \\ +x &\to x+ \qquad (x \in V) \\ a' &\to .\, \Lambda \\ \Lambda &\to a' \end{aligned}$$

If $\alpha = ara$, say, then the first production is

$$a'ara \to a + a'ra.$$

Apply the algorithm to the string *ararat*:

$$\begin{aligned} ararat &\to a'ararat \\ &\to a + a'rarat \\ &\to a + ra'arat \\ &\to a + ra + a'rat \\ &\vdots \\ &\to a + ra + rata' \\ &\to ar + a + rata' \\ &\vdots \\ &\to ararat + + a' \\ &\to ararat + + \end{aligned}$$

The next algorithm converts a string of tallies, $++\cdots+$, to a decimal count.

$$
\begin{aligned}
b'0+ &\to b'1 \\
b'1+ &\to b'2 \\
b'2+ &\to b'3 \\
&\vdots \\
b'8+ &\to b'9 \\
b'9+ &\to +b'0 \\
+b' &\to b'+ \\
b'+ &\to b'0+ \\
b' &\to .\,\Lambda \\
\Lambda &\to b'0
\end{aligned}
$$

Let the counting algorithm and the tally converter be denoted by M_1 and M_2, respectively. The two algorithms can be combined into an algorithm M_3, which carries out the process of M_1 until M_1 would terminate and then carries out the process of M_2 on the result. M_3 is constructed as follows:

1. Change M_1 to M_1' by substituting

$$
\begin{aligned}
+a' &\to a'+ \\
a' &\to b'0
\end{aligned}
$$

for the production

$$a' \to .\,\Lambda$$

2. Change M_2 to M_2' by deleting the last production ($\Lambda \to b'0$).
3. The combined algorithm is

$$
\begin{gathered}
M_2' \\
M_1'
\end{gathered}
$$

A general algorithm for combining two algorithms is much more complicated. The segment that carries out the process of the second algorithm has to come first in the combined algorithm, and a string has to be "protected" from its productions while the first process is being applied to the string.

ALGORITHM 4.1 Algorithm for combining two Markov algorithms M_1 and M_2 into a single algorithm M_3. Alphabet V_3 of M_3 is $(V_1 \cup V_1') \cup (V_2 \cup V_2')$, where V_1 and V_2 are the alphabets of M_1 and M_2, respectively, and V_1' and V_2' are the auxiliary alphabets. $V_3' = \{a', b', c', d', e'\}$.

1. Convert M_1 to M_1' by changing every terminal production $\alpha \to .\,\beta$ to $\alpha \to a'\beta$, and adding $\Lambda \to a'$ as the last production of M_1' (to take care of a stop by blocking).

2. Change the antecedent and the consequent of every production of M_2 as follows:

change Λ to c',
change $s_1 s_2 \cdots s_n$ to $c's_1c's_2c' \cdots c's_nc'$

(e.g., $\Lambda \rightarrow ab$ becomes $c' \rightarrow c'ac'bc'$). The modified productions are inapplicable to a string while the process of M_1 is carried out on it.

3. Convert the modified M_2 to M_2' by changing every terminal production $\alpha \rightarrow . \beta$ to $\alpha \rightarrow d'\beta$, and adding $c' \rightarrow d'$ as the last production of M_2'.

4. Introduce productions that change the string produced by M_1' to a form suitable for M_2' and do the final cleaning up. The complete M_3 is

$$
\begin{array}{rcll}
xa' & \rightarrow & a'x & (x \in V_1 \cup V_1') \\
a' & \rightarrow & b' & \\
b'x & \rightarrow & c'xb' & (x \in V_1 \cup V_1') \\
b' & \rightarrow & c' & \\
c'xd' & \rightarrow & d'c'x & (x \in V_2 \cup V_2') \\
d' & \rightarrow & e' & \\
e'c'x & \rightarrow & xe' & (x \in V_2 \cup V_2') \\
e'c' & \rightarrow . & \Lambda & \\
& M_2' & & \\
& M_1' & &
\end{array}
$$

Our final example is a Markov algorithm that appends either T or F to a string depending on whether or not the string is a K-formula. Symbol F is appended to the string at the start, and a pointer is moved along the string. The difference between the number of K-operators and the number of node symbols in the substring to the left of the pointer is given by tallies. If the number of tallies becomes zero before the pointer reaches the last symbol, or the number is nonzero when the last symbol is reached, then the string is not a K-formula. Otherwise the string is a K-formula, and the F is then changed to T. Here $V = N \cup \{*\}$, where N is the set of node symbols.

$$
\begin{array}{rcll}
Fx & \rightarrow & xF & (x \in V) \\
+x & \rightarrow & x+ & (x \in V) \\
a'* & \rightarrow & *+a' & \\
+a'x & \rightarrow & xa' & (x \in N) \\
a'xF & \rightarrow . & xT & (x \in N) \\
+ & \rightarrow & \Lambda & \\
a' & \rightarrow . & \Lambda & \\
\Lambda & \rightarrow & a'F &
\end{array}
$$

At this point we digress, and try to come to an understanding of what is meant by a *recursively solvable* (or *decidable*) *problem*. We define a problem as a function, and we say that a problem is recursively solvable if and only if the function is computable, i.e., if and only if there exists a mechanical procedure, which, applied to *any* member of the domain of the function, computes its image within a finite length of time. Take, for example, the problem of deciding whether or not a string is a K-formula. This is a function on the set of all strings into a set of two elements, say $\{T, F\}$. All K-formulas have the image T, all strings other than K-formulas map onto the element F. The Markov algorithm given above is the required mechanical procedure. The problem is therefore decidable.

A well-known abstract device for specifying mechanical procedures is the Turing machine, and it can be shown that a problem is recursively solvable if and only if there exists a Turing machine that computes the function. A mechanical device more powerful than the Turing machine, i.e., one that would compute a function that cannot be computed by a Turing machine, is not known and does not appear to exist. It can be shown also that the Turing machine for computing a function exists only if a Markov algorithm for computing the function exists. The two formalisms are therefore equivalent. Turing machines, important though they are in theoretical studies, are quite impracticable when it comes to describing procedures for solving practical problems. The usefulness of Markov algorithms, on the other hand, extends beyond the theory of decidability. They can provide reasonably efficient characterizations of operations on strings. We shall see in Section 12b that programs written in Comit, a programming language for processing strings, are very similar to Markov algorithms.

There are unsolvable problems relating to Turing machines and Markov algorithms themselves. For example, the problem of whether or not a Markov algorithm is applicable to a string is recursively unsolvable. It must be stressed that particular solutions can be found. Only the general procedure, which would work for *any* Markov algorithm and *any* string, cannot be devised.

4d. Languages and Grammars

There is very little that is interesting about a set V^* itself. We shall now look at subsets of V^* that we find interesting for some reason or other, and at ways of specifying rules for generating precisely those strings that are members of such subsets. Collections of rules that define subsets of V^* are known as grammars.

Traditionally we think of a grammar of a language as a set of rules a speaker of the language is supposed to obey. A grammar in the traditional sense defines "correct" usage. The modern approach is different. Instead of attempting to prescribe usage, the linguist accepts a language as used, and seeks a description of the language. He looks for a theory that specifies or predicts all sentences in the language, excluding strings that are not sentences. In other words, instead of defining a particular subset of the set of all strings over a vocabulary as *the* language, he takes a particular subset as given, and looks for rules that will generate precisely the members of this subset.

Natural languages, because of their complexity, have defied attempts at their complete specification. This is understandable. The use of a rich language is what makes us human beings, and a description of our linguistic competence would come close to a description of our existence. The methodology has, nevertheless, made significant contributions to important advances in other fields, particularly in computer science. One important source of our improved knowledge of programming languages has been the classification of subsets of V^* according to the form of the rules used in their generation. The classificatory approach has contributed also to a better understanding of the relation between languages and automata. Increased familiarity with linguistic techniques in the computing community has led to the investigation of a variety of structures by linguistic means. Analysis of pictures is an example; it no longer seems strange to speak of picture languages.

DEFINITION 4.18 Let V be an alphabet. Subsets of V^* are *languages* over V (generally denoted by L, possibly with subscripts). If $\alpha \in L$, then α is a *sentence* of L.

Definitions D.4.10 and D.4.11 are, of course, still applicable when sets A and B are interpreted as languages. In particular, if $L \subseteq V^*$, then $L^* = cl(L)$, the *closure* of language L.

DEFINITION 4.19 A *grammar* is a finite system of rules determining a language. The language determined by a grammar G will be denoted by $L(G)$. If $L(G_1) = L(G_2)$, then G_1 and G_2 are *equivalent*.

In D.4.19 the term *grammar* is left too vague to be of much use. We strengthen the term in the next definition, which introduces a specific class of grammars.

DEFINITION 4.20 A *constituent structure grammar* (or *phrase structure grammar*) is a quadruple $G = \langle V, V', P, S, \rangle$, where V and V' are an alphabet and an auxiliary alphabet, respectively, $S \in V'$, and $P \subseteq (V \cup V')^{*2}$ is a finite relation.

Members of P are called *production rules*. A rule $\langle \alpha, \beta \rangle$ is usually written as $\alpha \rightarrow \beta$. Members of V are called *terminal* symbols. Members of V' are called *nonterminals*, or *syntactic categories*, or *metalinguistic variables*, or simply *variables*. The term *constituent structure grammars* implies that these grammars, by means of variables, impose structure on constituents of sentences in their languages. We use *constituent structure* in preference to *phrase structure*, despite the wider acceptance of the latter term. Our reason is that *phrase*, because of its traditional place of importance in the terminology of linguistics as a study of natural languages, would evoke the false impression that natural languages will be our concern here. We will be concerned with formal languages or—to use a picturesque quote—chunks carved out of the free monoid V^*.

DEFINITION 4.21 Let $G = \langle V, V', P, S \rangle$ be a constituent structure grammar. We write $\sigma_i \Rightarrow \sigma_{i+1}$ if there exist $\sigma_i = \alpha\beta\gamma$ and $\sigma_{i+1} = \alpha\beta'\gamma$ in $(V \cup V')^*$ and $\beta \to \beta'$ is a member of P, and we write $\sigma_o \overset{*}{\Rightarrow} \sigma_t$ if either $\sigma_0 = \sigma_t$ or there exists a sequence $\sigma_0, \sigma_1, \ldots, \sigma_t$ such that $\sigma_i \Rightarrow \sigma_{i+1}$ for all $0 \leqq i < t$. The sequence is called the *σ_0-derivation* of σ_t; we denote it also by $\sigma_0 \Rightarrow \sigma_1 \Rightarrow \cdots \Rightarrow \sigma_t$.

DEFINITION 4.22 If $G = \langle V, V', P, S \rangle$ is a constituent structure grammar, then

$$L(G) = \{\alpha \mid \alpha \in V^* \quad \text{and} \quad S \overset{*}{\Rightarrow} \alpha\}$$

is the *constituent structure language generated* by G. (The distinguished metalinguistic variable S represents the class of sentences.)

Examples

1. Let N be a set of node symbols $\{a, b, c, \ldots, k\}$. Then $G_1 = \langle \{*\} \cup N, \{S\}, P_1, S \rangle$, where P_1 is the set of rules $\{S \to n \mid n \in N\} \cup \{S \to *SS\}$, is a grammar of K-formulas. We have, for example, $S \Rightarrow *SS \Rightarrow *S*SS \Rightarrow *S*bS \Rightarrow **SS*bS \Rightarrow **bS*bS \Rightarrow **ba*bS \Rightarrow **ba*bc$, and $S \overset{*}{\Rightarrow} **ba*bc$ is an S-derivation of the K-formula $**ba*bc$.

2. Let N be the set of Example 1. Then $G_2 = \langle \{*\} \cup N, \{K, \mathit{node}\}, P_2, K \rangle$, where P_2 is the set of productions $\{K \to \mathit{node}, K \to *KK\} \cup \{\mathit{node} \to n \mid n \in N\}$, is another grammar of K-formulas. G_1 and G_2 are equivalent: $L_1(G_1) = L_2(G_2)$. Obviously G_1 is the simpler grammar, but G_2 has the advantage that it differentiates explicitly between node symbols and the K-operator by assigning the node symbols to their own syntactic category.

3. Let $V = \{a, \mathit{apple}, \mathit{ate}, \mathit{bought}, \mathit{child}, \mathit{green}, \mathit{man}, \mathit{pear}, \mathit{the}\}$, $V' = \{S, A, N, NP, T, V, VP\}$, and let $G = \langle V, V', P, S \rangle$ be a grammar in which P contains the following rules:

$S \to NP.VP$	$N \to \mathit{apple}$	$T \to a$
$NP \to T.N$	$N \to \mathit{child}$	$T \to \mathit{the}$
$VP \to V$	$N \to \mathit{man}$	$V \to \mathit{ate}$
$VP \to V.NP$	$N \to \mathit{pear}$	$V \to \mathit{bought}$
$N \to A.N$	$A \to \mathit{green}$	

Let us derive a sentence of $L(G)$.

$$\begin{aligned} S &\Rightarrow NP.VP \\ &\Rightarrow T.N.VP \\ &\Rightarrow the.N.VP \\ &\Rightarrow the.child.VP \\ &\Rightarrow the.child.V.NP \\ &\Rightarrow the.child.V.T.N \\ &\Rightarrow the.child.V.a.N \\ &\Rightarrow the.child.V.a.A.N \\ &\Rightarrow the.child.V.a.green.N \\ &\Rightarrow the.child.ate.a.green.N \\ &\Rightarrow the.child.ate.a.green.pear \end{aligned}$$

It is only an accident that this sentence "makes sense." The string *a.apple.bought.the.green.green.green.pear* is also a sentence in $L(G)$. The symbol . indicates concatenation.

4. G is a grammar that generates all permutations of $\{a, b, c\}$: $G = \langle\{a, b, c\}, \{\pi, A, B, C\}, P, \pi\rangle$, where P consists of the productions $\pi \to ABC$, $AB \to BA$, $AC \to CA$, $BC \to CB$, $A \to a$, $B \to b$, $C \to c$. Then, for example, cba may be derived as follows: $\pi \Rightarrow ABC \Rightarrow BAC \Rightarrow BCA \Rightarrow CBA \Rightarrow cBA \Rightarrow cbA \Rightarrow cba$. Alternatively, we can define the simpler $G' = \langle\{a, b, c\}, \{\pi\}, P', \pi\rangle$, where now the productions are $\pi \to abc$, $ab \to ba$, $ac \to ca$, $bc \to cb$.

5. We can have a grammar $G = \langle V, V', P, S\rangle$ that contains no production of the form $S \to \alpha$, but the grammar is of little interest; $L(G)$ is vacuous. Likewise, there is no point in having a production rule whose consequent contains a nonterminal that does not appear in the antecedent of any other production of the grammar.

Production rules can be written in a compressed form, which has become known as the *Backus normal form* (*Bnf*). All productions with the same antecedent are written in a single line, the consequents being separated by vertical slashes. The sign ::= separates an antecedent from its consequents. To avoid ambiguity, members of V' are enclosed in angular brackets. In Bnf the productions of the grammar of Example 3 are written

$$\begin{aligned} \langle S\rangle &::= \langle NP\rangle\langle VP\rangle \\ \langle NP\rangle &::= \langle T\rangle\langle N\rangle \\ \langle VP\rangle &::= \langle V\rangle \mid \langle V\rangle\langle NP\rangle \\ \langle N\rangle &::= \langle A\rangle\langle N\rangle \mid apple \mid child \mid man \mid pear \\ \langle A\rangle &::= green \\ \langle T\rangle &::= a \mid the \\ \langle V\rangle &::= ate \mid bought \end{aligned}$$

The grammars defined by D.4.20 are known as *unrestricted* grammars. The definition is sufficiently general in a sense that we shall now clarify. It can be shown that any Turing machine can be represented directly as an unrestricted grammar, and conversely. Therefore, since the Turing machine is the most powerful mechanical device we know, any grammar that can be specified must be equivalent to some unrestricted grammar.

In order to facilitate the proof of equivalence of Turing machines and unrestricted grammars, definitions of the latter are usually tighter than our D.4.20. As an example of the constraints imposed, a definition might require that, for every $\langle \alpha, \beta \rangle \in P$, some substring of α be a member of V'. Under this constraint G' of Example 4 above is not a constituent structure grammar. Other constraints would exclude the grammars discussed in Example 5. The set of languages generated by the grammars of D.4.20 is, however, equal to the set generated by grammars subjected to the more usual constraints, and this is our justification for having opted for the simplest possible formulation in D.4.20.

Unrestricted grammars are too general to be of any practical use. All the more important questions about these grammars are undecidable. We shall now introduce restrictions on the production rules, which are not to be confused with the constraints of the paragraph above, that will define grammars of greater interest. These grammars are known as being of Type 1, Type 2, and Type 3; unrestricted grammars are of Type 0 under this classification. The languages generated by grammars of Type i are known as Type i languages. Every grammar of Type $i + 1$ is also of Type i, but the set of grammars of Type $i + 1$ is a proper subset of the set of grammars of Type i. Sets of the corresponding languages, too, obey the proper subset relation. It is reasonable to hold the opinion that Type 0 grammars are too amorphous to be called grammars. It is, indeed, preferable to call them unrestricted *rewriting systems*, and to reserve the term *grammar* for the restricted systems.

DEFINITION 4.23 A constituent structure grammar is of *Type 1* if the consequent of every production of the grammar contains at least as many symbols as the antecedent.

Example

G_2 of Example 2 of D.4.22 contains the production $node \to n$, but *node* is a single symbol, and G_2 is a Type 1 grammar. On the other hand, a system that contains the rule $\langle n \rangle + \langle n \rangle ::= \langle addition \rangle$, say, is not a Type 1 grammar; the antecedent contains three symbols, the consequent only one.

DEFINITION 4.24 A Type 1 grammar is of *Type 1'* only if each production of the grammar has the form $\alpha A \beta \to \alpha \gamma \beta$, where $\alpha, \beta \in (V \cup V')^*$, $A \in V'$, $\gamma \in (V \cup V')^* - \{\Lambda\}$.

Example

If we want to generate all permutations of n objects, it is, of course, possible to do so by means of a grammar that contains as many rules as there are permutations. The G of Example 4 of D.4.22 is an example of how one constructs a grammar with $\frac{1}{2}n(n+1)+1$ rules (i.e., fewer than $n!$ when $n > 3$). It is not a Type 1′ grammar, but a Type 1′ grammar can be constructed that generates all permutation and still contains fewer than $n!$ rules (except for small n). In this grammar every rule that has the form $AB \to BA$ in G is replaced by a set of four rules $\{AB \to AB', AB' \to A'B', A'B' \to BB', BB' \to BA\}$. The number of auxiliary symbols increases from n to n^2; each of the $\frac{1}{2}n(n-1)$ sets of four rules requires two additional auxiliaries.

Grammars of Type 1′ are called *context-sensitive* (or *context-dependent*) grammars. Although there exist Type 1 grammars that are not Type 1′, in one sense the two types are equivalent: If G is a Type 1 grammar, then there exists a Type 1′ grammar equivalent to G. Far too many problems in the theory of context-sensitive grammars are undecidable. Therefore these grammars are only slightly more interesting than the unrestricted rewriting systems.

DEFINITION 4.25 A grammar is of *Type 2* only if all of its rules take the form $A \to \gamma$, where $A \in V'$ and $\gamma \in (V \cup V')^* - \{\Lambda\}$.

Examples

1. The grammars of Examples 1, 2, and 3 of D.4.22 are of Type 2.
2. The major part of a definition of Algol or Fortran can be expressed in the form of Type 2 production rules. The following Type 2 grammar defines Fortran integers and real numbers. The definition is rather abstract in that there are no limits on the sizes of the numbers, and rightly so. Limits should only be imposed when the language is related to a particular computer; they would be determined by the size of the computer word.

⟨digit⟩ ::= 0 | 1 | 2 | 3 | 4 | 5 | 6 | 7 | 8 | 9
⟨sign⟩ ::= + | −
⟨unsigned integer⟩ ::= ⟨digit⟩ | ⟨unsigned integer⟩⟨digit⟩
⟨integer⟩ ::= ⟨unsigned integer⟩ | ⟨sign⟩⟨unsigned integer⟩
⟨unsigned F-number⟩ ::= . ⟨unsigned integer⟩ | ⟨unsigned integer⟩. |
  ⟨unsigned integer⟩. ⟨unsigned integer⟩
⟨F-number⟩ ::= ⟨unsigned F-number⟩ |
  ⟨sign⟩⟨unsigned F-number⟩
⟨exponent⟩ ::= E⟨integer⟩
⟨real⟩ ::= ⟨F-number⟩ | ⟨F-number⟩⟨exponent⟩ |
  ⟨integer⟩⟨exponent⟩

Type 2 grammars are known as *context-free* grammars. They have been widely studied, and a lot is known about them. For example, the problem whether the language generated by a grammar is empty, finite, or infinite is decidable for Type 2 grammars, but not for Type 1 grammars. However, the important problem of whether two Type 2 grammars are equivalent is undecidable.

DEFINITION 4.26 A grammar is of *Type 3* only if all of its productions are of the forms $A \to \alpha$ or $A \to \alpha B$, where $A, B \in V'$ and $\alpha \in V^* - \{\Lambda\}$, or, alternatively, they are all of the forms $A \to \alpha$ or $A \to B\alpha$.

Example

$G = \langle\{a, b, c\}, \{S, A\}, P, S\rangle$, where P consists of the productions

$$\begin{aligned}\langle S\rangle &::= ab\langle S\rangle \mid c\langle A\rangle \\ \langle A\rangle &::= ba\langle A\rangle \mid a\end{aligned}$$

is a Type 3 grammar. $L(G)$ is the complex product $(ab)^*c(ba)^*a$. Type 3 grammars are called also *one-sided linear*, or *finite state*, or *regular* grammars. G is a *right-linear* grammar because the nonterminal is written to the right of terminals in every consequent containing a nonterminal. Every right-linear grammar has an equivalent *left-linear* grammar. In the case of G, an equivalent left-linear grammar has the following productions:

$$\begin{aligned}\langle S\rangle &::= \langle A\rangle a \\ \langle A\rangle &::= \langle A\rangle ba \mid \langle B\rangle c \mid c \\ \langle B\rangle &::= \langle B\rangle ab \mid ab\end{aligned}$$

(B is an additional nonterminal symbol). The language $\{(ab)^k c(ba)^k a \mid k \geqq 0\}$, which is a proper subset of $L(G)$, cannot be produced by a Type 3 grammar.

4e. Languages and Automata

Type 3 grammars are called finite state grammars because they are related to *finite state automata*. Let us consider a finite state *acceptor*. This is a device that can be in one of a finite number of distinct *internal states* at any one time. It is equipped with a reading head, and under the reading head we pass a tape on which a string has been written. Depending on the state the automaton is in, and the symbol that it reads, it can remain in the same state or switch to a different state. In more formal terms, we have a set of states Q and a set of symbols S. The machine is described by a function f from $Q \times S$ into Q. One of the states is distinguished: $q_1 \in Q$ is the state in which the machine starts the scan of the tape and to which it should return after it has read the entire string presented to it. We then say that the automaton has *accepted* or *recognized* the string. Assume that the string contains a substring $s_{j-1}s_j$. If, on reading s_{j-1}, the machine goes into state q_i and $\langle q_i, s_j \rangle$ is not in the domain of f, the machine has to stop for lack of further instructions. We interpret this stop as nonacceptance of the string. The significant link between Type 3 languages and finite state acceptors is that for every Type 3 language there exists a finite state acceptor that will accept all sentences of the language and fail to accept strings that are not sentences. Moreover, if a language cannot be described by a Type 3 grammar, there exists no finite state acceptor for the language.

For every unrestricted rewriting system we can produce a Turing machine that will accept strings generated by this system, and no others. A Turing acceptor also has a finite number of internal states. It also reads from a tape, and the next action of the machine is a function from the Cartesian product of states and symbols. How then does it differ from a finite state acceptor? The difference is that a Turing machine has a tape of unbounded length on which it can write as much as it wants to, and that it can read what *it* has previously written. Therefore, in contrast to a finite state acceptor, a Turing acceptor has an infinite memory, i.e., the *finite* refers to memory capacity rather than to the number of internal states.

An automaton is very conveniently represented by a weighted digraph: Nodes represent internal states; arcs represent transitions from state to state. The digraph has an arc $\langle q_i, q_j \rangle$, with weight s_k, if and only if $\langle q_i, s_k, q_j \rangle$ is a member of the function describing the automaton. These digraphs are called *state graphs*. Figure 4.3 shows the digraph of a finite state automaton that

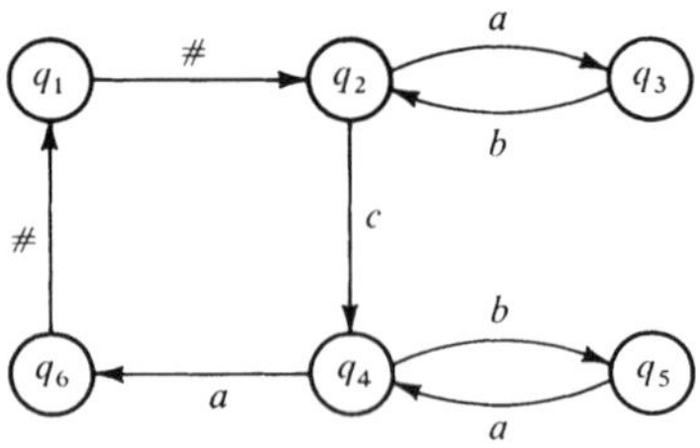

Figure 4.3

accepts the language $(ab)^*c(ba)^*a$. The symbol # is a delimiter. When a string is presented to an automaton, we use the # as special end markers. Thus, the string *ababcbaa* is written #*ababcbaa*# on the tape.

Instantaneous descriptions of an automaton help one understand the process by which an automaton accepts or rejects a string. An instantaneous description consists of the string and, in addition, a pointer to the symbol that is currently under the reading head. The description is not complete unless the current state of the automaton is indicated as well, and this is done by making the state symbol the pointer; it is placed on the left of the symbol being scanned. An initial instantaneous description is, for example, $q_1 \# ababcbaa \#$. For the automaton of Figure 4.3 $f(q_1, \#) = q_2$, and the transition to state q_2 is accompanied by a shift of the tape to the left. The next instantaneous description is therefore $\# q_2 ababcbaa \#$. The complete sequence of instantaneous descriptions depicting acceptance of *ababcbaa* by the automaton of Figure 4.3 is

$$
\begin{aligned}
q_1 \# ababcbaa \# &\rightarrow \# q_2 ababcbaa \# \\
&\rightarrow \# a q_3 babcbaa \# \\
&\rightarrow \# ab q_2 abcbaa \# \\
&\rightarrow \# aba q_3 bcbaa \# \\
&\rightarrow \# abab q_2 cbaa \# \\
&\rightarrow \# ababc q_4 baa \# \\
&\rightarrow \# ababcb q_5 aa \# \\
&\rightarrow \# ababcba q_4 a \# \\
&\rightarrow \# ababcbaa q_6 \# \\
&\rightarrow \# ababcbaa \# q_1 .
\end{aligned}
$$

The automaton halts in state q_1; i.e., the string is accepted. The next sequence of instantaneous descriptions illustrates nonacceptance of a string:

$$
\begin{aligned}
q_1 \# acabab \# &\rightarrow \# q_2 acabab \# \\
&\rightarrow \# a q_3 cabab \# .
\end{aligned}
$$

Since there is no $\langle q_3, c \rangle$ in the domain of f, the automaton stops.

Figure 4.4, our second example, shows an automaton that accepts the language a^*bc^*.

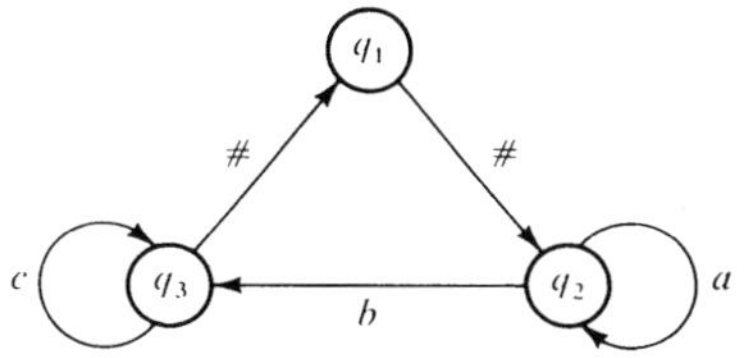

Figure 4.4

An alternative description of an automaton is in terms of a variable adjacency matrix. However, a more convenient way of representing an acceptor is by a *state-symbol matrix*. Let $Q = \{q_1, q_2, \ldots, q_n\}$ be the set of states of an automaton, and let $S = \{s_1, s_2, \ldots, s_m\}$ be its set of symbols (S includes the string end marker). The state-symbol matrix M is an $n \times m$

matrix in which the (i, j)th element is q_k if $f(q_i, s_j) = q_k$, or zero if $f(q_i, s_j)$ is undefined. A.4.2, a program that simulates finite state acceptors, is based on the use of the state-symbol matrix.

ALGORITHM 4.2 Finite state acceptor simulation program.

```
      INTEGER M(..,..),TAPE(..),STATE,SYMBOL,S,Q1,Q2
C1    SET UP MATRIX M - IN TRIPLES (Q1,S,Q2)INTEGERS
C1    1,2,3,... REPRESENT SYMBOLS
      DO 10  I = 1,...
      DO 10  J = 1,...
   10 M(I,J) = 0
      READ (5,100) N
      DO 15  J = 1,N
      READ (5,100) Q1, S, Q2
   15 M(Q1,S) = Q2
C2    READ TAPE AND PRINT IT FOR REFERENCE - PROGRAM
C2    STOPS WHEN THERE ARE NO MORE TAPES TO BE READ (A
C2    FEATURE PROVIDED BY MOST OPERATING SYSTEMS)
   20 READ (5,101) LENGTH, (TAPE(K), K = 1,LENGTH)
      WRITE (6,102)  (TAPE(K), K = 1,LENGTH)
C3    ACCEPTANCE TEST
      STATE = 1
      DO 25  NOW = 1,LENGTH
      SYMBOL = TAPE(NOW)
      STATE = M(STATE,SYMBOL)
   25 IF (STATE.EQ.0) GO TO 30
C4    BY CONVENTION AUTOMATON MUST STOP IN STATE 1
      IF (STATE.NE.1) GO TO 30
      WRITE (6,103)
      GO TO 20
   30 WRITE (6,104)
      GO TO 20
  100 FORMAT (3I4)
  101 FORMAT (I4/(40I2))
  102 FORMAT (1H0, 50I2/(1H ,50I2))
  103 FORMAT (1H0, 15HSTRING ACCEPTED)
  104 FORMAT (1H0, 15HSTRING REJECTED)
      END
```

Accepting power intermediate to those of Turing machines and finite state automata is provided by a pushdown store automaton. There is a very close relation between the abstract pushdown store automaton and a pushdown store. The latter has great practical importance in programming, and we shall discuss it in Section 10a. Discussion of the abstract pushdown store is outside the scope of this book.

Notes

The many worked examples and exercises of [Fa63] make it an excellent supplementary text for pursuing the subject of algebraic structures further. In our context the classic [Bi41] may be too thorough a text, but should be consulted for reference. For a basic introduction to the theory of groups see [Le49]. [Ca69], a very valuable survey of the use of computers in group theory, will mean little without extensive preliminary study of the theory. [Gr64], a very elementary work, connects group theory and the theory of graphs in a spirited fashion.

Introductions to the theory of Markov algorithms can be found in [Cu63, Gl64, Ko66]—[Cu63] is the more advanced. Our A.4.1 is adapted from [Cu63]; another composition algorithm can be found in [Ca63a]. [Ma51] and [Ma54] are primary sources. [Gu62] is a short, but highly informative, introduction to decidability. The solid [Da58] is a classic work on the theory of Turing machines and decidability problems, but [Mi67a] would serve better for an introductory text.

[Ba64a] is a good basic introduction to grammars in general; it is addressed to linguists and deals mainly with natural languages. [Gi66] is an advanced text on context-free languages. Our classification of grammars follows that of [Ch59], a slight modification having been suggested by the finer classification of [Ch63]. Decision problems associated with grammars are surveyed in [La64]. The Bnf notation is introduced in [Ba59a]; it has become widely known because of its use in the specification of Algol, see, e.g., [Na63]. The study of finite automata is best started by reading [Mc61], an extremely well written introductory survey, or the relevant sections of [Mi67a]. Then one can turn to the advanced texts, such as [Ne68a, Bo67a] ([Bo67a] stresses the state graph representation of automata).

Exercises

4.1 (a) Let R be the set of real numbers, and consider M_2: the set of all 2×2 matrices whose elements belong to R. Show that $\langle M_2, \cdot \rangle$, where $\cdot$ is matrix multiplication, is a semigroup with identity.

4.2 (a) Show that $\langle \mathscr{P}(A), +, \varnothing \rangle$, where A is a nonempty set and $+$ is the symmetric difference operation, is a semigroup. Is this algebra a group? Is it Abelian?

4.3 (a) Perform an analysis analogous to that of Example 3 of D.4.3 for an equilateral triangle and a regular pentagon. Find the "multiplication" tables of the two groups (they are the dihedral groups D_3 and D_5).

4.4 (a) The rigid motions of a regular polygon of n sides form the dihedral group D_n. Investigate the feasibility of writing a computer program that generates the "multiplication" table of D_n for any given n.

4.5 (a) Prove that every permutation can be expressed as a product of disjoint cycles.

4.6 (a) What are the 8-tuples that result when the permutation $p = (2\ 3\ 7\ 1)(4\ 5\ 8)$ is applied to 8-tuples $\langle 1, 2, 3, 4, 5, 6, 7, 8 \rangle$ and $\langle 8, 7, 6, 5, 4, 3, 2, 1 \rangle$? Express p as a product of transpositions. Is p even or odd?

4.7 (a) What pattern do the even and odd permutations follow in A.1.6?

4.8 (a) Prove that the algebras $\langle I, +, \cdot, 0, 1 \rangle$ and $\langle B, +, *, 0, 1 \rangle$ in Examples 1 and 2 of D.4.5 are rings.

4.9 (a) Show that $\langle [0, 1], max, \cdot, 0, 1 \rangle$, where $[0, 1]$ is the closed unit interval on the set of real numbers and $\cdot$ is arithmetic multiplication, is a Q-semiring.

4.10 (b) Let f and f_c be the functions of D.4.13. Find

(i) $f(America, A, me, f)$, (ii) $f(average, \Lambda, a, o)$,
(iii) $f_c(average, vera, \Lambda)$, (iv) $f_c(average, Vera, \Lambda)$,
(v) $f_c(assassin, n, nation)$, (vi) $f_c(f_c(assassin, sass, \Lambda), as, moccas)$.

4.11 (b) Let V be an alphabet, $A = (V^*)^3$, and $B = V^* \times \{\Lambda\} \times V^*$. Let $f_c: A \to V^*$ be a canonical substitution function (see D.4.13). Why is $\langle V^*, f_c \mid B \rangle$ not a semigroup?

4.12 (c) Given $V = \{u, v\}$ and the two composite Markov productions:

$$a'x \to xb'xa' \qquad (x \in V),$$
$$b'xy \to yb'x \qquad (x, y \in V).$$

Write the composite productions out in full.

4.13 (c) Give an example of a Markov algorithm that can never be blocked.

4.14 (c) Let $V = \{a, b\}$ and let S be any string on V. Let α, β, γ be *given* strings on V. Depending on whether $S = \alpha$ or $S \neq \alpha$, the following Markov algorithm gives a different result. What are the two results?

$$am \to ma$$
$$bm \to mb$$
$$ma \to m$$
$$mb \to m$$
$$m \to .\gamma$$
$$\alpha a \to m$$
$$\alpha b \to m$$
$$a\alpha \to m$$
$$b\alpha \to m$$
$$\alpha \to .\beta$$
$$\Lambda \to m$$

(Here, m is a marker.)

4.15 (c) Design a Markov algorithm that removes *all* occurrences of a specified substring from a given string. Test it on the string *ararat* for substring *ara*.

4.16 (c) Design Markov algorithms that change a string $a_1a_2 \cdots a_n$ on $V = \{u, v\}$ to

(i) $a_na_{n-1} \cdots a_1$,
(ii) $a_1a_2 \cdots a_na_na_{n-1} \cdots a_1$.

4.17 (c) Combine the two algorithms of Exercise 4.16 by means of the technique of A.4.1.

4.18 (c) Given two *binary* integers $a_1a_2 \cdots a_n$ and $b_1b_2 \cdots b_m$ as the string $a_1a_2 \cdots a_n + b_1b_2 \cdots b_m$. Design a Markov algorithm that finds the sum of the two integers as a binary integer.

4.19 (c) Design a Markov algorithm that converts a Boolean form to its arithmetic representation (see Exercise 2.26).

4.20 (c) What facilities should be added to the formalism of Markov algorithms to convert the formalism into a programming language?

4.21 (d) Which of the following strings are sentences in $L(G)$ of Example 3 of D.4.22?

(i) *the.child.ate*
(ii) *The.child.ate.the.apple*
(iii) *the.child.bought.an.apple*
(iv) *the.green.pear.ate.a.apple*

4.22 (d) Rewrite the production rules of G_1 of Example 1 of D.4.22 in Bnf.

4.23 (d) Write grammars for the following languages:

(i) a^*bcc^*,
(ii) $\{a^kba^k \mid k \geqq 0\}$,

(iii) $\{n \mid n \text{ is a Fortran variable name}\}$,
(iv) $\{a^k b^k a^k \mid k \geqq 1\}$,
(v) $\{n \mid n \text{ is a Boolean form}\}$.

What are the types of these grammars?

4.24 (d) Describe the languages generated by grammars that have the following production rules:

(i)
$\langle S\rangle ::= \langle B\rangle\langle A\rangle \mid b$
$\langle A\rangle ::= a$
$\langle B\rangle ::= \langle C\rangle\langle S\rangle$
$\langle C\rangle ::= c$

(ii)
$\langle S\rangle ::= a\langle A\rangle c \mid b$
$\langle A\rangle ::= a\langle S\rangle c \mid b$

(iii)
$\langle S\rangle ::= \langle A\rangle\langle A\rangle$
$\langle A\rangle ::= \langle A\rangle\langle B\rangle \mid a$
$\langle B\rangle ::= b$

(iv)
$\langle S\rangle ::= ad\langle A\rangle da \mid a\langle S\rangle a \mid aca$
$\langle A\rangle ::= b\langle A\rangle b \mid bd\langle S\rangle db$

4.25 (d) Find right- and left-linear grammars equivalent to the Type 2 grammar $\langle\{a, b, c\}, \{A, S\}, P, S\rangle$, where P consists of the productions

$$\langle S\rangle ::= a\langle A\rangle c$$
$$\langle A\rangle ::= \langle A\rangle bb \mid b$$

4.26 (d) Consider a Type 1 grammar $\langle V_i, V_i', P_i, S\rangle$ that generates all permutations of a set of i symbols ($i \geqq 2$), and in which the antecedent of every production belongs to the star of V_i'. Show that P_i need contain at most $\frac{1}{2}i(i+1)+1$ production rules. (See Example 4 of D.4.22.)

4.27 (e) Draw state graphs of acceptors for the following languages:

(i) a^*bcc^*;
(ii) $\{ab^{2n+1}c \mid n \geqq 0\}$;
(iii) $(ab)^* \cup \{bc, a\}^*b$.

4.28 (e) Implement A.4.2 and test the acceptor for the language of Part (iii) of Exercise 4.27 on the following strings:

(i) #ababab#
(ii) ##
(iii) #b#
(iv) #abcb#
(v) #bcbcaaabcb#
(vi) #abb#
(vii) #bcababb#
(viii) #bcabab#

Strings (i)–(v) alone belong to the language.

PART II

APPLICATIONS OF STRUCTURES

CHAPTER 5

Trees

5a. Trees as Grammatic Markers

Let us turn back to Example 3 of D.4.22 and look at the derivation of *the.child.ate.a.green.pear*. If, instead of the derivation

$$S \Rightarrow NP.VP \Rightarrow T.N.VP \Rightarrow the.N.VP \Rightarrow \cdots,$$

we followed the sequence

$$S \Rightarrow NP.VP \Rightarrow NP.V.N \Rightarrow T.N.V.N \Rightarrow \cdots,$$

we could still arrive at the same sentence. There are in fact 32 different S-derivations of this sentence in the given grammar G, but the derivations are not *essentially* different. Equivalence of derivations is brought out very clearly when derivations are represented by directed trees. Figure 5.1 shows the derivation tree for our sentence. Each node carries a label—the label is a terminal symbol if the node is terminal; otherwise it is an auxiliary symbol. Here order is important. The labels on the terminal nodes, taken in a left-to-right sequence, must give *the.child.ate.a.green.pear* and not, for example, *ate.green.pear.a.the.child.*

In constructing a derivation tree we start with an isolated node labeled S. Application of $S \rightarrow NP.VP$ corresponds in the construction process to two arcs being dropped from the isolated node, the terminal nodes of these arcs

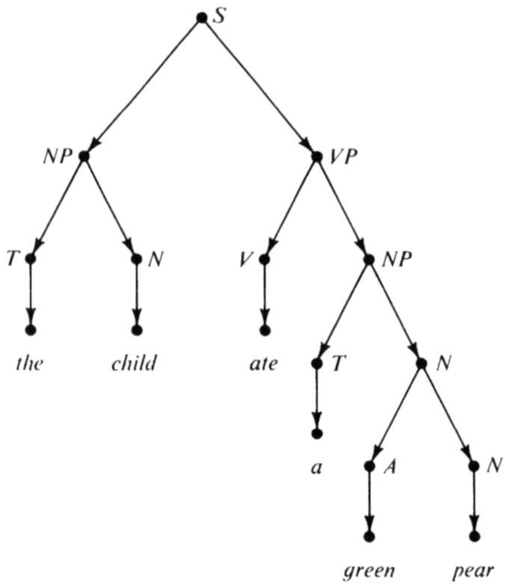

Figure 5.1

receiving labels *NP* and *VP*. If next we apply $NP \rightarrow T.N$, then arcs are suspended from the node labeled *NP*. If, on the other hand, rule $VP \rightarrow V.NP$ is used next, arcs are suspended from the node labeled *VP* and nothing happens at the node labeled *NP*. In general, if a derivation has produced at some point a string $c_1c_2 \cdots a \cdots c_n$, then in the construction of the tree we have at this stage a structure in which the terminal nodes carry labels $c_1, c_2, \ldots, a, \ldots, c_n$, reading from left to right. If the derivation now proceeds with application of a production $a \rightarrow b_1b_2 \cdots b_m$, then the corresponding action in the construction of the tree consists of suspending m arcs from the terminal node labeled a, and labeling the terminal nodes of these arcs (from left to right) $b_1, b_2, \ldots, b_m$. The important thing is that all 32 derivations of our example ultimately result in the same tree; i.e., a derivation tree displays the structure of a sentence without reflecting minor details of the derivation process. Note, however, that the representation of structure by derivation trees is effective only with context-free grammars (Types 2 and 3).

It is possible that a sentence has two or more essentially different derivations. We say then that the grammar (and language) is *ambiguous*. A context-free grammar is ambiguous if and only if at least one sentence generated by the grammar has more than one derivation tree. Consider grammar $G = \langle\{a, b\}, \{S, A, B\}\ P, S,\rangle$, where P is the set of rules,

$$\begin{aligned} \langle S\rangle &::= \langle A\rangle\langle B\rangle \\ \langle A\rangle &::= a\langle A\rangle \mid \langle A\rangle b \mid a \\ \langle B\rangle &::= \langle B\rangle b \mid b. \end{aligned}$$

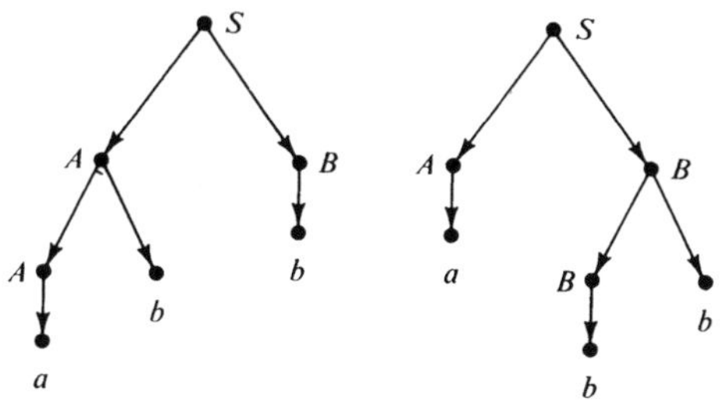

Figure 5.2

Figure 5.2 shows that the sentence *abb* has two derivation trees. Grammar *G* is therefore ambiguous.

Can one tell whether a grammar is ambiguous? Unfortunately, as regards the grammars we defined in Section 4d, the ambiguity problem is recursively solvable only for Type 3 grammars. This means that even in the case of Type 2 grammars there exists no algorithm that will decide for every grammar of this type whether it is ambiguous.

Undecidability of the ambiguity problem has very important effects on the design of programming languages. Type 3 grammars are too primitive for programming languages, and grammars of most programming languages approximate to Type 2 grammars. The best known example is Algol. The major part of the document defining Algol consists of production rules; they are productions of a Type 2 grammar. Examples of essentially different derivations have been found in Algol, and the set of derivations has been changed to remove essential ambiguities (with no absolute guarantee, however, that the grammar is unambiguous as it now stands). We have a very good reason for a preoccupation with ambiguity. In some contexts it is sufficient merely to recognize a string as a sentence in the language defined by the grammar. Not so with programming languages. A compiler is not only a recognizer; it also generates code. The code generated depends on the derivation tree, and two different structural interpretations of a sentence could give rise to two different sequences of instructions. A programmer who wants one of the sequences is not happy when the compiler inserts the other in his object program.

Undecidability of the ambiguity problem in general does not exclude the possibility of being able to find procedures that could test *particular* grammars for ambiguity, but search for the particular is bad scientific practice. It is easy to prove that a given language has an infinite number of grammars generating it. Frequently only some of the grammars are ambiguous, and then the language is said to be essentially unambiguous. A language is *essentially ambiguous* if and only if every grammar that generates it is ambiguous. The designer of a programming language should design an essentially unambiguous language and an unambiguous grammar to go with it. In the past languages have been specified first, and grammars fitted to the languages afterward in a manner that has been *ad hoc* to a greater or lesser extent. Instead, we should look for classes of grammars of sufficient power (subclasses of Type 2 grammars, but not as restrictive as Type 3) for which algorithmic tests for ambiguity exist. If we have such tests, then the design of a language can proceed side by side with the specification of its grammar in an iterative manner, until the language satisfies the requirements put on it, and its grammar is unambiguous.

Unfortunately the difficulty of finding algorithms for anything that has to do with grammars increases very rapidly as the power of the grammar increases. Consider a very simple class of grammars known as *parenthesis grammars*. A parenthesis grammar is a context-free grammar in which all productions have the form $A \to (\alpha)$, where α contains no occurrences of (or of). If no two productions in a parenthesis grammar have the same consequent, the grammar is said to be *backward-deterministic*. A backward-deterministic parenthesis grammar is unambiguous. Let $G' = \langle\{a, b, (\ ,\ )\},$ $\{S, A, B\}, P, S\rangle$ be a parenthesis grammar with production rules,

$$\begin{aligned}\langle S\rangle &::= (\langle A\rangle\langle B\rangle),\\ \langle A\rangle &::= (a\langle A\rangle) \mid (\langle A\rangle b) \mid (a),\\ \langle B\rangle &::= (\langle B\rangle b) \mid (b).\end{aligned}$$

Under G' derivations analogous to those of Figure 5.2 produce two distinct sentences: $(((a)b)(b))$ and $((a)((b)b))$. This is so because the sentences carry the structure of their derivation trees with them as patterns of parentheses. A decision procedure for telling whether two parenthesis grammars are equivalent exists, but, although parenthesis grammars are rather simple, derivation of this procedure is very involved.

The strings produced by parenthesis grammars suggest a linear representation for derivation trees. Let us write $\langle A, a_1\rangle, \langle A, a_2\rangle, \ldots, \langle A, a_n\rangle$ for the arcs originating at a node labeled A, and terminating at nodes labeled a_1, $a_2, \ldots, a_n$. This sequence of arcs is represented by $A(a_1a_2 \cdots a_n)$. The linear string that represents the derivation tree of Figure 5.1 is

$$S(NP(T(the)N(child))VP(V(ate)NP(T(a)N(A(green)N(pear))))).$$

A *syntactic chart* displays the entire set of productions of a grammar. This is a structure resembling a directed tree in that the order of arcs is sometimes important. Nonterminal nodes are labeled either & or *OR*; terminal nodes have terminal symbols for labels. Some arcs are labeled with auxiliary symbols; others carry no labels. To see how the chart is constructed, consider the productions

$$\langle VP\rangle ::= \langle V\rangle \mid \langle V\rangle\langle NP\rangle$$

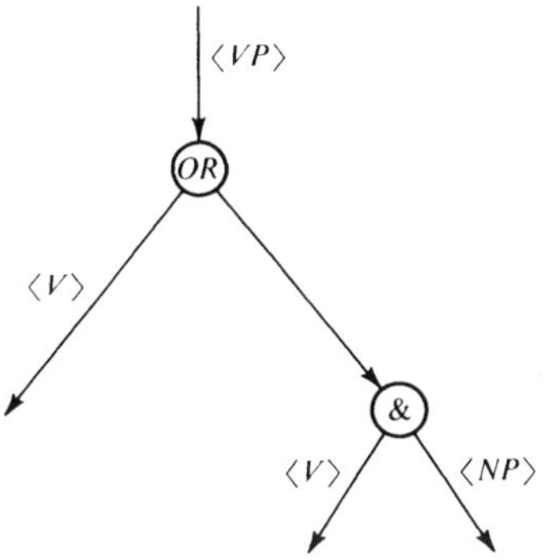

Figure 5.3

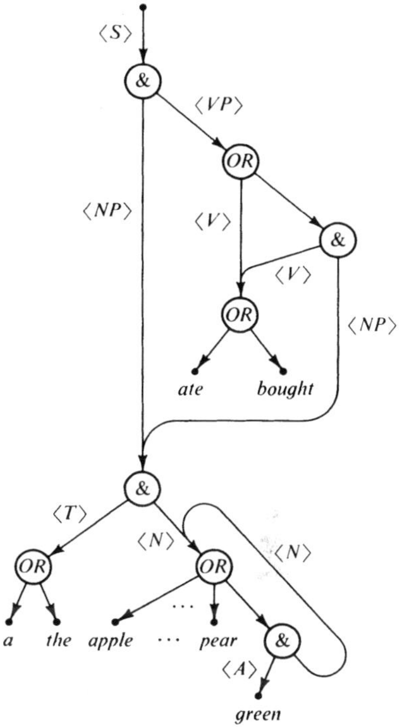

Figure 5.4

of Example 3 of D.4.22. They are represented by the structure shown in Figure 5.3. The arc labeled $\langle VP \rangle$ terminates at a node that carries the label *OR*. This label indicates choice: $\langle VP \rangle$ can be rewritten in as many ways as there are arcs originating from this node. Here their number is 2. The first of the arcs carries the label $\langle V \rangle$, and the path through the node and down this arc represents the production $VP \rightarrow V$. The other arc carries no label. Its sole purpose is to lead to the node labeled &. The label & symbolizes necessity: If the second production is chosen, then all symbols carried by the arcs originating from the & node must be used in the rewriting of *VP*. This is where the need for order comes in. The left-to-right order of the arcs specifies the order that the labels carried by the arcs have in the production rule. Here the left-to-right order is $\langle V \rangle \langle NP \rangle$. This branch therefore represents the production $VP \rightarrow V.NP$.

Figure 5.4 shows the complete syntactic chart of the grammar of Example 3 of D.4.22. Let us see how the chart is used. On entering the chart, we come at once to an & node, and we are told that $\langle S \rangle$ must be rewritten as $\langle NP \rangle \langle VP \rangle$. No choice is given in the rewriting of $\langle NP \rangle$ either, and the string becomes ⟨*T*⟩⟨*N*⟩⟨*VP*⟩. With the rewriting of ⟨*VP*⟩ we are given a choice: It can be rewritten as ⟨*V*⟩ or as ⟨*V*⟩⟨*NP*⟩ but, if the second alternative is taken, there is no choice in the rewriting of ⟨*NP*⟩. We take the second alternative, and the complete string to this stage becomes

$$\langle T\rangle\langle N\rangle\langle V\rangle\langle T\rangle\langle N\rangle.$$

Every auxiliary in this string can be rewritten in more than one way. For the first ⟨*N*⟩ we choose ⟨*A*⟩⟨*N*⟩, and rewrite the ⟨*N*⟩ in this string as a further ⟨*A*⟩⟨*N*⟩, giving ⟨*A*⟩⟨*A*⟩⟨*N*⟩. The complete string is now

$$\langle T\rangle\langle A\rangle\langle A\rangle\langle N\rangle\langle V\rangle\langle T\rangle\langle N\rangle.$$

We can replace every auxiliary in this string by a terminal symbol: For the ⟨*T*⟩ we can choose *a* or *the*, the ⟨*A*⟩ must be rewritten as *green*, as must the second ⟨*A*⟩; the ⟨*N*⟩ can be rewritten as *apple*, or *child*, or *man*, or *pear*; and so on. One sentence in the language specified by the grammar is *a.green.green.apple.bought.the.child.* If a chart contains a cycle, as it does in our example ($N \to A.N$), no bound can be put on the longest sentence. Here we can make the substring *green.green.green.* ··· as long as we wish.

For a second example we take the right-linear grammar of the example of D.4.26. Its syntactic chart is shown in Figure 5.5.

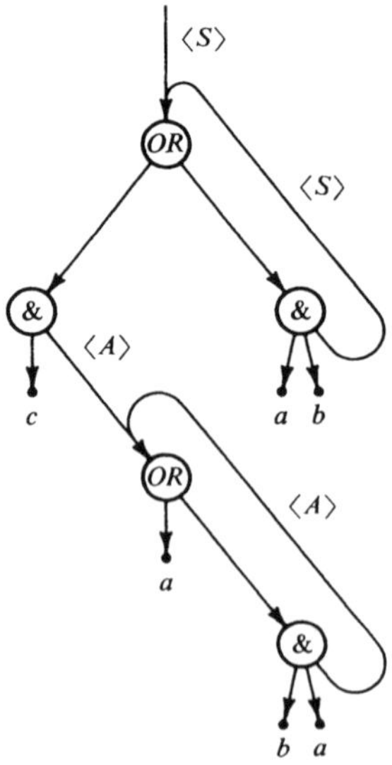

Figure 5.5

5b. Representation of Prefix Formulas

The order of evaluation of an arithmetic expression such as

$$a + bc + d/e - f \tag{5.1}$$

is determined by convention. Evaluation proceeds from left to right, but multiplication or division is performed before addition or subtraction. The precedence imposed by the hierarchy of operations can be overruled by means of parentheses: The whole expression enclosed in a pair of parentheses is considered a single operand. For example, we can use parentheses to enforce strict left-to-right evaluation:

$$((a + b)c + d)/e - f. \tag{5.2}$$

If every expression of the form operand–operator–operand is enclosed in parentheses, and we stipulate that evaluation proceeds outward from the innermost parentheses, then the hierarchy of operations does not have to be specified—it is implicit in the structure. An example of a completely parenthesized expression:

$$(((a + b)(c + d))/(e - f)). \tag{5.3}$$

In completely parenthesized form expressions (5.1) and (5.2) become $(((a + (bc)) + (d/e)) - f)$ and $(((((a + b)c) + d)/e) - f)$, respectively.

Arithmetic expressions can be represented by trees. Every operation corresponds to a labeled nonterminal node. The label specifies the operation, and arcs originating from the node lead to the operands. Figure 5.6 shows the trees corresponding to the three expressions (5.1), (5.2), and (5.3).

Now, starting from the top and keeping to the left, we trace round a tree, stringing together the symbols as they are encountered, see Figure 5.7. This procedure generates the expression,

$$- + + a \times bc/def. \tag{5.1a}$$

From the other trees of Figure 5.6 we obtain

$$-/+ \times +abcdef, \tag{5.2a}$$

$$/\times +ab+cd-ef. \tag{5.3a}$$

These expressions are *prefix* forms of (5.1), (5.2), and (5.3). Each operator precedes its two operands. Take the multiplication operator in (5.2a). It needs two operands. The + cannot be an operand; it is itself an operator. Therefore, in order to obtain the first operand for the multiplication, we have to evaluate $+ab$ before we can perform the multiplication. The second

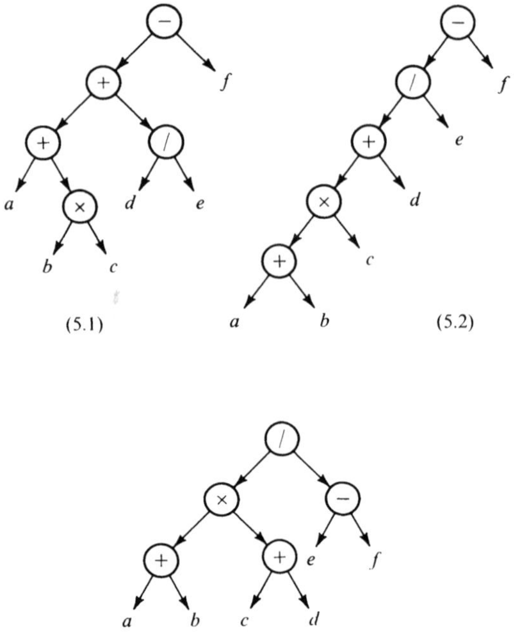

(5.1) (5.2) (5.3)

Figure 5.6

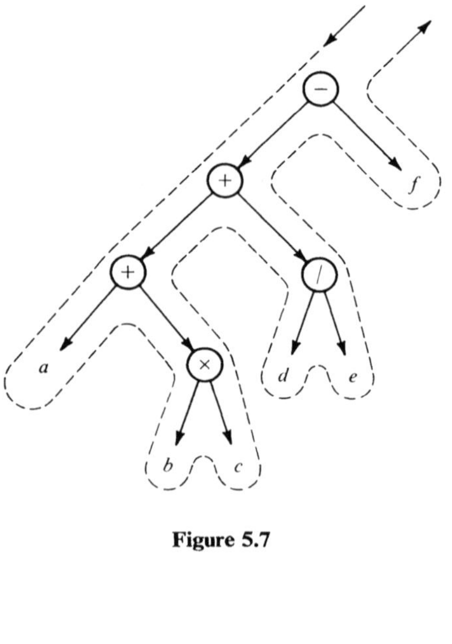

Figure 5.7

operand of × is *c*. Prefix expressions have the same advantage that completely parenthesized expressions have; one does not have to specify a hierarchy of operations. Their further advantage over the completely parenthesized expressions is that they are shorter; their disadvantage is that we are not as used to the prefix form operator–operand–operand as to the *infix* sequence operand–operator–operand.

The hierarchy of operations in arithmetic is so well established that there is no real danger that the intentions of the writer of an arithmetic expression will be misinterpreted by the reader. When we turn to expressions in the statement calculus the situation is not as satisfactory. The functionally complete set of operations $\{\vee, \wedge, \neg\}$ comprises all the logical operations available in Fortran (as `.OR.`, `.AND.`, and `.NOT.`). In logic the usual convention requires evaluation of $\wedge$ before $\vee$, with parentheses again being used to override the precedence imposed by this convention. The unary operator $\neg$ takes precedence over both $\wedge$ and $\vee$, but parentheses are important: The forms $\neg p \wedge q$ and $\neg(p \wedge q)$ require different evaluation sequences. In the first case $\neg p$ is evaluated first, and then $\neg p \wedge q$. In the second case $p \wedge q$ is evaluated first; the truth value of the whole expression is then the complement of the truth value of $p \wedge q$. In Standard Fortran the expression

```
.NOT. A1 .AND. A2.OR. A3 .AND. A4
```

is interpreted to correspond to the tree of Figure 5.8, but the Fortran compiler supplied with the CDC 3200 computer interprets the formula as having the structure in Figure 5.9; i.e., the conventions of logic are not universally

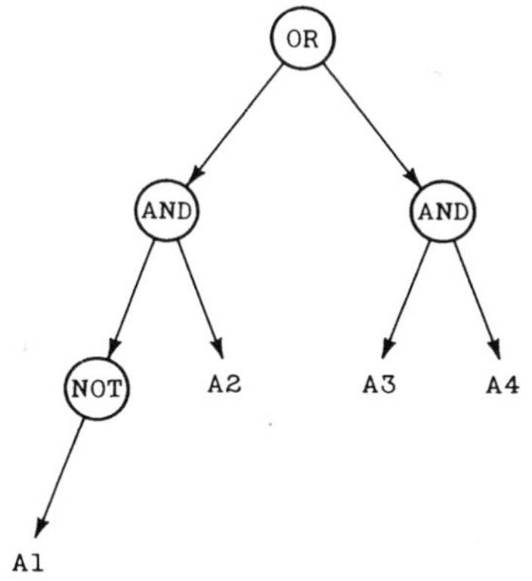

Figure 5.8

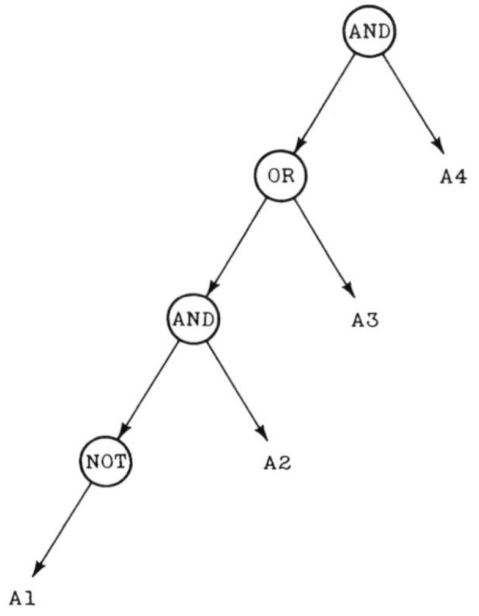

Figure 5.9

accepted. This emphasizes the need for an unambiguous representation, such as the one provided by prefix notation. The prefix formulas corresponding to the two trees are distinct. Using conventional symbols for the logical operations, they are

$$\vee \ \wedge \ \neg \ \text{A1 A2} \ \wedge \ \text{A3 A4}, \tag{5.4}$$

$$\wedge \ \vee \ \wedge \ \neg \ \text{A1 A2 A3 A4}. \tag{5.5}$$

In the interpretation of formulas (5.4) and (5.5) it is still necessary to know that $\neg$ is a unary operation, and that $\vee$ and $\wedge$ are binary operations. If the same symbol can stand for a unary and a binary operation, as, for example, $+$ and $-$ do in arithmetic, prefix formulas can be ambiguous. Both of the trees in Figure 5.10 correspond to the prefix formula $--ab$, but the first

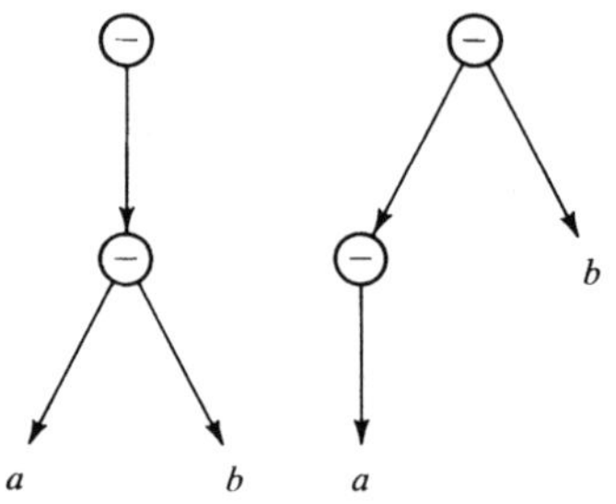

Figure 5.10

tree represents $-(a-b) = -a+b$, while the second stands for $(-a)-b = -a-b$. We had similar difficulty with the tree of Figure 5.1. In writing down the linear string corresponding to the tree (and this linear string is a prefix formula) we had to use parentheses to avoid ambiguity.

Sometimes a tree can be simplified. For example, tree (a) of Figure 5.11,

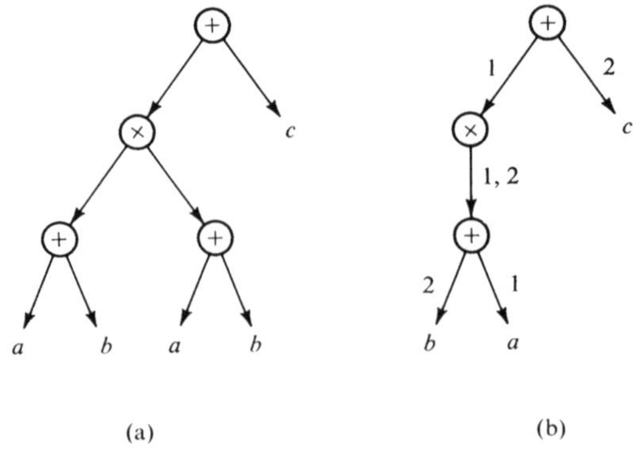

Figure 5.11

which corresponds to $+ \times +ab+abc$, can be drawn in the simpler form (b). Labels on arcs of tree (b) specify left-to-right order of the operands; this tree is therefore a true labeled weighted digraph.

We shall discuss conversion algorithms for prefix formulas in Section 10b. If a data structure can be represented as a tree, then the choice between leaving it in this form and converting it to a prefix formula is a matter of convenience, largely determined by the programming language that is to be used in implementing the processing algorithms.

5c. Sort Trees and Dictionaries

Quite often a computer program has to generate a list of all distinct words in a text. This can be accomplished by reading in the entire text, and producing the list by normal sorting procedures. There are, however, serious objections to this approach. First, if the number of distinct words (types) is small compared with the number of all words (tokens), memory space is wasted in storing redundant information. Sometimes, particularly when the length of the input text is unknown, this results in an unnecessarily complicated program; provision has to be made for transfer of data to and from auxiliary storage when, if some other approach were taken, all relevant data could be accommodated in core storage. Second, total processing time improves when input operations are made to overlap with straight computing. It is therefore desirable to generate the list of types while the text is being read. This becomes essential rather than just desirable when reduction of processing time is a major concern of the programmer, e.g., in writing nonexperimental compilers. In many assemblers, compilers, and interpreters the creation of tables of variable names (symbol tables) is one of the most important activities. Third, the table may have to go into full use before it is completed. This is the case with symbol tables in incremental compilers.

The simplest table of types is an unordered list. A word is simply adjoined to the end of the list if it is not already in the list. In the worst case, i.e., when a word is not in the list, the word has to be compared against every listed word. Alternatively, the list can be kept in alphabetic order. Then the time required to establish whether or not a word is already in the list is greatly reduced (see Section 11a), but insertion of a new word requires shifting of all words that come after it. The simplest arrangements are therefore the least economical. We need a structure that can be searched through in less time than an unordered list, but one that also permits new words to be readily added to it. A sort tree is one such structure.

ALGORITHM 5.1 Algorithm for creating a sort tree. A *sort tree* is a labeled binary tree. The labels make up a table of types. If a node is labeled, it is called *filled*; if not, it is called *free*. Let w_i stand for the ith word in the text, and assume that the text contains n words.

1. Initiate the tree as a single free node (this remains the root of the tree throughout), and set $i = 0$.
2. Set $i = i + 1$. If $i > n$, stop.
3. Enter the tree at the root.
4. If the node reached is free, make w_i the label of this node, drop two arcs from the node, thus creating two new free nodes, and go to 2.
5. A filled node has been reached. If w_i is equal to the label, go to 2.
6. If w_i is "smaller" than the label, proceed to the next node by taking the left arc down; if "larger", take the right arc down. Go to 4.

Example

Figure 5.12 shows the tree generated from the text,

Friends, Romans, countrymen, lend me your ears;
I come to bury Caesar, not to praise him.

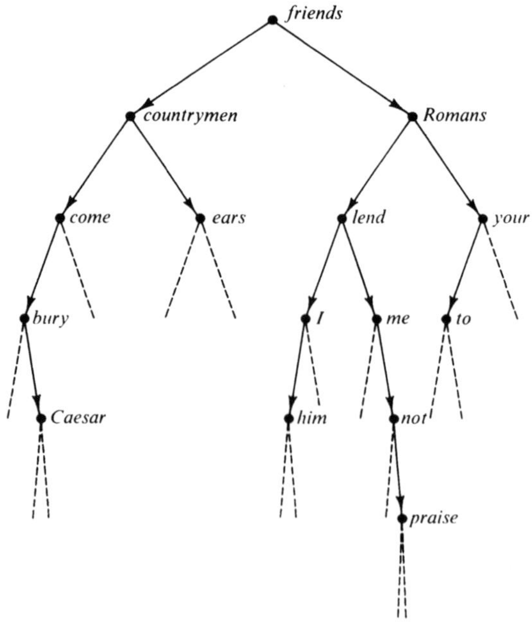

Figure 5.12

(The tree contains many arcs that terminate at free nodes, and are, therefore, superfluous. This is a minor technical detail. The algorithm can be made to suppress creation of redundant arcs, but the reformulated algorithm would not show the salient features of the procedure quite as clearly as A.5.1.)

An alphabetized list of the labels of nodes in a sort tree is produced by tracing round the tree. The procedure is similar to that of Figure 5.7, but here a label is added to the list when it is first reached in *upward* motion. Figure 5.13 illustrates the procedure. The labels are added to the list in the order of the dots on the line traced around the tree. The alphabetized list is *bury, Caesar, come, countrymen, ears, friends, him, I, lend, me, not, praise, Romans, to, your*.

For simplicity we have made the labels single items of data. In most practical applications of sort trees labels are ordered pairs $\langle s_k, a_k \rangle$, but only the first coordinates (the symbols s_k) enter the comparisons of A.5.1. The set of labels is then a function f, with $a_k = f(s_k)$. Assume that the tree is used to generate frequency counts of words in a text. Then, at Step 4 of A.5.1, $\langle w_i, 1 \rangle$ instead of just w_i is made the label, and at Step 5, if w_i is equal to the first coordinate of the label, the second coordinate is incremented by 1. After the algorithm has stopped, the a_k of each $\langle s_k, a_k \rangle$ is the number of occurrences of the type s_k in the text. For example, with the short text of the example of A.5.1, all labels, except the label $\langle$*to*, 2$\rangle$, would have the form $\langle s_k, 1 \rangle$. If the tree is a symbol table of an incremental compiler, then the second coordinates of labels might be addresses assigned to the variables (first coordinates).

The trees of Figures 5.12 and 5.13 are equivalent in the sense that the alphabetization algorithm produces the same list when applied to them. The

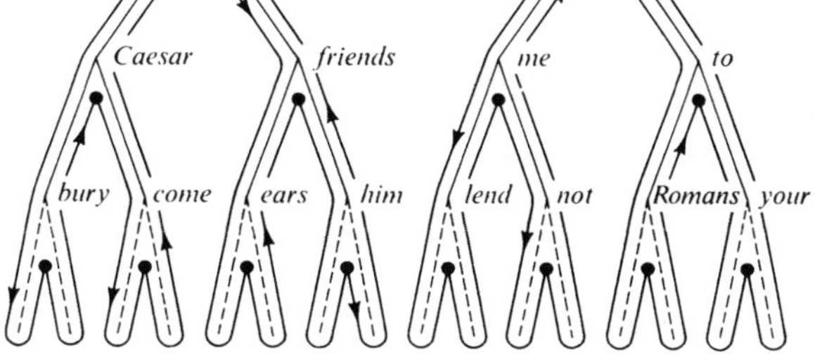

Figure 5.13

tree of Figure 5.13 is balanced, that of Figure 5.12 is not. Assume that we continue processing more text, and that the words in the text have an equal probability of occurrence. Then the expected number of comparisons that the next word of input has to undergo is $(1 \times 1 + 2 \times 2 + 4 \times 3 + 4 \times 4 + 3 \times 5 + 1 \times 6)/15 = 3.6$ for the unbalanced tree and $(1 \times 1 + 2 \times 2 + 4 \times 3 + 8 \times 4)/15 = 3.27$ for the balanced tree. Processing efficiency can be improved if a sort tree is converted to the equivalent balanced tree after 15 or 31 nodes have been filled. However, if the text is short, the subsequent gains in processing time may be too small to justify the time spent on rearranging the tree. An input text that is already in alphabetic order is a calamity.

Sort trees can be easily implemented using Fortran or a similar language. In our example the input will be a string of `N` integers. For simplicity it will be assumed that the integers are already in store, in array `D`. An array of pointers `P` will be generated by A.5.2. Element `P(J,1)` is a pointer to the node reached by following the left arc originating from `J`, and element `P(J,2)` is

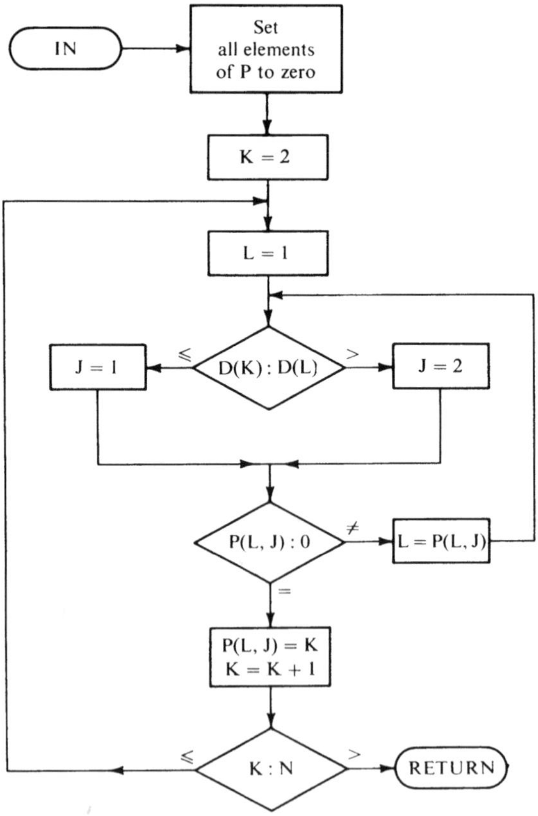

Figure 5.14

a pointer to the node reached by taking the right arc. If a pointer is zero, the corresponding arc would lead to a free node, i.e., would be redundant. Duplication will be permitted. In terms of A.5.1 this means deletion of Step 5 and change of *smaller* in Step 6 to *less than or equal.*

ALGORITHM 5.2 Figure 5.14 is an algorithm for implementation of a sort tree.

Example

Let the input string (array D) be

26 23 26 27 22 25 28 26 21 28 23 24.

Then array P of Figure 5.15 is the array generated by A.5.2. The sort tree is an exact representation of the arrays D and P, and vice versa. The nodes have been given identifying numbers 1, 2, ..., 12, which correspond to subscripts in D. The numbers in parentheses are the labels; the label of node J is D(J).

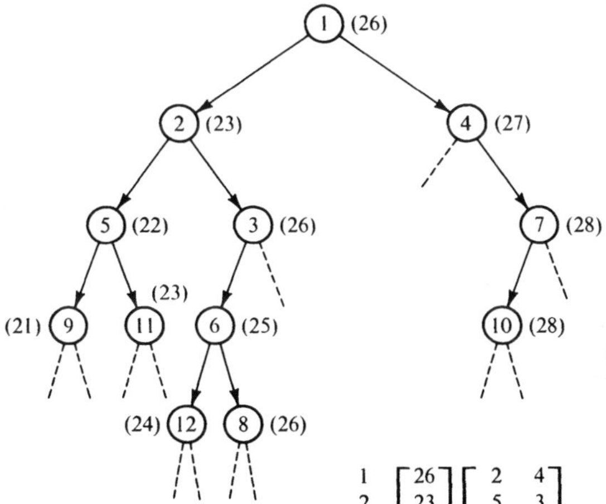

	Array D	Array P	
1	26	2	4
2	23	5	3
3	26	6	0
4	27	0	7
5	22	9	11
6	25	12	8
7	28	10	0
8	26	0	0
9	21	0	0
10	28	0	0
11	23	0	0
12	24	0	0

Figure 5.15

The alphabetization algorithm applied to the tree of Figure 5.15 produces the sorted string,

21 22 23 23 24 25 26 26 26 27 28 28.

As the example shows, A.5.2 can be used as the first stage of a sorting procedure. The second stage, conversion of array D to a sorted array, will be discussed in Chapter 10 as an example of application of pushdown stores. Some gain is to be had in going through the second stage if the sort tree is to be used extensively after its construction. Search efficiency is improved in the case of an unbalanced tree, and array P is made available for storing other data. The extravagant storage requirements do, however, dctract from the usefulness of the algorithm as a sorting technique. Therefore, unless there is a special reason for constructing a sort tree to begin with, other sorting procedures can be found that are more practicable. One such procedure follows.

ALGORITHM 5.3 Subroutine TREEUP sorts an array of N elements. In the worst case approximately $2N(log_2 N - 1)$ comparisons and $N(log_2 N - 1)$ interchanges have to be made. The algorithm is most easily followed if array A is *pictured* as a directed tree rooted at A(1) with *imaginary* arcs joining elements. An arc terminating at A(J) originates from A(JHALF), where JHALF = J/2, and the operation is Fortran integer division; e.g., arcs terminating at A(4) and A(5) originate from A(2). Figure 5.16 shows the structure of an array with six elements.

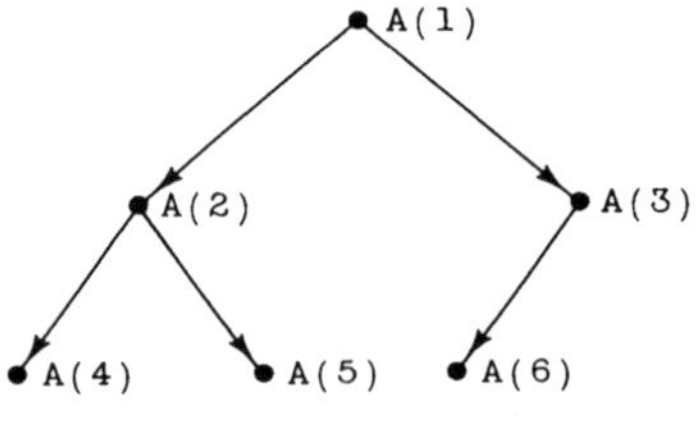

Figure 5.16

The segment starting with statement 20 (the shift procedure) is the mainstay of the program. The first loop (ending with statement 5) uses it to rearrange the elements so that the relation A(JHALF) ≧ A(J) holds for JHALF = 2,3,...,N/2. When the shift procedure is entered with the elements so ordered, the largest element of a subtree specified by I and L is shifted to the root node of the subtree. In the loop ending with statement 15 the shift procedure is used to shift the largest of the N elements into A(1). This element is then interchanged with A(N). Then the largest of the remaining N-1 elements is shifted into A(1) and interchanged with A(N-1), and so on. After N-1 iterations the array is sorted.

```
      SUBROUTINE TREEUP (A,N)
      DIMENSION A(N)
      ITOP = N/2
      ASSIGN  5 TO NN
      L = N
      DO  5  K = 2,ITOP
      I = ITOP + 2 - K
      GO TO 20
    5 CONTINUE
      ASSIGN 10 TO NN
      L = N
      DO 15  K = 2,N
      I = 1
      GO TO 20
   10 COPY = A(1)
      A(1) = A(L)
      A(L) = COPY
   15 L = L - 1
      RETURN
C  THE SHIFT PROCEDURE THAT FOLLOWS CAN BE WRITTEN
C  AS A SEPARATE SUBROUTINE
   20 COPY = A(I)
   25 J = 2 * I
      IF (J - L) 28, 30, 35
   28 IF (A(J+1).GT.A(J)) J = J + 1
   30 IF (A(J).LE.COPY) GO TO 35
      A(I) = A(J)
      I = J
      GO TO 25
   35 A(I) = COPY
C  SOME COMPILERS MAY OBJECT TO THE FOLLOWING TRANSFER
C  ALTHOUGH ASA STANDARD FORTRAN PERMITS IT
      GO TO NN, (5,10)
      END
```

A sort tree can function as a dictionary. Then the s_k of a label $\langle a_k, s_k \rangle$ may be an English word, and the a_k its equivalent in some other language, e.g., $\langle$*friends, amis*$\rangle$ in an English–French dictionary, or $\langle$*Freunde, friends*$\rangle$ in a German–English dictionary. Another way in which trees can be used in the construction of dictionaries is exemplified by the following scheme. For simplicity we shall assume that the entries are single words rather than pairs. We construct a tree with 27 arcs originating from the root, the first 26 arcs leading to nodes labeled $a, b, c, \ldots, z$. From each of these 26 nodes we suspend another tree having exactly the same structure; i.e., we have a tree with $27 + 26 \times 27$ arcs. Each terminal node has a list attached to it, which consists of words whose first two letters are the labels of the nodes in the path from the root to the list. For example, by going from the root to the node labeled c, and then to the node labeled o, we reach a list of words that begin with *co*: *come*, *countrymen*, ... (see Figure 5.17). Since this list is reached by a unique path, the first two letters of every word are superfluous, and can be chopped off to give *me*, *untrymen*, ... The loss of the first letters presents difficulties with short words, such as *a*, *an*, *me*. This is where the 27th arc (the ω-arc) comes in. Words only one letter long are stored at the node reached by taking the ω-arc from the root. In the case of a word that consists of two letters we go from the root to the node indicated by the first letter of the word, and take the ω-arc from there, e.g., *me* is stored as *e* at the node reached by first taking the *m*-arc from the root, and then the ω-arc.

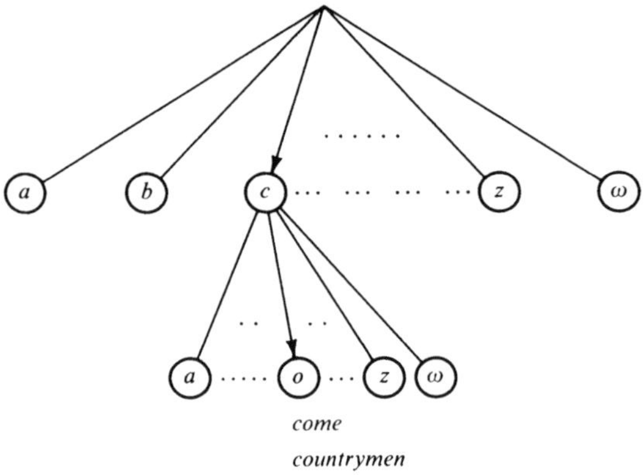

Figure 5.17

A similar application arises with arrays. A multidimensional array can be stored as a linear sequence of elements, and the position of an element in the array computed from the declared bounds of the array (Section 9b). Alternatively, the array can be set up as a tree in which the number of arcs originating from a node on level i is equal to the $(i + 1)$th bound. For example, if a 3-dimensional array has bounds $\langle 5, 4, 10 \rangle$, the array is represented by a tree with 5 arcs originating from the root, and 4 arcs originating from each of the 5 nodes on level 1. The 20 nodes on level 2 are terminal nodes, and a vector of 10 elements is attached to each of them. Let us see now how a particular element in this array is reached, say $\langle 2, 3, 5 \rangle$. The second of the arcs from the root leads to the appropriate node on level 1, and the third arc originating from this node terminates at a node that has the vector containing element

⟨2, 3, 5⟩ attached to it. The required element is the fifth element in the vector; it is accessed by the machine language technique of indexing.

5d. Decision Trees and Decision Tables

Flowcharts have undeniable merits as devices for displaying the logic of computer programs. Sometimes, however, a situation can become too complex for flowcharts. The greater the number of conditional statements in a program, the more confusing the corresponding flowchart. Finally, it can become so abstruse that the tracing of a particular execution sequence is no better than running a maze. This is where decision tables can restore clarity. Note, however, that only a very limited class of programs, arising mainly in business applications, and only parts of programs in this class, can be represented by decision tables.

Let us take a simple example based on a check being presented to a bank. The bank asks a number of questions:

q_1—*Is the check covered?*
q_2—*Is the account blocked?*
q_3—*Is the balance greater than* $300?

Answers to the questions form triples of yeses and noes. The triple ⟨*Y*, *N*, *Y*⟩ causes the bank to take an action different from that elicited by ⟨*N*, *N*, *N*⟩. The possible actions are:

A1—*Do not accept the check.*
A2—*Accept unconditionally.*
A3—*Accept, but charge a handling fee.*

There are eight possible triples. If the triples are tabulated, together with the appropriate actions, the result is an *extended-entry decision table*. The decision table for our example is Table 5.1. Each triple and the corresponding action, i.e., each column in the table, is a *decision rule*.

TABLE 5.1

EXTENDED-ENTRY DECISION TABLE

Rule		1	2	3	4	5	6	7	8
Condition	q_1	Y	Y	Y	Y	N	N	N	N
	q_2	Y	Y	N	N	Y	Y	N	N
	q_3	Y	N	Y	N	Y	N	Y	N
Action		A1	A1	A2	A3	A1	A1	A1	A1

Here we selected a very simple example to keep explanation of techniques simple. It is not unknown to have situations involving as many as 100 conditions. The extended-entry table for 100 conditions has 2^{100} rules! Fortunately we can combine rules to arrive at a *limited-entry* decision table. The conversion of an extended-entry table to a limited-entry table is best followed in a *sequential decision tree*. This is a binary tree in which arcs carry labels Y and N, nonterminal nodes have the conditions for labels, and the terminal nodes are labeled with the appropriate actions. Figure 5.18 shows the sequential decision tree corresponding to Table 5.1. The tree is, of course, a flowchart in an unfamiliar guise.

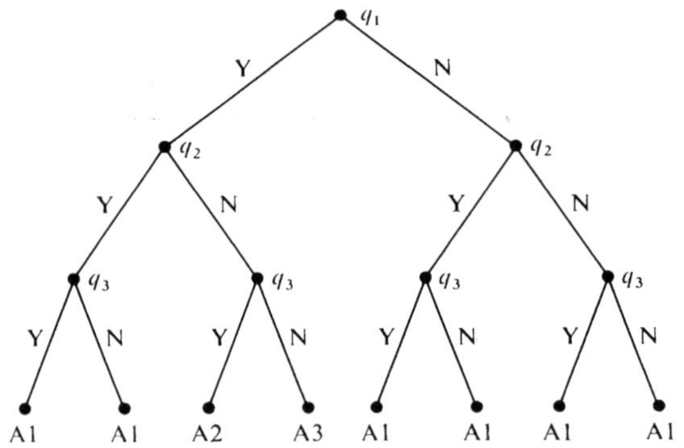

Figure 5.18

From the tree we can see at once that if q_1 and q_2 have both brought the response Y, then q_3 does not have to be asked; the outcome is independent of the response to this question. If the answer to q_1 is N, then the action is independent of the other conditions. We can therefore prune the tree down to that of Figure 5.19. The latter can be converted back to a decision table. This table, Table 5.2, is of the limited-entry variety. There are only four rules. The absence of an entry in a limited-entry table signifies a "don't care" condition.

TABLE 5.2

LIMITED-ENTRY DECISION TABLE

Rule	1	2	3	4
Check covered?	Y	Y	Y	N
Account blocked?	Y	N	N	–
Balance > $300?	–	Y	N	–
Action	A1	A2	A3	A1

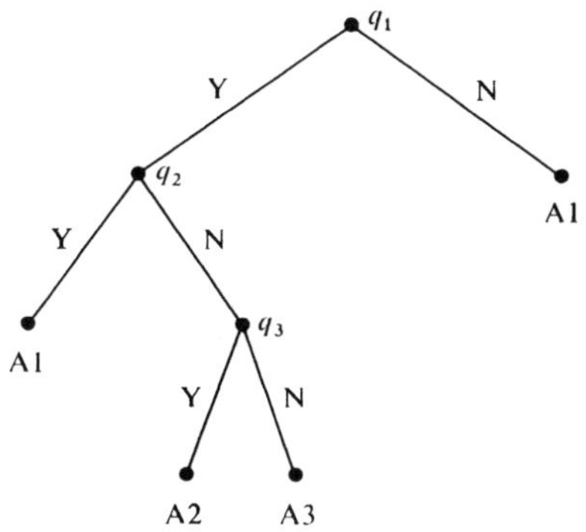

Figure 5.19

When there are 100 conditions one cannot very well construct the complete sequential decision tree. Pruning away of redundant branches has to take place while the tree is being constructed. For instance, in our rather oversimplified situation (we have not allowed for overdrafts for one thing) we know from the start that an uncovered check must evoke action A1.

In a computer a decision table is set up as a matrix of bits and a vector of addresses, called *table matrix* and *transfer vector*. In the table matrix 1 stands for Y, and 0 for N or a "don't care" condition. Decision rule n is represented by the nth column of this matrix and the nth element of the transfer vector. Table 5.1 becomes

$$\begin{bmatrix} 1 & 1 & 1 & 1 & 0 & 0 & 0 & 0 \\ 1 & 1 & 0 & 0 & 1 & 1 & 0 & 0 \\ 1 & 0 & 1 & 0 & 1 & 0 & 1 & 0 \end{bmatrix}$$

$$[\text{A1} \quad \text{A1} \quad \text{A2} \quad \text{A3} \quad \text{A1} \quad \text{A1} \quad \text{A1} \quad \text{A1}]$$

In processing a particular check we first set up a *data vector* of three elements (or, in general, as many elements as there are conditions). If, for this particular check, q_1 is answered with a *yes*, the first element of the data vector is 1; otherwise it is 0. The second element is 1 if q_2 is answered with *yes*, and 0 otherwise; and so forth. The data vector is compared with successive columns of the table matrix. If the vector matches a column, then the corresponding address in the transfer vector is used to transfer control to the program segment that performs the required action. Assume that a check has given rise to the data vector

$$\begin{bmatrix} 0 \\ 1 \\ 0 \end{bmatrix}.$$

It matches the sixth column of the table matrix, and the sixth element of the transfer vector is therefore used to transfer control to the program segment that prints a message "Check cannot be accepted" or performs some similar action.

The limited-entry decision table of our example becomes

$$\begin{bmatrix} 1 & 1 & 1 & 0 \\ 1 & 0 & 0 & 0 \\ 0 & 1 & 0 & 0 \end{bmatrix}$$

$$[\mathrm{A1} \quad \mathrm{A2} \quad \mathrm{A3} \quad \mathrm{A1}]$$

but

$$\begin{bmatrix} 0 \\ 1 \\ 0 \end{bmatrix}$$

does not match any column in this table matrix. The way out of the impasse is to suppress a 1 in the data vector whenever the 1 would be compared with a 0 that stands for a "don't care." The masking out is done by means of a *mask matrix* in which every Y or N entry of the decision table is represented by 1, and a "don't care" by 0. In our case the mask matrix is

$$\begin{bmatrix} 1 & 1 & 1 & 1 \\ 1 & 1 & 1 & 0 \\ 0 & 1 & 1 & 0 \end{bmatrix}.$$

Now, before the data vector is compared with column n in the table matrix, it is modified by use of the nth column of the mask matrix. If an element in the data vector is 1, but the corresponding element in the nth column of the mask matrix is 0, the entry in the data vector is made 0. This modification is achieved by taking the $\wedge$ of each element of the data vector and the corresponding element of the mask matrix ($1 \wedge 1 = 1$, $1 \wedge 0 = 0 \wedge 1 = 0 \wedge 0 = 0$). In our case this means that

$$\begin{bmatrix} 0 \\ 1 \\ 0 \end{bmatrix}$$

remains unchanged for the first three comparisons (in the first three columns of the mask matrix the second element is 1), but that it is changed to

$$\begin{bmatrix} 0 \\ 0 \\ 0 \end{bmatrix}$$

for the comparison with the fourth column. This procedure is called the *rule mask technique.*

If the cost of setting up the data vector is high, then an *interrupted* rule mask technique should be used. Let us rearrange the mask matrix and the table matrix with its associated transfer vector so that the numbers of zeros in columns of the mask matrix are in descending order. The result is Table 5.3. Applicability of the first rule in the rearranged table (Rule 4) is determined by condition q_1, and applicability of the second rule (Rule 1) by conditions q_1 and q_2 alone. Evaluation of the data vector can therefore be interrupted after its first element is computed (and the remaining elements set to zero). No more is needed for comparison with the first rule. If a match results, action A1 is performed. If not, evaluation of the data vector resumes, and the second element is computed. This partial data vector is then compared with the second column of the table matrix. The third element is computed only if there is still no match.

TABLE 5.3

REPRESENTATION OF A DECISION TABLE

	Mask matrix				Table matrix				
q_1	1	1	1	1	0	1	1	1	
q_2	0	1	1	1	0	1	0	0	
q_3	0	0	1	1	0	0	1	0	
					[A1	A1	A2	A3]	Transfer vector
					4	1	2	3	Rule number

A concise representation of the sequence of interruptions is provided by a $2 \times (c + r)$ *sequencing matrix*, where c is the number of conditions and r the number of rules. The top row contains numbers representing conditions; the bottom row numbers of rules. The conditions follow the order of their evaluation. One element in each column is the number of either a condition or a rule; the other is zero. In our example the matrix is

$$\begin{bmatrix} 1 & 0 & 2 & 0 & 3 & 0 & 0 \\ 0 & 4 & 0 & 1 & 0 & 2 & 3 \end{bmatrix}$$

A sequencing matrix is interpreted as follows: Starting with element $\langle 1, 1\rangle$, one moves along the top row evaluating conditions until a zero is reached; then one switches to the bottom row and tests rules until a zero is encountered in this row; then switches back to the top row and resumes evaluation of conditions until a further zero is encountered; and so on. The sequencing matrix of the original limited-entry table is

$$\begin{bmatrix} 1 & 2 & 3 & 0 & 0 & 0 & 0 \\ 0 & 0 & 0 & 1 & 2 & 3 & 4 \end{bmatrix}.$$

In practice there is no need to rearrange the mask and table matrices bodily. The arrangement of columns in the sequencing matrix gives enough information for the process to be performed as if the matrices were rearranged.

If the mask matrix is, for example,

$$\begin{bmatrix} 1 & 0 & 1 & 1 & 1 & 1 & 1 \\ 0 & 1 & 1 & 0 & 1 & 1 & 1 \\ 0 & 1 & 1 & 1 & 0 & 1 & 1 \\ 1 & 0 & 0 & 1 & 1 & 1 & 1 \end{bmatrix},$$

the order of evaluation of the first three conditions can be q_1 and q_4 followed by q_2, or q_1 and q_4 followed by q_3, or q_2 and q_3 followed by q_1. If the cost of evaluation is the same for all conditions, then it would not seem to matter which order of evaluations is adopted. If, however, the costs differ, they should be taken into account.

Evaluation costs are not alone in affecting the efficiency of a program derived from a decision table. One must also consider the probabilities associated with the conditions. Returning to the example of Table 5.3, let us associate with each q_i a cost c_i and a probability of success (*yes* answer) p_i. To make the probabilities realistic we shall leave undefined the context in which the table is used. A set of values is given in Table 5.4. The interrupted rule mask procedure illustrated by Table 5.3 is exactly equivalent to the

TABLE 5.4

COSTS AND PROBABILITIES OF CONDITIONS

Condition q_i	Cost c_i	Probability p_i
q_1	40	0.85
q_2	15	0.25
q_3	25	0.60

sequential decision tree of Figure 5.19. We have redrawn the tree as Figure 5.20, adding costs and probabilities. The expected cost of interrupted processing of a data vector can be easily computed by reference to the tree. It is $40 + 0.85(15 + 0.75 \times 25) = 68.6875$. By contrast, if the interrupting procedure were not followed, the cost would be $40 + 15 + 25 = 80$.

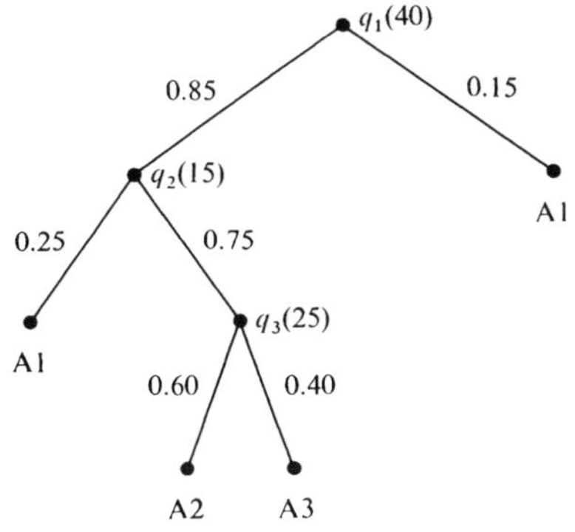

Figure 5.20

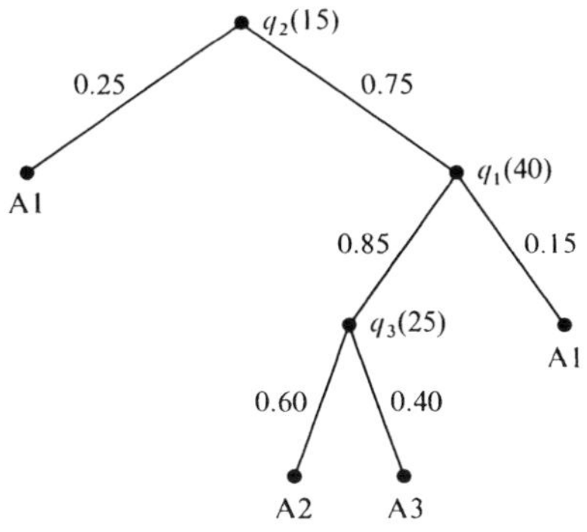

Figure 5.21

In the tree of Figure 5.20 we see that if q_1 is answered with a *yes*, and q_2 is answered with a *yes*, action A1 is taken. If q_1 is answered *no*, A1 is again taken, irrespective of what the answer to the unasked q_2 might be. In other words, A1 is taken if q_2 is answered with a *yes*, no matter how q_1 is answered. This means that we can rearrange the tree, putting q_2 at the root. Figure 5.21 shows the rearranged tree. Here the expected cost is $15 + 0.75(40 + 0.85 \times 25) = 60.9375$, which is a further reduction.

In a practical application one would rarely be able to assume that the probabilities associated with the conditions q_i are independent of each other. For example, if our conditions were

q_1—*Older than 40 years?*
q_2—*Older than 50 years?*

then the probability that an individual is older than 50 is greater for people over 40 than for the total population, and it is zero for people whose age does not exceed 40. Therefore, the optimization of decision tables should be based on the probabilities of applicability of the rules rather than on the probabilities associated with conditions.

Decision tables have some definite advantages as a form of documentation over sequential decision trees. Unless conditions are reduced to simple symbols, a tree is in danger of being overcluttered. Just the opposite holds for a table; spelling out the conditions actually enhances the readability of the table. Moreover, on account of the geometry, the information content of a rectangular table inscribed in the rectangle of a page of a document is greater than the information carried by a triangular tree inscribed in a page of the same size. If the document is generated by a computer, then the printer is adequate for producing tables. The program for generating trees is more complex, and has to use a plotter for output. Nevertheless, as some of our examples have shown, trees can sometimes be very useful for gaining insight in the structure of decision tables.

A further advantage of decision tables is that a small extended-entry table can be directly coded in Fortran or some other higher-level language. Conversion of a tree to a Fortran program would be a much more devious process. Assume that every Y in an extended-entry table and the data vectors is coded as 1, and every N as 0. In every row of the table substitute 2^{m-1} for every 1, where m is the number of the row. Then compute the column totals. In the case of Table 5.1 they are 7, 6, ..., 1, 0, i.e., column totals identify rules uniquely. Now, given a data vector of n elements, perform the same valuation, and obtain a value between 0 and $2^n - 1$, which is equal to the column total of the column that the data vector matches. This value (increased by 1) can therefore be used in a computed `GO TO` to effect a transfer of control to the program segment in which the appropriate action is performed. The program for Table 5.1 is simply

```
INDEX = 1 + D(1) + 2*D(2) + 4*D(3)
GO TO (N1,N1,N2,N3,N1,N1,N1,N1), INDEX
```

where `D` is the data vector and N1, N2, N3 stand for statement numbers associated with program segments in which actions A1, A2, A3, respectively, are performed.

Notes

In practice one is not interested in generating all sentences of a language. Rather, one wants to know whether a given sentence belongs to a particular language. This requires a parser. In [In66], which deals with the writing of compilers for programming languages, the syntactic chart of a grammar is used in the construction of a parser. Parsing of programming languages is surveyed in [Fe68]. Of the 226 items in the bibliography of this survey [Gr65] may be singled out; its distinguishing feature is the comparison of the efficiencies of a number of parsing strategies for context-free languages on the basis of the performance of Turing machine recognizers equivalent to the parsing strategies. Parsing of natural languages is surveyed in [Bo63a, Sa67]. [Ha67b] is a text on computational linguistics; it contains parsing algorithms. Automated natural language processing in general is surveyed in [Si66a, Bo67b]. McNaughton has shown that a backward-deterministic parenthesis grammar is unambiguous, see [Mc67], and Knuth proves in [Kn67a] that the problem of whether a Type 2 language has a parenthesis grammar can be decided.

The sort tree of Figure 5.12 is adapted from [Bo60], which contains a thorough analysis of search efficiency in the sort tree. A more recent analysis can be found in [Ar69a]. Similar techniques are described in [Br59, Fr60, Wi60, Sc63, Ka65, Su63, Cl64, Pa69b]. Alogrithm 5.3 derives from [Fl64]. It is of interest to note that trees have been used in the analysis of sorting procedures on a number of occasions, see [Oe57, Bu58, Hi62a.] For the representation of arrays by trees see Section 2.3 of [Kn68]. Our Exercise 5.13, in which a tree is used to depict the inclusion relation, was suggested by the practical applications described in [Bo63b, Bo67c].

The literature on decision tables is surveyed in [Ki67] to the end of 1966. Of the numerous articles on the techniques we single out [Fi66], a description of a large problem made interesting by the inclusion of a quiz to test the reader's attentiveness. [Mc68] is an expository monograph; unfortunately it tends to gloss over the subtler difficulties. One such difficulty is ambiguity in limited-entry tables, discussed in the very important [Ki68]. A very thorough analysis of the conversion of limited-entry decision tables to computer programs so as to minimize processing time and storage requirements is carried out in [Re66, Re67]. Compression of decision tables by means of a parsing technique is described in [Ch67]. The technique of the computed `GO TO` described in the final paragraph of Section 5d is taken from [Ve66].

Exercises

5.1 (a) Write a program that determines, given the adjacency matrix of a digraph D, whether or not D is a directed tree, and, if it is, finds its root.

5.2 (a) Sentence *a.green.man.bought.the.green.apple* has been generated by the grammar of Example 3 of D.4.22. Draw the derivation tree of this sentence and represent the tree by a linear string.

5.3 (a) Show that $G = \langle\{a, b\}, \{S, A\}, P, S\rangle$, where P consists of

$$\langle S\rangle ::= ab\langle S\rangle b \mid a\langle A\rangle b \mid a$$
$$\langle A\rangle ::= b\langle S\rangle$$

is an ambiguous grammar. What is $L(G)$? Find an unambiguous grammar G_1 such that $L(G) = L(G_1)$.

5.4 (a) Draw syntactic charts of the grammars of Exercises 4.24 and 4.25.

5.5 (a) Find a Type 3 grammar that generates the language of Part (iii) of Exercise 4.27 and draw the syntactic chart of the grammar. Would you find the syntactic chart of assistance in the design of an acceptor for the language?

5.6 (b) Given infix formula `R = ((A+B)*(A+C))/X + Y`. Draw the tree corresponding to this formula and use the tree to write down the corresponding prefix formula.

5.7 (b) Define a parenthesis grammar that generates completely parenthesized arithmetic expressions in variables $a, b, \ldots, z$, and binary operators $+, -, /, \times, =$.

5.8 (b) Although multiplication in the set of real numbers is an associative operation, it is not associative in the set of computer representations of real numbers. For example, `A*(B*C)` and `(A*B)*C` may have very different values if floating point underflow occurs. Therefore, it may be quite realistic to have to write an extended product as `((A*(B*C))*(D*E))*F`. Draw the tree of this expression, and find its prefix form. Then draw the tree of the K-formula $***a*bc*def$, and rewrite the K-formula in infix form. Find infix forms of the following K-formulas:

(i) $**a***a*bc**dcbef$;
(ii) $***a***bbcd***cbdee$;
(iii) $*a**d*b*cd*ea$.

Does the prefix or the infix form afford a better insight in the structure of the digraphs represented by the formulas? What effect does change in the left-to-right order of arcs in the tree of a K-formula have?

5.9 (c) Construct tree dictionaries for the following passages:

(i) The only problem to be faced in using the foregoing algorithm to obtain a solution to the traveling salesman problem for an arbitrarily large number of cities is a storage problem.

(ii) Assuming that subsequent text will be composed entirely of words from this dictionary and that all words will occur with equal probability, compute the average number of comparisons for this dictionary and for an equivalent balanced tree dictionary.

Apply Part (ii) to both trees.

5.10 (c) Set up array `P` (see the example of A.5.2) for the sort tree of Part (i) of Exercise 5.9.

5.11 (c) Assume that we have a procedure A for converting a sort tree stored as arrays `D` and `P` (see Figure 5.15) to an array of numbers sorted in ascending order. Modify A.5.2 so that it produces array `P` in such a form that procedure `A` now puts the numbers into descending order. Test your algorithm on the data of the example of A.5.2.

5.12 (c) The two trees of Exercise 5.9 are to be combined into a single sort tree. One way would be to send the first passage down the second tree. Since a considerable effort has gone into the construction of the first tree, it may be possible to find a more efficient procedure that utilizes the structure of this tree. Investigate.

5.13 (c) Figure 5.22 is a Venn diagram showing 11 elements, which belong to five sets. Devise a tree representation of the situation depicted by the diagram. What structure would you need to represent the situation if a sixth set $F = \{b, c, k, m\}$ were introduced?

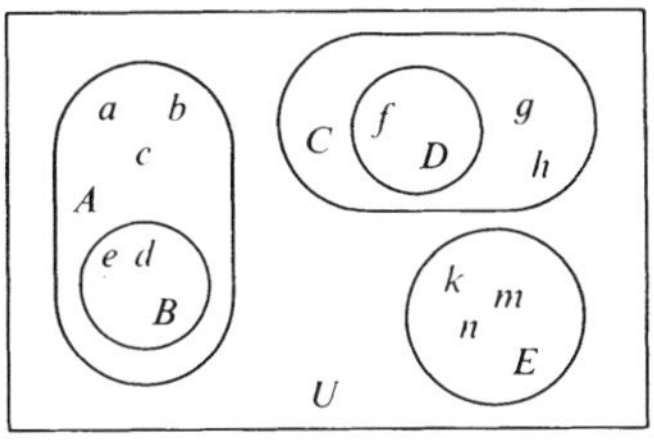

Figure 5.22

5.14 (d) In Exercise 2.16 consider the inputs as conditions, and the presence or absence of an output signal as actions A and B, respectively. Set up extended-entry decision tables corresponding to Parts (ii) and (iv) of Exercise 2.16, and find limited-entry tables equivalent to the extended-entry tables.

5.15 (d) In processing driving license applications of drivers who hold licenses from other states all applicants are given a quiz on road rules. In addition, a driving test is given to applicants under the age of 21, or over the age of 65, and to applicants who have had licenses suspended. Moreover, applicants over 65 have to undergo a medical examination, and applicants under 21 who have had licenses suspended have to attend a course on road safety. Tabulate the data as an extended-entry table, and convert this to a limited-entry table.

5.16 (d) Table 5.5 is an extended-entry decision table in which the top row gives the probability of a data vector matching the corresponding decision rule. Draw a sequential decision tree that corresponds to the table and reduce it to two different trees of four terminal nodes each (rooted at q_1 and q_3, respectively). Assuming that the data vector is passed across the table matrix from the left, arrange the columns in the table matrices of the corresponding limited-entry tables in such a way that the expected number of comparisons is minimized. Which arrangement is the more efficient?

TABLE 5.5

	0·10	0·25	0·15	0·20	0·05	0·05	0·05	0·15
q_1	Y	Y	Y	Y	N	N	N	N
q_2	Y	Y	N	N	Y	Y	N	N
q_3	Y	N	Y	N	Y	N	Y	N
Action	A1	A3	A1	A2	A1	A1	A1	A1

5.17 (d) Set up interrupted rule mask procedures to correspond to the two reduced decision trees of Exercise 5.16. If the evaluation costs of conditions q_1, q_2, and q_3 are, respectively, 1.5, 4, and 2 units, what is the cost of producing a data vector when the interrupted rule mask technique is not used? Which of the interrupted rule mask procedures is more efficient?

5.18 (d) Assume that each probability in Table 5.5 is $\frac{1}{8}$ (this assumption would have to be made if actual probabilities were not known). Which of the interrupted rule mask procedures of Exercise 5.17 is now the more efficient?

5.19 (d) Consider the procedure that you selected as the more efficient in Exercise 5.18. With each of the three conditions there is associated an estimate of the probability that the condition will have to be evaluated. If for a sufficiently long time counts are kept of the number of times each condition is in fact evaluated, one might find that the assumption of Exercise 5.18 is wrong. It is even possible that the other interrupted rule mask procedure is more efficient. Can any conclusion be made about the relative efficiencies of the two procedures from the counts?

CHAPTER 6

Paths and Cycles in Digraphs

6a. Shortest Path Problems

A road map is a labeled graph. Towns are nodes and roads between towns edges (or, if we want to interpret the map as a digraph, pairs of arcs); mileages are weights. If two towns in a densely populated area are three or four hundred miles apart, very many routes lead from one to the other. The one with the least total mileage is the shortest path. To a motorist setting out from town A, and wishing to reach town B as early as possible, finding the shortest path is a very real practical problem.

Still keeping to the example of roads, the path that is shortest in the sense of least traveling time is not always the path of shortest distance; it may take twice as long to cover 10 miles on a country lane as to go 20 miles on a freeway. Traveling times can therefore be more suitable weights than distances. Costs can also function as weights. In fact, we can take any nonnegative number (under certain conditions even a negative number) for a weight, and define the shortest path problem as the problem of finding a path for which the sum of weights is a minimum. Solutions of the problem are independent of physical interpretation of the weights. The unweighted digraph is merely a special case, interpreted as a weighted digraph with every weight equal to 1. The shortest path given by the general algorithm must then be the path of fewest arcs.

We shall first give an algorithm for the more fundamental problem of the existence of paths. By Corollary 2 of Th.3.6, if $Q = X + X^2 + \cdots + X^n$, where X is the adjacency matrix of a digraph with n nodes, then an element in the path matrix P of the digraph is 1 if and only if the corresponding element of Q is nonzero. Use of Boolean operations $\wedge$ and $\vee$ gives P more directly ($1 \wedge 1 = 1$, $1 \wedge 0 = 0 \wedge 1 = 0 \wedge 0 = 0$; $0 \vee 0 = 0$, $0 \vee 1 = 1 \vee 0 = 1 \vee 1 = 1$). We define Boolean matrix operations $C = A \wedge B$ and $D = A \vee B$ in a set of square matrices of order n by

$$c_{ij} = \bigvee_{k=1}^{n} (a_{ik} \wedge b_{kj}),$$

$$d_{ij} = a_{ij} \vee b_{ij}.$$

Then $P = X \vee X^2 \vee \cdots \vee X^n$, where $X^k = X^{k-1} \wedge X$. A very efficient algorithm for computing P follows.

ALGORITHM 6.1 Given a square matrix X of order n with elements in $\{0, 1\}$.

1. Set $X^* = X$.
2. Set $j = 1$.
3. Set $i = 1$.
4. If $x_{ij}^* = 1$, then set $x_{ik}^* = x_{ik}^* \vee x_{jk}^*$ for all k from 1 to n.
5. Set $i = i + 1$. If $i \leqq n$, go to 4.
6. Set $j = j + 1$. If $j \leqq n$, go to 3; else stop.

THEOREM 6.1 (Warshall's theorem) If X is the adjacency matrix of a digraph, then the X^* generated by A.6.1 is the path matrix.

Proof. We show first that $x_{ik}^* = 1$ implies $p_{ik} = 1$. If $x_{ik} = 1$, then certainly $x_{ik}^* = 1$, and $p_{ik} = 1$. If $x_{ik} \neq 1$, then x_{ik}^* is set to 1 in Step 4. This means that at some stage of the process $x_{ij}^* = x_{jk}^* = 1$. Each of these, similarly, comes from X itself or from a previous application of Step 4; i.e., x_{ij}^* was set to 1 because $x_{ij} = 1$ or by virtue of $x_{ij'}^* = x_{j'j}^* = 1$ having held, and similarly for x_{jk}^*. Since the process is finite, the sequence of applications of Step 4 that finally leads to $x_{ik}^* = 1$ must have started out with X containing a finite set of elements $x_{ii_1} = x_{i_1i_2} = x_{i_2i_3} = \cdots = x_{i_mk} = 1$, but this means that $p_{ik} = 1$. Next prove that $p_{ik} = 1$ implies $x_{ik}^* = 1$. If $p_{ik} = 1$, then an expression of the form $x_{ii_1} = x_{i_1i_2} = \cdots = x_{i_mk} = 1$ must hold; i.e., $x_{ii_1}^* = x_{i_1i_2}^* = \cdots = x_{i_mk}^* = 1$ holds initially. Treating this expression as a string, our purpose is to justify a series of replacements of substrings $x_{aj}^* = x_{jb}^*$ by x_{ab}^* that finally leaves the string $x_{ik}^* = 1$. The iteration with $j = 1$ in A.6.1 produces in Step 4 $x_{i'k'}^* = 1$ for all i' and k' such that $x_{i'1}^* = x_{1k'}^* = 1$, and this justifies substitution of substring $x_{i'k'}^*$ for every substring $x_{i'1}^* = x_{1k'}^*$. The iteration with $j = 2$ similarly justifies substitution of $x_{i'k'}^*$ for every substring $x_{i'2}^* = x_{2k'}^*$. The substitutions are continued with increased j until $j = t$, where t is the largest of the subscripts $i_1, i_2, \ldots, i_m$. Clearly, at the end of the iteration, the string has been reduced to $x^*_{ik} = 1$. Consequently $p_{ik} = 1$ implies $x^*_{ik} = 1$.

ALGORITHM 6.2 (generalized Warshall algorithm) Let $\langle Q, +, \cdot, 0, 1\rangle$ be a Q-semiring, and define operations in the set of square matrices of order n whose elements are members of Q, $C = AB$ and $D = A + B$, as follows:

$$c_{ij} = a_{i1}b_{1j} + a_{i2}b_{2j} + \cdots + a_{in}b_{nj}$$
$$d_{ij} = a_{ij} + b_{ij}$$

Define also $X^k = X^{k-1}X$. Then $X + X^2 + \cdots + X^n$ is equal to the X^* produced by A.6.1 with Step 4 replaced by

4. Set $x^*_{ik} = x^*_{ik} + (x^*_{ij} \cdot x^*_{jk})$ for all k from 1 to n.

Examples

1. Consider $\langle R^\infty_+, min, +, \infty, 0\rangle$, the Q-semiring of Example 2 of D.4.6. In this case the algorithm can be expressed as the following program:

```
      SUBROUTINE PATHLS (X, N, RINF)
      DIMENSION X(N,N)
      DO 10  J = 1,N
      DO 10  I = 1,N
      IF (X(I,J).EQ.RINF) GO TO 10
      DO  5  K = 1,N
    5 X(I,K) = AMIN1(X(I,K), X(I,J)+X(J,K))
   10 CONTINUE
      RETURN
      END
```

The subroutine is entered with a matrix of weights (for example, mileages) associated with the arcs. If no arc joins nodes i and j, then the element `X(I,J)` is set to `RINF`, which is a very large number, say 10^{15}. `RINF` is our "infinity." On return `X(I,J)` is still equal to `RINF` if there is no path from i to j; otherwise `X(I,J)` contains the length of the shortest path from i to j.

2. In the case of the Q-semiring (Boolean algebra) $\langle\{0, 1\}, \vee, \wedge, 0, 1\rangle$ A.6.2 becomes A.6.1.

Knowledge of the length alone of a shortest path is rarely sufficient; one has to know the path explicitly. A.6.3 specifies all simple paths and cycles in a digraph. It is then an easy matter to find the shortest paths. The algorithm is based on a matrix operation which we shall call symbolic multiplication.

Let a square matrix of order n have strings for its elements. The symbolic product $C = AB$ in a set of such matrices is defined by $c_{ij} = a_{i1}b_{1j} + a_{i2}b_{2j} + \cdots + a_{in}b_{nj}$, where $+$ denotes juxtaposition and $a_{ik}b_{kj}$ denotes concatenation if neither a_{ik} nor b_{kj} is null, but $a_{ik}b_{kj} = \Lambda$ if $a_{ik} = \Lambda$ or $b_{kj} = \Lambda$. Juxtaposition is associative and commutative; it distributes over concatenation as follows: $\alpha(\beta + \gamma) = \alpha\beta + \alpha\gamma$, $(\beta + \gamma)\alpha = \beta\alpha + \gamma\alpha$. The null string is neutral in juxtaposition: $\Lambda + \alpha = \alpha + \Lambda = \alpha$.

ALGORITHM 6.3 Let the arcs of a digraph with n nodes be identified by symbols $a, b, c, \ldots$, and let the symbols be assembled into a variable adjacency matrix V, in which nonexistence of an arc is indicated by the null string Λ.

1. Define the matrix V_1: for $i, j = 1, 2, \ldots, n$ set $(v_{ii})_1 = \Lambda$ and, if $i \neq j$, set $(v_{ij})_1 = v_{ij}$.
2. Define the vector D_1 by $(d_i)_1 = v_{ii}$, $i = 1, 2, \ldots, n$.
3. Set $q = 1$.
4. If $q = n$, stop; else form the symbolic matrix products $V_L = V_1 V_q$ and $V_R = V_q V_1$.
5. Define V_{q+1}: for $i, j = 1, 2, \ldots, n$ set $(v_{ii})_{q+1} = \Lambda$ and, if $i \neq j$, make $(v_{ij})_{q+1}$ the juxtaposition of the terms that occur in *both* $(v_{ij})_L$ and $(v_{ij})_R$.
6. Define D_{q+1} by $(d_i)_{q+1} = (v_{ii})_L = (v_{ii})_R$, $i = 1, 2, \ldots, n$.
7. Set $q = q + 1$; go to 4.

Element $(v_{ij})_t$, $i \neq j$ represents all simple paths of length t from node i to node j, and $(d_i)_t$ represents all simple cycles of length t through i.

Example

Figure 6.1 shows a digraph and the matrix V associated with the digraph. The sequence of matrices and vectors that follows illustrates application of A.6.3 to this digraph.

$$V_1 = \begin{bmatrix} \Lambda & a & \Lambda & b \\ \Lambda & \Lambda & c & \Lambda \\ \Lambda & \Lambda & \Lambda & e \\ f & g & h & \Lambda \end{bmatrix} \qquad D_1 = \begin{bmatrix} \Lambda \\ \Lambda \\ d \\ \Lambda \end{bmatrix}$$

$$(V_L)_2 = (V_R)_2 = \begin{bmatrix} bf & bg & ac + bh & \Lambda \\ \Lambda & \Lambda & \Lambda & ce \\ ef & eg & eh & \Lambda \\ \Lambda & fa & gc & fb + he \end{bmatrix}$$

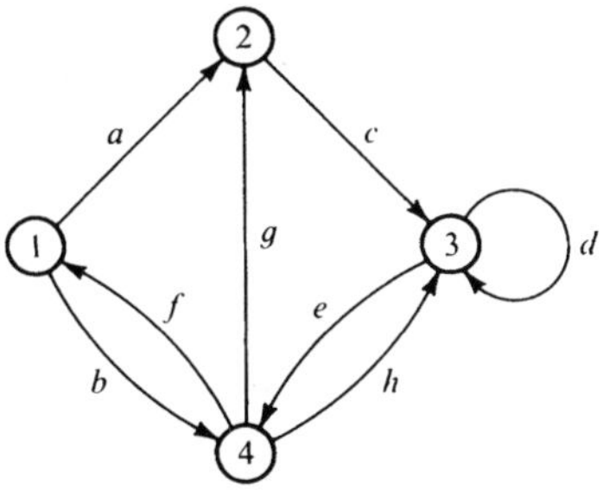

$$V = \begin{bmatrix} \Lambda & a & \Lambda & b \\ \Lambda & \Lambda & c & \Lambda \\ \Lambda & \Lambda & d & e \\ f & g & h & \Lambda \end{bmatrix}$$

Figure 6.1

$$V_2 = \begin{bmatrix} \Lambda & bg & ac + bh & \Lambda \\ \Lambda & \Lambda & \Lambda & ce \\ ef & eg & \Lambda & \Lambda \\ \Lambda & fa & gc & \Lambda \end{bmatrix} \qquad D_2 = \begin{bmatrix} bf \\ \Lambda \\ eh \\ fb + he \end{bmatrix}$$

$$(V_L)_3 = \begin{bmatrix} \Lambda & bfa & bgc & ace \\ cef & ceg & \Lambda & \Lambda \\ \Lambda & efa & egc & \Lambda \\ hef & fbg + heg & fac + fbh & gce \end{bmatrix}$$

$$(V_R)_3 = \begin{bmatrix} \Lambda & \Lambda & bgc & ace + bhe \\ cef & ceg & ceh & \Lambda \\ \Lambda & efa & egc & efb \\ \Lambda & \Lambda & fac & gce \end{bmatrix}$$

$$V_3 = \begin{bmatrix} \Lambda & \Lambda & bgc & ace \\ cef & \Lambda & \Lambda & \Lambda \\ \Lambda & efa & \Lambda & \Lambda \\ \Lambda & \Lambda & fac & \Lambda \end{bmatrix} \qquad D_3 = \begin{bmatrix} \Lambda \\ ceg \\ egc \\ gce \end{bmatrix}$$

$$(V_L)_4 = \begin{bmatrix} acef & \Lambda & bfac & \Lambda \\ \Lambda & cefa & \Lambda & \Lambda \\ \Lambda & \Lambda & efac & \Lambda \\ gcef & hefa & fbgc & face \end{bmatrix} \qquad (V_R)_4 = \begin{bmatrix} acef & aceg & aceh & bgce \\ \Lambda & cefa & \Lambda & cefb \\ \Lambda & \Lambda & efac & \Lambda \\ \Lambda & \Lambda & \Lambda & face \end{bmatrix}$$

$$V_4 = \begin{bmatrix} \Lambda & \Lambda & \Lambda & \Lambda \\ \Lambda & \Lambda & \Lambda & \Lambda \\ \Lambda & \Lambda & \Lambda & \Lambda \\ \Lambda & \Lambda & \Lambda & \Lambda \end{bmatrix} \qquad D_4 = \begin{bmatrix} acef \\ cefa \\ efac \\ face \end{bmatrix}$$

Since there can be no simple paths of length n, the purpose of the final iteration is merely to detect Hamiltonian cycles. For this it is sufficient to evaluate just the diagonal elements of $(V_L)_n$ or $(V_R)_n$. Since $(V_L)_2 = (V_R)_2$, the first iteration can be treated as a special case also. In all other iterations it is necessary to compute both products in order to eliminate nonsimple paths.

Two undesirable features of A.6.3 should be noted. First, the algorithm has to be programmed in a language with string processing capabilities. Second, if one is interested in shortest paths alone, the algorithm takes an unnecessarily long time. A modified Warshall algorithm that uses a matrix of paths finds explicit shortest paths much faster than A.6.3.

ALGORITHM 6.4 Modified Warshall algorithm for finding shortest paths in a digraph with n nodes. Let X be the matrix of weights defined in Example 1 of A.6.2, and let V be a variable adjacency matrix as in A.6.3. The operation on elements of V^* in Step 6 is concatenation.

1. Set $X^* = X$ and $V^* = V$.
2. Set $j = 1$.
3. Set $i = 1$.
4. If $x_{ij}^* = \infty$, go to 8.
5. Set $k = 1$.
6. If $(x_{ij}^* + x_{jk}^*) < x_{ik}^*$, then
   a. set $x_{ik}^* = x_{ij}^* + x_{jk}^*$;
   b. set $v_{ik}^* = v_{ij}^* v_{jk}^*$.
7. Set $k = k + 1$. If $k \leq n$, go to 6.
8. Set $i = i + 1$. If $i \leq n$, go to 4.
9. Set $j = j + 1$. If $j \leq n$, go to 3; else stop.

Example

Let the arcs of the digraph of Figure 6.1 be assigned weights according to the following scheme:

a	b	c	d	e	f	g	h
1.0	0.2	0.4	0.2	0.6	0.3	0.7	1.5

Then

$$X = \begin{bmatrix} \infty & 1.0 & \infty & 0.2 \\ \infty & \infty & 0.4 & \infty \\ \infty & \infty & 0.2 & 0.6 \\ 0.3 & 0.7 & 1.5 & \infty \end{bmatrix} \qquad V = \begin{bmatrix} \Lambda & a & \Lambda & b \\ \Lambda & \Lambda & c & \Lambda \\ \Lambda & \Lambda & d & e \\ f & g & h & \Lambda \end{bmatrix}$$

and A.6.4. stops with

$$X^* = \begin{bmatrix} 0.5 & 0.9 & 1.3 & 0.2 \\ 1.3 & 1.7 & 0.4 & 1.0 \\ 0.9 & 1.3 & 0.2 & 0.6 \\ 0.3 & 0.7 & 1.1 & 0.5 \end{bmatrix} \qquad V^* = \begin{bmatrix} bf & bg & bgc & b \\ cef & ceg & c & ce \\ ef & eg & d & e \\ f & g & gc & fb \end{bmatrix}.$$

There may be more than one shortest path $(i, \ldots, k)$ from node i to node k. Element v_{ik} specifies just one of the paths, but the algorithm can be modified to find all shortest paths (Exercise 6.5). Whereas A.6.4 still depends on manipulation of strings, the algorithm that follows is purely numerical.

ALGORITHM 6.5 Let X be a matrix of weights as in A.6.4. A matrix M is constructed as follows:

$$\begin{aligned} m_{ik} &= k, \quad \text{if } x_{ik} \neq \infty; \\ &= 0, \quad \text{if } x_{ik} = \infty. \end{aligned}$$

Element m_{ik} is replaced by m_{ij} whenever, in the course of computation of X^*, x_{ik}^* is set equal to $x_{ij}^* + x_{jk}^*$. At the end of the computation m_{ik} contains the name (identifying number) of the first intermediate node on a shortest path from i to k, e.g., if this path is $(i, t, \ldots, k)$, then $m_{ik} = t$. Moreover, m_{tk} contains the first intermediate node on a shortest path from t to k, and so on. Hence a shortest path from i to k is defined in a reasonably explicit fashion. The `RINF` of our subroutine has the same purpose as in Example 1 of A.6.2.

```
      SUBROUTINE SHORTP (X,XSTAR,NODES,N,RINF)
      DIMENSION X(N,N), XSTAR(N,N), NODES(N,N)
      DO  5  I = 1,N
      DO  5  J = 1,N
      XSTAR(I,J) = X(I,J)
      NODES(I,J) = 0
    5 IF (X(I,J).NE.RINF) NODES(I,J) = J
      DO 15  J = 1,N
      DO 15  I = 1,N
      IF (XSTAR(I,J).EQ.RINF) GO TO 15
      DO 10  K = 1,N
      T = XSTAR(I,J) + XSTAR(J,K)
      IF (T.GE.XSTAR(I,K)) GO TO 10
      XSTAR(I,K) = T
      NODES(I,K) = NODES(I,J)
   10 CONTINUE
   15 CONTINUE
      RETURN
      END
```

Example

Take the digraph of Figure 6.1, with matrix X as in the example of A.6.4. Then XSTAR is the X^* of that example, and

$$\text{NODES} = \begin{bmatrix} 4 & 4 & 4 & 4 \\ 3 & 3 & 3 & 3 \\ 4 & 4 & 3 & 4 \\ 1 & 2 & 2 & 1 \end{bmatrix}$$

on exit from SHORTP.

A traveler wishing to go from place a to place b is not interested in the shortest path from c to d, unless c and d happen to lie on his shortest route from a to b. A.6.5 is an uneconomical procedure for solving the traveler's problem, and we now give two procedures that are more efficient than the general algorithm. The first finds shortest paths from a given node to every other node that is reachable from this node; the second finds just the one shortest path from the given node to another given node.

ALGORITHM 6.6 (Dantzig's algorithm) Given a node a_1 in a digraph $\langle A, R\rangle$. Let A_n be the set of all nodes reachable from a_1. For each $a_j \in A_n$ the algorithm finds the shortest path $(a_1, \ldots, a_j)$ by iteration. Let $c(a_j)$ be the length of this shortest path. We denote by A_i the set of all nodes a_k for which $c(a_k)$ is known at the start of the ith iteration. The process starts with $A_1 = \{a_1\}$ and finds in turn $A_2, \ldots, A_n$. Note that $A_1 \subset A_2 \subset \cdots \subset A_n$.

1. Set $A_1 = \{a_1\}$, $c(a_1) = 0$, $i = 1$. Construct a digraph T consisting of a single node a_1.
2. For every arc $\langle a_k, a_t \rangle$ such that $a_k \in A_i$ and $a_t \notin A_i$ compute $d_{kt} = c(a_k) + w_{kt}$, where w_{kt} is the (positive) weight of the arc.
3. Set $A_{i+1} = A_i$ and determine $d = min\ d_{kt}$. For every arc $\langle a_k, a_t \rangle$ for which $d_{kt} = d$ (there may be more than one such arc):
   a. Set $c(a_t) = d$.
   b. Set $A_{i+1} = A_{i+1} \cup \{a_t\}$.
   c. In T draw the arc $\langle a_k, a_t \rangle$. (T already contains node a_k; node a_t may also be already in T.)
4. If $od(A_{i+1}) = 0$, stop; else set $i = i + 1$ and go to 2.

Digraph T of A.6.6 is acyclic. If all shortest paths $(a_1, \ldots, a_j)$ are unique, it is, moreover, a directed tree. Every path in T corresponds to a shortest path in the digraph $\langle A, R \rangle$.

Example

We apply the algorithm to the digraph of Figure 6.2.

$i = 1$. $A_1 = \{1\}$, $c(1) = 0.0$.
$d_{12} = 0.0 + 0.1 = 0.1$,
$d_{13} = 0.0 + 0.7 = 0.7$.
$d = 0.1$, $c(2) = 0.1$, $A_2 = \{1, 2\}$.

$i = 2$. $d_{13} = 0.7$,
$d_{24} = 0.1 + 0.2 = 0.3$,
$d_{25} = 0.1 + 0.2 = 0.3$.
$d = 0.3$, $c(4) = c(5) = 0.3$, $A_3 = \{1, 2, 4, 5\}$.

$i = 3$. $d_{13} = 0.7$,
$d_{46} = 0.3 + 0.8 = 1.1$,
$d_{47} = 0.3 + 0.4 = 0.7$,
$d_{53} = 0.3 + 0.5 = 0.8$,
$d_{57} = 0.3 + 0.5 = 0.8$.
$d = 0.7$, $c(3) = c(7) = 0.7$, $A_4 = \{1, 2, 3, 4, 5, 7\}$.

$i = 4$. $d_{46} = 1.1$,
$d_{76} = 0.7 + 0.3 = 1.0$,
$d_{78} = 0.7 + 0.1 = 0.8$.
$d = 0.8$, $c(8) = 0.8$, $A_5 = \{1, 2, 3, 4, 5, 7, 8\}$.

$i = 5$. $d_{46} = 1.1$,
$d_{76} = 1.0$.
$d = 1.0$, $c(6) = 1.0$, $A_6 = A$.

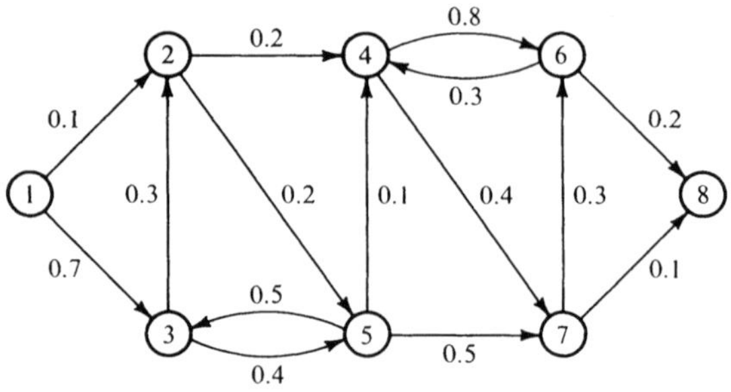

Figure 6.2

$c(1) = 0.0$
$c(2) = 0.1$
$c(3) = 0.7$
$c(4) = 0.3$
$c(5) = 0.3$
$c(6) = 1.0$
$c(7) = 0.7$
$c(8) = 0.8$

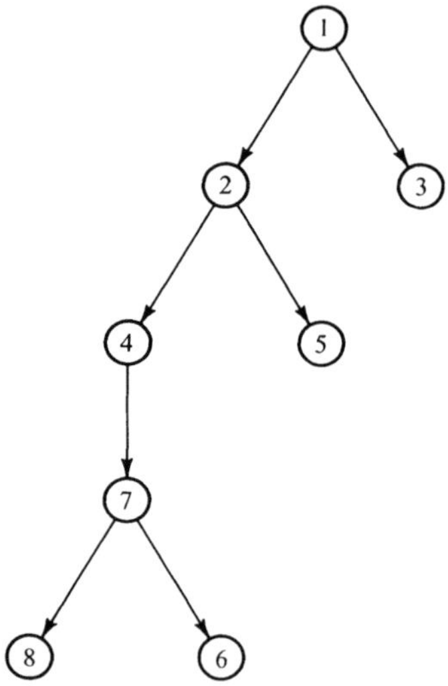

Figure 6.3

The results are displayed in Figure 6.3. The tree is a partial subdigraph of the digraph of Figure 6.2.

If a digraph contains n nodes, then, in the ith iteration of A.6.6, at most $i(n - i)$ values of d_{kt} have to be searched through for a minimum; i.e., there are at most $i(n - i) - 1$ comparisons. The greatest total number of comparisons that can arise (for a complete symmetric digraph) is therefore $\sum_{i=1}^{n-1} [i(n - i) - 1]$, and it can be shown that this is approximately equal to $n^3/6$. By contrast, the number of comparisons in A.6.5 is n^3 in the worst case.

One application of A.6.6 is in the design of a distribution network for some utility (gas, water, etc.). If node a_1 stands for a supply station, and nodes $a_2, a_3, \ldots$ represent localities to which the utility is to be supplied, then A.6.6 can be used to produce the most economical layout of supply lines. In this application digraph T must be a tree. The algorithm should therefore be slightly modified so that it finds only one shortest path from a_1 to a node a_j when more than one shortest path to a_j exists.

ALGORITHM 6.7 Given nodes a_1 and a_n, the following algorithm finds a shortest path $(a_1, \ldots, a_n)$: Apply Dantzig's algorithm in the forward direction from a_1, but simultaneously apply a similar procedure in a backward direction from a_n, computing lengths of shortest paths $c'(a_j)$ from nodes a_j to node a_n. When $c(a_m)$ and $c'(a_m)$ have both been found for some node a_m, then we have a "forward" set of nodes A, and a "backward" set A'. The shortest path is the path $(a_1, \ldots, a_i, a_j, \ldots, a_n)$, where $a_i \in A$ and $a_j \in A'$ (possibly $a_i = a_j = a_m$), for which $c(a_i) + c'(a_j) + d_{ij}$ is a minimum.

Example

Use A.6.7 to find the shortest path from node 1 to node 8 in the digraph of Figure 6.2. The algorithm proceeds as follows.

$i = 1$. Forward computations as in the example of A.6.6.
$A'_1 = \{8\}$, $c'(8) = 0.0$.
$d'_{68} = 0.2 + 0.0 = 0.2$,
$d'_{78} = 0.1 + 0.0 = 0.1$.
$d' = 0.1$, $c'(7) = 0.1$, $A'_2 = \{7, 8\}$.

$i = 2$. Forward computations as in the example of A.6.6.
$d'_{68} = 0.2$,
$d'_{47} = 0.4 + 0.1 = 0.5$,
$d'_{57} = 0.5 + 0.1 = 0.6$.
$d' = 0.2$, $c'(6) = 0.2$, $A'_3 = \{6, 7, 8\}$.

$i = 3$. Forward computations as in the example of A.6.6 give $c(7)$, and $c'(7)$ is already known. We have $A = \{1, 2, 3, 4, 5, 7\}$, $A' = \{6, 7, 8\}$. The shortest path is $(1, 2, 4, 7, 8)$, of length $c(7) + c'(7) = 0.8$.

Another interesting problem is that of finding a *longest* simple path from one node to another. Although the problem appears to be related to the shortest path problem, Warshall's algorithm cannot be used to solve it in the general case. The minimization process of the algorithm for shortest paths necessarily produces simple paths. Suppose now that a try is being made to derive a longest path by maximation. If there exists a simple path $(a, \ldots, c, \ldots, b)$, but there is a cycle through c, then the "length" of the path $(a, \ldots, b)$ can be made arbitrarily large by going round and round the cycle (this is the reason why we have not formally defined the length of a nonsimple path, see D.3.7). Another difficulty is this: In the shortest path problem, if $(a, c, \ldots, b)$ is a shortest path from a to b, then the subpath $(c, \ldots, b)$ is a shortest path from c to b, a property on which Warshall's algorithm is based; if, on the other hand, $(a, c, \ldots, b)$ is a longest simple path from a to b, there may well exist a simple path $(c, a, \ldots, b)$ that is longer than the subpath $(c, \ldots, b)$ of $(a, c, \ldots, b)$.

Acyclic digraphs provide an exception. In matrix X of A.6.5 make all weights greater than zero, letting $x_{ij} = 0$ indicate nonexistence of arc $\langle i, j\rangle$. Define semigroups $\langle R_+, max, 0\rangle$ and $\langle R_+, *\rangle$, where R_+ is the set of non-negative real numbers and operation $*$ is defined as follows: $a * b = a + b$ if neither a nor b is zero, but $a * 0 = 0 * a = 0$. We can find no identity element for $\langle R_+, *\rangle$, i.e., R_+ cannot be a semiring. This is a consequence of the difficulties mentioned in the paragraph above. But, if the digraph with adjacency matrix X is acyclic, then, with operations $+$ and $\cdot$ interpreted as *max* and $*$, respectively, A.6.2 gives lengths of longest paths, and a program similar to A.6.5 can be used to find these paths (Exercise 6.9).

6b. Cycles

A.6.1 is an algorithm for detecting the presence of cycles in a digraph: Node i lies on a cycle if $x_{ii}^* = 1$ in the path matrix. Computation of an $n \times n$ path matrix by A.6.1 may take as many as n^3 operations. A method can be derived from A.6.7 that may take somewhat less time than A.6.1 to decide for just one node a_i whether or not this node lies on a cycle. The forward and backward procedures of A.6.7 are both applied to a_i. If a node a_k can be found for which both $c(a_k)$ and $c'(a_k)$ exist, then a_i lies on a cycle. Moreover, the procedure defines explicitly an elementary cycle of least distance through a_i. Now, if every arc on the cycle is loaded with a very large weight, and the process is repeated, a different elementary cycle through a_i is found (unless there is just the one cycle). The process can be repeated until it generates a cycle that has been generated before. At this stage all elementary cycles through a_i have been found.

One application of the set of cycles through a given node arises with K-formulas. Let $\{k_1, k_2, \ldots, k_m\}$ be a minimal set of K-formulas of a digraph. The leading nodes of these K-formulas define a node base of the digraph. Let a_i be the leading node of k_i, and let A_i be the set of nodes that lie on cycles through a_i (we assume that a_i lies on a cycle of zero length in any case, i.e., that A_i contains at least a_i). Set A_i can be generated by taking the subdigraph defined by k_i, and applying the procedure based on A.6.7 to leading node a_i. Let us generate $A_1, A_2, \ldots, A_m$. These sets define all node bases of the digraph. The number of different node bases is $\prod_{i=1}^{m} |A_i|$, and a node base is found by taking exactly one node from each A_i.

We shall now show that cycles can be found by operations on K-formulas themselves. First we discuss how they are related to K-formulas.

THEOREM 6.2 If C is a cycle in a digraph D, and K is a minimal set of K-formulas of D, then the K-formulas $*b_ib_j$ for all arcs $\langle b_i, b_j\rangle$ in C have been absorbed in just the one member of K.

Proof. The theorem follows from the requirement of Step 3 in A.3.4 that every substitution must give rise to the longest possible K-formula. One of the $*b_ib_j$ must be, of course, the first to be absorbed. If it is absorbed by K-formula k_i, then k_i continues to be this longest possible K-formula for at least as long as there remain K-formulas of arcs in C still to be absorbed.

THEOREM 6.3 Let k_i be a member of a minimal set of K-formulas of digraph D. Let C be a cycle in D, and assume that k_i has absorbed the K-formulas of all members of C. Then k_i contains at least two occurrences of the symbol of some node through which C passes.

Proof. Of all K-formulas of arcs in C, some K-formula $*ab$ is absorbed last. Since $\langle a, b\rangle$ lies on a cycle, there are also arcs $\langle \ldots, a\rangle$ and $\langle b, \ldots\rangle$ in C, and their K-formulas have already been absorbed. This means that k_i contains symbols a and b. Now, the absorption of $*ab$ is achieved in one of two ways. If b is the leading node of k_i at this moment, then the current k_i can be absorbed into $*ab$, and k_i contains at least two occurrences of a thereafter. Alternatively, $*ab$ can be substituted for an occurrence of a in k_i. Then k_i contains at least two occurrences of b.

For what follows we introduce the notion of levels of K-formulas. A minimal set of K-formulas consists of K-formulas on level 0. If a K-formula k_i in this set can be written as $** \cdots *a_i\alpha_1\alpha_2 \cdots \alpha_n$, where the α_j are again K-formulas, then the α_j are K-formulas on level 1. They are said to *derive* from k_i. If an α_j can in turn be expressed in terms of K-formulas, then these K-formulas are on level 2, and so forth. Consider the K-formula

$$***ac*b***c*db*edae,$$

which represents the digraph of Figure 6.4. Here the K-formulas on level 1 are $\alpha_1 = c$, $\alpha_2 = *b***c*db*eda$, $\alpha_3 = e$. The second level K-formula that derives from α_2 is $***c*db*eda$, and from this K-formula we can derive third level K-formulas, $*db$, $*ed$, a. The first two of these give rise to fourth level K-formulas b and d, respectively. The level structure is illustrated by the level tree of Figure 6.4.

THEOREM 6.4 If within a K-formula of a node a the symbol a occurs again, and the distance between the symbols cannot be reduced by applications of the switch rule, then the node symbols between the inclusive bounds of the two occurrences of a define a cycle.

Proof. Leading nodes of K-formulas on level $i + 1$ are all reachable from the leading node of the ith level K-formula from which the $(i + 1)$th level K-formulas derive. It follows that any path in a level tree represents a path in

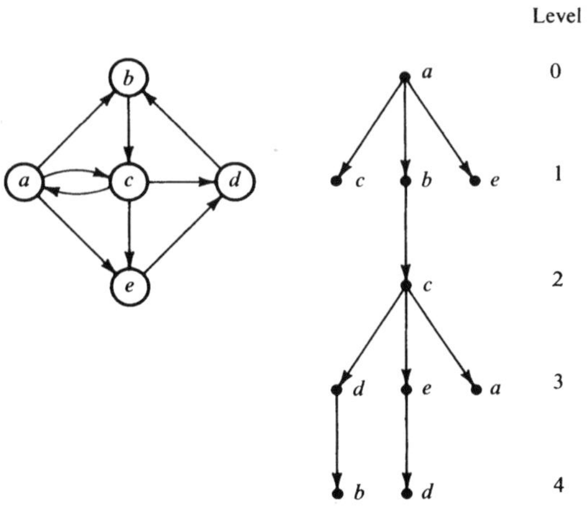

Figure 6.4

the corresponding digraph, and any path in the digraph that is composed of leading nodes of K-formulas on levels i, $i + 1$, $i + 2$, ..., where each of these K-formulas derives from its predecessor, is represented by a path in the level tree. In particular, if a K-formula of a contains another occurrence of a, then there is a path $(a, \ldots, a)$ in the level tree and, hence, in the digraph. This path is, of course, a cycle. An application of the switch rule to the K-formula corresponds in the level tree merely to a rearrangement of the arcs originating from a particular node. Paths in the tree are, however, independent of the order of arcs. This completes the proof. By way of explication we note that applications of the switch rule that bring the second a as close as possible to the first correspond in the level tree to a rearrangement of the subtree rooted at the node representing the first a in such a way that the path $(a, \ldots, a)$ becomes the leftmost path from the root of this subtree.

Examples

1. Consider ****ac*b***c*db*edae*. Substring **b***c*db*eda* is a K-formula of b in which there is a second occurrence of b. The second b cannot be switched any nearer to the first. Therefore (b, c, d, b) is a cycle in the digraph of Figure 6.4. The proof of Th.6.4 makes it obvious that there is a cycle through the leading node of a K-formula if the symbol of the node occurs again somewhere in the K-formula. This is the case with the a here. Two applications of the switch rule produce in turn ****a*b***c*db*edace* and ****a*b***ca*db*edce*. The latter tells us that (a, b, c, a) is a cycle.

2. Only one application of the switch rule is possible in **c**db*e*dc*. It produces **c**d*e*dcb*, and (c, d, e, d, c) is therefore a cycle. This cycle is not elementary, but it is easy to see that it is the union of elementary cycles

(c, d, c) and (d, e, d). Substrings $**d*e*dcb$ and $*d*e*dc$ are K-formulas, and cycle (d, e, d) can be derived directly from either of them.

No amount of switching in the K-formula of Example 1 of Th.6.4 can produce additional cycles. By inspection of the digraph we see, however, that it contains cycles (a, c, a), (a, e, d, b, c, a), and (b, c, e, d, b) as well. The more radical interchange rule has to be used to find these cycles. An application of the interchange rule corresponds in the level tree to the interchange of subtrees whose roots carry the same labels (i.e., the tree is transformed into a truly different tree). For example, in the tree of Figure 6.4, we can detach the subtree rooted at c, and reattach it to the terminal node labeled c. Existence of the cycle (a, c, a) can be deduced from the new tree (interchange transforms the K-formula to $***a***c*db*eda*bce$, and the switch rule then gives $***a***ca*db*ed*bce$). Cycle (a, e, d, b, c, a) is more difficult to detect. First one has to detach the arc originating from the node labeled e, and attach it to the terminal node that carries the label e. Then the arc originating from the node labeled d is transferred to the terminal node labeled d. Finally the subtree suspended from the node labeled b is attached to the terminal node labeled b. The order in which the three interchanges are carried out does not matter; in any case the result is a transformed tree containing a path through nodes labeled a, e, d, b, c, a.

THEOREM 6.5 Given a minimal set of K-formulas representing a digraph. For any elementary cycle in the digraph, application of switch and interchange rules to some K-formula belonging to the minimal set produces a sequence of node symbols that defines the cycle in accordance with Th.6.4.

Proof. By Th.6.2 only one K-formula of the minimal set has to be considered. Denote it by k. Th.6.3 tells us that the symbol of some node on the cycle occurs at least twice in k. Assume that this symbol is a, and write the cycle as $(a, b_1, b_2, \ldots, a)$. The K-formula $*ab_1$ has been absorbed in k. Node a is still the leading node of some K-formula in k, and so is b_1. If the K-formula of a is on level i, the one of b_1 is on level $i + 1$ and must derive from that of a. Therefore there is certainly a path corresponding to (a, b_1) in the level tree. Now assume that there is a path corresponding to $(a, b_1, \ldots, b_n)$ and show that this path can be made into one corresponding to $(a, b_1, \ldots, b_n, b_{n+1})$. K-formula $*b_nb_{n+1}$ of arc $\langle b_n, b_{n+1}\rangle$ has also been absorbed in k, and the same argument that establishes the existence of a path corresponding to (a, b_1) tells us that there is a path corresponding to (b_n, b_{n+1}) in the level tree. Now, if this path is not already a continuation of that corresponding to $(a, b_1, \ldots, b_n)$, it can be made such. At least two nodes then have labels b_n; one is the terminal node of the path corresponding to $(a, b_1, \ldots, b_n)$, another is the node from which the path corresponding to (b_n, b_{n+1}) originates. We interchange the subtrees suspended from these two nodes. Since an elementary cycle in a finite digraph has finite length, the process that builds up the path corresponding to $(a, b_1, \ldots, a)$ is finite. Each interchange in the level tree is accompanied by the appropriate interchange in k. Finally, after the path corresponding to the cycle has been built up, it is possible to transform the appropriate K-formula of a in k to the form required by Th.6.4 by a finite number of applications of the switch rule.

Th.6.5 tells us that a sequence of applications of switch and interchange rules is an algorithm for finding all cycles in a digraph, but does not specify the sequence. Construction of the algorithm is a rather difficult research topic. Recall that all simple paths can be found also by A.6.3. More research is needed even to decide whether an algorithm based on Th.6.5 would be superior to A.6.3.

The algorithm discussed next can definitely be superior to A.6.3 (as long as A.6.3 is used for the sole purpose of finding cycles). Assume that we are given the adjacency matrix A and the path matrix P of a digraph with n nodes. Now, if $a_{ik}p_{ki} = 1$, then there exist paths (i, k) and $(k, \ldots, i)$; i.e., there exists a cycle $(i, k, \ldots, i)$. If, however, $a_{ik}p_{ki} = 0$ for all $k = 1, 2, \ldots, n$, then node i does not lie on any cycle. In principle the algorithm makes use of the products $a_{ik}p_{kj}$ to construct something similar to the level trees of K-formulas. The process is initiated by drawing a node and labeling it 1. Then an arc is drawn for every i such that $a_{1i}p_{i1} = 1$ $(i = 1, 2, \ldots, n)$, and the terminal node of the arc is labeled i. For the purposes of the argument assume that $a_{1i}p_{i1} = 1$ for some i; i.e., assume that we have a nontrivial tree. Take any terminal node labeled k, say, where $k \neq 1$. Draw an arc from this node for every j such that $a_{kj}p_{j1} = 1$ $(j = 1, 2, \ldots, n)$, and assign label j to the terminal node of the arc. Note, however, that the arc is *not* drawn if $j \neq 1$ and j is already the label of some node on the path from the root of the tree to the node labeled k. Continue this process until all terminal nodes carry the label 1. There is now a path from the root of the tree to a terminal node for every elementary cycle through node 1 in the digraph. The cycles are defined by labels on nodes in the paths. Next generate submatrices A_2 and P_2 by deleting the first row and column of A and of P. An analogous procedure now finds cycles that go through node 2, but not through node 1. In the third stage delete the first two rows and columns of A and P, and use the resulting submatrices A_3 and P_3 to find cycles through node 3, etc. In effect, this is what A.6.8 does.

ALGORITHM 6.8 Algorithm for finding all cycles in a digraph, given the adjacency and path matrices of the digraph. The program, as written, can cope with a digraph having at most 50 nodes.

```
      SUBROUTINE CYCLES (A,P,N)
      LOGICAL A(N,N), P(N,N)
      LOGICAL USED(50)
      INTEGER ROOT, REACHJ(50), PATH(51)
      DO 35  ROOT = 1,N
C  INITIATE TREE
      DO  5  K = ROOT,N
      REACHJ(K) = ROOT
    5 USED(K) = .FALSE.
      LEVEL = 1
      PATH(1) = ROOT
      I = ROOT
C  TEST WHETHER PATH CAN BE EXTENDED
   10 JMIN = REACHJ(I)
      IF (JMIN.GT.N) GO TO 20
      DO 15  J = JMIN,N
   15 IF (A(I,J).AND.P(J,ROOT).AND..NOT.USED(J))
     X       GO TO 30
C  BACKTRACK IN TREE, RESETTING REACHJ AND USED
   20 REACHJ(I) = ROOT
   25 USED(I) = .FALSE.
      LEVEL = LEVEL - 1
      IF (LEVEL.EQ.0) GO TO 35
      I = PATH(LEVEL)
      GO TO 10
C  EXTEND PATH
   30 USED(J) = .TRUE.
      REACHJ(I) = J + 1
      LEVEL = LEVEL + 1
      PATH(LEVEL) = J
      I = J
      IF (J.NE.ROOT) GO TO 10
C  PRINT PATH
      WRITE (6,100) (PATH(NODES), NODES = 1,LEVEL)
  100 FORMAT (1H , 40I3 / 1H , 15X, 11I3)
      GO TO 25
C  END OF MAIN LOOP - REACHED WHEN LEVEL=0
   35 CONTINUE
      RETURN
      END
```

Example

Figure 6.5 shows a digraph, its adjacency matrix, and the trees constructed by the algorithm for nodes 1, 3, and 4. Since the first rows in A_2 and A_5 are zero, there are no trees for nodes 2 and 5. The trees for nodes 3 and 4 illustrate a phenomenon that we shall discuss in detail further on; it suffices to note here that the constructions do not represent cycles. Since the digraph contains a Hamiltonian cycle, $p_{ij} = 1$ for all $i, j = 1, 2, \ldots, n$. The program outputs the cycles in lexicographic order:

1 3 4 5 1

1 3 4 5 2 1

1 3 5 1

1 3 5 2 1

1 4 5 1

1 4 5 2 1

1 5 1

1 5 2 1.

Note that in the tree for node 1 the four subtrees rooted at the nodes labeled 5 are identical. So are the two subtrees rooted at the nodes labeled 4. Processing speed can be greatly improved if the program takes identity of subtrees into account. The improved program would be more complicated, and would require much additional storage.

One can become sidetracked in the second and subsequent stages of A.6.8. Assume that in the second stage of the process some $a_{ik}p_{k2}$ is equal to 1, but that $p_{k2} = 1$ by virtue of the existence of a path $(k, \ldots, j, 1, \ldots, 2)$ alone. In the tree this means that time is wasted on the construction of a path, with nodes labeled $2, \ldots, i, k, \ldots, j$, which, because it cannot be extended any further, does not give rise to a cycle. The trees for nodes 3 and 4 in Figure 6.5 illustrate this phenomenon.

The sidetracking is avoided if the process is carried out in reverse. If there has been no reason for computing path matrix P in advance, the backward process should be faster. Here A_k is the same submatrix as before, but P_k is the path matrix for the subdigraph on nodes $k, k+1, \ldots, n$ alone. We start with A_n and P_n and find the cycle through n in the subdigraph on the single node n; if it exists, it is the sling (n, n). Next take A_{n-1}, compute P_{n-1} from A_{n-1} and P_n, and find cycles through node $n-1$. In general, having found cycles through node $k+1$, take A_k, compute P_k from A_k and P_{k+1} (see Exercise 6.10), and find cycles through node k.

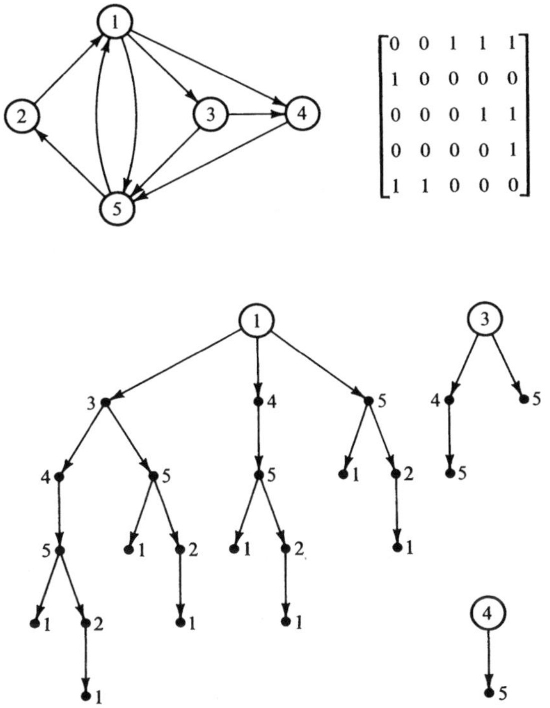

Figure 6.5

We end this section by noting that an algorithm for finding cycles can be used to find the longest elementary path from node a to node b. If the digraph does not contain arc $\langle b, a\rangle$ already, we add it to the digraph. Elementary cycles $(a, \ldots, b, a)$ then define elementary paths $(a, \ldots, b)$, and the longest of the cycles defines the longest elementary path from a to b.

6c. A Scheduling Problem

All but the most trivial production processes are composed of a number of separate activities. The activities are not independent, and in most practical cases their interdependence can be very complicated indeed. Perhaps the most effective visual aid to the understanding of a schedule of activities is a display of the activities and their relations in the form of a network. But the network of a complicated process can itself be very complicated. Mechanical procedures have therefore been developed for highlighting certain critical sections of the network. Proper managerial supervision of the project is then much easier to achieve. The information derived from the network by the analysis program enables management to allocate resources to a better effect.

The analysis techniques have become very important tools in operations research. Some very sophisticated systems have been developed: PERT (Program Evaluation Review Technique), CPA or CPM (Critical Path Analysis or Critical Path Method), RAMPS (Resource Allocation and Multi-Project Scheduling), and so forth. Here we can only introduce the basic terminology and some of the more fundamental principles.

Instead of choosing an impressive example and showing how to save a few million dollars, we shall consider a very simple task, the writing and mailing of a letter. The decomposition of this "production process" might be as follows:

A. Get paper and envelopes.
B. Get pen.
C. Write letter.
D. Address envelope.
E. Put letter into envelope and seal envelope.
F. Get stamp.
G. Affix stamp.
H. Mail letter.

Let us assume that three members of a family are involved in these activities. Mrs. S. is the instigator of the project, she has a letter to write. Since her writing is not as legible as her husband's, she has Mr. S. addressing the envelope. Mrs. S. also has trouble remembering where she last put her pen. Mrs. S. then assigns activities B and D to her husband, keeping activities A, C, and E for herself. Junior is called in for the more strenuous activities; he is made responsible for F, G, and H. Figure 6.6 shows the sequence of activities for the three persons involved.

Figure 6.6 shows some of the relations between activities, but it does not

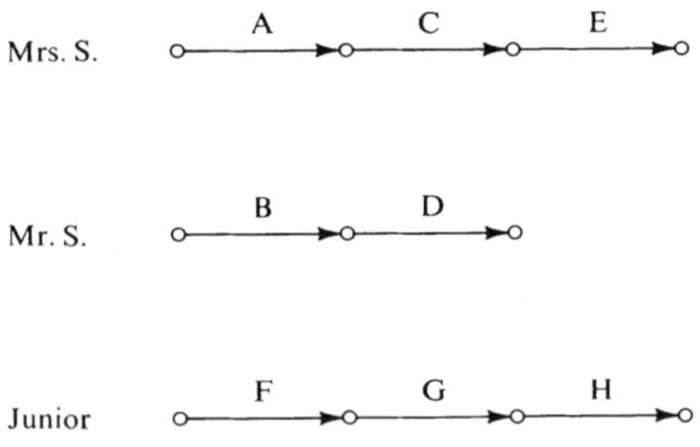

Figure 6.6

describe the process completely. The following relations are not made explicit:

(a) Activities A, B, and F can be concurrent.

(b) One of C or D must be completed before the other is started (assuming that there is only one pen). Assume that D precedes C.

(c) Activities A and B must both be completed before D can be started.

(d) Activity G should not be started before D is completed (if writing of the address is not successful at first try, and G has been completed, then a stamp is lost); activities E and G cannot be concurrent, but C and G can.

(e) Activity E must precede activity H.

The five drawings of Figure 6.7 show successive stages in the construction of the network. For example, Drawing (c) represents the network after Conditions (a), (b), and (c) have been incorporated. Dependence of the start of D on completion of A is indicated by a *dummy activity*, represented by the broken line. Another dummy activity is introduced in Stage (d); it indicates that G cannot start before D is finished. The convention adopted throughout

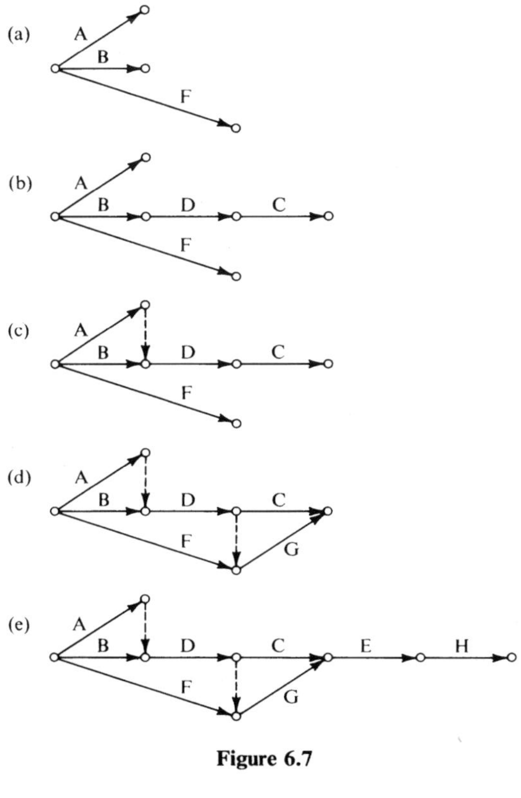

Figure 6.7

is that an activity represented by an arc originating from a node a cannot start before all activities represented by arcs terminating at a have been completed.

Let us now identify the nodes. Then we can dispense with activity symbols in the network itself (as long as we keep a reference list of arcs and the corresponding activities). Finally let us estimate durations of activities. The duration of an activity is an estimated time difference between the finish and the start of the activity. Dummy activities have zero durations. The duration of an activity is assigned as a weight to the arc representing the activity. In Figure 6.8, which shows the complete network, durations are given in minutes.

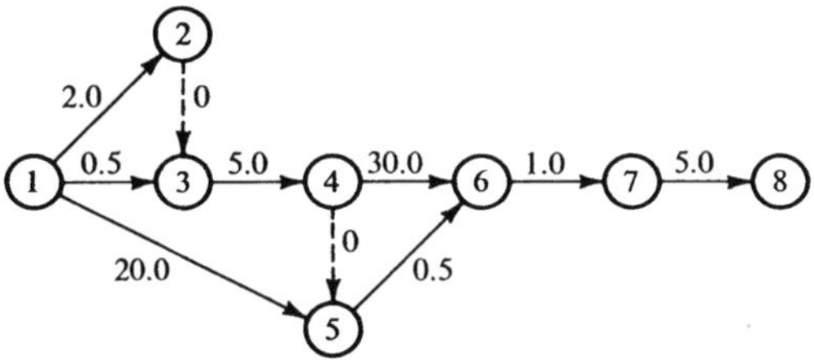

Figure 6.8

6d. Critical Path Scheduling

Let us now give a rigorous definition of what we mean by a scheduling network.

DEFINITION 6.1 A *scheduling network* is a 6-tuple $\langle E, e_s, e_t, A, D, W\rangle$, where

E is a set of numbered nodes—$E = \{1, 2, \ldots, n\}$,
e_s is the only node with zero indegree,
e_t is the only node with zero outdegree,
A is a subset of $E \times E$,
D is a set of values (weights),
W is a function $W: A \to D$,

and the digraph defined by $\langle E, A\rangle$ is acyclic. Members of E, A, and D are called, respectively, *events*, *activities*, and *durations*. If $W(i, j) = 0$, then $\langle i, j\rangle$ is a *dummy activity*. Events e_s and e_t are called *start* and *termination*, respectively. If, for every $\langle i, j\rangle \in A$, the relation $i < j$ holds, then the activities are said to be in *topological order*.

The network of Figure 6.8 is in topological order; a network constructed from the digraph of Figure 6.2 by removing arcs $\langle 5, 3\rangle$ and $\langle 6, 4\rangle$ is not.

The requirement that the node with zero indegree or zero outdegree be unique is not unduly restrictive. If there is more than one node having zero indegree to begin with, one simply selects one of these nodes for the start and joins the others to it by dummy activities. Multiple terminations are dealt with similarly. If a network is in topological order, then $e_s = 1$ and $e_t = n$.

Scheduling networks cannot contain cycles because of the interpretation of activities incident with an event; an activity originating from a node cannot start before all activities terminating at the node are completed. This is best seen in the case of a sling. Then we have the absurd situation that an activity cannot start before it is completed! In a small network cycles can be detected by visual inspection, but in a practical problem involving hundreds or even thousands of events and activities the network has to be checked for absence of cycles mechanically. Note that a scheduling network is acyclic if all activities are in topological order; i.e., an algorithm that orders the activities also detects cycles.

TABLE 6.1

PRECEDENCE TABLE FOR CONSTRUCTION OF FIGURE 6.7(e)

Activity	Immediate successors
A	D
B	D
C	E
D	C, G
E	H
F	G
G	E
H	–

Normally construction of the network proceeds from a *precedence table* of activities. Table 6.1 is a precedence table for the example of Section 6c. Generation of precedence tables is a highly skilled operation requiring intimate knowledge of the production process. In our example the choice of C as a successor of D is determined by knowledge of availability of machinery (only one pen) and manpower (overall performance is made more efficient if the addresser of the envelope is released to some other project as early as possible). The scheduling network itself does not contain all the information that has gone into its construction. Therefore, if a preliminary analysis of the network suggests that it should be rearranged, the rearrangements that are made are only partly determined by the results of the analysis. Whether or not one has to rearrange depends greatly on the skill with which the original table was constructed.

Construction of precedence tables and their possible rearrangement will remain a manual activity for some time to come. The data that have to be taken into account are often qualitative rather than quantitative, and they vary greatly in type. Our techniques for dealing with such data in a computer are as yet too crude to justify mechanization of this stage. The next stage, the drawing of the network with dummy activities included and events numbered, or, what is equivalent to actually drawing it, numbering of events and setting up of a table of arcs, can be programmed, but is still most often done manually. Assume that we have constructed a network as far as Stage (e) in the particular example of Figure 6.7. The manual algorithm A.6.9 numbers events in such a way that activities are in topological order.

ALGORITHM 6.9 (Fulkerson's rule) X is the set of nodes numbered at any particular stage in the application of the algorithm to an acyclic digraph.

1. Set $i = 1$.
2. If there is no unnumbered node with zero indegree, then go to 5.
3. Assign number i to an unnumbered node with zero indegree.
4. Set $i = i + 1$; go to 2.
5. If all nodes are numbered, restore all removed arcs and stop.
6. If $od(X) \neq 0$, remove all arcs originating from X and go to 2.
7. Error condition: Existence of an unnumbered node and $od(X) = 0$ imply existence of a cycle in the network.

DEFINITION 6.2 Let $\langle E, 1, n, A, D, W \rangle$ be a topologically ordered scheduling network. With each node $i \in E$ there are associated two times, the *earliest event time* $t^-(i)$ and the *latest event time* $t^+(i)$, defined as follows:

(a) $t^-(1) = 0$.
(b) $t^-(k) = max_{i \in E}[t^-(i) + W(i, k)], \qquad k \neq 1.$
(c) $t^+(n) = t^-(n)$.
(d) $t^+(i) = min_{k \in E}[t^+(k) - W(i, k)], \qquad i \neq n.$

Example

Table 6.2 is a list of earliest and latest event times for the network of Figure 6.8. Let us look at the computations for $t^-(5)$ and $t^+(1)$. $W(1, 5)$ and $W(4, 5)$ are defined in the case of $t^-(5)$. The maximum of $t^-(1) + W(1, 5) = 0 + 20.0$ and $t^-(4) + W(4, 5) = 7.0 + 0$ is 20.0. In the case of $t^+(1)$ we have $W(1, 2)$, $W(1, 3)$, and $W(1, 5)$ defined. The values of $t^+(k) - W(1, k)$ are, respectively, $2.0 - 2.0 = 0$, $2.0 - 0.5 = 1.5$, and $36.5 - 20.0 = 16.5$. The minimum is 0.

TABLE 6.2

EARLIEST AND LATEST EVENT TIMES FOR THE NETWORK OF FIGURE 6.8

Event	Earliest event time	Latest event time
1	0	0
2	2.0	2.0
3	2.0	2.0
4	7.0	7.0
5	20.0	36.5
6	37.0	37.0
7	38.0	38.0
8	43.0	43.0

DEFINITION 6.3 Let $\langle E, 1, n, A, D, W\rangle$ be a topologically ordered scheduling network. With each activity $\langle i, j\rangle \in A$ we associate a time, called the *float* of the activity, defined

$$float(i, j) = t^+(j) - t^-(i) - W(i, j).$$

An activity with zero float is called a *critical activity*, and a path $(1, \ldots, n)$ consisting entirely of critical activities is a *critical path*.

Example

Table 6.3 lists the floats of all activities in the network of Figure 6.8. The path (1, 2, 3, 4, 6, 7, 8) is critical. Every scheduling network contains at least one critical path; there may be more than one.

TABLE 6.3

FLOATS OF ACTIVITIES IN THE NETWORK OF FIGURE 6.8

Activity	Float
$\langle 1, 2\rangle$	0
$\langle 1, 3\rangle$	1.5
$\langle 1, 5\rangle$	16.5
$\langle 2, 3\rangle$	0
$\langle 3, 4\rangle$	0
$\langle 4, 5\rangle$	29.5
$\langle 4, 6\rangle$	0
$\langle 5, 6\rangle$	16.5
$\langle 6, 7\rangle$	0
$\langle 7, 8\rangle$	0

THEOREM 6.6 A critical path in a scheduling network is a longest path from start to termination.

Proof. Exercise 6.25.

We shall now examine the significance of critical path analysis. First we note that the total time to complete the project is 43.0 min in our example. Let us look at Event 5, the only event for which $t^- \neq t^+$. The earliest event time, 20.0 min, is the earliest time by which we *can* reach the event. The latest event time is the time by which we *may* reach the event without upsetting the schedule. There is, therefore, nothing gained by urging Junior to hurry up in getting to and from the post office. Indeed, Junior may as well have an ice cream on his way, and he may spend up to 16.5 minutes at the drug store. If, however, he stays there for 18 minutes, say, then the duration of activity $\langle 1, 5\rangle$ becomes 38 minutes, the new critical path is (1, 5, 6, 7, 8), and the project completion time jumps to 44.5 minutes. (Activities $\langle 1, 2\rangle$, $\langle 2, 3\rangle$, $\langle 3, 4\rangle$, $\langle 4, 6\rangle$ are then no longer on the critical path. If the duration of $\langle 1, 5\rangle$ is exactly 36.5 minutes, then both (1, 2, 3, 4, 6, 7, 8) and (1, 5, 6, 7, 8) are critical.)

But the purpose of critical path analysis is not to give Junior a reason for indulging. Rather it is to reduce the task completion time. If the letter has to be in the mailbox in less than 43 minutes, then critical path analysis tells us not to worry about activities $\langle 1, 3\rangle$, $\langle 1, 5\rangle$, and $\langle 5, 6\rangle$. The efficiency of the *critical* activities must be improved. In our example most activities are critical, so an attempt to improve the efficiency of every activity would not be too costly. But in practice, in a realistic large-scale network with perhaps 5000 activities, only 10%, perhaps, of the activities would be found on critical paths. By concentrating on these 500 activities, rather than all 5000, and reallocating resources so that the critical activities are completed in shorter time, the project completion time can be greatly reduced.

We must always remember that the initial durations are only estimates, and that the network has to be periodically reviewed. A drastic difference between an actual duration and its estimate may significantly change the pattern of critical paths.

The method is fairly new, dating from about 1957, but its success has been spectacular. It is thought that the Polaris missile program, which took five years to complete, would have taken seven years without critical path analysis.

Notes

A.6.1 was discovered independently by Roy [Ro59], and by Warshall [Wa62]. An Algol implementation can be found in [In 62]. Our proof of Th.6.1 is based on the proof in [Wa62]. More than one digraph may have the same path matrix: The inverse problem of finding the digraph of fewest arcs corresponding to a given path matrix is solved in [Si65, Ba69a, Mo69]. The generalization of Warshall's algorithm to Q-semirings is due to Robert and Ferland [Ro68], but Floyd discovered the particular application described in Example 1 of A.6.2 much earlier—[Fl62] is an Algol procedure for finding lengths of shortest paths. Floyd's algorithm remained unknown in operations research circles, and too many papers have since been devoted to the description and justification of less efficient methods. A decomposition strategy for finding shortest paths in very large digraphs is described in [Hu68].

A.6.3 is due to Ponstein [Po66]. A similar procedure, based on prime number coding, can be found in [Gi60]. A.6.5 is suggested by [Ro68]. A.6.6 is described in [Be62, Bu65]. The technique of A.6.7 is suggested in [Be62]. A similar procedure is described in [Ni66]; [Bo67d] is a corresponding Algol program. One procedure for finding longest paths in an acyclic digraph is described in [Ch66]. The longest path problem is closely related to the traveling salesman's problem, see [Ha62a, Pa64]. The traveling salesman's problem is surveyed in [Be68b]; we add [Ro66] to the bibliography of the survey.

A proof of Th.6.4 can be found in [Kr64], but our proof is new. A.6.8 is adapted from Figure 7 of [Fl67].

A nonannotated bibliography of the extensive literature on critical path methods can be found in [Bi62], supplemented by [Le66]. To a computer scientist publications that deal specifically with computers in critical path scheduling have greater relevance than the more general works. Of the latter we mention only a few: [Lo64] is an elementary introduction; its companion volume [Lo66] contains exercises; [Ba67b] is a more advanced work (its bibliography contains 197 items), our A.6.9 is adapted from it; Chapter 2 of [El66], a solid work on systems analysis, deals with scheduling networks. [Ka63] is a somewhat dated survey of the organization of computer programs used for analyzing activity networks. Algorithms for topological ordering are given in [La61, Ka62]. The RAMPS system is described in [Mo63], and the critical path algorithm built into RAMPS in [Kl67] (the program of [Kl67] accepts input that is not topologically ordered; another example of this is the program of [La65a]). [Mo67] contains a detailed description of internal representation of data in a program for very large problems. In [Ki66] a very interesting application of PERT to the design of a magnetic core memory is described. Automatic plotting of networks is described in [Ho67], and [Fi68] contains an algorithm for the automatic construction of the network from precedence relations in the set of activities. [Sc68] is another paper dealing with construction of networks. [Ho63b] deals with the somewhat similar problem of assembly line balancing (a Fortran program is included; the procedure is based on the adjacency matrix of a digraph representation of the assembly line).

Exercises

6.1 (a) Denote the digraph of Figure 6.9 by D. Write a program that finds the adjacency matrix of the cycle digraph of D. How many strong components does D have?

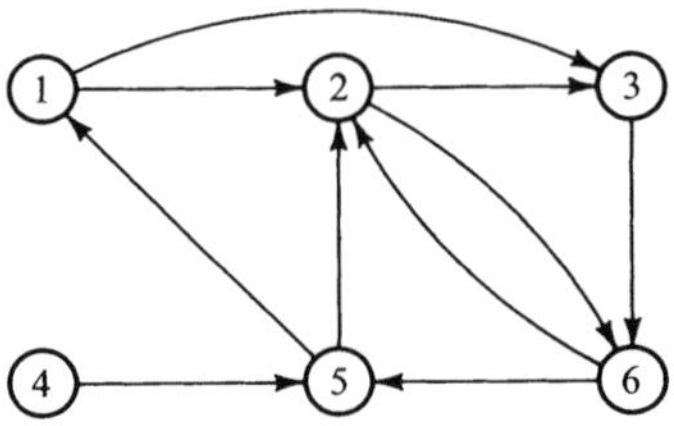

Figure 6.9

6.2 (a) The input to subroutine PATHLS of Example 1 of A.6.2 is

$$X = \begin{bmatrix} 10^{15} & 0.1 & 10^{15} \\ 10^{15} & 10^{15} & 0.1 \\ 0.1 & 10^{15} & 10^{15} \end{bmatrix}.$$

What is the output? Is the output consistent with D.3.9? If it is not, what changes should be made to the input to make the output consistent with D.3.9?

6.3 (a) In A.6.1 interchange iteration indices j and k so that Steps 2, 4, and 6 become

2. Set $k = 1$.
4. If $x_{ij}^* = 1$, then set $x_{ik}^* = x_{ik}^* \vee x_{jk}^*$ for all j from 1 to n.
6. Set $k = k + 1$. If $k \leq n$, go to 3; else stop.

What effect does the interchange have on the algorithm? What effect would interchange of j and i have?

6.4 (a) Define a variable adjacency matrix for the digraph of Figure 6.9 and apply A.6.3 to it. (If you are familiar with a programming language that has string processing capabilities, then you should use a computer for this. If not, then the drudgery of this exercise should provide incentive to learn such a language.)

6.5 (a) Modify A.6.4 so that it finds all shortest paths.

6.6 (a) Given matrix `NODES`, as generated by A.6.5, write a subroutine that returns the shortest path (`I`, ..., `J`) for specified `I` and `J` as a vector of node numbers.

6.7 (a) Express A.6.6 as a Fortran subroutine. Test the subroutine on the digraph of Figure 6.2.

6.8 (a) Dantzig's algorithm does not in general give the shortest paths from a given node when some of the weights are negative. Try to find a reason for this. One would expect that a maximization procedure analogous to Dantzig's algorithm would find longest paths from a given node in an acyclic digraph. This same reason makes it impossible to do so. Investigate in the acyclic digraph that results when arcs $\langle 5, 3\rangle$ and $\langle 6, 4\rangle$ are removed from the digraph of Figure 6.2.

6.9 (a) Write a program similar to A.6.5 to define longest paths in an acyclic digraph. Use this program and the program of Exercise 6.6 to find the longest path from node 1 to node 8 in the digraph of Figure 6.2 with arcs $\langle 5, 3\rangle$ and $\langle 6, 4\rangle$ removed.

6.10 (a) Let $D = \langle A_n, R\rangle$, where $A_n = \{a_1, a_2, \ldots, a_n\}$, be a digraph with adjacency matrix X, and let $A_k = \{a_1, a_2, \ldots, a_k\}$ ($A_k \subset A_n$). Given X, and the path matrix of the subdigraph $\langle A_k, R \cap (A_k \times A_k)\rangle$, find the path matrix of $\langle A_{k+1}, R \cap (A_{k+1} \times A_{k+1})\rangle$.

6.11 (a) Construct a weighted digraph that represents the maze of Figure 6.10 and use A.6.7 to find the shortest path from A to B.

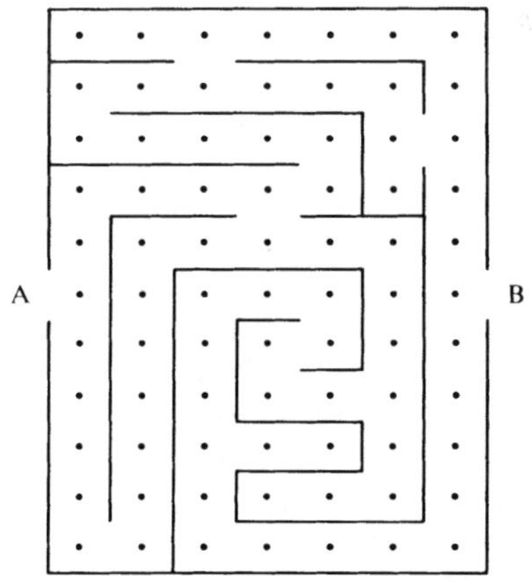

(The distance between adjacent dots is 1 unit)

Figure 6.10

6.12 (b) The number of cycles in a digraph can be quite large. Evaluate

$$\sum_{i=2}^{n} (\sum_{j=1}^{n} (\prod_{k=1}^{i-1} (j-k)))$$

for $n = 5$, and show that this expression gives the number of elementary cycles (excluding slings) in a complete symmetric digraph on n nodes.

6.13 (b) Draw level trees of the following K-formulas:

(i) $***a*b*c*d***eabdcd$,
(ii) $*1**2*64*3**425$,
(iii) $***a***bb***ebdedce$.

Is there a simple way of determining the number of levels in a level tree from the form of its K-formula?

6.14 (b) Draw the digraph corresponding to the K-formula of Part (i) of Exercise 6.13. Find all cycles in the digraph by inspection, and then, for every cycle, transform the K-formula by means of switch and interchange rules until a sequence of node symbols in the K-formula defines the cycle in accordance with Th.6.4.

6.15 (b) Write a program for finding cycles in a digraph by the "backward" variant of the procedure of A.6.8.

6.16 (b) Use A.6.8 to find the longest elementary path from node 1 to node 8 in the digraph of Figure 6.2 (disregard weights, i.e., take all weights to be equal to 1).

6.17 (c) A production process comprises activities *a*, *b*, *c*, and *d*. Activity *c* cannot start before *a* is completed, and *d* cannot start before *a* and *b* are both completed. Give a graphical representation of the process.

6.18 (c) Smith, Jones, and Brown are gathering information separately at three production plants; it takes them 8, 6, and 7 days, respectively, to complete this task. Smith and Brown pass their data to Jones, who then proceeds to work out a feasible production schedule for all three plants. It takes Jones 6 days to draw up the schedule. Meanwhile Brown is contacting suppliers and collating data on availability of supplies; this takes 10 days. Both the production schedule and the report on supplies are needed by Jones to work out delivery dates, and by Brown to advise suppliers of the dates on which supplies should reach the production plants. These activities take 2 and 3 days, respectively. Jones hands his schedule of deliveries to Smith, who then organizes transport, taking 3 days to do so. The planning is completed when Smith and Brown have completed their respective tasks. Give a graphical representation of this planning process.

6.19 (d) Write a program that checks, given the adjacency matrix of a digraph, that there is exactly one node with zero indegree and exactly one node with zero outdegree in the digraph. (In CPA parlance the program checks for "holes" in the activity network.)

6.20 (d) Find the critical path or paths in the network of Figure 6.11.

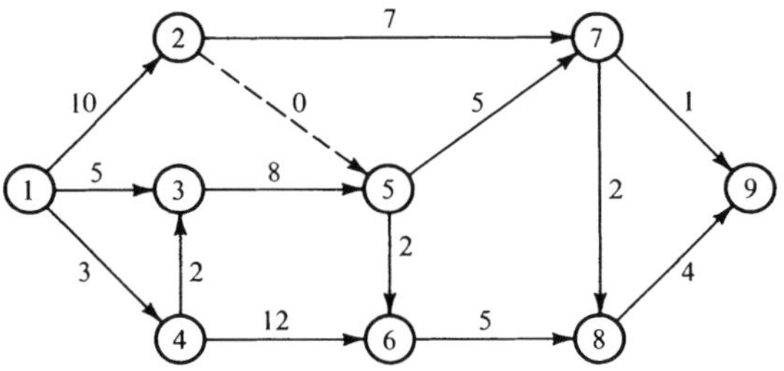

Figure 6.11

6.21 (d) In what respects does the network of Figure 6.11 change when durations of activities $\langle 4, 6\rangle$ and $\langle 5, 6\rangle$ are halved?

6.22 (d) Find the critical path or paths in the network of Exercise 6.18.

6.23 (d) It is discovered that Smith is supposed to take a holiday after he finishes gathering information at his plant and before he starts organizing transport. Amend the network of Exercise 6.18 so that it contains Smith's holiday as a new activity. What is the greatest number of days that Smith can be away without affecting the completion time of the planning process?

6.24 (d) Assign the n numbers 1, 2, ..., n to nodes in a network. Write a program based on A.6.9 that takes the set of arcs in the network for its input and renumbers the nodes in such a way that the arcs are topologically ordered.

6.25 (d) Prove Th.6.6.

6.26 (d) A network has been checked and found to contain no holes (see Exercise 6.19) and no cycles. The program of Exercise 6.9 is used to find a longest path $(1, \ldots, n)$. How can one tell whether this longest path is a unique longest path from 1 to n?

6.27 (d) Use the matrix of lengths and the matrix of nodes generated when the procedure of Exercise 6.9 is applied to an activity network to find *all* critical paths in the network.

6.28 (d) In what respects is A.3.3 superior to A.6.9 for detecting cycles in a network?

CHAPTER 7

Digraphs of Programs

7a. Flowchart Digraphs

Normally an algorithm comprises a preamble, which defines the objects that are to be operated on, and a list of operations, presented in such a manner that the order of their execution is unambiguously prescribed. A computer program that performs the way one wants it to perform is, of course, an algorithm. In a program one finds three types of statements. Declarations are statements of the first type. They are the preamble. Statements of the second type have to do with the actual computation, and statements of the third type effect transfers of control.

Every statement, regardless of type, is just a sequence of symbols, and normally one defines a computer program as a string or a sequence of strings, well formed according to a set of rules, which constitute the syntax of the language in which the program is written. Syntax alone, however, leaves a language unintelligible, and semantic rules must be added to turn a language into an effective medium of communication. In the case of a programming language some of the semantic rules deal with the sequence in which execution of the program is to proceed. They give an interpretation to statements of Type 3, but, since these statements look very much like statements of the other types, it is not easy to visualize the flow of control in a program by just looking at its listing.

If a grid of transfers of control specifying the execution sequence is superimposed on a program, our understanding of the program is greatly aided. Flowcharting provides the grid, but a flowchart is both a digraph of flowlines and a specification of the statements of the program (except Type 1 statements). In the next two sections we shall be interested in the digraph alone. Therefore we want a definition of a program in which the digraph is considered part of the program, but abstraction of the digraph is made easy.

DEFINITION 7.1 A computer program is a quintuple $\langle S, A, E, T, N\rangle$, where $S = \{s_1, s_2, \ldots, s_n\}$ is a nonempty set of executable statements, N is a set of declarative statements (possibly empty), $E \subseteq S$ is a nonempty set of entry points, $T \subseteq S$ is a nonempty set of terminal statements, and A is a square matrix of order n satisfying the following conditions:

(a) $a_{ij} = 1$ if control may pass from s_i to s_j;
$a_{ij} = 0$ otherwise.

(b) Column i of A is zero if and only if $s_i \in E$; row j of A is zero if and only if $s_j \in T$.

(c) For every element s_i in $(S - E)$ there exists a sequence of non-zero elements of A, $a_{k_1k_2}, a_{k_2k_3}, \ldots, a_{k_uk_{u+1}}$, such that $s_{k_1} \in E$, $s_{k_{u+1}} \in T$, and one of $k_2, k_3, \ldots, k_{u+1}$ is i.

At first sight this looks like a case of excessive formalization for its own sake, but the appearance changes when we interpret the s_i as nodes in a digraph having A for its adjacency matrix. By taking A on its own we have abstracted the digraph of possible transfers of control. We call this the *flowchart digraph* of the program. In terms of the flowchart digraph, Condition (b) of D.7.1 becomes $E = \{s_i \mid id(s_i) = 0\}$ and $T = \{s_i \mid od(s_i) = 0\}$, and Condition (c) states that every executable statement in the program must be reachable from an entry point and that a terminal statement must in turn be reachable from this statement.

Let us now look at the forms particular Fortran statements take in a flowchart digraph. Formats, dimension statements, and other nonexecutable statements are ignored in the construction of a flowchart digraph. Every executable statement must be given an identification number. We shall find it necessary to introduce auxiliary nodes; i.e., a flow chart digraph will contain more nodes than there are statements in the corresponding program. Figure 7.1 shows subdigraphs of the program segments that follow. The 5, 6, 5a, SUB5, etc. in Figure 7.1 are unique identifying names of nodes. They are not labels in the sense of labels in labeled digraphs.

(a)
```
(5) X = T
(6) T = Y
```

(b)
```
(5) IF (X.NE.Y) X = Y
(6) ...
```

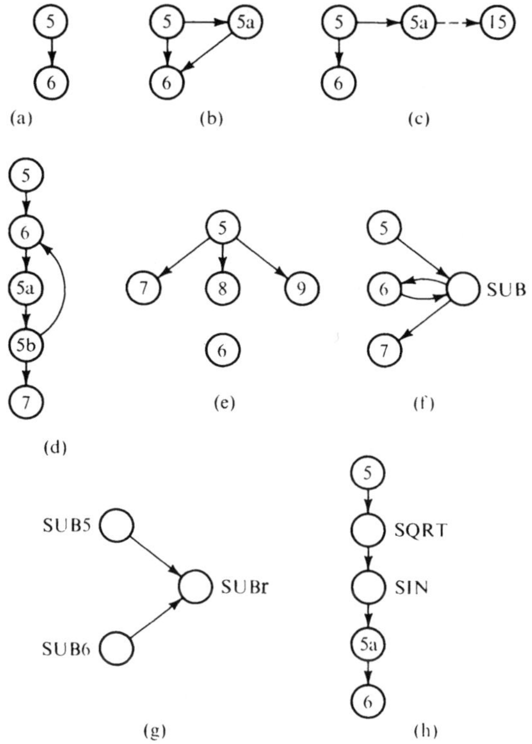

Figure 7.1

Here an auxiliary node 5a has to be introduced. It corresponds to the statement `X = Y`.

(c)

```
(5) IF (X.NE.Y) GO TO 15
(6) ...
```

The `GO TO 15` is an executable statement. Therefore, for the sake of consistency, we have introduced the auxiliary node 5a, although here we could have drawn the direct pointer ⟨5, 15⟩ instead.

(d)

```
(5) DO  6  K = 1,N
 6  A(K) = 0
(7) ...
```

Two auxiliary nodes are required. It does not matter whether they are named 5a and 5b, or 6a and 6b, or something else altogether. The first stands for the incrementation of `K`, the second for the test of `K` against `N`.

(e)

```
(5) GO TO (7,7,8,9), J
 6  ...
```

Node 6 is isolated only as far as our subdigraph is concerned. There must be some statement transferring control into 6, and some way out of 6 as well.

(f)

```
(5)  CALL SUB (A,B)
(6)  CALL SUB (C,D)
(7)  ...
```

A decision has to be made whether to consider every routine as a separate program or to draw a single program flowchart for the entire program comprising a main routine and its subprograms. Since one is not normally interested in the structure of library routines, the former approach appears to be more sensible. We introduce an auxiliary node carrying the name of the subprogram. Then, if the structure of the subprogram is of interest, its flowchart digraph is produced, but separately, and we assume that this digraph *can* be substituted in place of the auxiliary node.

(g) Assume that in subroutine SUB we have the program segment

```
(5)  RETURN
       :
(6)  RETURN
```

Note now that there are only single arcs ⟨SUB, 6⟩ and ⟨SUB, 7⟩ in digraph (f). Therefore, to permit substitution of digraph SUB for the node SUB in digraph (f), we must have a single terminal node in SUB, and we introduce the auxiliary collective return node SUBr. Then, if we make the substitution, ⟨SUB, 6⟩ and ⟨SUB, 7⟩ are replaced by ⟨SUBr, 6⟩ and ⟨SUBr, 7⟩, respectively.

(h)

```
  5    X = SIN(SQRT(X)) + Y
(6)    ...
```

A node named 5 has to precede nodes that stand for function subprograms in case some GO TO statement in the program transfers control to 5. Node 5a represents the computation steps that remain to be done after the function calls.

7b. Detection of Programming Errors

Some programming errors and irregularities can be detected in the flowchart digraph on its own. Others require the flowchart digraph to be used in conjunction with the program or, at least, to have a naming convention for the nodes that assigns particular significance to certain nodes in the flowchart digraph.

Discontinuities can be detected with great ease. Consider the program segment

```
      GO TO 15
  20  ...
```

where the program contains no Type 3 statement giving access to statement 20. By D.7.1 we require that every executable statement be reachable from some entry node. Let us first find the path matrix P of the program digraph (as a logical array). Then, assuming that $E = \{1\}$ and that the digraph contains N nodes,

```
      I = 1
      DO 50  K = 2,N
      NOPATH(K-1) = 0
      IF (P(1,K)) GO TO 50
      NOPATH(I) = K
      I = I + 1
  50  CONTINUE
```

assembles in vector NOPATH numbers of nodes that are not reachable from the entry node. If, on exit from the loop, NOPATH(1) is zero, then every node is reachable from the entry node. An analogous procedure detects nodes from which no terminal statement can be reached.

A less frequent error condition arises when a Type 3 statement transfers control to a nonexecutable statement, e.g., a format statement. This type of error can be detected during construction of the flowchart digraph. If, for example, the Type 3 statement is

```
  64  GO TO (69,69,70,71), J
```

where 70 is the number of a format statement, and a list of numbers of executable statements has already been generated, then the setting up of arc $\langle 64, 70\rangle$ is prevented by 70 not being in the list. If, on the other hand, nodes are numbered while the flowchart digraph is being set up, then the arc is generated, but there is no arc originating from 70, and the error is detected in the search for nodes from which there is no path to a terminal node.

Next let us consider DO-loops. If node numbers corresponding to the incrementation step and the exit test are distinguished in some manner (e.g., nodes representing exit tests, and only those nodes, have numbers from a given range, say 100–199, assigned to them), then cycles corresponding to DO-loops can be distinguished from other cycles. Assume that cycle $(a_1, a_2, \ldots, a_1)$ represents a DO-loop, that b_k is the DO-statement and a_1 the statement following it, and that there exists an arc $\langle b_j, a_i \rangle$ $(\neq \langle b_k, a_1 \rangle)$ such that a_i is on the cycle, but b_j is not. The DO-loop is then entered from outside. Unless there exists another cycle $(a_1, \ldots, b_j, a_i, \ldots, a_1)$, in which case b_j represents a statement belonging to a legitimate extension of the DO-loop, this is an error condition.

Superfluous transfers, such as

```
      GO TO 43
         ⋮
43    GO TO 47
```

can be detected if names of nodes corresponding to Type 3 statements are distinguished. Then, if nodes a and b on a path (a, b, c) represent Type 3 statements, the program is simplified if arcs $\langle a, b \rangle$ and $\langle b, c \rangle$ are replaced by $\langle a, c \rangle$. The greatest significance of this example is that it shows the limited practical utility of simplification schemes. Transfers of control during execution of a program take very little time; search for the paths (a, b, c) uses a lot of it. The net result of this "simplification" is an increase in total processing time.

Detection of programming errors is an essential part of compilation. Although techniques that are not explicitly based on the digraph representation may in general be faster and use less storage than the digraph techniques, the latter would be the faster if a flowchart of the program were to be constructed as well. Then the flowchart digraph and a list of its cycles would have to be generated in any case, and the error detection techniques described above could be implemented at little additional cost.

7c. Segmentation of Programs

At times a computer program cannot be accommodated in its entirety in the main store of the computer. The program has to be segmented, and there arises the problem of selecting cutting points to minimize the number of segment interchanges between different levels of storage.

The segmentation problem is extremely difficult. Let the executable statements (and auxiliary statements) in a program be numbered consecutively $1, 2, \ldots, n$. If a cut is made between statemnets i and $i + 1$, the cut produces two segments, comprising, respectively, statements $1, 2, \ldots, i$ and $i + 1, \ldots, n$. In the simplest approach to the problem cuts are made at points in the program that are spanned by the least number of program loops.

Let us call the two segments A and B. In the flowchart digraph there may be more than one arc $\langle i, j \rangle$ such that the statements associated with i and j are in A and B, or in B and A. The natural representation of a segment is a subdigraph of the flowchart digraph defined on the set of nodes representing statements in the segment *and* nodes that do not belong to this set, but lie at the other end of arcs originating or terminating in the set. If the subdigraph is looked upon as a flowchart digraph in its own right, then these latter nodes are entry or terminal nodes of the digraph. Figure 7.2 illustrates segmentation of the following skeleton program, with the cut having been made between statements 9 and 10.

```
 1   IF (...) GO TO 15
 3   ... = ...
 4   ... = ...
 5   IF (...) GO TO 19
 7   DO 10 ...
 8   DO 10 ...
 9   ... = ...
10   ... = ...
15   ... = ...
16   ... = ...
17   IF (...) GO TO 4
19   STOP
```

A few terms have to be defined now. If two paths $(a, \ldots, b)$ in a flowchart digraph have only nodes a and b in common, then the two paths are *parallel.* If P and Q are paths from an entry node to a terminal node, P is simple and $P \subset Q$, and Q does not contain a path that is parallel to a subpath of P, then $(Q - P)$ is a *return path.* By Th.3.1 every cycle in a flowchart digraph contains a return path. The return path of a cycle corresponding to a DO-loop contains just one arc. Return paths of other cycles may contain more than one arc. Paths $(5, 7, \ldots, 19)$ and $(5, 6, 19)$ are parallel in the complete flowchart digraph of our example; return paths are $(12, 9)$, $(14, 8)$, and $(17, 18, 4)$. If $(a, \ldots, b, c)$ represents a return path, then $\langle b, c \rangle$ is the *final arc* of the return path.

The criterion that a cut should be made at a point spanned by the least number of loops becomes, in terms of the flowchart digraph, that the segmentation process should sever the least number of return paths. Let us take adjacency matrix C of the cycle digraph of a program (A.3.3). Then

$$L_i = \sum_{j=1}^{i} \sum_{k=i+1}^{n} c_{kj}$$

seems to define the number of return paths severed by a cut after statement i. This is not strictly true, as the following skeleton program shows.

```
1   DO   5 ...
2   GO TO 4
3   GO TO 5
4   GO TO 3
5   ... = ...
8   STOP
```

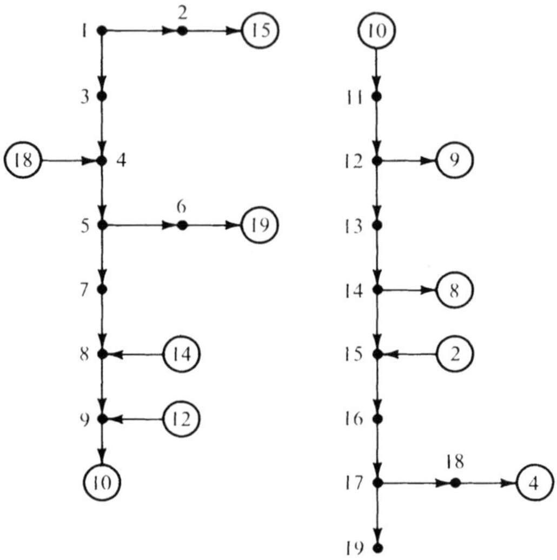

Figure 7.2

Here $L_3 = 2$, although there is only the one cycle (2, 4, 3, 5, 6, 7, 2). The L_i become reliable counts if the flowchart digraph is made acyclic by removing the final arc of every return path, the remaining arcs are put into topological order, and the removed arcs restored.

Even when the L_i give the true number of severed return paths there is the difficulty that one must have a knowledge of the iteration counts of the loops for optimal segmentation. These will rarely be available. Segment interchanges depend also on the location of data. Flow of data across a cut has as much importance as the flow of control. Significant work on this aspect of segmentation has barely started. Summarizing, we are quite far from having sound procedures for optimizing segmentation. It would appear that the selection of cutting points will have to be a dynamic process, the cuts being selected during execution of the program on the basis of information gathered while the program is being executed.

7d. Automatic Flowcharting

In trying to understand a program one is primarily interested in the flow of control. We have already seen that a listing of the program is of little help because one string of symbols looks to us very much like any other. In a flowchart sequencing of statements for execution is indicated by flowlines, and the statements themselves by boxes. A line is one dimensional, a box two dimensional, and this difference we can discern at a glance. The use of flowcharts as aids to the design of programs and for documentation has been with us since the very earliest days of electronic computing. Here we shall consider the use of the theory of digraphs in the automatic preparation of flowcharts from programs. The utility of automated documentation should be obvious. Flowcharts prepared while a program is still being debugged can be equally important—for drawing attention to unintended irregularities in the flow of control.

It is easy enough to write a procedure that encloses statements of a program in boxes, superimposes a grid of arbitrarily intersecting flowlines on the column of boxes, and writes *flowchart* at the bottom. To deserve this name, however, the drawing must enable one to trace the flow of control readily, and identify with ease points at which decisions are made. Therefore a flowchart must be two dimensional, with the order of statements of the original program rearranged where rearrangement simplifies the pattern of flowlines. In quite a normal program we may well find a hundred statements interposed between statements x and y, where statement x tells that control may next go to y. A two-dimensional disposition of the boxes allows boxes for statements that are executed one after the other to be placed near each other. For the sake of clarity the number of intersections of flowlines should be kept as low as possible.

Moreover, particularly if the flowchart is being generated as a documentation record to be filed, it must be separated into pages of standard size by making cuts at fixed intervals of 10.5 or 12.5 in., say. The product will hardly be acceptable unless the flowchart generator has already segmented the flowchart into pages of the required size. The paging requirement imposes a limit on the width of the flowchart as well (generally the width is limited in any case by the nature of the output device).

The quality of output of existing automatic flowchart generators varies greatly, and is far from perfect even with the better generators. This is probably due to the approaches to the problem having been largely *ad hoc*, the inherent lack of generality of such approaches preventing a continuous growth of knowledge. If a problem is formulated in sound mathematical terms, then an existing store of knowledge can be put to use, and partial solutions, instead of being isolated achievements, add to this store. The appropriate mathematical formulation of the flowcharting problem should proceed from some definition similar to our D.7.1.

The paging problem in flowchart generation is related to the segmentation problem for programs, but the two problems are distinct. In both cases one operates on a flowchart digraph. Ideally segments should be generated by cutting arcs that do not lie on cycles. Unfortunately this solution is inadequate when the segments turn out too large. A cut across a loop can have a very serious effect on the execution time of the program, and it is therefore necessary to exercise great care in selecting cuts.

In the preparation of a flowchart one's main concern is that the logical pattern of the program should be easy to follow, and it is as important to draw parallel paths originating from a decision node side by side on the same page as it is to avoid cutting loops. In the first instance the flowchart digraph has to be segmented in such a way that all arcs belonging to a cycle are confined to a single segment, and also all paths that are parallel to each other. Segments that are too small can be reassembled into page-sized segments without difficulty. Unfortunately, compared to an admissible program segment in most multilevel storage systems, a page of a flowchart, which cannot take more than 25 or so flowchart symbols (nodes), is quite small; and, too frequently for one's liking, flowchart segments have to be subdivided further. We have no adequate theory to guide us in this process, but cutting of loops is not as critical as in the case of program segmentation, and the lack of a theory is not much of a handicap. Despite this, a list of all cycles in the flowchart digraph is still needed, for guidance in the subdivision of segments that are too large and for designing the layout of a page.

We call an arc of a digraph a *separator* of the digraph if it does not lie on a cycle or on a path that is parallel to some other path. It is easy to see that deletion of separators produces a partial digraph whose connected components are the segments of the flowchart digraph. Exercise 7.8 requires one to find the separators in a flowchart digraph that has a single entry node and a single terminal node. The solution to this exercise suggests the approach that can be taken in designing an algorithm for the much more difficult case of multiple entry and terminal nodes (Exercise 7.9).

Assume now that the nodes to go on a page have been selected. The next problem relates to layout. An axis has to be selected within a page of the flowchart. This will be called the main flowpath. Other paths are drawn either side of the main flowpath. An automatic flowchart generator cannot be expected to identify the main flowpath with the path having greatest logical significance (often even the writer of the program would find this a difficult task), and some other criterion has to be used. In the program segment corresponding to the subdigraph that is being fitted to the page there will be one or more points at which control passes into the segment or out of it.

The corresponding nodes are entry nodes and terminal nodes, respectively. An empirical main flowpath is obtained by finding the longest elementary path originating at an entry node. Preferably the path should originate at an entry node that is the last node of the main flowpath relative to the page immediately preceding this page, and it should end at a terminal node, but there is little loss of clarity if the flowpath originates at some other entry node and ends at a node that is not terminal. The reason is that an automatic flowcharter should produce flowcharts on at least two levels of detail. On the lower level the output is a conventional flowchart, separated into pages. On the higher level a chart is generated that represents the logical sequence and interrelation of the pages.

If A.6.3 was used to find the cycles, then we have now also a list of all simple paths in the flowchart digraph, and one of these paths is the longest simple path in the subdigraph that we are fitting to the page. If, on the other hand, the cycles were found by the more accessible A.6.8, then the longest simple path still remains to be found. We break all cycles by removing final arcs of return paths. Then the longest simple path in the page can be found by the procedure of Exercise 6.9, and the removed arcs restored.

Finally there remains the problem of minimizing the number of intersections of flowlines in the drawing. One could analyze the undirected graph corresponding to the flowchart digraph; the layout problem then becomes the planarity problem (Section 3h). A practical algorithm for finding how near a graph is to being planar does not appear to have been discovered. This certainly means that we do not have a good general algorithm that will rearrange edges in the geometrical representation of the graph until the number of intersections is a minimum, but, since paging reduces the number of flowlines (explicit flowlines are replaced by connectors), the problem becomes tractable when a page is considered on its own. As the longest elementary path is the axis of the page and all cycles are known, a trial-and-error procedure is then probably as efficient as any other that could be discovered.

Notes

A survey of applications of graph theoretical models of programs made prior to 1960 can be found in [Ma60]. [Ka60] is an early paper describing the use of properties of digraphs in the detection of programming errors and elimination of redundancies. Later work on optimization of programs is reported in [Cl66a, Al69, Lo69a, Ba69b]. The procedures are explicitly based on or at least influenced by graph theory (except [Cl66a]); [Al69] contains the most extensive treatment.

Ramamoorthy applies graph theory to analysis of programs in a number of papers: [Ra66a] establishes the terminology and gives some basic algorithms (a variant of the algorithm of Exercise 7.3 is an example), [Ra66b] deals with segmentation of programs for a multiprogrammed computer, and [Ra67] extends the work to maintenance of hardware. The indices L_i were introduced by Schurmann [Sc64]. Estimation of running times for specific sections of a program is an important precondition for effectiveness of segmentation. Because of its difficult nature, little work has been done on this problem, but some results may be found on pp. 364–369 of [Kn68], and in [Ha66, Ma67a, Kr68]. Efficiency of program segmentation depends also on data flow across segment boundaries; [Ku67] deals with optimization of programs with respect to data flow.

Analysis of time sharing and multiprogramming by means of graph theoretical models is carried out in a number of papers; we recommend, in rough order of increasing difficulty, [Lo69b, Re68, Ka66].

Automatic flowcharting is considered in [Sc58, Ha59, Kn63, Kr64, An65, Sh66, Be69a]. Automatic construction of flowcharts specifically for debugging is discussed in [St65].

Exercises

7.1 (a) Draw flowchart digraphs of the programs of A.1.5 and A.5.3.

7.2 (b) Write a program that detects in a flowchart digraph nodes from which no terminal statement can be reached. Assume that there may be more than one terminal statement. (The algorithm should be similar to that given in Section 7b for finding nodes that are not reachable from an entry node.)

7.3 (b) Assume that nodes $n, n-1, \ldots, n-k$ represent terminal statements in a flowchart digraph. The following algorithm detects nodes from which no terminal statement can be reached:

1. Transfer column n of adjacency matrix A of the flowchart digraph to vector T.
2. For $j = n-1, \ldots, n-k$ and for $i = 1, 2, \ldots, n$ set $t_i = t_i \vee a_{ij}$ (for definition of operation $\vee$ see A.6.1).
3. For $j = 1, \ldots, n-k-1$ and for $i = 1, 2, \ldots, n$: if $a_{ij} = 1$, then set $t_i = t_i \vee a_{ij}$.
4. Repeat Step 3 until no more changes take place in T.

If now a t_i is zero $(1 \leqq i < n-k)$, then no terminal statement can be reached from node i. Discuss the relative efficiencies of this algorithm and that of Exercise 7.2.

7.4 (c) Use A.3.3 to find the cycle digraph of the program of A.5.3 and compute the loop indices L_i defined in Section 7c.

7.5 (c) Consider the following definition of a return path in a flowchart digraph: "Every arc that lies on a simple path from an entry node to a terminal node (a forward path) is a forward arc; an arc that is not forward is a return arc. The return arcs define return paths: A path $(b, \ldots, a)$ made up entirely of return arcs, such that nodes a and b lie on a forward path, is a return path." Show by an example that this definition is *not* equivalent to our definition in Section 7c.

7.6 (c) Is the number of return paths in a flowchart digraph necessarily equal to the number of cycles? Is the number of final arcs of return paths necessarily equal to the number of cycles?

7.7 (c) Identify return paths in the flowchart digraph of the program of A.5.3 by visual inspection and remove the final arcs of these paths. Write a program (or use a program that you may have already written) to find the longest path in the resulting structure.

7.8 (d) If a flowchart digraph has a single entry node and a single terminal node, then the finding of separators is a relatively easy matter. Develop an algorithm that finds the separators in this special case.

7.9 (–) (major project) Write a subroutine that finds the separators in a flowchart digraph.

7.10 (–) (major project) Given an adjacency matrix of order 25 or less that defines a subdigraph of a flowchart digraph. Develop an algorithm that places nodes of the subdigraph in such a way relative to each other that the flowchart boxes corresponding to the nodes would be properly laid out on a page of the flowchart. The output might be a matrix D, defined as follows:

$$d_{ij} = a \times 3^0 + b \times 3^1,$$

where $a = -1$ if node i is below node j,
$= 0$ if the nodes are on the same vertical level,
$= 1$ if i is above j,
and $b = -1$ if node i is to the left of node j,
$= 0$ if the nodes are on the same horizontal level,
$= 1$ if i is to the right of j.

7.11 (–) (research project) Computer programs are nearly always such that the procedure of Exercise 7.9 gives segments that are much too large to fit on a page. Develop (in outline) a complete solution of the paging problem.

7.12 (–) (research project) Write an algorithm that finds final arcs of all return paths in a flowchart digraph. (The definition of a return path given in

Section 7c will be of little use in this endeavor. However, the purpose of Exercises 7.5 and 7.7 has been to give an intuitive understanding of what a return path is, and the constructive definition of Exercise 7.5 is, in fact, equivalent to the definition of Section 7c in most cases. One approach to the solution of the problem would be to find a constructive definition that is equivalent to our definition in all cases, but the possibility that a useful definition of final arcs could be found without having to define return paths should not be ignored.)

CHAPTER 8

Other Applications of Graphs

8a. Flow Problems

DEFINITION 8.1 A *capacitated network* (*capacitated transportation network*) is a quadruple $\langle A, Q, c, R_+ \rangle$, where $\langle A, Q \rangle$ is a digraph, R_+ is the set of nonnegative real numbers, and $c\colon Q \to R_+$ is a function. We denote $c(i,j)$ by c_{ij}, and call it the *capacity* of arc $\langle i,j \rangle$.

Example

The weighted network of Figure 8.1 is a capacitated network. Only the interpretation given to weights c_{ij} distinguishes the capacitated network of Figure 8.1 from other weighted digraphs. A capacity c_{ij} can be thought of as the greatest amount of some commodity that can be moved along $\langle i,j \rangle$ in a unit of time. A fundamental problem in the theory of networks is to find the greatest amount of a commodity that can be conveyed from one given node to some other given node in the network. This is the maximal flow problem.

DEFINITION 8.2 Let $\langle A, Q, c\ R_+ \rangle$ be a capacitated network. A *flow* f of value v from node s to node t is a function $f\colon Q \to R_+$ such that

(a) $0 \leqq f(i,j) \leqq c_{ij}$ for all $\langle i,j \rangle \in Q$,

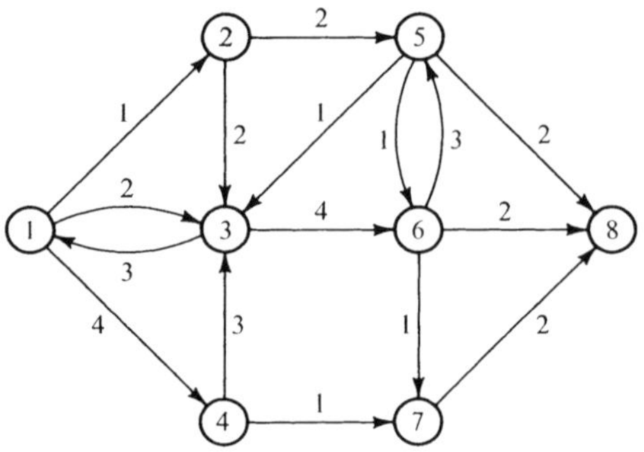

Figure 8.1

(b) $\sum_{Q^+_{\{p\}}} f(p, i) - \sum_{Q^-_{\{p\}}} f(i, p) = v$ for $p = s$,
$= -v$ for $p = t$,
$= 0$ for $p \in A - \{s, t\}$.

Nodes s and t are called, respectively, *source* and *sink* of the network with respect to flow f. (Q^+ and Q^- are defined by D.3.6.)

Conditions (a) of D.8.2 state that nowhere can the flow along an arc exceed the capacity of the arc. Conditions (b) are conservation equations. A commodity is pumped into the network by way of the source at some set rate. At t as much of the commodity must come out as goes in at s. At all intermediate points the inflow is equal to the outflow. The problem is to find the greatest value of v that is consistent with the conditions.

DEFINITION 8.3 Let $\langle A, Q, c, R_+ \rangle$ be a capacitated network with source s and sink t. Let X be a subset of A such that $s \in X$, $t \notin X$. Then the set of arcs Q^+_X is a *cut* in the network, and the sum of capacities of all arcs in Q^+_X is the *capacity of the cut*, denoted $C(X)$.

Example

In the capacitated network of Figure 8.1 let node 1 be the source and node 8 the sink. The cut corresponding to $X = \{1, 2, 4, 7\}$ is the set of arcs $\{\langle 1, 2\rangle, \langle 2, 3\rangle, \langle 2, 5\rangle, \langle 4, 3\rangle, \langle 7, 8\rangle\}$. The cut corresponding to $\{1, 2, 3, 4\}$ is $\{\langle 2, 5\rangle, \langle 3, 6\rangle, \langle 4, 7\rangle\}$. Capacities of the two cuts are $2 + 2 + 2 + 3 + 2 = 11$ and $2 + 4 + 1 = 7$, respectively.

THEOREM 8.1 If X defines a cut in a capacitated network $\langle A, Q, c, R_+ \rangle$, and v is the value of a flow in the network, then $v \leq C(X)$.

Proof. Let f' be an extension of f to the domain $A \times A$, defined by

$$f'(i, j) = f(i, j) \quad \text{for} \quad \langle i, j\rangle \in Q,$$
$$= 0 \quad \text{for} \quad \langle i, j\rangle \notin Q.$$

From Conditions (b) of D.8.2, in terms of the extension f', we obtain

$$\sum_{i \in A} f'(p, i) - \sum_{i \in A} f'(i, p) \begin{matrix} = v & \text{for} & p = s, \\ = 0 & \text{for} & p \in A - \{s, t\}. \end{matrix}$$

Hence, summing over all members of X,

$$\sum_{p \in X,\, i \in A} f'(p, i) - \sum_{i \in A,\, p \in X} f'(i, p) = v.$$

But, by symmetry,

$$\sum_{p \in X,\, i \in X} f'(p, i) - \sum_{i \in X,\, p \in X} f'(i, p) = 0.$$

Hence

$$\sum_{p \in X,\, i \in A - X} f'(p, i) - \sum_{i \in A - X,\, p \in X} f'(i, p) = v,$$

and, dropping the zero terms,

$$v = \sum_{Q_X^+} f(p, i) - \sum_{Q_X^+} f(i, p).$$

Hence

$$v \leqq \sum_{Q_X^+} f(p, i) \leqq \sum_{Q_X^+} c_{pi} = C(X).$$

THEOREM 8.2 (Ford and Fulkerson's max-flow min-cut theorem) Let $\langle A, Q, c, R_+ \rangle$ be a capacitated network with v_{max} the greatest possible flow from source s to sink t. Let C_{min} be the smallest capacity of a cut in the network. Then $v_{max} = C_{min}$.

Proof. Let the maximal flow v_{max} from s to t be defined by function f'. Define a set X' recursively as follows:

(i) $s \in X'$;
(ii) if $p \in X'$, and $f'(p, q) < c_{pq}$ or $f'(q, p) > 0$ (or both), then $q \in X'$.

We show that making t a member of X' leads to the contradiction that the flow can be improved. From the definition of X', if $t \in X'$, then there exists a semipath $S = (s, \ldots, t)$. Now, if $\langle p, q \rangle \in S$ is in the direction from s to t, define $f_{pq} = f'(p, q) - c_{pq}$. If $\langle p, q \rangle$ is in the reverse direction, put $f_{pq} = f'(p, q)$. Finally set $v' = min_S f_{pq}$. Now the flow in each "forward" arc can be increased by v' and the flow in each "backward" arc decreased by v' without affecting the conditions of D.8.2. But then the maximal flow becomes $v_{max} + v'$, and this is an obvious contradiction. Hence $t \notin X'$, i.e., set X'

defines a cut. An arc with only one node in X' cannot have satisfied Part (ii) of the definition of X', and we must have

$$\begin{aligned} f'(p, q) &= c_{pq} \quad \text{for every} \quad \langle p, q\rangle \in Q_{X'}^{+}, \\ f'(q, p) &= 0 \quad \text{for every} \quad \langle q, p\rangle \in Q_{X'}^{-}. \end{aligned}$$

But then, from the proof of Th.8.1, putting v_{max} for the v there,

$$\begin{aligned} v_{max} &= \sum_{Q_{X'}^{+}} f'(p, q) - \sum_{Q_{X'}^{-}} f'(q, p) \\ &= \sum_{Q_{X'}^{+}} c_{pq} - 0 \\ &= C(X'). \end{aligned}$$

Since $v_{max} \leqq C(X)$ in general, $C(X') = C_{min}$.

Direct use of the theorem is not feasible. Let $|A| = n$. Then the number of candidates for the subset of A that would define a min-cut set is

$$\sum_{k=0}^{n-2} C(n-2, k) = 2^{n-2}.$$

Even for a relatively small network with 30 nodes 2^{28} still amounts to 268,435,456. But the *proof* of Th.8.2 is constructive in that it defines a min-cut set, and this definition is the basis for the following algorithm.

ALGORITHM 8.1 (Ford–Fulkerson algorithm) Given a capacitated network $\langle A, Q, c, N_+ \rangle$, where N_+ is the set of nonnegative integers and c is now a function into N_+. Flow function f is integer valued also. The following algorithm finds a set $X' \subset A$ that defines a minimal cut. Labels are attached to the nodes as part of the procedure.

1. Assign arbitrary flows $f(p, q)$ that are consistent with D.8.2 to all arcs; e.g., make the flows all zero.
2. Make $h(s) = \infty$ and attach label $\langle -, h(s)\rangle$ to s. Select node s for the p of Steps 3 and 4.
3. For every arc $\langle p, q\rangle$ such that q is not labeled compute $d(p, q) = c_{pq} - f(p, q)$, and, if $d(p, q) > 0$, attach to q the label $\langle p^{+}, h(q)\rangle$, where $h(q) = min[h(p), d(p, q)]$.
4. For every arc $\langle q, p\rangle$ such that q is still not labeled and $f(q, p) > 0$ attach to q the label $\langle p^{-}, h(q)\rangle$, where $h(q) = min[h(p), f(q, p)]$. Node p is now processed.
5. If t is labeled, then go to 7.
6. Select a labeled node p that has not been processed and go to 3. If none remain to be processed, stop. Then the labeled nodes constitute set X' and the $f(p, q)$ define the greatest flow.

7. The algorithm has produced a semipath on labeled nodes $(s, \ldots, t)$, where the first coordinate of the label on a node names the node that precedes it in the semipath. Adjust the flow: Working backward from t, until s is reached, for every labeled node q on semipath $(s, \ldots, t)$ set $f(p, q) = f(p, q) + h(t)$ or set $f(q, q) = f(q, p) - h(t)$, depending on whether the label of q is $\langle p^+, h(p)\rangle$ or $\langle p^-, h(q)\rangle$, respectively.
8. Discard all labels and go to 2.

Example

Tables 8.1 and 8.2 summarize computation of the greatest flow from node 1 to node 8 in the capacitated network of Figure 8.1. Initially all flows are assumed zero; final flows are as shown in Figure 8.2 (where arcs in which the flow is zero have been omitted). Here $X' = \{1, 3, 4\}$, but note that this is not the only set defining a minimal cut.

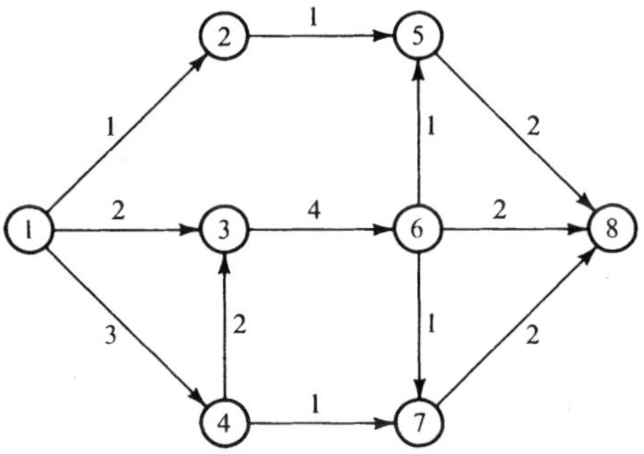

Figure 8.2

TABLE 8.1

NODE LABELS

Node	Iteration 1	2	3	4	5	6
1	$\langle -, \infty\rangle$	$\langle -, \infty\rangle$	$\langle -, \infty\rangle$	$\langle -, \infty\rangle$	$\langle -, \infty\rangle$	$\langle -, \infty\rangle$
2	$\langle 1^+, 1\rangle$					
3	$\langle 1^+, 2\rangle$	$\langle 1^+, 2\rangle$	$\langle 4^+, 3\rangle$	$\langle 4^+, 2\rangle$	$\langle 4^+, 2\rangle$	$\langle 4^+, 1\rangle$
4	$\langle 1^+, 4\rangle$	$\langle 1^+, 4\rangle$	$\langle 1^+, 4\rangle$	$\langle 1^+, 3\rangle$	$\langle 1^+, 2\rangle$	$\langle 1^+, 1\rangle$
5	$\langle 2^+, 1\rangle$		$\langle 6^+, 2\rangle$	$\langle 6^+, 1\rangle$	$\langle 6^+, 1\rangle$	
6	$\langle 3^+, 2\rangle$	$\langle 3^+, 2\rangle$	$\langle 3^+, 2\rangle$	$\langle 3^+, 1\rangle$	$\langle 3^+, 1\rangle$	
7	$\langle 4^+, 1\rangle$	$\langle 4^+, 1\rangle$	$\langle 4^+, 1\rangle$	$\langle 4^+, 1\rangle$	$\langle 6^+, 1\rangle$	
8	$\langle 5^+, 1\rangle$	$\langle 6^+, 2\rangle$	$\langle 5^+, 1\rangle$	$\langle 7^+, 1\rangle$	$\langle 7^+, 1\rangle$	

TABLE 8.2

ADJUSTED FLOWS

Arc	Iteration					
	1	2	3	4	5	6
⟨1, 2⟩	1	1	1	1	1	
⟨1, 3⟩		2	2	2	2	
⟨1, 4⟩			1	2	3	
⟨2, 3⟩						
⟨2, 5⟩	1	1	1	1	1	
⟨3, 1⟩						
⟨3, 6⟩		2	3	3	4	
⟨4, 3⟩			1	1	2	
⟨4, 7⟩				1	1	
⟨5, 3⟩						
⟨5, 6⟩						
⟨5, 8⟩	1	1	2	2	2	
⟨6, 5⟩			1	1	1	
⟨6, 7⟩					1	
⟨6, 8⟩		2	2	2	2	
⟨7, 8⟩				1	2	

A network that contains multiple sources $s_1, s_2, \ldots, s_k$ can be converted to a network with a single source by adding a collective source node s' and augmenting the network with arcs of infinite capacity from s' to each of the s_i. The problem becomes one of finding the maximal flow from s' to t. Multiple sinks can be dealt with in a similar fashion. Another generalization involves capacities. So far we have assumed that the lower bound on the flow in any arc is zero. In the general case two explicit flow bounds c_{low} and c_{high} ($c_{low} \leqq c_{high}$) are assigned to each arc, and a flow in the arc is required to lie in the interval $[c_{low}, c_{high}]$. The bounds can be negative, a negative $f(p, q)$ being interpreted as a flow from q to p of magnitude $|f(p, q)|$. In the most common situation $c_{low} = -c_{high}$ for every arc; flow in an arc can be in both directions, a common bound restricting its magnitude in either direction. One way of dealing with this special case is to add arc $\langle q, p \rangle$ for every $\langle p, q \rangle$ in the network. Both pairs of bounds can then be set to $[0, c_{high}]$, and we have again the capacitated network of D.8.1. If both $\langle p, q \rangle$ and $\langle q, p \rangle$ are already in the network, having bounds $[-c_1, c_1]$ and $[-c_2, c_2]$, respectively, simply assign capacity $c_1 + c_2$ to each of the two arcs. (In the general case it is possible to construct a network that supports no flow, e.g., a network with $Q = \{\langle s, 2\rangle, \langle 2, t\rangle\}$, where the bounds are [1, 2] and [3, 4].)

A maximal flow need not be unique, and one may be required to select a particular maximal flow that satisfies some additional criterion. Consider a function $m: Q \to R_+$ that assigns to each arc $\langle p, q\rangle \in Q$ a value $m(p, q)$, which we interpret as the cost of sending one unit of a commodity through $\langle p,q\rangle$. The total cost associated with a flow f is $\sum_{\langle p,q\rangle \in Q} m(p,q)\cdot f(p,q)$ and the problem is to find a maximal flow for which the total cost is a minimum. This is known as the minimal cost problem. Investigation of the general bounded network and minimal cost problems is beyond the scope of this book.

The max-flow min-cut theorem is related, rather unexpectedly, to a very important result of combinatorial mathematics concerning the existence of distinct representatives of subsets of a set.

DEFINITION 8.4 Let S be a set, and let $\mathscr{S} = \{S_1, S_2, \ldots, S_m\}$ be a family of subsets of S (not necessarily disjoint). A set of *distinct* elements $\{a_1, a_2, \ldots, a_m\}$ such that $a_i \in S_i$ and $a_i \notin S_j$ $(j \neq i)$ is a *system of distinct representatives* for $\mathscr{S}$.

THEOREM 8.3 (P. Hall's theorem) Let

$$\mathscr{S} = \{S_i \mid i \in I\}, \qquad \text{where} \qquad I = \{1, 2, \ldots, m\},$$

be a family of subsets of a set S. A system of distinct representatives for $\mathscr{S}$ exists if and only if $|\bigcup_{i \in I_k} S_i| \geqq k$ for every I_k, where I_k is a subset of I having k elements, and for every $k = 1, 2, \ldots, m$.

Proof. Necessity is obvious. To prove sufficiency construct a network with nodes s and t, and a node for each element $x_j \in S$ and each subset $S_i \in \mathscr{S}$, i.e., a total of $2 + |S| + m$ nodes. Draw arc $\langle s, S_i\rangle$ for every node S_i, arc $\langle x_j, t\rangle$ for every node x_j, and arc $\langle S_i, x_j\rangle$ whenever $x_j \in S_i$. Complete the construction by assigning capacity 1 to each arc. Now construct a flow f. If (s, S_i, x_j, t) is a path, then $x_j \in S_i$, and x_j can be a representative of S_i. Now, if this path carries a nonzero flow, we interpret the flow in the path as making x_j the representative of subset S_i. Since the inflow at S_i cannot exceed 1, only one of the arcs originating from S_i can carry the flow out. Moreover, the outflow from x_j cannot exceed 1 either. Therefore, the only flow into x_j must come from S_i. Every f selects representatives of some of the subsets, and a maximal flow selects representatives of the greatest possible number of subsets. Therefore, $\mathscr{S}$ has a system of distinct representatives if and only if the value of the maximal flow from s to t is m. Suppose now that $\mathscr{S}$ has no system of distinct representatives. Then the value of the maximal flow from s to t must be less than m, and by Th.8.2 the capacity of a minimal cut is less than m. Now assume that the cut contains k_1 arcs that originate at s. Therefore, the set X' of Th.8.2 contains exactly $m - k_1$ nodes of type S_i. Denote the set of these nodes by S', and let X be the set of elements in the union of all the members of $\mathscr{S}$ that belong to S'. Partition X into $X_1 = \{x_j \mid x_j \in X'\}$ and $X_2 = \{x_j \mid x_j \notin X'\}$, and assume that $|X_1| = k_2$. Then the cut contains k_2 arcs that terminate at t. Consequently, arcs originating from S' and terminating in X_2, which make up the rest of the cut, must be fewer than $m - k_1 - k_2$ in number. This means that $|X_2| < m - k_1 - k_2$. But then $|X| = |X_1| + |X_2| < m - k_1$, and, since $|S'| = m - k_1$, we have shown that the nonexistence of a system of distinct representatives implies the existence of a collection of k subsets the union of which has fewer than k elements, i.e., the theorem has been proven by proving its contrapositive.

Example

An association has a number of committees; a member of the association may serve on any number of them. At the end of the year each committee selects one of its members to report on its activities. The problem is to choose the reporters in such a way that no individual reports on more than one committee. This problem can be solved by setting up a network as in the proof of Th.8.3. The S_i stand for committees, the x_j for members of committees. If there are k committees, and the maximal flow in the network has the value k, then the problem has a solution, and the flow defines the solution. If more than one flow is maximal, there is more than one solution.

8b. Graphs in Chemistry

There is probably no science in greater need of mechanized information retrieval than chemistry. Millions of chemical compounds are known; new ones are produced at an ever faster rate. The chemist has two main problems: First, he wants to find out whether the substance in his test tube is already known; second, given a substance, he wants to know the properties of similar substances. Both problems reduce to a matching process; a description of the given substance has to be matched against descriptions of substances that make up the data base of the retrieval system. A precondition for a satisfactory retrieval system is a standard representation of chemical compounds. The representation must be unambiguous, amenable to classification so as to facilitate search, and reasonably compact. Quite a number of systems of representation are in use. The great diversity of possible search requests ensures that almost any system has some feature that makes it superior to others for dealing with some particular aspect of the retrieval problem.

Chemical formulas are very compact, but they are ambiguous. Consider the hydrocarbons of the paraffin series. Their general formula is C_kH_{2k+2} (e.g., CH_4, C_2H_6). Each carbon atom is linked to four other atoms; a hydrogen atom can be linked to only one other atom, which has to be carbon. The total number of links in C_kH_{2k+2} is $\frac{1}{2}(4k + 2k + 2) = 3k + 1$. Let us interpret the links as edges in a graph having the atoms for nodes. Then the theory of Section 3h tells us that the graph contains no circuits ($m - n + 1 = 3k + 1 - 3k - 2 + 1 = 0$), i.e., that the graph is a tree. The subgraph of carbon atoms must be a connected tree, and more than one tree of the carbon atoms can be constructed for every $k \geq 4$. Each distinct tree on the k carbon atoms represents an *isomer* of C_kH_{2k+2}. Since isomers have different properties, the formular representation is inadequate. For example, when $k = 6$, there are 5 isomers, all corresponding to the single formula C_6H_{14}. Enumeration of isomers is an important problem. Cayley (1875) interpreted the isomerism problem for the paraffin series as an enumeration problem for trees, and solved it in this form; his solution is one of the classics of graph theory.

In a diagrammatic representation each isomer has a different structural diagram. Figure 8.3 shows the structural diagrams of isomers of H_4C_{10}.

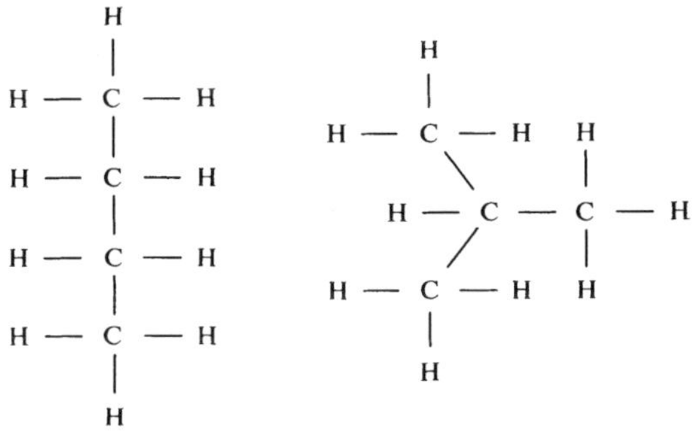

Butane Isobutane

Figure 8.3

Structural diagrams are very often drawn with the hydrogen atoms stripped off. Then, if, for example, there are b edges incident with a carbon atom, we know that $4 - b$ is the number of hydrogen atoms bonded to the carbon atom. Sometimes two atoms are linked by more than one bond. Either diagram of Figure 8.4 is an example. (In these diagrams, knowing that oxygen has 2 bonds and nitrogen 3, we can easily work out that the molecules contain three hydrogen atoms.) Structural diagrams are labeled graphs with parallel edges; such graphs are known as *multigraphs*. Labeled multigraphs give a reasonably adequate description of chemical substances. It must be remembered, however, that the graphs are projections of three-dimensional structures into the plane. For stereochemistry, which distinguishes different arrangements of the atoms of a molecule in space as stereoisomers, the planar representation is still inadequate.

The matching problem can be dealt with very well in terms of chemical structures. It becomes the graph isomorphism problem, considerably simplified by the labels carried by the nodes. The procedure that we are going to describe, Sussenguth's algorithm, is based on two principles. First, if graphs G and G^* are isomorphic, then the subset of nodes of G that exhibit some property must correspond to the subset of nodes of G^* that exhibit this same property. Second, if the subsets of nodes of G and G^* that are characterized by some property do not have the same number of elements, then the two graphs cannot be isomorphic.

Consider the structural diagrams of Figure 8.4 in which subscripts are node identifiers. The matching procedure starts with generation of subsets of nodes that represent the same type of atom; nodes that represent oxygen

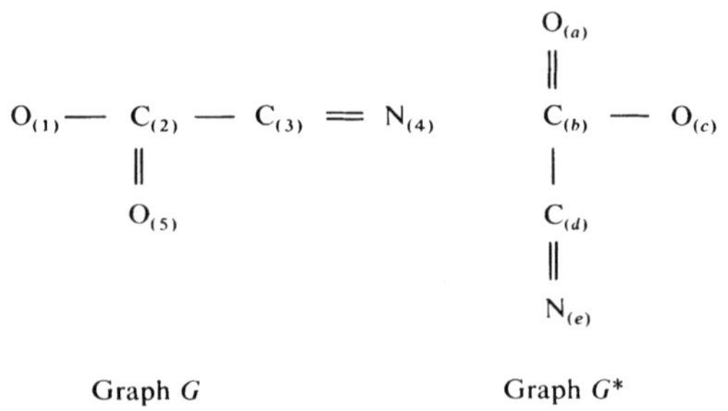

Figure 8.4

atoms, say, in graph G must correspond to nodes representing oxygen atoms in G^*. These correspondences are shown by lines 1 to 3 in Table 8.3. Next the subsets of nodes that are joined by the same type of bond are made to correspond (lines 4 and 5). At this point a partitioning procedure takes over. The purpose of partitioning is to reduce the number of nodes of G^* to which a node of G can correspond. The purpose is achieved by taking intersections. For example, from the correspondence of $\{1, 5\}$ and $\{a, c\}$ (line 3), we know merely that node 1 of G corresponds to one of a or c in G^*, but

$$\{1, 5\} \cap \{1, 2, 3\} = \{1\} \quad \text{and} \quad \{a, c\} \cap \{b, c, d\} = \{c\},$$

implying that 1 corresponds to c. Partitioning gives lines 6 and 7 in the table. Three of the nodes are now matched, but correspondence of elements in sets $\{2, 3\}$ and $\{b, d\}$ is still unresolved. We define the *degree* of a node as the number of edges incident with it (counting sets of parallel edges linking two nodes as a single edge), and generate subsets of nodes that have the same degree (lines 8 to 10). There is no need to resume the partitioning procedure; all nodes are matched. The correspondence of nodes in the two graphs is

Graph G:	1	2	3	4	5
Graph G^*:	c	b	d	e	a

TABLE 8.3

MATCHING OF CHEMICAL STRUCTURES

Basis for subset generation		Subset of G	Subset of G^*	Line
Node label:	C	{2, 3}	$\{b, d\}$	1
	N	{4}	$\{e\}$	2
	O	{1, 5}	$\{a, c\}$	3
Bond:	single	{1, 2, 3}	$\{b, c, d\}$	4
	double	{2, 3, 4, 5}	$\{a, b, d, e\}$	5
Partition:	lines 3, 4	{1}	$\{c\}$	6
	lines 3, 5	{5}	$\{a\}$	7
Degree:	1	{1, 4, 5}	$\{a, c, e\}$	8
	2	{3}	$\{d\}$	9
	3	{2}	$\{b\}$	10
Neighbors:	line 2	{3}	$\{d\}$	8a
	line 6	{2}	$\{b\}$	9a

If some nodes were still not matched, new subsets would have to be generated. Let us define the *order* of a node as the number of edges in the smallest circuit through this node. Subsets of nodes having the same order would be put in correspondence. (All nodes in the graphs of Figure 8.4 have the same order, namely 0, and the concept of order is of no value here.) Another concept that can be used is one of neighborhood. If a node of G is already matched to a node of G^*, then nodes that are the immediate neighbors of the identified nodes form corresponding subsets. If, instead of using the degrees of nodes to generate subsets, we had made use of neighborhoods, the matching procedure would have continued as shown in Table 8.3 below the dividing line.

In general, the algorithm terminates when every node in G has been paired off with a node in G^*, or when two corresponding subsets of nodes of G and G^* are found to differ in the number of nodes they contain. If the former is the case, graphs G and G^* are isomorphic; if the latter, isomorphism is impossible. Occasionally, however, the algorithm exhausts all subset generating properties before either of the two conditions is satisfied. This happens when more than one isomorphism is possible between the two graphs, or when the subset generating properties are incomplete in the sense that some property that would establish isomorphism or the lack of it has been neglected in the design of the algorithm.

For the sake of the argument assume that the algorithm cannot take us beyond line 7 of Table 8.3. Nodes in subset {2, 3} have remained unmatched, but we know that their possible correspondents must belong to $\{b, d\}$. Therefore we postulate the two correspondences

$$A(1) = \begin{bmatrix} 1 & 2 & 3 & 4 & 5 \\ c & b & d & e & a \end{bmatrix} \quad \text{and} \quad A(2) = \begin{bmatrix} 1 & 2 & 3 & 4 & 5 \\ c & d & b & e & a \end{bmatrix}$$

and carry out a node-by-node comparison of the two graphs. If both assignments are valid, there are two isomorphisms; if only one is valid, there is a single isomorphism; if both assignments result in contradictions, then the graphs are not isomorphic.

After the nodes have been partitioned according to their labels (lines 1 to 3 of Table 8.3), only the structure of the graphs determines validity of the arrangements. The structure can be represented by a *connectivity matrix* C in which $c_{ij} = n$, where n is the number of times edge $\{i, j\}$ is drawn in the multigraph. We have

$$C_G = \begin{bmatrix} 0 & 1 & 0 & 0 & 0 \\ 1 & 0 & 1 & 0 & 2 \\ 0 & 1 & 0 & 2 & 0 \\ 0 & 0 & 2 & 0 & 0 \\ 0 & 2 & 0 & 0 & 0 \end{bmatrix} \quad \text{and} \quad C_{G*} = \begin{bmatrix} 0 & 2 & 0 & 0 & 0 \\ 2 & 0 & 1 & 1 & 0 \\ 0 & 1 & 0 & 0 & 0 \\ 0 & 1 & 0 & 0 & 2 \\ 0 & 0 & 0 & 2 & 0 \end{bmatrix}.$$

(In C_{G*} subscripts 1, ..., 5 refer to nodes $a, \ldots, e$, respectively.) From C_{G*} generate the connectivity matrices corresponding to the two assignments. They are

$$C_{A(1)} = \begin{bmatrix} 0 & 1 & 0 & 0 & 0 \\ 1 & 0 & 1 & 0 & 2 \\ 0 & 1 & 0 & 2 & 0 \\ 0 & 0 & 2 & 0 & 0 \\ 0 & 2 & 0 & 0 & 0 \end{bmatrix} \quad \text{and} \quad C_{A(2)} = \begin{bmatrix} 0 & 0 & 1 & 0 & 0 \\ 0 & 0 & 1 & 2 & 0 \\ 1 & 1 & 0 & 0 & 2 \\ 0 & 2 & 0 & 0 & 0 \\ 0 & 0 & 2 & 0 & 0 \end{bmatrix}.$$

Since $C_G = C_{A(1)} \neq C_{A(2)}$, there exists a single isomorphism.

A structure diagram is completely specified by its connectivity matrix and a vector of labels E. In our example the vectors are

$$E_G = [\mathrm{O} \quad \mathrm{C} \quad \mathrm{C} \quad \mathrm{N} \quad \mathrm{O}] \qquad \text{and} \qquad E_{G*} = [\mathrm{O} \quad \mathrm{C} \quad \mathrm{O} \quad \mathrm{C} \quad \mathrm{N}].$$

For the input of chemical structures one can use a special chemical typewriter. The typewriter produces a paper tape, and a conversion program uses the information on the tape to set up the connectivity matrix and label vector.

An analogous procedure can be used to identify a given graph G as a subgraph of another graph G^*. Defining properties are again used to generate corresponding subsets. Denote by S and S^* the subsets of nodes of G and G^*, respectively, that are generated by the same defining property. The two principles on which the matching algorithm is based now become: If G is a subgraph of G^*, then $S \subseteq S^*$; if $|S| > |S^*|$, then G cannot be a subgraph of G^*.

8c. Graphs in Information Retrieval

Information retrieval consists of two disciplines: document retrieval and fact retrieval. A scientist, before he sets out to perform some piece of research, wants a list of references to all work in his chosen problem area. The literature search is done by a document retrieval system. If his examination of the retrieved publications shows that what he wants to do is unlikely to have been done before, then the scientist starts on his project. Soon he finds that he needs answers to highly specific questions; e.g., he may want to know the number of isomers of $C_{10}H_{22}$. For this information he turns to a fact retrieval system. If the system cannot supply the answer (Answer: $C_{10}H_{22}$ has 75 isomers), then he has to appeal again to the document retrieval system for a list of references on enumeration of isomers.

A retrieval system can mean many things. It may mean the scientist and his personal file of index cards, or the scientist and a few bibliographies, or the scientist and a roomful of handbooks, or a little request card and a computer with a huge data base and a battery of complex document and fact retrieval procedures. The fully automatic system of the last example does not exist. Semiautomatic document retrieval systems, which require the scientist to do some work himself, are, however, in existence. Automatic fact retrieval systems also exist, and in some narrow areas, e.g., retrieval of stock market quotations and chemical retrieval, they are fully automatic; but these systems do rarely more than search through a few lists of data. A system with a *wide* data base, capable of making *complex* inferences, is still only a hope for the future. This is due partly to the sheer size of the required data base, partly to linguistic problems. We shall discuss only the simpler, but still very difficult, problem of document retrieval.

With a primitive semiautomatic system the user supplies a set of key words, e.g. {*graph*, *network*, *tree*}, and the system retrieves documents that have one or other of these words or some derivative of it in the title. The *semi* refers to the work that is likely to go on afterwards. First, the set of titles retrieved may contain "Programming *graphic* devices," "Handbook of mathematical functions with formulas, *graphs*, and mathematical tables," "The place of a small library in the national *network*," etc. These references have to be weeded out manually, but one has to be careful not to assume offhand, for example, that "The enumeration of *trees* by weight and diameter" deals with forestry. Second, the system will fail to retrieve documents that are relevant, but do not have one of the given key words in their titles, e.g., "The four-color problem."

Both types of shortcomings can be dealt with reasonably effectively. One takes a dictionary of concepts, known as a *thesaurus*, and selects terms from this dictionary that will describe a particular document. They become the *index terms* associated with the document; their selection is the process of *indexing*. The user of the system selects words or phrases from the same thesaurus to define the categories of documents that he wants retrieved. These we shall call *search terms*. The number of elements in the intersection of the sets of index and search terms determines the relevance of a document. Both index and search terms can be weighted, and the system can use the weights to evaluate a finer measure of relevance of a document with respect to the search request. The retrieved documents are graded in order of relevance. Under this scheme "The four-color problem" and "The enumeration of trees by weight and diameter" are assigned high relevance; the other titles quoted above sink to the bottom of the list. The system is still semiautomatic in that the selection of index terms has to be done manually, or, at best, semiautomatically.

This has been an exceedingly naïve account of the retrieval problem. Continuing in the same naïve vein, let us take a superficial look at the structure of a thesaurus. The semantic value of a word is unknown unless we have some idea how this word is related to other words. A concept is not a single word (or phrase), but a word and other words related to it, words that are, in turn, related to the words to which the first word is related, and so on, to any suitable distance. A thesaurus is a dictionary in which related words are grouped into clusters. Consider *Roget's Thesaurus*, which contains 1000 clusters, each cluster defining a particular concept. For example, cluster No. 792 defines the concept *thief*; 65 terms are listed in this cluster. A conventional thesaurus does not, however, indicate the strength of the relation between two words in a cluster. The cluster of our example contains the terms *fence* and *viking*. It is hard to see any relation between the terms, let alone define its strength, until other terms of the cluster are used to set up a relational chain *fence—receiver of stolen goods—thief—robber—pirate—viking*. The length of the chain between two terms can be used as a measure of the relatedness between the terms. This suggests that a graph gives a more effective representation of a concept than an unstructured list. The concept graphs can be combined into a single graph representing the entire thesaurus; it is connected because each term that belongs to more than one cluster is still represented by a single node. The term *fence* alone, because it occurs in 6 of the 1000 clusters, connects 6 concept graphs.

Let us now take a new approach to the retrieval problem. With an effectively constructed thesaurus it is possible to permit the user to supply only a few search terms. The system then uses the thesaurus to generate a larger internal set for the comparisons with index sets of documents. The internal set could be the union of clusters in which more than one or two, say, of the supplied terms have been found. If the thesaurus defines few concepts, as is the case with *Roget's Thesaurus*, the concepts are very broad, and the internal set is much too large. Moreover, to conserve storage, redundancy should be eliminated by making the entire thesaurus a single graph, but then the individual concepts lose their identity. In this single thesaurus graph a procedure for constructing the internal set could be based on the following, or some similar, definition: The internal set contains all the search terms supplied, and every term that is joined by an edge to at least four other terms in the set, of which at least two must belong to the initial search set. Figure 8.5 shows a subgraph in which the term *arc* is represented in three conceptual senses (as in *arc light*, *arc in a network*, *arc of a circle*). In terms of the search set {*arc*, *path*} and this graph, the only internal set consistent with the definition is {*arc*, *graph*, *network*, *circuit*, *path*}. The probability that all

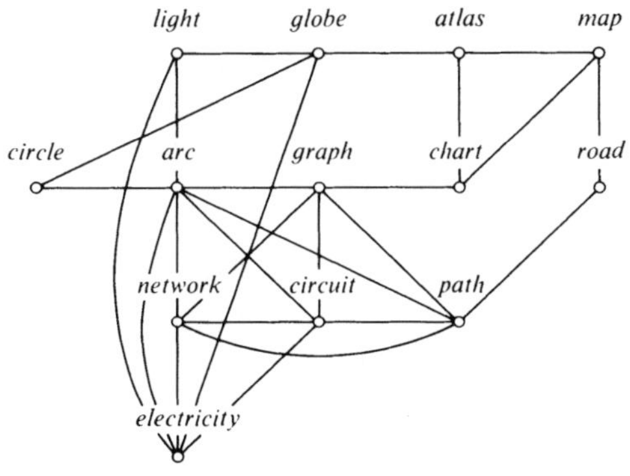

Figure 8.5

relevant documents will be retrieved is much greater with this set than with the original set {*arc*, *path*}.

Another graph that can be of interest in the retrieval of documents is the *citation digraph* $\langle D, R\rangle$, where D is a set of documents and R is the set $\{\langle d_i, d_j\rangle \mid d_i \text{ cites } d_j\}$. If we have already selected a subset X of D that consists of relevant documents, and we have a document $d_k \notin X$, then the number of arcs in the intersection $R_X^+ \cap R_{\{d_k\}}^-$ (see D.3.6) is the number of documents in X that cite the document d_k. This number can be used in the evaluation of the relevance of d_k to the search request.

A further use of the citation digraph is in establishing the absolute importance of a document. Libraries are becoming very short of space. It is, therefore, important to determine the importance of documents in their collections. Documents that are in little demand can be cleared from the open shelves to give readier access to those for which a strong demand exists. One measure of the importance of a document d_k is the number of other documents that cite it, directly or indirectly. In terms of the citation digraph this is the number of nodes from which node d_k is reachable.

Notes

For further study of flow problems one can turn to Chapter 7 of [Bu65], Part II of [Be62], and the classic [Fo62]. Our formulation of the Ford–Fulkerson algorithm (A.8.1) follows closely the exposition given in [Be64a]. Our proof of Th.8.3 derives from that given in [Ha65b].

Sussenguth's matching procedure for chemical structures is described in [Sa64] and, in greater detail, in [Su65]. References to chemical notations, and to the part they play in the retrieval of chemical information can be found in [Ta67]. One should consult current volumes of the *Annual Review of Information Science and Technology* for up-to-date surveys of various aspects of mechanized information retrieval. Salton's SMART system is perhaps the most advanced operational system for experimentation with automated information retrieval procedures. Salton's textbook on information retrieval [Sa68a] describes retrieval principles in relation to this existing system. For this reason it has a sustained vitality that warrants a very warm recommendation of the book. The graph of our Figure 8.5 relates words in the simplest possible way. For an indication of further possibilities inherent in this approach one should study the sophisticated relational scheme devised by Quillian, see [Qu67]. Quillian's work has influenced Simmons and his co-workers in the design of their question answering system Synthex—see [Si66b, Si66c], and [Ca67] for an amusing illustration of the lack of perfection that systems of this type still tend to exhibit.

Exercises

8.1 (a) Interpret the networks of Figures 3.6 and 6.8 as capacitated transportation networks and find, respectively, the value of maximal flow from node c to node f and from node 1 to node 8.

8.2 (a) If all capacities in the network of Figure 8.1 are (i) doubled, and (ii) halved, what is the value of the maximal flow from node 1 to node 8?

8.3 (a) If in the network of Figure 8.1 we change c_{58} to 3, then the value of the maximal flow can be increased to 7 by altering the capacity of any one of two other arcs. Identify these arcs.

8.4 (a) In the network of Figure 8.1 let nodes 1 and 2 be two sources and node 8 the sink. Find the maximal flow from the sources to the sink.

8.5 (a) One generalization of the capacitated network is a network in which (some or all) nodes are capacitated as well. For example, although the sum of capacities of arcs terminating at a node may be 15, say, and the sum of capacities of arcs originating from this node may also be 15, the total flow through the node may be restricted to values not exceeding 12, say. A node-capacitated network may be converted to an equivalent network in which arcs alone are capacitated by the addition of dummy nodes and arcs. Describe the conversion process.

8.6 (a) Committees of an association are composed as follows: $A = \{a, b, d\}$, $B = \{b, d, e, g\}$, $C = \{c, d\}$, $D = \{a, b, c\}$, $E = \{a, f, g\}$, $F = \{c, d, g\}$, $G = \{a, b, c, f\}$. Show that it is possible to choose reporters from all committees in such a way that no individual reports on more than one committee.

8.7 (a) A Department of Computer Science wants to give the following courses in Spring Term: CS1, CS12, CS13, CS248, CS293, CS31, CS328. Faculty members are qualified to teach the courses as follows:

Brown	CS1, CS13, CS248;
Douglas	CS1, CS31;
Evans	CS1, CS248, CS31;
Harris	CS12, CS293;
Jones	CS1, CS13;
Kelly	CS12, CS293, CS328;
Smith	CS1, CS13, CS31.

Show that the department cannot give all seven courses if each faculty member is to teach only one course.

8.8 (b) Graph *S* of Figure 8.6 is a subgraph of Graph *G*. Use the matching procedures described in Section 8b to prove this. (The leftmost line in the drawing of *S*, which is not an edge, serves to indicate the degree of the node to the right of it.)

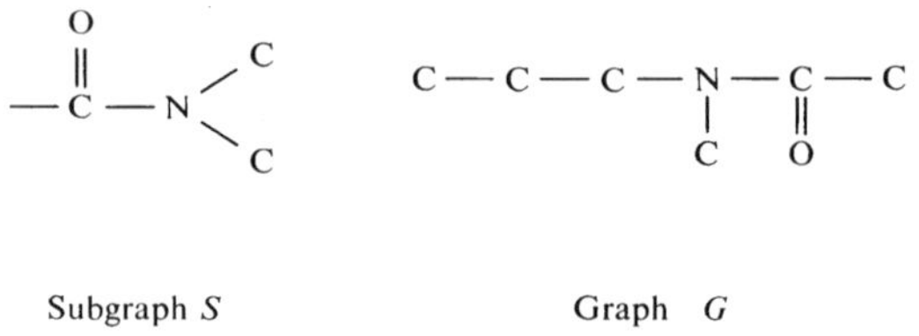

Figure 8.6

8.9 (b) (nontrivial project) Implement Sussenguth's algorithm for matching chemical structures as a computer program.

PART III

COMPUTER REPRESENTATION OF STRUCTURES

CHAPTER 9

Arrays

9a. Storage Media and Their Properties

In this section, which serves for an introduction to the remainder of the book, we shall endeavor to come to a limited understanding of some storage media used in modern computers. It is not our purpose to delve into the physical principles underlying the different storage mechanisms; we shall consider only those properties that may have direct relevance to programming.

The most important storage unit in a modern computer is its magnetic core store. The cores, each of which stores one bit, are arranged in planes with 32, or 64, or 128, or 256 cores in both directions (the numbers are powers of 2 to enable full use to be made of the bits assigned to specification of addresses in machine language commands), and the planes are stacked one on top of the other. Usually the number of cores in a plane determines the number of words in the memory unit, and the number of planes determines the length of a word. For example, a 128 by 128 unit 36 planes deep stores 16 384 (or 16 K for short) 36-bit words. The time of access to a word, called the *cycle time*, is the time required to enter an address in the address register, read the information from the selected location, and, if the reading operation is destructive, write the information back into the same location. The IBM 360/50 is a typical modern computer of average size. Depending on the model, the number of 32-bit words in its core memory can vary from 16K to 131K; the cycle time is 2 μsec. The important feature of core memory is that access time is independent of the location of a word. The time to access 100 words stored sequentially is the same as the time to access 100 words that have their addresses selected at random. For this reason core memory is called a *random access memory*.

Sometimes core storage is arranged on two levels. There is a *main* store and a slower *mass* or *bulk* store. The fast main store of the IBM 360/50 can be supplemented with a bulk store having a cycle time of 8 μsec and a capacity of 262K or 524K 32-bit words. Access is still fully random, and a word in the mass core store functions exactly like a word in the main store. Therefore, the mass store is simply an extension of the main store. A different approach is taken by CDC with their 6000 series machines, which have a cycle time of 1 μsec and main core memory ranging in size from 32K to 131K 60-bit words. Here information has to be transferred from the bulk store into the main store before it can be operated on, but transfer is very fast. Most efficient transfer is in blocks of 8 60-bit words, and transfer rates of 0.1 to 0.8 μsec for a word can be achieved (these figures depend on the sizes of the main store and the bulk store—the latter can vary in size from 126K to 2015K words).

The accessing mechanism of a magnetic core store is electronic. This means that the storage unit and the components that give access to it remain stationary during reading and writing. In the other storage devices that we shall consider the storage medium is a thin surface layer of magnetic material. Information passes between the computer proper and the storage device through a *read–write head*. Access to a particular region of the surface is gained by bringing the head in near contact with the region by mechanical means. Consequently access times for magnetic surface storage are longer than for core storage.

A magnetic *disk store* consists of a number of disks, their flat surfaces coated with magnetic material, in continuous rotation about an axis through their centers. The surfaces are divided into annular regions, called *tracks*. Access to a location in a disk store involves selection of disk, track, and the bit position at which the location starts. Some early disk stores had a single head, which could be a long distance from where it was needed. Nowadays there is at least one head for every disk; there may even be several heads permanently assigned to the one disk to reduce the amount of mechanical motion. The head nearest to the selected track is positioned over the track, and data transfer occurs when the motion of the disk carries the appropriate section of the track past the head. For our example we take the CDC 808 disk store, designed for use in CDC 6000 series computer systems. It provides storage for 13×10^6 60-bit words on 128 disk surfaces. The head positioning time varies from 20 to 110 msec. There is also a *latency time* to consider. This is the time one has to wait for the storage location to move up to the head after the head is in position. The maximum latency time is the time for one revolution of the disk. For the 808 this is 50.8 msec. In our example a head gives access to six adjacent tracks at once, and the maximal transfer rate of 168K words per sec is achieved when four heads transfer information at once.

Both access times and costs of *magnetic drums* are intermediate to those of bulk core and disk stores. The magnetic material is carried by the curved surface of a rigid rotating cylinder. The geometry permits more reading heads to be attached to a drum than to a disk of equivalent capacity, and more heads can be made to transmit information simultaneously. Hence waiting times and transfer rates are better than for disk stores. A typical drum (CDC 863) has a capacity of 4×10^6 6-bit characters; the average waiting time is 17 msec, and a maximal transfer rate of 2×10^6 characters per sec can be achieved. Some early computers, e.g., the IBM 650, had drums for their main stores, and were called "drum computers." Those drums, however, were very slow by modern standards. Because disks and drums are provided with addressing mechanisms they are sometimes called *random access devices*, but they are not as "random" as core memory.

The storage medium of a *magnetic tape* is supported by a flexible ribbon, up to about 2400 ft long. A typical tape unit, the IBM 729VI, moves tape at 112.5 in. per sec, and a 1 in. strip of tape can accommodate 800 rows of information, each row storing a 6-bit character. A maximum of 10×10^6 characters could be stored on a reel 2400 ft long, and the entire reel could be read in about 4 min. However, the maximal capacity cannot be achieved because information is stored in blocks, and gaps some $\frac{3}{4}$ inches long must be left between blocks to allow for stopping and starting between blocks. The combined stopping and starting time may be of the order of 5 msec. With large blocks, comprising, for example, 2000 36-bit words, the effect of the gaps on either capacity or transfer rate is barely noticeable (5% decrease in capacity, and a similar decrease in the maximal transfer rate of 15K words per sec).

In contrast to disks and drums, magnetic tape devices have no addressing mechanism. One simply sets the tape in motion and identifies a required block by counting the blocks that pass the reading head, or by reading identifying data stored within the blocks themselves. In either case this is a matter of programming, and the access time is determined by the sequence in which the blocks were originally stored on the tape. For this reason magnetic tape storage is called *sequential storage*. If data have to be accessed in more or less random manner, latency times of the order of minutes make tapes completely useless, but they serve very well as an inexpensive medium for permanent storage of bulky records and for some types of temporary storage.

A multiprogramming system processes several programs at the same time, and each of these programs may individually require more storage space than the main core store can provide. The users of the system certainly cannot be expected to cope with the intricate details of transfers between the main store and auxiliary storage devices in this environment, and the transfers must be scheduled and carried out by the system. There is, therefore, no obstacle to improving the well-being of the user by letting him assume that he has the computer all to himself, and that the main store at his disposal is nearly unlimited in size. Multiprogramming systems support this illusion by some variant of the *paging concept*. The entire memory of the computer is divided into pages of, say, 512 words, and the programmer can use a large number of these pages. The pages go in and out of core memory as required, but there cannot be too much of a delay between the request for a page and its arrival in core storage. The illusion would soon be shattered if magnetic tapes were used for the auxiliary memory.

9b. Storage of Arrays

Consider a function f with domain $A_1 \times A_2 \times \cdots \times A_t$. Let $|A_k| = n_k$, and denote the members of A_k by $a_{k1}, a_{k2}, \ldots, a_{kn_k}$. The $1, 2, \ldots, n_k$ are *subscripts* or *indices*, and we denote $f(a_{1i}, a_{2j}, \ldots, a_{tw})$ in terms of indices by $f_{ij\cdots w}$. Take a very simple case, a function f defined on $A \times B$, where $|A| = n$ and $|B| = m$. This function can be represented by an $n \times m$ matrix of its values, i.e., by an array of nm "boxes" arranged in the plane:

$$\begin{bmatrix} f_{11} & f_{12} & \cdots & f_{1m} \\ f_{21} & f_{22} & \cdots & f_{2m} \\ \vdots & \vdots & \vdots & \vdots \\ f_{n1} & f_{n2} & \cdots & f_{nm} \end{bmatrix}.$$

The indices determine the position of the box that stores f_{ij}; we find it at the intersection of the ith row and the jth column. The function with values $f_{ij\cdots w}$ has $n_1 n_2 \cdots n_t$ members, and it is represented by a t-dimensional array in t-space. The box that contains $f_{ij\cdots w}$ in this space is found at the intersection of vectors defined by subscripts $i, j, \ldots, w$ parallel to the axes of the space.

Fortran does permit us to refer to f_{ij} by the name `F(I,J)`, but all conventional computers still require a multidimensional array to be stored as a linear sequence of elements. The compiler must, therefore, contain a procedure that computes the actual address of an element in an array from its specification in terms of indices. The mapping from the multidimensional array to the linear array is not unique; here we shall quite arbitrarily assume that elements of the multidimensional array are stored in lexicographic order of their indices, e.g., $f_{111}, f_{112}, \ldots, f_{11n_3}, f_{121}, \ldots, f_{n_1n_2n_3}$. We want a *storage mapping function*, *loc*, defined on the indices as follows:

$$loc(i, j, \ldots, w) = loc(1, 1, \ldots, 1) + c_0 + c_1 i + c_2 j + \cdots + c_t w,$$

where the $c_0, c_1, \ldots, c_t$ are constants. The problem is to find the constants. To see how it is done let us consider an array A, dimensioned (10, 15, 7), where we want the address of element `A(5,7,4)`. This element comes fourth in the seventh row of the fifth plane of the array; i.e., it is preceded by 4 planes, 6 rows, and 3 elements. But a plane contains 15 × 7 elements, and a row 7 elements. Hence

$$loc(5, 7, 4) = loc(1, 1, 1) + 4(15 \times 7) + 6(7) + 3,$$

and for three-dimensional arrays in general we have

$$\begin{aligned} loc(i, j, k) &= loc(1, 1, 1) + n_2 n_3 (i - 1) + n_3 (j - 1) + (k - 1) \\ &= loc(1, 1, 1) - (n_2 n_3 + n_3 + 1) + n_2 n_3 i + n_3 j + k. \end{aligned}$$

It is now easy to see that in the general case of a t-dimensional array

$$\begin{aligned} c_0 &= -(n_2 \cdots n_t + n_3 \cdots n_t + \cdots + n_t + 1), \\ c_1 &= n_2 \cdots n_t, \\ c_2 &= n_3 \cdots n_t, \\ &\vdots \\ c_{t-1} &= n_t, \\ c_t &= 1. \end{aligned}$$

The part played by array bounds $n_1, n_2, \ldots, n_t$ in storage mapping is one reason why most programming languages that provide subscripted variables require the actual dimensions of arrays to be declared. Fortran permits up to three dimensions; in Algol the number of dimensions is unlimited, but implementers of the language must, of course, set some limit on this number. Some compilers, e.g., the Watfor Fortran compiler, generate a code that tests during execution of a program whether a reference to a subscripted variable lies outside the declared array bounds. If it does, and were to remain undetected, information might get written into locations that house other data or even the program itself, and the result would be a rather erratic behavior of the program. The increase in execution time due to these tests is well worth it in a compiler that is used mainly for program testing.

The standard higher-level languages (Fortran, Algol, PL/I) require all arrays to have a rectangular structure. Very often this results in considerable waste of storage space. For example, if a matrix A is *upper triangular*, i.e., if $a_{ij} = 0$ whenever $j < i$, then only elements with $j \geq i$ need to be stored. Similarly, nearly half the information contained in a *symmetric* matrix A is redundant because $a_{ij} = a_{ji}$, and again only the upper (or lower) triangle of the values has to be stored. Another example is the *tridiagonal* matrix in which only the elements a_{ij} such that $j = i$ or $j = i \pm 1$ are nonzero:

$$\begin{bmatrix} a_{11} & a_{12} & 0 & 0 & 0 & \dots \\ a_{21} & a_{22} & a_{23} & 0 & 0 & \dots \\ 0 & a_{32} & a_{33} & a_{34} & 0 & \dots \\ & & \vdots & \vdots & \vdots & \vdots \end{bmatrix}.$$

Nonzero elements of an upper triangular matrix A, dimensioned (N, N), can be stored in vector UPT of dimension M = N*(N+1)/2 as follows:

```
UPT(1)  UPT(2)  UPT(3)  UPT(4)  ...  UPT(M)
A(1,1)  A(1,2)  A(2,2)  A(1,3)  ...  A(N,N)
```

Element A(I,J) is in location UPT(K), where K = J*(J-1)/2 + I. A tridiagonal matrix T, dimensioned (N,N), can be stored in vector TRD of dimension M = 3*N - 2 as follows:

```
TRD(1)  TRD(2)  TRD(3)  TRD(4)  TRD(5)  ...  TRD(M)
T(1,1)  T(2,1)  T(1,2)  T(2,2)  T(2,3)  ...  T(N,N)
```

Element T(I,J) is in location TRD(K), where K = 2*J + I - 2. Alternatively, a tridiagonal matrix can be stored in three vectors:

```
        [T(1,2)  T(2,3)  T(3,4)  ...  T(N-1,N)]
[T(1,1)  T(2,2)  T(3,3)  T(4,4)  ...  T(N,N)]
        [T(2,1)  T(3,2)  T(4,3)  ...  T(N,N-1)]
```

Two upper triangular matrices A and B, both dimensioned (4, 4), say, can be stored as a single matrix AWITHB, dimensioned (5, 4), as follows:

$$\text{AWITHB} = \begin{bmatrix} \text{A(1,1)} & \text{A(1,2)} & \text{A(1,3)} & \text{A(1,4)} \\ \text{B(1,1)} & \text{A(2,2)} & \text{A(2,3)} & \text{A(2,4)} \\ \text{B(1,2)} & \text{B(2,2)} & \text{A(3,3)} & \text{A(3,4)} \\ \text{B(1,3)} & \text{B(2,3)} & \text{B(3,3)} & \text{A(4,4)} \\ \text{B(1,4)} & \text{B(2,4)} & \text{B(3,4)} & \text{B(4,4)} \end{bmatrix}.$$

Then A(I,J) = AWITHB(I,J), but B(I,J) = AWITHB(J+1,I).

9c. Sparse Matrices

In Example 3 of D.3.4 we represented a matrix X by a network. The network can be represented, in turn, by a different matrix, which we call Q (to make the example more interesting we have added elements $x_{51} = 7$ and $x_{54} = 3$):

$$Q = \begin{bmatrix} 1 & 1 & 2 & 2 & 3 & 4 & 4 & 5 & 5 & 5 & 5 \\ 1 & 2 & 1 & 3 & 5 & 1 & 4 & 1 & 3 & 4 & 5 \\ 4 & 3 & 2 & 4 & -1 & 8 & -2 & 7 & 1 & 3 & 2 \end{bmatrix}.$$

The first two elements of a column in Q define an arc, the third is the weight associated with the arc, or, in the terminology of arrays, a column stores the indices and the value of a nonzero element of X. Generalizing, let a matrix X have n rows and m columns, and let k be the number of nonzero elements in X. Normally the matrix would be stored in nm locations, but when $3k < nm$, a matrix such as Q gives a more economical representation. We call a matrix with many zero elements a *sparse* matrix. Unfortunately, if we had to multiply two sparse matrices stored in this fashion, the representation would prove rather awkward. It is reasonably easy to find all elements belonging to a given row of X because they occupy contiguous columns in Q, but the elements belonging to a given column are scattered throughout Q.

Therefore, instead of the representation of Figure 3.7, we shall consider a digraph in which there is a node for every nonzero element of the sparse matrix, and triples ⟨*row number*, *column number*, *value*⟩ are assigned as labels to the nodes. All nonzero elements belonging to the same row or column in the matrix are represented by adjoining nodes on a path in the digraph. Figure 9.1 shows the matrix of our example under this representation.

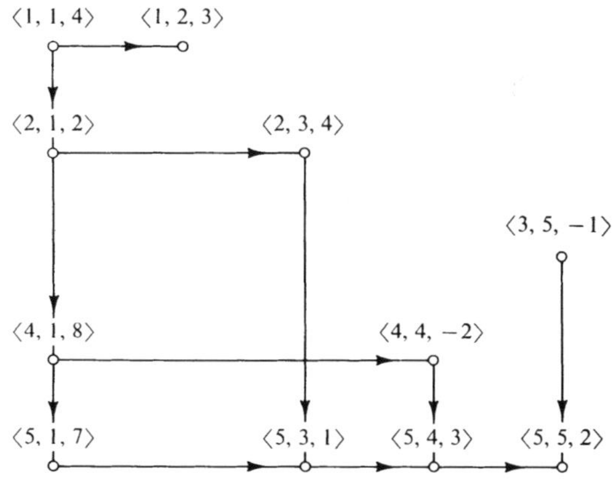

Figure 9.1

Now we can represent matrix X (or the digraph of Figure 9.1) by a 4×11 matrix QQ in which the first three rows are a copy of Q, but a fourth row contains pointers linking elements that belong to the same column in X:

$$\text{QQ} = \begin{bmatrix} 1 & 1 & 2 & 2 & 3 & 4 & 4 & 5 & 5 & 5 & 5 \\ 1 & 2 & 1 & 3 & 5 & 1 & 4 & 1 & 3 & 4 & 5 \\ 4 & 3 & 2 & 4 & -1 & 8 & -2 & 7 & 1 & 3 & 2 \\ 3 & 0 & 6 & 9 & 11 & 8 & 10 & 0 & 0 & 0 & 0 \end{bmatrix}.$$

We shall call this the four-row representation of the matrix. Let us look at column 1 of X. Elements that belong to this column are identified by a 1 in the second row of QQ. They are stored in columns 1, 3, 6, and 8 of QQ, and the entries in the fourth row of QQ do send us from the first column to the third, then to the sixth, and finally to the eighth column. The zero in QQ(4,8) indicates that the element stored in QQ(3,8) is the last nonzero element belonging to column 1 of X. We supplement QQ with two "entry" vectors NC and NR:

$$\text{NC} = [1 \quad 2 \quad 4 \quad 7 \quad 5]$$
$$\text{NR} = [1 \quad 3 \quad 5 \quad 6 \quad 8 \quad 12].$$

The value in NC(J) points to the column of QQ that contains the first nonzero element of column J in X; if NC(J) = 0, then all elements of column J are zero. NR(I) contains the number of the column in QQ that stores the first nonzero element of row I in X and, since NR(I+1) contains a pointer to the first nonzero element of row I+1, there are NR(I+1) - NR(I) contiguous columns storing elements of row I. If all elements of row I are zero, then NR(I) = NR(I+1). This convention makes it necessary for NR to contain one more element than there are rows in X; this last element contains a number that is 1 greater than the number of columns in QQ.

The normal procedure for forming a matrix product C = AB is

```
      DIMENSION A(N,K), B(K,M), C(N,M)
        ⋮
      DO 5 I = 1,N
      DO 5 J = 1,M
      C(I,J) = 0.
      DO 5 L = 1,K
    5 C(I,J) = C(I,J) + A(I,L)*B(L,J)
```

We shall now give a program for finding C = AB when matrices A and B have four-row representation.

ALGORITHM 9.1 Formation of the matrix product $C = AB$. The N × K matrix A with KA nonzero elements is specified by arrays QA, NCA, NRA. The K × M matrix B with KB nonzero elements is specified by QB, NCB,

NRB. Matrix C has the normal representation. Values of NP and KP are greater by 1 than those of N and K, respectively. For convenience elements of the matrices are assumed integers.

```
      SUBROUTINE ABYB (QA,NCA,NRA,QB,NCB,NRB,C,KA,KB,
     1                 N,NP,K,KP,M)
      INTEGER QA(4,KA), QB(4,KB), C(N,M)
      DIMENSION NCA(K), NCB(M), NRA(NP), NRB(KP)
      DO 50  I = 1,N
      DO 10  J = 1,M
   10 C(I,J) = 0
C   TEST FOR EMPTY ROW IN A
      IF (NRA(I).EQ.NRA(I+1)) GO TO 50
      DO 40  J = 1,M
      KANOW = NRA(I)
      KBNOW = NCB(J)
C   TEST FOR EXHAUSTED COLUMN IN B OR ROW IN A
   15 IF (KBNOW.EQ.0.OR.KANOW.EQ.NRA(I+1)) GO TO 40
C   FIND K SUCH THAT A(I,K), B(K,J) BOTH NONZERO
      IF (QA(2,KANOW)-QB(1,KBNOW)) 20,25,30
   20 KANOW = KANOW + 1
      GO TO 15
   25 C(I,J) = C(I,J) + QA(3,KANOW)*QB(3,KBNOW)
      KANOW = KANOW + 1
   30 KBNOW = QB(4,KBNOW)
      GO TO 15
   40 CONTINUE
   50 CONTINUE
      RETURN
      END
```

The four-row representation of matrices cannot be implemented by normal means when we lack a prior knowledge of the number of nonzero elements in a matrix. In the next section we shall develop techniques that will enable us to store arrays without dimensioning them.

9d. Storage Allocation at Execution Time

Fortran requires constant array bounds to be specified before execution of a program. Adjustable dimensions are permitted in subprograms, but the actual arrays that take the place of the dummy "place holders" during execution of a subprogram must have been given constant dimensions. For example, if the program contains the statement

```
CALL SUB (X,5)
```

and the definition of SUB starts with

```
SUBROUTINE SUB (ARRAY,N)
DIMENSION ARRAY(N,N)
```

then somewhere X must have been given constant bounds X(5,5) In A.9.1 matrix C was stored in the conventional form because we did not have the information that would have enabled us to declare a four-row array to take this matrix. In Algol and PL/I programs array bounds can be specified and storage allocated to the arrays during execution of a program, but this facility still does not solve our problem. We want to avoid declaring some arrays altogether.

One way of solving the problem is to store all arrays of unknown dimensions in a large vector, and to use a second array for bookkeeping. For example, if we declare the arrays by

```
DIMENSION STORE(10000), INFO(100,2)
```

then we have 10,000 locations reserved for storing arrays, and the 100 rows of INFO can be used to store information about 100 arrays at a time. We do not refer to an array by a name, but by the number K of the row in INFO that contains information pertaining to it. INFO(K,1) contains the subscript of the element of STORE at which the stored array begins, and INFO(K,2) contains its size. Further, there is a location INDEX, which contains a pointer to the first element in STORE that has not been used up.

To see how the scheme works assume that two arrays are to be read from cards and put into STORE in the conventional representation of matrices, that their product is to be formed, also in STORE, and that finally the product is to be converted to the four-row representation. We assume that the first card of the input contains two numbers, which specify the number of rows and columns, respectively, of the first matrix, and that this card is followed by cards containing the elements of the matrix punched in lexicographic order of the indices. This set of cards is followed by a second set that defines the second matrix. We have to assume that the number of columns in the first matrix is equal to the number of rows in the second (otherwise the product would be undefined). The program that follows reads the two matrices and forms their product, but the conversion to the four-row representation is left as an exercise (Exercise 9.9).

```
      COMMON INDEX, STORE(10000), INFO(100,2)
      DIMENSION NSTORE(10000)
      EQUIVALENCE (STORE(1), NSTORE(1))
C   THE EQUIVALENCE PERMITS US TO REFER TO THE STORAGE
C   ARRAY BY TWO NAMES.  THIS GETS AROUND THE
C   DIFFICULTY CREATED BY THE FORTRAN NAMING
C   CONVENTION FOR INTEGERS AND REALS.
      INDEX = 1
      CALL READER (N,K,1)
      CALL READER (K,M,2)
      INFO(3,1) = INDEX
      INFO(3,2) = N*M
      DO 10  I = 1,N
      DO 10  J = 1,M
      INOW = INFO(1,1) + (I-1)*K
      JNOW = INFO(2,1) + (J-1)
      STORE(INDEX) = 0.
      DO  5  L = 1,K
      STORE(INDEX) = STORE(INDEX)
     X                  + STORE(INOW)*STORE(JNOW)
      INOW = INOW + 1
    5 JNOW = JNOW + M
   10 INDEX = INDEX + 1
      :
            (Conversion to four-row representation)
      :
      SUBROUTINE READER (I,J,K)
      COMMON INDEX, STORE(10000), INFO(100,2)
      READ (5,100) I,J
      INFO(K,1) = INDEX
      INFO(K,2) = I*J
      IHI = INDEX + INFO(K,2) - 1
      READ (5,101) (STORE(II), II = INDEX, IHI)
      INDEX = IHI + 1
      RETURN
  100 FORMAT (2I5)
  101 FORMAT (8F10.0)
      END
```

Assume that N=10, K=5, M=20, and that there are 15 nonzero elements in the product matrix. Then, after the four-row representation has been found, STORE and INFO appear as shown in Figure 9.2.

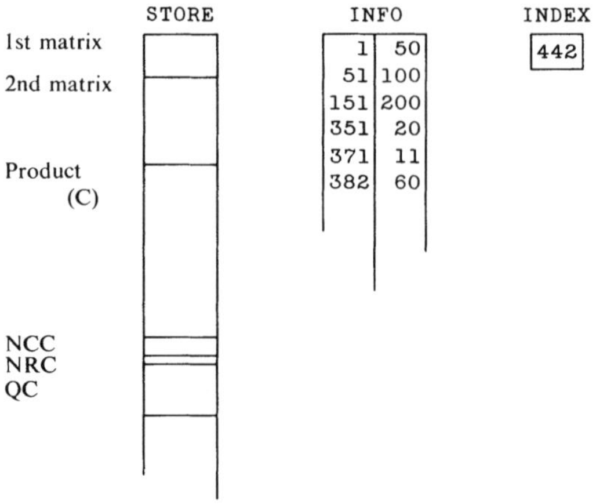

Figure 9.2

Assume now that we no longer require the first three matrices in STORE. The storage space occupied by these arrays can be used again. We leave as an exercise (Exercise 9.10) the writing of the program that deletes the three arrays and shifts the remaining arrays to the beginning of STORE. Figure 9.3 shows the appearance of the storage regions after the deletions.

The dynamic storage allocation scheme described above is too specific. We would prefer to relegate much of the processing to general subprograms. For example, we would like to have a subroutine that compacts STORE by deleting all arrays that are no longer required. This subroutine could be called by other routines whenever the storage capacity of STORE is exceeded. A second unsatisfactory feature of the present scheme is that arrays are specified by row numbers of INFO instead of mnemonic names.

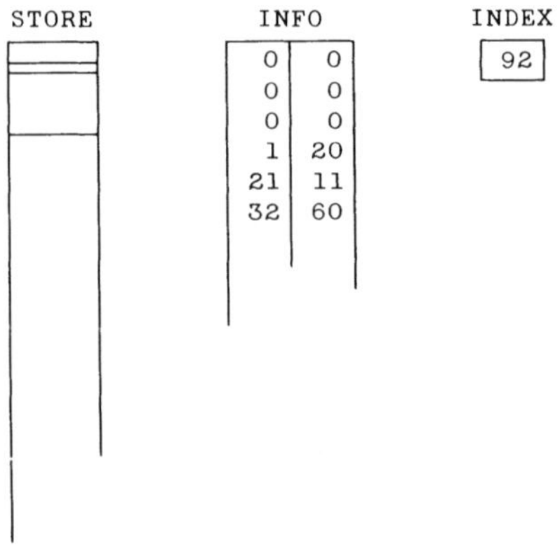

Figure 9.3

Let us analyze the compaction problem. Assume that a third column is added to INFO, and that INFO(k, 3) contains 1 if array k is still required, -1 if it is no longer in use, and 0 if there is no such array. The programmer has the responsibility of putting -1 in the appropriate location in INFO when he has finished with an array. The compaction subroutine, which we shall call GRBAGE, can then compact STORE by deleting arrays marked -1, moving all arrays marked 1 to the head of the store, and making appropriate changes in INFO. Looking again at the situation depicted in Figure 9.3, assume that a new array has been read into STORE. The first row of INFO was empty, and we assume that the information pertaining to the new array has been stored there. The first row of INFO then corresponds to the fourth array in STORE. But, if the order of arrays in STORE differs from the order in which information about them is stored in INFO, we are in serious difficulties with the design of GRBAGE; the compacting procedure cannot then simply go down INFO, compacting as it moves from row to row. The easiest solution is to require that each row of INFO be used only once, and that one cannot use row I+1 before row I has been used. Under this scheme information about the new array would be put into row 7 of INFO. Another solution is for GRBAGE to shift rows of INFO as well, so that the position of an array in STORE always agrees with the number of the row in INFO that contains information about it. But how will the main program and the other routines know what changes have been made to INFO?

We take the second alternative. However, instead of worrying how to pass information about changes made to INFO out of the subroutine, we change the whole scheme so that the main program is no longer concerned with INFO. This is done by introducing mnemonic names for the dynamically stored arrays, and specifying all operations on these arrays by means of subprograms. The main program then becomes just a sequence of calls to the subprograms. We shall use integer variables for naming arrays in STORE. If variable NN refers to an array, then it is made to contain the number of the row in INFO that contains information about this array. INFO is augmented with yet another column; this column contains references to the names of arrays. Assume that array NN is the 27th array in STORE. Then the location NN contains the value 27, and INFO(27,4) contains a reference to location NN. Now assume that the markers in INFO(25,3) and INFO(26,3) are both -1, and that GRBAGE is called. If all the other 24 arrays preceding NN are still in use, then GRBAGE makes NN the 25th array in STORE, transfers contents of the 27th row of INFO to the 25th row, and changes the value in location NN to 25. Establishing access to variable NN is the trickiest part in the design of the scheme because Fortran does not provide the address of NN. This part of the problem is solved by asking the programmer to put names of all arrays that at some time will occupy STORE into COMMON, as in the following example:

```
      INTEGER A, B, C, XA, XB, ......
      COMMON A, B, C, N1, N2, N3, XA, XB, ......
```

Let us now write an initiating routine START, which is to be called by the main program before anything else is done:

```
      SUBROUTINE START
      COMMON NAMES(250)
      COMMON/INTERN/INDEX,INFROW,STORE(10000),
     X               INFO(100,4)
      DO 1 K = 1,250
    1 NAMES(K) = K
      DO 2 I = 1,100
      DO 2 J = 1,4
    2 INFO(I,J) = 0
      INDEX = 1
      INFROW = 1
      RETURN
      END
```

Variable N3 has the same location as NAMES(6), and, after return from START, this location contains the value 6. Straight after the call to START we may decide to make a call to a subroutine that reads in an array from cards. Assume that the array has size 50, and that we have decided to call it N3. Then the input routine puts (1, 50, 1, 6) into the first row of INFO and changes the 6 in NAMES(6) to 1. After N3 has been stored INDEX contains the value 51, and INFROW, which points to the row of INFO that is to be used next, contains 2. If next we read matrix N1 (size 64) from cards, the second row of INFO is made (51, 64, 1, 4), and NAMES(4) receives the value 2. INDEX and INFROW become 115 and 3, respectively. Taking this example still further, indicate that array N3 is not to be saved by making INFO(N3,3) equal to -1, and then call GRBAGE. This subroutine takes the 6 from INFO(1,4) and restores it to NAMES(6); after this the name N3 may be used to refer to another array. Then it shifts array N1 to locations STORE(1) to STORE(64), sets INDEX equal to 65 and INFROW equal to 2, puts (51, 64, 1, 4) into the first row of INFO, clears the second row to zero, and changes the 2 in NAMES(4) to 1.

An important consideration in the design of the system is what to do with dimensions of arrays. They could be stored with the arrays themselves—the block occupied by an array in STORE then contains several locations additional to those that store the elements of the array; these locations contain the dimensions of the array. In what follows we shall assume that this course has been taken. Of course, some implementers may prefer to transmit dimensions in argument lists of calls. The processing routines are then simpler, and this may outweigh the inconvenience of having to clutter up argument lists with dimension data.

For our final example we assume that the calls

```
      CALL CONVRT (C,C4)
      NUMBER = KOUNT (C4)
      CALL DELETE (C)
```

convert a matrix C to its four-row representation C4, determine the number of nonzero elements in the matrix, and indicate that C need not be saved after this. Function KOUNT and subroutine DELETE are simply

```
      FUNCTION KOUNT(NNN)
      COMMON/INTERN/INDEX,INFROW,STORE(10000),
     X              INFO(100,4)
      KOUNT = INFO(NNN,2)/4
      RETURN
      END

      SUBROUTINE DELETE(NNN)
      COMMON/INTERN/INDEX,INFROW,STORE(10000),
     X              INFO(100,4)
      INFO(NNN,3) = -1
      RETURN
      END
```

GRBAGE has to be called whenever the value in INDEX exceeds 10,000, or the value in INFROW exceeds 100. Therefore, before a start is made on the actual conversion, subroutine CONVRT calls GRBAGE if INFROW exceeds 100:

```
      IF (INFROW.GT.100) CALL GRBAGE
      IF (INFROW.GT.100) GO TO 111
      :
  111 WRITE (6,100)
      CALL EXIT
  100 FORMAT(30H TOO MANY ARRAYS — EXIT CALLED)
      :
```

Assume that we build up the final value of INDEX by adding 4 whenever a nonzero element is transferred from C to C4. If the value exceeds 10,000, then one calls GRBAGE, tests INDEX again and terminates if it still exceeds 10,000, adjusts some parameters (C4 has certainly been shifted to a new region in STORE, and C may also have been shifted), and resumes the conversion process. It is probably more convenient to have the error exits in GRBAGE itself.

This sketchy introduction to the implementation of dynamic storage allocation facilities in Fortran has taken us from a relatively simple technique to a complex system of subprograms. We have given only a few examples of the subprograms that a complete system would contain. Although the system can be written entirely in Fortran, and the calls to the subprograms are Fortran statements, we have, in effect, a new programming language. The system extends the power of Fortran by giving a non-Fortran interpretation to the variables that refer to arrays in STORE. In Fortran I, J, and K are integer variables, and K = I + J is a meaningful statement. If, however, we use I, J, and K to represent dynamically stored arrays, then the variables are no longer integer variables in the usual sense (K = I + J is then meaningless). The powerful technique for creating new programming languages by assigning new interpretations to certain constituents of an existing language is known as *embedding*; we shall discuss the technique further in Section 12c.

Calls to subprograms are the statements of our new language; the language is, in fact, defined by the subprograms. Once the representation of data and the mechanism for access to the data have been decided on, and a few basic routines that implement the access mechanism have been written, one is at liberty to write as few or as many additional subprograms as one wishes. Even a fairly complex system of some 30 subprograms should not take an experienced programmer more than a few days to write. It is a simple matter to add new facilities; one simply writes a few new subprograms.

Notes

Apart from the highly technical [Ri67] (with a bibliography of some 1000 entries), I know of no book that I would be prepared to recommend for reference on storage media and their properties. This is so because in the rapidly advancing science and technology of computing the fastest changes take place in the technology of storage devices. Consequently periodic literature alone is capable of keeping up with the changes. *Scientific American* carries an occasional survey article. Information on particular devices can be found in *Datamation*, *Computers and Automation*, and in the "Products" section of *CACM*.

Addressing of elements in multidimensional arrays is discussed in [He62, Hi62b]. A compaction method for sparse arrays is given in [Ba63]. [Ga64] contains an Algol procedure for the control of a system that changes array sizes dynamically. A Fortran-embedded programming system that constitutes one solution of the dynamic storage allocation problem is described in [Sa68b].

Exercises

9.1 (b) A Fortran array `X` is dimensioned (10, 10, 10). Assuming that elements of arrays are stored in lexicographic order of their indices, find displacements of the locations of `X(5,6,7)`, `X(10,9,8)`, and `X(1,5,9)` relative to the location of `X(1,1,1)`.

9.2 (b) In the definition of the storage mapping function we assumed that every subscript has 1 for its lower bound. In Algol the lower bound of a subscript can be any integer. Define the storage mapping function for this general case.

9.3 (b) Assume that we have a computer in which the word size is 36 bits, and that an array A of dimensions (L, M, N), which has all of its elements belonging to $\{0, 1\}$ is stored in a compact form, 36 elements to a word, in a linear array `LOGIC`. `LOGIC(1)`, `LOGIC(2)`, `LOGIC(3)` contain the dimensions, and the remaining words of `LOGIC` contain the elements of A. Assembler language subprograms for storing and fetching elements of A can be written, and these subprograms can be referenced by a Fortran program:

```
CALL STORE (NUMBER,LOGIC,I,J,K)
```

causes `NUMBER` to be stored as a_{ijk}, and the function reference

```
NFETCH (LOGIC,I,J,K)
```

supplies the value of a_{ijk}. Both these subprograms require a subroutine `PLACE`, which, given arguments `LOGIC, LDIM, I, J, K,` where `LDIM` is the dimension of `LOGIC`, returns the subscript of the element of `LOGIC` in which a_{ijk} is to be found, and its bit position. Write this subroutine in Fortran. What should be stored in the first three words of `LOGIC` in preference to the dimensions for more efficient performance of `PLACE`?

9.4 (b) Given matrices

$$A = \begin{bmatrix} 1 & 2 & 3 & 4 \\ 0 & 3 & 4 & 5 \\ 0 & 0 & 5 & 6 \\ 0 & 0 & 0 & 7 \end{bmatrix} \quad \text{and} \quad B = \begin{bmatrix} 1 & 1 & 3 & 5 \\ 0 & 3 & 3 & 5 \\ 0 & 0 & 5 & 5 \\ 0 & 0 & 0 & 7 \end{bmatrix},$$

write a program that reads the two matrices into a compact 5×4 array and finds the matrix products AB and BA, which are also upper triangular matrices, in two vectors (each of dimension 10).

9.5 (b) Apply the principle illustrated by Figure 5.17 to the design of a storage scheme for arrays. Describe your scheme fully, and give a detailed analysis, with reference to a particular computer with which you are familiar, of its advantages and disadvantages as compared to a scheme based on the storage mapping function.

9.6 (c) Give four-row representations, including the entry vectors, for matrix X of Example 1 of D.3.12 and for matrix X^3 (see the example of Th.3.6).

9.7 (c) Given four-row representations of matrices A and B, subroutine ABYB of A.9.1 finds the product of the two matrices. Write similar routines for finding

(i) the sum $A + B$,
(ii) the transpose A', where $(a')_{ij} = a_{ji}$.

9.8 (c) Given an N × M matrix MAT with K nonzero elements. Write a subroutine that converts MAT to its four-row representation. The subroutine starts as follows:

```
SUBROUTINE FORROW (MAT,N,M,MATFOR,K,NC,NR,NP)
DIMENSION MAT(N,M), MATFOR(4,K), NC(M), NR(NP)
```

9.9 (d) Incorporate the conversion procedure of Exercise 9.8 in the matrix multiplication program of Section 9d.

9.10 (d) With reference to the storage scheme depicted in Figure 9.2, write a subroutine that deletes the first K arrays stored in STORE, shifts all remaining arrays to the head of STORE, and adjusts INFO and INDEX accordingly.

9.11 (d) (major project) Write a set of subprograms to provide Fortran with dynamic storage allocation facilities. (The proper design of the system is more important than the coding of the routines.)

CHAPTER 10

Pushdown Stores, Lists and List Structures

10a. Pushdown Stores

Basically a pushdown store is a set of storage locations, which are initially empty. As data are stored in the pushdown store, the store "remembers" the order in which they were stored. When a datum is to be fetched from a conventional store, a copy of the datum is actually moved. The datum itself remains in the store and can be fetched again and again. In the case of a pushdown store a fetch instruction moves a datum right out of the store. Only one datum is accessible in a pushdown store at any one time; in the case of a *last in–first out* (LIFO) store this is the most recently stored item.

We can picture a LIFO store as a stack of buttons on which information is inscribed. LIFO stores are in fact very often called *stacks*. Putting a button on an empty table constitutes the first storing operation in our illustration. The next storing operation places a second button on top of the one already on the table. Each subsequent storing operation increases the height of the stack of buttons. To get a datum from the stack we must take the topmost button, i.e., the one that was stored last. There is no direct access to the first button stored; pulling it out would collapse the stack. The only way of getting at this button is by removal of the buttons above it, one by one, until the last button is exposed. Its removal leaves the stack empty. The term *pushdown store* originates from a similar illustration. In some parts of the world bus conductors are provided with coin dispensers that operate as follows. The dispenser is a hollow tube closed at the bottom and partly closed by a semi-circular rim at the top. In an empty dispenser a disk is held against the rim by a spiral spring occupying the entire tube. The dispenser is filled by pushing coins in under the rim, thus pushing down the disk. The pressure of the spring behind the disk holds the stack of coins in place between the rim at the top and the disk at its bottom. A coin being pushed into the dispenser pushes down the coins already there and becomes the topmost coin. In dispensing the topmost coin is slid out from under the rim. Since it is at the top, the coin that went in last comes out first. The remaining coins are pushed upward by the spring and the coin that was immediately below the dispensed coin pops into the topmost position.

One advantage that a stack has over a conventional store is that its user does not have to worry about addresses. Only two instructions are needed: a storing instruction, which we shall designate PUSH, and a fetching instruction to which we give the name IPOP. When stacks are considered in the context of the theory of automata they have unlimited depth, but in practice we must decide on a workable limit. Let us take 100 for the limit here. Then an array, say N(1) ··· N(100), can be set up as a stack simply by providing a pointer, which we shall call IP, and by making sure that IP points at all times to the topmost datum in the stack. Initially IP is given the value zero. Instruction PUSH increments the value of IP and inserts a datum in the location to which IP then points. Instruction IPOP fetches the item to which IP points and then decrements IP. Figure 10.1 gives a trace of the changes in the appearance of the array N during execution of a sequence of PUSH and IPOP instructions. The appearance of the array and the value of IP *after* execution of an instruction are given in the same row as the instruction. The contents of the stack are indicated by shading.

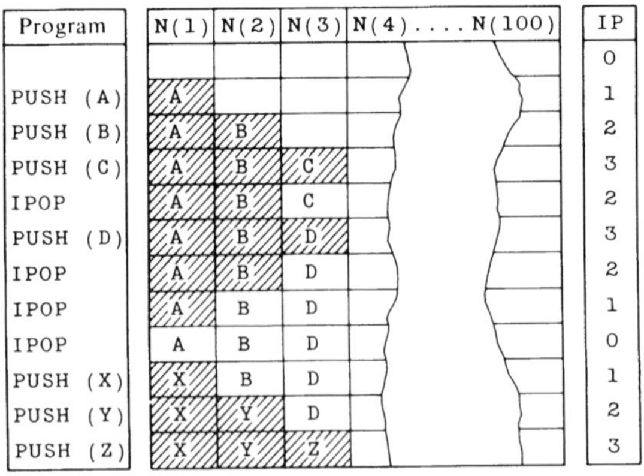

Figure 10.1

It is very easy to set up a Fortran array as a working stack. All that is needed is an initiating routine that sets IP to zero, and two routines that take the functions of PUSH and IPOP, respectively.

```
      SUBROUTINE SKOPEN
      COMMON N(100), IP
      IP = 0
      RETURN
      END

      SUBROUTINE PUSH (IN)
      COMMON N(100), IP
      IF (IP.EQ.100) GO TO 999
      IP = IP + 1
      N(IP) = IN
      RETURN
  999 WRITE (6,100)
      CALL EXIT
  100 FORMAT (1H1,28HSTACK OVERFLOW - EXIT CALLED)
      END

      FUNCTION IPOP (IT)
C   BOTH IPOP AND IT RETURN THE DATUM
      COMMON N(100), IP
      IF (IP.EQ.0) GO TO 999
      IT = N(IP)
      IPOP = IT
      IP = IP - 1
      RETURN
  999 WRITE (6,100)
      CALL EXIT
  100 FORMAT (1H1, 29HSTACK UNDERFLOW - EXIT CALLED)
      END
```

Note that the main program must have access to IP to enable it to test the stack for emptiness. Access is through COMMON. Therefore, initiation can be performed in the main program itself; i.e., we can make do without a separate initiating routine.

Perhaps the simplest example of the use of the stack is the inversion of the order of elements in a vector. Suppose that we are given an unknown nonzero number of nonzero elements (<100) in the low end of an array LOT, delimited by a zero value. The elements are to be stored in reverse order in the same array. The following sequence of Fortran statements will do the job:

```
      DO 20 K = 1,100
      IF (LOT(K).EQ.0) GO TO 25
20    CALL PUSH (LOT(K))
25    K = K - 1
      DO 30  J = 1,K
30    LOT(J) = IPOP (IT)
```

The order in which data go into the stack determines the order in which they will come out, and this is the reverse of the input order. Therefore, the ease with which the stack can be made to deal with the reversal problem should cause no surprise. Indeed, all applications that involve the use of stacks are concerned with change of order.

In another type of restricted access device, which can still be called a pushdown store, data are pushed down at one end, but access to the stored data is only from the other end of the store. This type of store has come to be known as a *queue*. It is called also a *first in–first out* (FIFO) store. A Fortran implementation of a queue is more difficult than the implementation of a stack. Still more difficult to implement is a scheme in which storage and access is at either end of the data area.

10b. Prefix, Postfix, and Infix Formulas

In Section 5b we showed how arithmetic expressions could be represented by trees. We shall now deal with the compilation of arithmetic expressions. The use of stacks in this context has achieved the distinction of a classic in computer science. In Section 5b we described what is meant by a completely parenthesized expression. For example, the Fortran statement

```
A = B*C + D*(E-F)
```

becomes

```
(A=((B*C)+(D*(E-F))))
```

when completely parenthesized. The purpose of the parentheses is to avoid having to specify a hierarchy of operations. At the other extreme we have parenthesis-free notations. The above statement becomes

```
=A+*BC*D-EF
```

in *prefix* notation (see Section 5b), or

```
ABC*DEF-*+=
```

in *postfix* notation. Again the hierarchy of operations does not have to be specified.

Let us now see how a postfix formula would be compiled. A.10.1 sets up a table of entries $n_1 \# n_2$, where n_1 and n_2 stand for variable names (or locations) and # stands for an operator. The result of the operation of row i is stored in an auxiliary location $T.i$.

ALGORITHM 10.1 Write the statement to be converted to an operand–operator–operand table as $s_1 s_2 \cdots s_k$.

1. Set $i = 1$; set $J = 1$.
2. Set $S = s_i$. If S is an operator, go to 4.
3. Push down S; set $i = i + 1$; go to 2.
4. Pop up n_2; pop up n_1; enter $n_1 S n_2$ in row J of table.
5. If stack is empty, stop.
6. Set $S = T.J$; set $J = J + 1$; go to 3.

Example

Applied to `ABC*DEF-*+=` the algorithm produces Table 10.1. It is then easy to compile the statement: Simply replace each line of the table by an appropriate "chunk" of machine code. There is really no need to generate the table as such. In Step 4, instead of entering $n_1 S n_2$ into a table, the appropriate machine commands can be generated and put out directly.

TABLE 10.1

OPERAND–OPERATOR–OPERAND TABLE

Row	Operand–operator–operand		
1	B	*	C
2	E	−	F
3	D	*	$T.2$
4	$T.1$	+	$T.3$
5	A	=	$T.4$

There still remains the question: How does one get the statements into postfix form? We observe that the order of the variable names is the same in all notations. Only the order of the operators and their locations change. We shall see that the reordering of the operators can be performed by a stack with the aid of a table of priorities.

ALGORITHM 10.2 The following algorithm for converting an infixed arithmetic statement to postfix form uses a priority table:

Operator—	(	)	=	+ or −	* or /	**
Priority—	0	1	2	3	4	5

The priority of a symbol s will be denoted by $p(s)$, e.g., $p(+) = 3$. Write the statement as $s_1 s_2 \cdots s_k$. We assume that the statement does not contain + or − as unary operators.

1. Set $i = 1$.
2. Set $S = s_i$. If S is (, then go to 9.
3. If S is a variable name, transfer it to the output string and go to 10.
4. If stack is empty, go to 9.
5. Pop up symbol n from the stack.
6. If $p(n) \geqq p(S)$, transfer n to output string and go to 5.
7. If S is) and n is (, then go to 10.
8. Stack n.
9. Stack S.
10. Set $i = i + 1$. If $i \leq k$, then go to 2.
11. Empty stack into output string and stop.

Example

Table 10.2 illustrates the conversion of the statement

```
A = B*C + D*(E - F)
```

to postfix form. In the table the stack status and the additions to the output string refer to the state of affairs on reaching Step 10.

TABLE 10.2

GENERATION OF POSTFIX FORMULA

s_i	Stack status	Addition to output string
A		A
=	=	
B	=	B
*	=*	
C	=*	C
+	=+	*
D	=+	D
*	=+*	
(	=+*(	
E	=+*(	E
−	=+*(−	
F	=+*(−	F
)	=+*	−
	Emptying of stack	*
		+
		=

Finally we consider again the representation of a sort tree created by A.5.2. By putting arrays D and P of Figure 5.15 through a procedure that uses a stack we can print elements of D in ascending order.

ALGORITHM 10.3 Figure 10.2 is a flowchart of a program for the output of elements in a sort tree in ascending order. D is the array of data, and P is the array of pointers (see Figure 5.15).

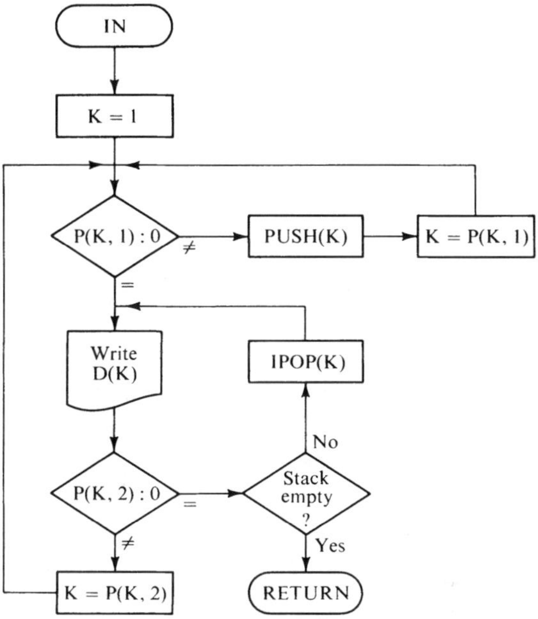

Figure 10.2

10c. Storage Levels for a Pushdown Store

In the theory of automata the concept of a stack calls for unlimited depth. Although we cannot achieve this ideal, a finite stack looks to the user just like an infinite one until it overflows. The deeper we make our finite stack, the smaller the probability that overflow will occur. For the practical applications discussed in the preceding section we certainly do not need deep stacks. However, we might find a stack of very great depth useful for experiments on automata. Therefore, we may have to set ourselves the design criterion that our stack should be deep. But if we try to satisfy this criterion by assigning, say, 10,000 fixed locations to the stack, and the number of symbols in the stack does not exceed 10, say, throughout a particular run, storage space is tied up unnecessarily.

Storage space is conserved if the stack is made extensible by changing from static to dynamic storage allocation. In effect, we provide two stacks, a fixed foreground stack and a dynamic background stack. The foreground stack consists of, say, 200 fixed contiguous locations. If a PUSH instruction would cause overflow of the fixed stack, the 100 symbols at the bottom of this storage area are moved out and stacked as a block in the dynamic stack. Then the top 100 symbols are moved to the bottom of the fixed stack and the pointer adjusted accordingly. Alternatively, they may be left physically in place, but made to look to the system as if they had been shifted. To achieve this effect one needs to divide the 200 locations into two blocks of 100 locations and provide a rather elaborate switching scheme for pointers.

If now an IPOP instruction finds the fixed stack empty, the topmost block of 100 symbols is popped into it from the background stack. Since the dynamic stack is so organized that no space is tied up that is not in actual use, this at once releases the 100 locations for other use. The background stack is then simply an extension of the fixed stack, and a user need not even know of its existence. We have suggested a block size of 100 rather than 200 to avoid extensive shifting of blocks.

The blocks of the background stack can be left in core storage, or they can be transferred to disk or drum. If the operating system under which the program is run provides file management capabilities, then the transfers of blocks can be effected by calling the file management system into action. The details of the organization of the background stack will be discussed in Section 10i.

10d. Lists—Introductory Concepts

Very often the storage needs of a program cannot be predicted in advance. One solution of the problem was given in Section 9d. For example, instead of basing the design of a stack on a foreground stack and a background stack of overflow blocks, as we did in the preceding section, we would place the entire stack in an array storage area of the type described in Section 9d. However, implicit in the design of the system of Section 9d is the assumption that only one array is "active" at any one time. If, then, we wished to make use of more than one stack simultaneously, this dynamic storage arrangement would be inapplicable. The system discussed in the preceding section would not be very satisfactory either. Assume that we are going to have at most n stacks in use at any one time. Then we could set aside a matrix of 200 rows and n columns for the stack storage area, and, instead of having a single pointer, provide a vector of n pointers for the n stacks. The permanent assignment of $200 \times n$ locations to the stacks is rather wasteful of space. Our purpose now is to look at more economical schemes for organizing an area of store in such a way that on demand more storage space can be assigned to any one of a number of sets of data at any stage of execution of a program.

For inspiration we look to the IBM 650, a rather antique computer. In most present day computers instructions have in general the format

(Operation code)(Address of operand),

and normally instructions are executed in the order in which they are stored unless explicit transfer instructions override the normal sequence. The IBM 650 format of instructions was

(Op. code)(Addr. of operand)(Addr. of next instruction).

The storage medium was a rotating magnetic drum. Since the address of the next instruction was always given explicitly, the instructions could be distributed over the drum in any order. The purpose was optimization: The instructions were to be so placed that, by the time the current instruction had been executed, the drum would have brought up the next instruction.

To give another illustration of the concept let us look at the Fortran program segment

```
IT = N(IP)
IPOP = IT
IP = IP - 1
```

Here the results depend on the order of the instructions. If, however, we rewrite the segment as in Figure 10.3, i.e., if we provide statement labels and pointers (the `GO TO` statements), then the boxed statement pairs can be shuffled at will without affecting the results of the program.

It does not take long to realize that if a program can be stored in nonconsecutive locations, the same can be done with data, i.e., that a scheme of

```
      GO TO 5
   +--------------+
   | 5 IT=N(IP)   |
   |   GO TO 6    |
   +--------------+
   +--------------+
   | 6 IPOP=IT    |
   |   GO TO 7    |
   +--------------+
   +--------------+
   | 7 IP=IP-1    |
   |   GO TO 8    |
   +--------------+
    8 .........
```

Figure 10.3

pointers can represent the logical ordering of data just as well as the conventional interpretation that consecutive memory locations contain consecutive items of data. A data storage scheme in which the logical order of data is determined by pointers is known as a *list storage* scheme, and a vector of data stored under such a scheme is called a *list*.

To see how a list storage scheme works let us take an array of dimensions 100×2, say. We shall call the first cell in a row the *data cell*, and the second the *pointer cell*. Let us now link up the rows by inserting in the pointer cell of each row the number of the row that follows it. The last row, of course, has nothing following it. Here a zero is inserted in the pointer cell. This structure is a list (Figure 10.4). At the start all data cells are empty; i.e., the list is a *list of available space* (LAVS list) or *free list*. We also have a *list name* (here it is LAVS), which contains the number of the first row occupied by the list (here this number is 1).

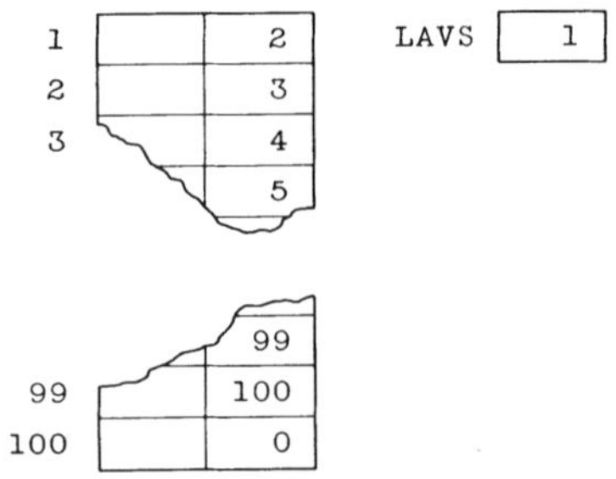

Figure 10.4

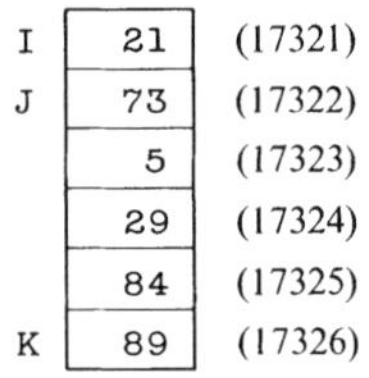

Figure 10.5

Consider now a set of data I = 21, J = [73, 5, 29, 84], K = 89. Under a conventional static storage scheme these data might be stored as shown in Figure 10.5. Throughout the program every instruction referring to the datum I then contains the address 17321 in its address field. Similarly K is referred to by the address 17326, and J(3) by a base address 17322 and an increment of 2.

Let us now set up *lists* I, J, and K. Lists in isolation are useless; the whole purpose is to have them operated on by programs, which must know where to find them. Therefore, each list must have a name, which will contain the number of the first row occupied by the list.

ALGORITHM 10.4 The following algorithm stores a vector of values $[d_1, d_2, \ldots, d_n]$, where n need not be known in advance, as list L. The list storage area is the 2-column matrix LIST.

1. Set L = LAVS.
2. Set $i = 1$.

3. Set LIST(LAVS,1) $= d_i$.
4. Set ITEMP = LAVS.
5. Set LAVS = LIST(LAVS,2).
6. If d_i is the last element of a vector, set LIST(ITEMP,2) = 0 and stop.
7. Set $i = i + 1$, go to 3.

In Step 1 of A.10.4 lists L and LAVS are made to coincide. This is the only place where L is explicitly referred to. In Step 3 a datum is inserted in the list. Note that the sequence of pointers is not changed except at the very end. In Step 6 the end of list L is marked by placing 0 in the last pointer cell. This breaks the sequence of pointers; i.e., list L is separated from LAVS.

Figure 10.6 shows the list store after lists I, J, and K have been stored in it by A.10.4. At first sight there does not seem to be any advantage to this

1	21	0
2	73	3
3	5	4
4	29	5
5	84	0
6	89	0
7		8
…		
99		100
100		0

LAVS	7
I	1
J	2
K	6

Figure 10.6

scheme. The first six elements of the first column of LIST look exactly like Figure 10.5. What, then, is the purpose of the pointers? To show their purpose assume that I does not contain a conventional piece of data, but, instead, that I is a stack. Assume now that a datum IDATUM is to be pushed down. The following Fortran subroutine will do the job.

```
      SUBROUTINE PUSH (STACK,IN)
      COMMON LIST(100,2), LAVS
      INTEGER STACK
C   CHECK FOR LIST STORE OVERFLOW
      IF (LAVS.EQ.O) GO TO 999
      LIST(LAVS,1) = IN
      ITEMP = LAVS
      LAVS = LIST(LAVS,2)
      LIST(ITEMP,2) = STACK
```

```
      STACK = ITEMP
      RETURN
 999  WRITE (6,100)
      CALL EXIT
 100  FORMAT (1H1,28HLAVS EXHAUSTED - EXIT CALLED)
      END
```

Assume that IDATUM contains the value 58. Figure 10.7 gives a schematic representation of the list (stack) I after execution of

```
      CALL PUSH (I,IDATUM)
```

Row 7 now contains the topmost element of the stack in its data cell.

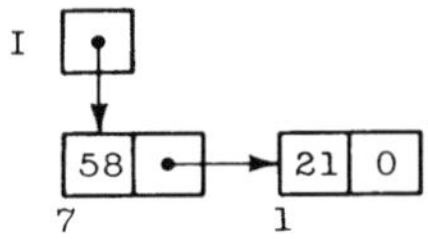

Figure 10.7

Next, assume that I, J, and K are all made to function as stacks, and that the following sequence of calls is executed

```
      CALL PUSH (J,38)
      CALL PUSH (I,17)
      CALL PUSH (K,46)
```

The schematic configuration of the three stacks and the appearance of the list store are shown in Figure 10.8.

Under this arrangement for stacks function IPOP becomes

```
      FUNCTION IPOP (STACK)
      COMMON LIST(100,2), LAVS
      INTEGER STACK
C   CHECK FOR STACK UNDERFLOW
      IF (STACK.EQ.0) GO TO 999
      IPOP = LIST(STACK,1)
      ITEMP = STACK
      STACK = LIST(STACK,2)
      LIST(ITEMP,2) = LAVS
      LAVS = ITEMP
      RETURN
 999  WRITE (6,100)
      CALL EXIT
 100  FORMAT (1H1, 29HSTACK UNDERFLOW – EXIT CALLED)
      END
```

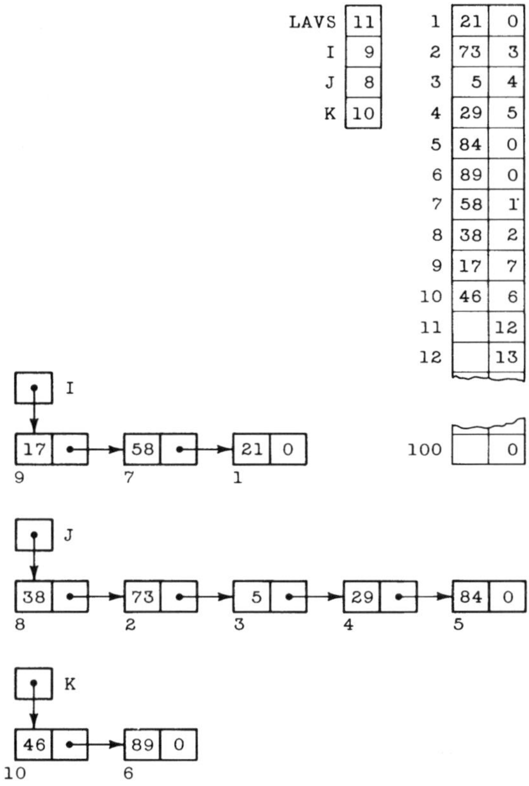

Figure 10.8

Note that each row of `LIST` is returned to `LAVS` when it is released by the `IPOP` operation. When the last datum is removed from the stack, the pointer in its name (e.g., location `I` for stack `I`) is made 0. A list with 0 in its name is an *empty* or *undefined* list.

List storage schemes have their disadvantages as well as advantages. One of the disadvantages is that access to a specified datum is slow. Since the kth datum is found by tracing through a sequence of k pointers, the average access time increases with k. In an array the access time is independent of k. Another disadvantage is that a list store can accommodate much less data than a conventional store of the same size. We shall look at list storage schemes that have better storage utilization characteristics in the next section, but there is no list storage scheme as efficient as a conventional store in this respect. Conventional storage arrangements have, however, the disadvantage of being static.

Operators can be grouped into two classes according to the type of data they operate on. The operators of arithmetic operate on numbers and belong to the first class. Rearrangement operators operate on symbolic data and belong to the second class. There is some overlap of the two classes; comparison operators can belong to either class. It is important to realize that numerals can function as numbers and as symbols. Thus, when we add two numerals, we are adding numbers, but when we sort a set of numerals, or insert a numeral in a sequence of numerals, or push a numeral into a stack, then we operate on the numerals as symbols. We speak of two branches of computation: numerical and symbolic. For numerical computation conventional static storage is normally more efficient. Symbolic computation, which involves extensive rearrangement of data, is often performed more effectively if the data are stored as lists.

10e. Formats of List Elements

A list *element* is composed of a datum field, one or more pointer fields, and, possibly, one or more marker fields, the purpose of which will be discussed further on. Under the scheme developed in the preceding section an element occupies one row in a two-column matrix. The first cell of a row is the datum field, and the single pointer field occupies the second cell of the row. We shall now see how we can save some space by compacting the fields of a list element. These schemes have to be implemented at assembler language level. Therefore, in what follows, pointer fields will be assumed to contain actual machine addresses rather than row numbers of an array.

A word in the IBM 7090 and related machines is 36 bits long, but only 15 bits are needed for a pointer. This leaves 21 bits for other uses, of which 3 bits might be assigned to a marker field, and the remaining 18 bits for storage of symbols. The 7090 character code uses 6 bits for the representation of a character. These 18 bits can then be used to store a string of three characters or a numerical symbol in the range 0–262,143 ($262,143 = 2^{18} - 1$). This "packed" representation can be used with any machine that has a similar word structure, but the details will be determined by the instruction set of the particular machine considered. For example, IBM 7090 and 7040 are very similar computers, but, while the 7040 has instructions giving direct access to character fields, access to individual characters in the 7090 is by the shifting operations. In the 7090 best efficiency may result if a word in the list store is formatted as shown in drawings (a) or (b) of Figure 10.9. In the 7040, on the other hand, formats (c) or (d) may lead to better execution times. Of course, these formats show only a few of the possible designs.

Basic storage units of IBM System/360 computers are individually addressable 8-bit bytes. The design allows for extension of core storage to as many as 16,777,216 bytes. Since $16,777,215 = 2^{24} - 1$, an address has to be 24 bits

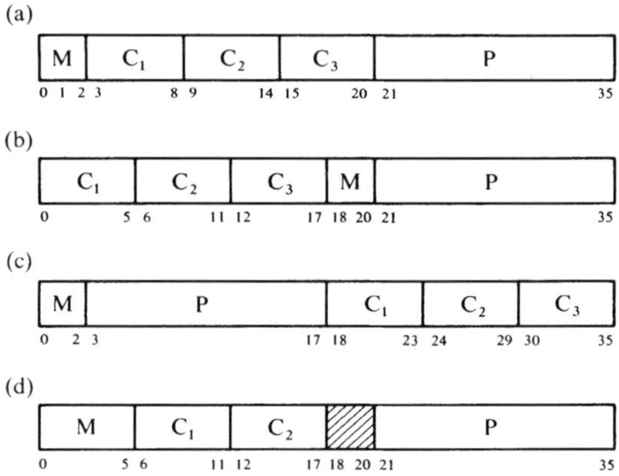

Figure 10.9 M, marker; C, character; P, pointer. In (d) bits 18–20 are not used; they may be used for a second marker field.

long. Some economy is achieved by regarding an address as the sum of a base address and a displacement from the base. The base address is loaded into a general purpose register, which then becomes a *base register*. In an instruction a storage location is completely specified by the displacement, which may range from 0 to 4095, and the address of the appropriate base register, which can range from 1 to 15 (a sixteenth register, register 0, cannot be used as a base register). When the instruction is decoded, the actual address is computed by adding the displacement to the contents of the designated base register. The displacement is specified by 12 bits and the base register by 4 bits; i.e., an effective address requires only 16 bits, a considerable improvement on 24 bits. Four consecutive bytes form a 32-bit word. The store can then be pictured as consisting of 1024-word blocks, a block being accessed by its base address, and a word within the block by the displacement from the base.

We shall restrict list elements to single words, using two bytes for the pointer field and the remaining two bytes for the datum field. Since the System/360 character code requires one byte for each character, this design provides for the storage of two characters in each list element. There is more than one form that the pointers can take. The most obvious choice is to use the form by which an effective address is specified in an instruction. Format (a) of Figure 10.10 shows this arrangement. The B-field specifies the base register, and the D-field the displacement. Unfortunately there are only the 16 general purpose registers. In addition to being used as base registers, they serve as index registers and accumulators. If more than 8 registers, say, were assigned to the list store as base registers, the reduction of overall efficiency due to a shortage of registers for other purposes might well become

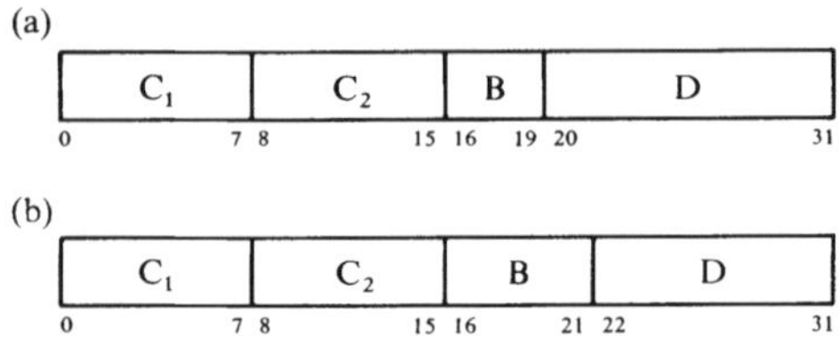

Figure 10.10

intolerable. But, if we use only 8 base registers, then the largest store that we can have has only 8192 words. For most practical purposes a somewhat larger store is needed.

Another possibility is to use the actual address of the next element. Prior to the execution of an instruction that refers to this element the address of the element is loaded from the pointer field into the base register to which the instruction refers. The displacement field of the instruction is zero. This scheme has the disadvantage that three bytes are now used by the pointer, leaving only one byte of a word for data. Its good feature is that only one base register need be used for all operations on elements of the list store.

Our next scheme combines the advantages of the two preceding schemes. In common with the first scheme a list element contains a 2-byte datum field, a B-field, and the displacement field. The B-field, however, has a different interpretation. It contains a pointer to an entry in a 16-word directory. The entries of the directory are base addresses. Under the first scheme, before an instruction that refers to the next element in the list is executed, the B-field and the displacement are both inserted in the instruction. Under the present scheme the instruction already contains the address of a base register, and the one base register can, of course, be used for all operations on the lists. Only the displacement field is now inserted in the instruction itself, and the base register is loaded from the directory. This scheme is slower than the first two, but it represents a considerable improvement on the storage efficiency of the second scheme and permits a list store that is twice as large as that permitted by the first scheme.

If we were to accept still slower access to elements in a list, then the maximum size of the list store can be increased to 65,536 words. Since the address in the displacement field refers to only one byte in every four; i.e., words rather than bytes are addressed, the lowest two bits of the displacement field are redundant. This means that the D-field can be shortened by two bits, and these two bits used to extend the B-field; a list element then has format (b) of Figure 10.10. The B-field can now contain values up to 63, permitting a directory of 64 entries. The displacement is reconstructed by two left shifts of the D-field.

Our designs have not provided for a marker field, but, if such a field is necessary, one can modify the last scheme. Instead of using the two bits saved by the design to extend the B-field, one or both of the bits can be used as a marker field.

10f. List Structures

In Section 5a we saw that derivation trees can be represented by parenthesized linear strings. For example, the tree of Figure 5.1 can be represented by

$$S(NP(T(the)N(child))VP(V(ate)NP(T(a)N(A(green)N(pear))))).$$

Similarly, the tree of Figure 10.11 can be represented by $a(bc(fg(j))d(hi)e)$.

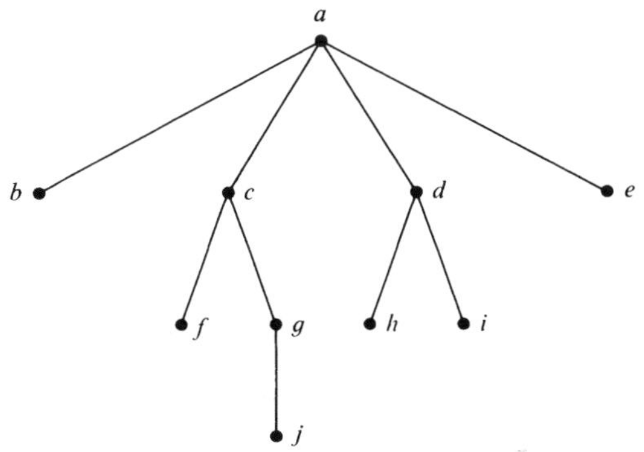

Figure 10.11

However, often it is more convenient to have a representation that corresponds more directly to the way we picture a tree. A tree structure can be represented in this more explicit fashion by a set of linear lists if the lists are provided with a mechanism for branching from one list to another. Since anything at all can be put in the data field of a list element, the datum can be a pointer to another list, *designated as such by the setting of a marker.* A set of lists can then be linked by inserting pointers in the data fields and setting markers. If we have lists X and Y, and there exists a link *from* list X *to* list Y, then Y is a *sublist* of X. Unless sublists have to be independently referenced they need not have their own names.

The tree of Figure 10.11 can be represented by linked lists as in Figure 10.12. We have assumed format (a) of Figure 10.9, with the convention that the first bit of the marker field, which is the sign bit of a 7090 word, is 1 (minus) if the data field contains a pointer and 0 (plus) if it contains a symbol.

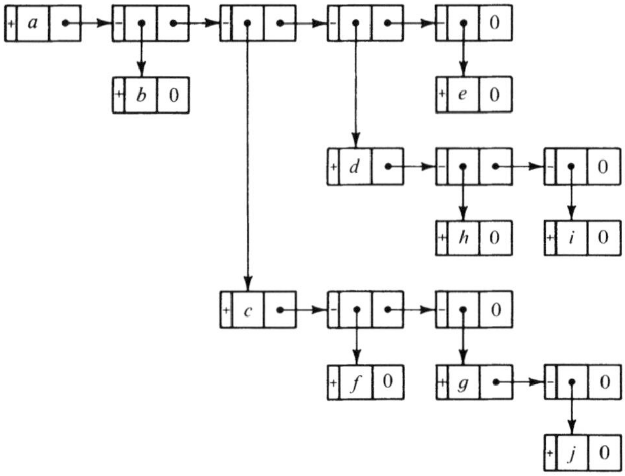

Figure 10.12

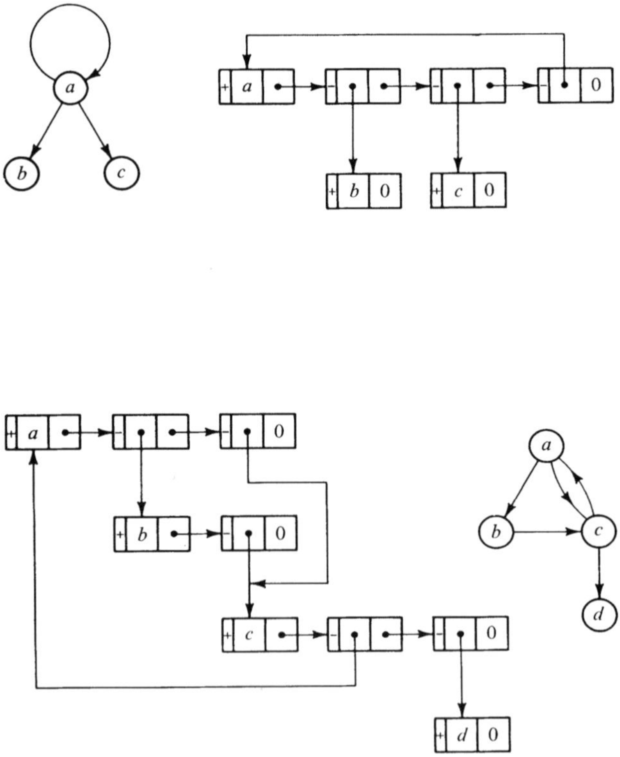

Figure 10.13

Bits 1 and 2 are not used. If the datum is a pointer, i.e., an address, it is stored in bit positions 3–17. This field, called the *decrement field* in the 7090 terminology, is directly accessible to some machine instructions.

Lists can be used to represent more complicated structures than trees. Figure 10.13 shows two digraphs and their representation by lists. Conversely, a set of linked lists can be represented by a digraph in which each list is represented by a node and each element that contains a pointer in its data field by an arc. A set of linked lists is a *list structure* if its digraph is connected. It is a *reentrant* list structure if its digraph contains cycles. In particular, a list that is a sublist of itself, i.e., that contains a pointer to itself, is a *reentrant list*. The digraph of a reentrant list is a sling. If two lists have the same list as a sublist, then they are said to *share* the sublist. In the digraph the indegree of a node representing a shared list is equal to at least 2.

Let us now suppose that we have stored a tree as a list structure and that we want to produce a parenthesized string from it. As we traverse the list structure we may come to a pointer to a sublist. We have to go down the sublist, but we have to mark in some way how far down the main list we have

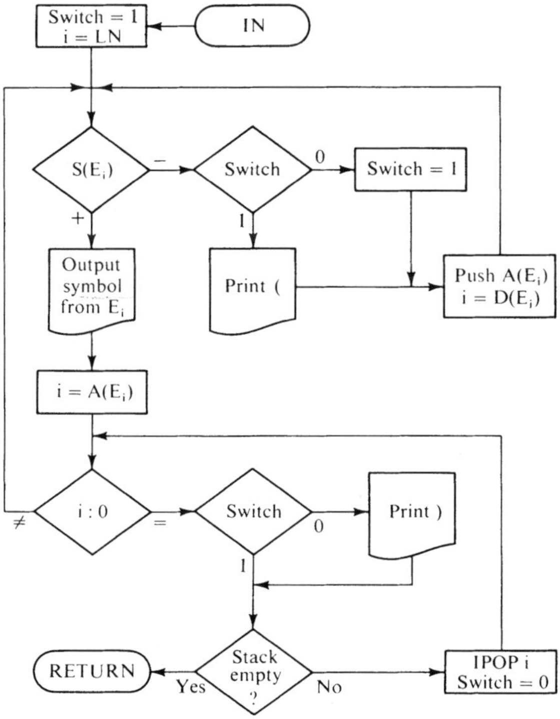

Figure 10.14

come; in other words, we have to remember the address of the next element in the main list. The remembering is done by a stack.

ALGORITHM 10.5 Figure 10.14 shows an algorithm for generating a parenthesized expression from a tree stored as a list structure. E refers to a list element, and E_i has location i. The list name (LN) contains the address of the first element of the list. The format has been chosen with the IBM 7090 in mind; S(E) is the sign bit of element E, D(E) is the decrement field (bits 3–17), A(E) is the address field (bits 21–35).

Let us now turn again to the sort tree of Figure 5.15 and consider the implementation of the sort tree as a list structure on the IBM 7090, say. Arrays D and P are implicitly a list structure, but, since they are fixed arrays, we cannot extend the structure. By using a pair of words for each list element we can create an extensible list structure as shown in Figure 10.15 that uses only 25 words of store (including the list name) as against the 36 words used by the arrays. This particular representation is possible only because 2 is the greatest number of arcs that originate from any node of the tree.

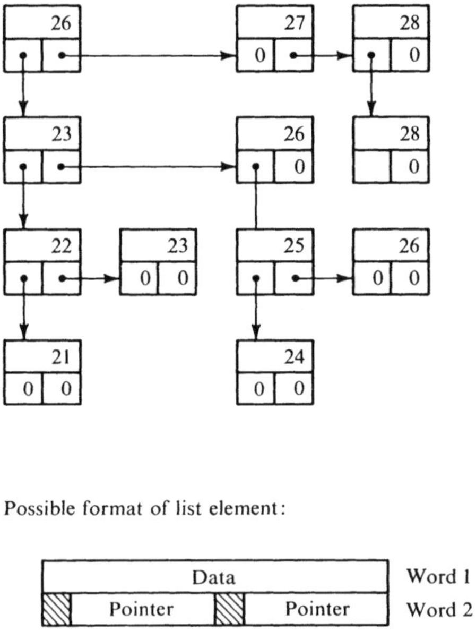

Figure 10.15

10g. Threaded and Symmetric Lists

If, instead of putting a zero in the pointer field of the last element of a list, we point back from this element to where the list starts, then we have a *threaded* list (also called a *circular* list). We shall call the pointers in the last elements *return addresses*. Figure 10.16 shows a threaded list structure, which represents the tree of Figure 10.11. To avoid making Figure 10.16 incomprehensible by too many lines, return lines have not been drawn, but are implied by return addresses in the appropriate pointer fields. In the case of the main list the return address is still zero, and this zero indicates the end of the structure; the return address of a sublist refers to the element that establishes a link to this sublist in the list of which it is a sublist. Because of the return addresses a threaded list structure can be traversed without the aid of a stack. Marker fields are essential: Not only do we have to differentiate between data elements and elements that represent structural features of the list structure; now the last element of a list has to be distinguished as well.

ALGORITHM 10.5a Figure 10.17 shows an algorithm for generating a parenthesized expression from a tree stored as a threaded list structure. The notation convention is the same as in A.10.5. However, since now the marker field occupies two bit positions, the first bit is called `S₁(E)`, and

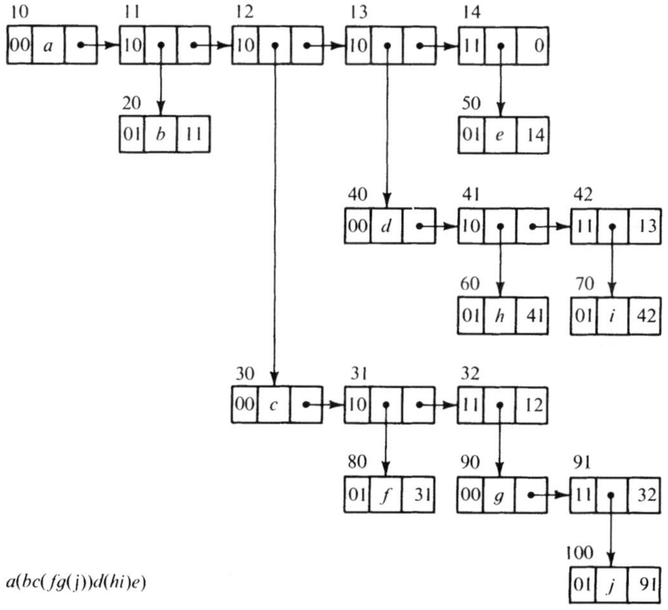

Format of marker: Bit 1 — 0 indicates a data element;
1 indicates a pointer element.

Bit 2 — 0 indicates that the element is not last in the list;
1 indicates the last element.

Figure 10.16

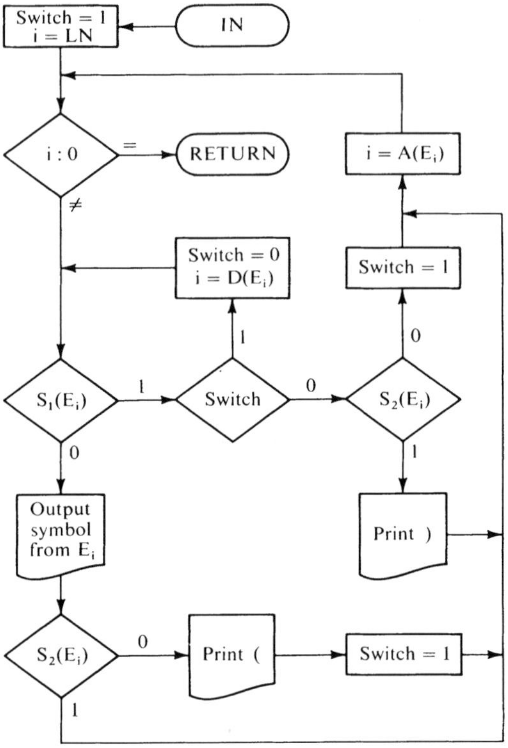

Figure 10.17

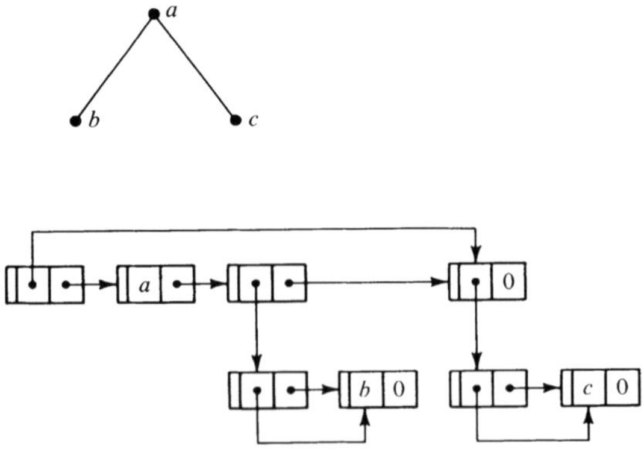

Figure 10.18

the second $S_2(E)$. The switch indicates whether a particular pointer element is reached for the first or second time.

The convention adopted in Figure 10.16 does not allow for sharing of sublists. Although it is quite easy to make sharing possible by a change in the format of the marker fields to permit more than one "last" element, and, consequently, more than one return address, nothing is gained. The one good feature of threaded list structures is that they can be traversed without having to use a stack. Multiple return addresses would make it again necessary to keep a record of the history of a traversal; i.e., a stack would have to be used.

Both conventional and threaded lists can be traversed in one direction only. This is not always convenient. For example, in order to append a new element to the end of a list, the entire list has to be traversed to get to its end. The particular problem can be solved by providing the first element of a list with a pointer to the end of the list, as in Figure 10.18, but a more general solution is obtained when every element in a list is provided with two pointers: a forward pointer and a backward pointer. The list is then a *symmetric* list.

In most computers an element of a symmetric list has to occupy at least two words. One of the words is used for storing a symbol or a pointer to a sublist. The other contains the forward and backward pointers, and a marker field. Examples of symmetric list structures are shown in Figures 10.19 to 10.21. Since any one of a number of feasible designs is as good as another, the format of marker fields and, in Figure 10.20, the contents of the pointer fields of elements representing terminal nodes of the tree have been left unspecified.

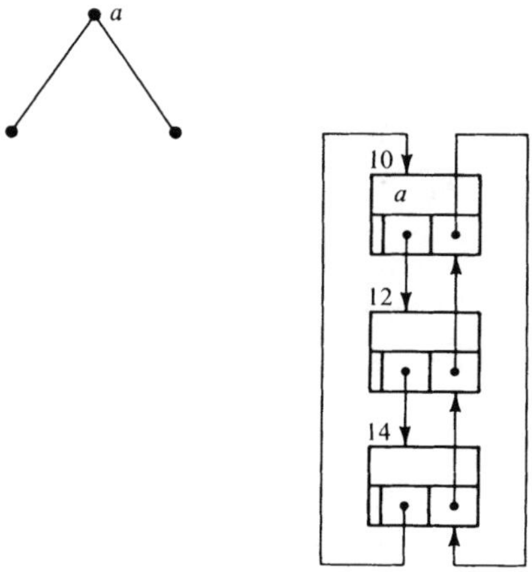

Figure 10.19

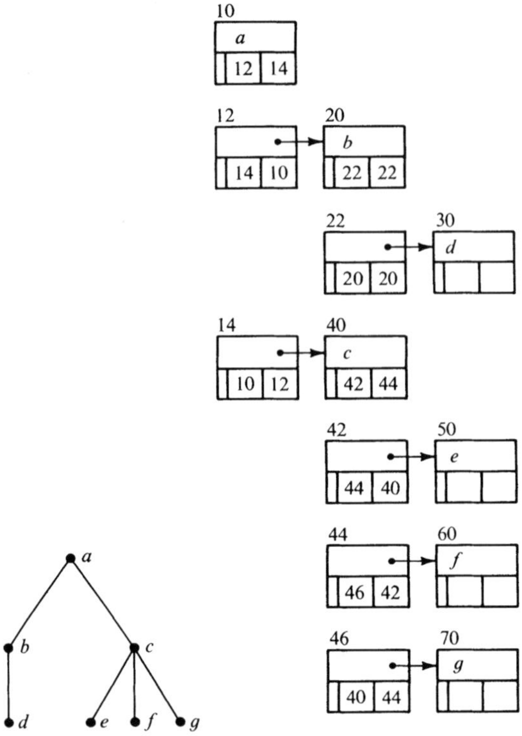

Figure 10.20

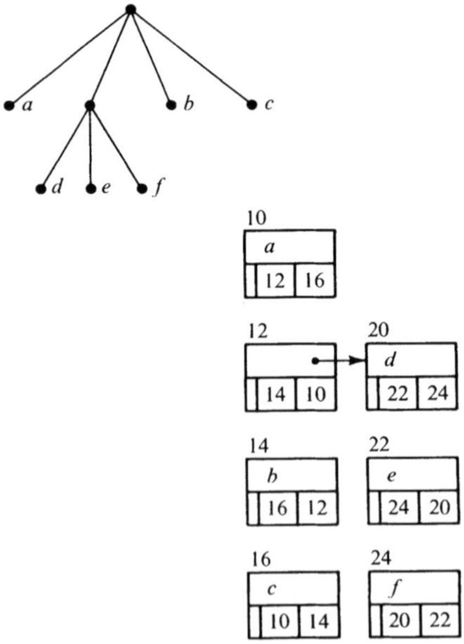

Figure 10.21

10h. Representation of Digraphs as List Structures

We have already seen digraphs represented as list structures; notably in Figure 10.13. A feature of this representation is that lists are shared, and this means that threaded list structures cannot in general represent digraphs.

An alternative representation to that of Figure 10.13 keeps symbols and structure separate, and the digraph is represented by a set of lists. In this representation some lists consist entirely of pointers, others entirely of symbols. A marker *in the list name* indicates the type of the list, and—since all elements in the one list are of the same type—it is not necessary to provide individual elements with markers. A set of lists that represents the labeled digraph of Figure 10.22 is shown in Figure 10.23. SYMBOL is the list of all

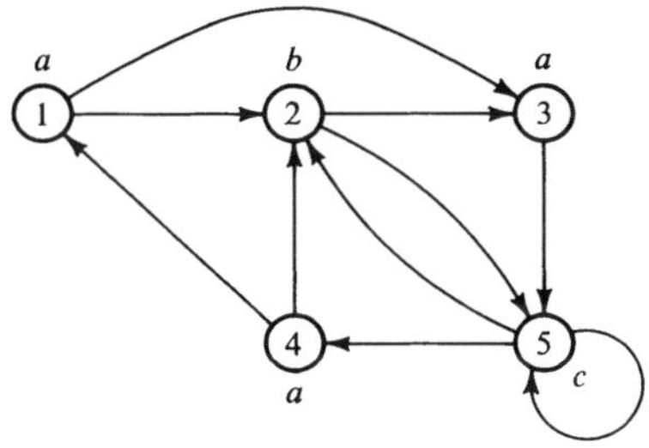

Figure 10.22

node labels. Each node is represented by a list of addresses of the names of lists representing other nodes. These lists indicate the nodes to which there are arcs from the given node; i.e., the list associated with a node is, in effect, a list of arcs originating from this node. Since each of the lists has its own name, we have a set of independent lists rather than a list structure.

Changes that affect structure alone or symbols alone are somewhat easier to make when there is separation of symbols and structure than when the two are intermingled. This advantage, however, is insignificant compared to that brought about by elimination of markers. At the end of Section 5c (Figure 5.17) we considered dictionaries arranged as trees. The important data in this design are the lists of dictionary entries suspended from terminal nodes, and the success of the system depends on the speed with which an entry can be retrieved. Unless structure is separated from data, too much time would be spent examining markers. In general, one should try to keep symbols and structure separate whenever one of these types of information dominates the other. Also, elimination of markers is very much a necessity in list processing on an IBM System/360 computer. In an IBM 7090 word a marker can be accommodated in a 3-bit field for which it would be difficult to find any other use (Figure 10.9). There is no such field to spare in the System/360 word

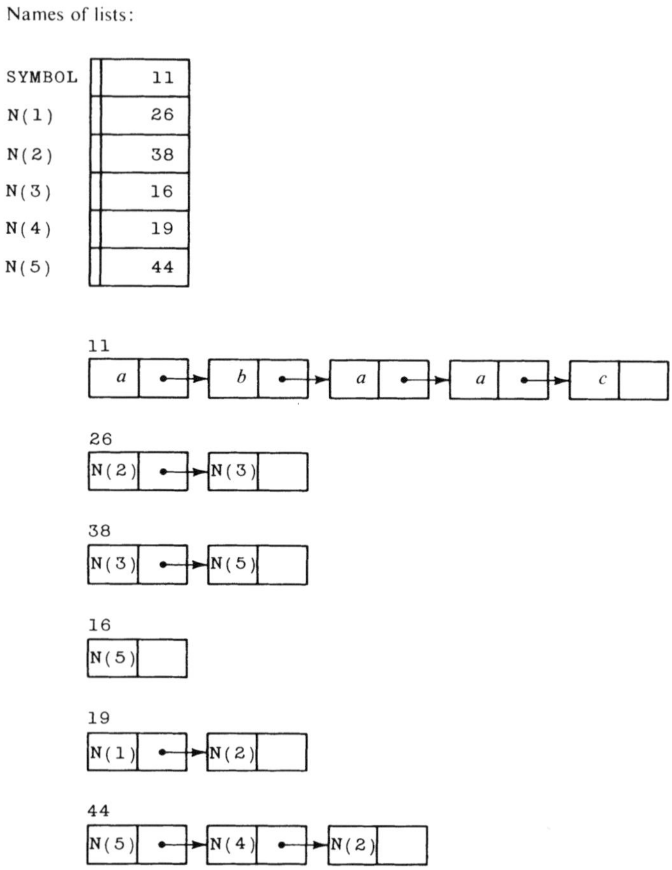

Figure 10.23

(Figure 10.10). Consequently, use of markers can become very costly in terms of storage space.

Referring back to Figure 10.23, one may object to this representation on the grounds that in contrast to a list structure, which may be specified by a single external name, the set of lists requires as many external names as there are lists in the set. If the number of nodes of the digraph is large, a considerable block of static storage has to be set aside for list names. This becomes awkward when the representation of a digraph is being generated without prior knowledge of the number of nodes (we then have precisely the situation that made us investigate list storage schemes in the first place). The only solution is to do away with the block of static storage and store the names themselves as a list, with a single external name referring to this list.

The implementation of such an indirect addressing scheme presents a number of technical problems, but they are fairly easy to solve.

10i. Multiword List Elements

In processing symbols the main operations are insertion, deletion, and rearrangement. Each of these operations calls for changes in structure, and these changes are best made when each symbol is linked into a list as a separate entity, i.e., wholly occupies a single list element. For such applications the schemes that we have been discussing are very good. But there are other application areas requiring dynamic storage in which these schemes do not function as well. In Section 10c we discussed a stack in which the active part is in a static storage region and inactive deeper sections have been transferred to a dynamic store in blocks of 100. The blocks can be stored in a list store that has pairs of words for its elements, but this arrangement is very inefficient, in terms of both time and space. Since each of the 100 symbols has to be transferred separately, the transfer rate is low. On the other hand, while the block is in the list store, no changes are made to it. This means that the pointer words paired to the data words have no functional use.

Efficiency is improved by use of multiword elements. Presume that the list store is divided into 101-word elements. As in the 2-word element, one of the words contains pointers, but now a single pointer word services 100 words of data. However, the list store will probably be used for a variety of purposes, and it would not be good policy to divide the entire list store into elements of the same size. One possibility is to use elements whose size may be one of a restricted set of possible sizes, say $\{2, 4, 8, 16, \ldots, 256\}$. Then the 100 symbols of a background block can be stored in a 128-word element, with 27 words going to waste, or they may be distributed over a set of linked smaller elements: a 64-word element, a 32-word element, and an 8-word element, reducing the number of words wasted. Under this scheme the size of the block must be stored in the pointer word of the block (external directories holding the sizes of blocks have been used as well, but they reduce the flexibility of the scheme). Available space can be arranged as a set of lists, one for each size, or as a single list. Greater efficiency is achieved with a set of lists. If no element of a required size is available, it is created by splitting a larger element; this is the reason for making each size half the next larger size. Periodically all unused elements are shifted to one end of the list store and restructured into larger elements.

An alternative is a scheme in which the sizes of elements are completely unrestricted. Initially the entire list store is a single element of available space. If a stack block has to be stored, 101 cells are taken from this element, reducing its size by this amount. Again, if a list structure representation of the tree *a*(*bc*(*fg*(*j*))*d*(*hi*)*e*) is to be created, then 19 elements of size 2 are taken from the element of available space. As lists are erased, they are returned to available space, and the available space region now becomes a multielement list. If at some later stage a stack block is to be stored, and a search through the list of available space fails to produce an element of size 101 or larger, all unused elements are shifted to one end of the list store and merged to form a single element of available space from which the 101-word element can then be sliced off.

10j. Management of List Stores

In several places we have written in passing that lists and list structures are erased and then returned to the list of available space (`LAVS`). Let us now discuss mechanisms for their return. Consider the tree of Figure 10.11 and its representation in Figure 10.12. The tree, as a list structure, has an external name containing a pointer to the first element of the structure (the element containing the symbol *a*). The complete tree is erased by setting the pointer in its name to zero. If this is done, then all list elements that make up the tree become *inaccessible*. One might, on the other hand, merely delete the subtree rooted at *d*. This would be done by shifting the second pointer from the fourth element in the first row of Figure 10.12 to the second pointer field of the third element in this row, and then the six elements that represent the subtree would become inaccessible.

Return of elements to `LAVS` after each erasure is difficult because of the possible interference between the return operation and the operation in progress at that time. Instead, most list processing systems postpone the retrieval of inaccessible elements until `LAVS` is nearly or completely exhausted. A procedure that retrieves all inaccessible elements, called the *garbage collector*, is then brought into action. Garbage collection proceeds in two stages. First, the external names are examined. Whenever a name contains a pointer, all paths that have their origin defined by this pointer are traced through, and every element encountered in tracing the paths is marked. After this has been done for every name all accessible elements are marked and all inaccessible elements remain unmarked. The marked elements define lists that are *active* at this time. In the second stage the unmarked elements are strung together to form a new list of available space, and the markers removed from the accessible elements.

Sharing of sublists presents no difficulty. If two structures have shared a sublist, and the sublist has been deleted from one of the structures, the elements of the sublist are still accessible in the other structure. The requirement that at least one bit in the marker field of every element has to be set aside for exclusive use in garbage collection is generally of minor consequence. There is, however, one major difficulty. We have seen that the traversal of list structures requires one to use a stack (except in the case of threaded list structures). Unfortunately, there is no available space for the stack in the list store when garbage collection takes place. One solution is to provide the system with a stack external to the list store and inaccessible to the programmer. A more attractive solution is to use a reversal-of-pointers technique, which enables one to retrace paths to their origins. The following marking algorithm makes use of this technique.

ALGORITHM 10.6 Marking algorithm for a garbage collector. The algorithm marks all elements that belong to a list structure defined by a nonzero pointer in its name (LN). NPLACE is the address of LN. In each element we assume a marker field comprising locations S_1, S_2, and S_3—S_1 is 0 if the list element contains data and 1 if it contains a pointer to a sublist; S_2 is used to keep track of the reversal of pointers; the algorithm sets $S_3 = 1$ in all elements accessible from LN. The pointers A(E) and D(E) have the same interpretation as in A.10.5. Note that the algorithm works for reentrant list structures, and that the lists may consist of multiword elements.

1. Set I = LN, J = NPLACE.
2. If $S_3(E_I) = 1$, go to 9.
3. Set $K = A(E_I)$, $A(E_I) = J$, $S_3(E_I) = 1$, J = I, I = K.
4. If I $\neq$ 0, go to 3.
5. If $S_1(E_J) = 0$, go to 8.
6. If $S_2(E_J) = 1$, go to 9.
7. Set $K = D(E_J)$, $D(E_J) = A(E_J)$, $A(E_J) = I$, $S_2(E_J) = 1$, I = K; go to 2.
8. Set $K = A(E_J)$, $A(E_J) = I$; go to 10.
9. Set $K = D(E_J)$, $D(E_J) = I$.
10. Set I = J, J = K; if K $\neq$ NPLACE, go to 5.
11. Stop.

Comments. If a list is already marked—it may be a sublist of another structure that has already been processed, or it may belong to a reentrant structure—then the processing of the list is bypassed (Step 2). In Steps 3 and 4 one moves down the list, marking elements and reversing pointers, until the end of the list is reached. Then the path along the list is retraced backwards, pointers being restored to their original settings (Steps 8 to 10). Sublists are dealt with in this stage. An element containing a pointer to a sublist is visited twice during the backward traversal. Step 6 distinguishes between the two instances. The first time $S_2(E)$ is 0, and one interrupts the restoring of pointers to move into the sublist. The action shifts to Step 2 via Step 7, but Step 6 is reached again after processing of the sublist, all of its sublists, and so forth has been completed. This time $S_2(E)$ is 1, and restoring of pointers in the list is resumed.

If all list elements have the same size, the creation of the new list of available space in the second stage of the garbage collection process is very simple. Complications arise when multiword elements of various sizes have to be compacted into a single element of available space. Active list structures must be relocated, and this involves updating of pointers. As part of the process one may wish to do some compacting of the active lists as well. This is an extremely difficult task, and an intermediate transfer of the list structures to auxiliary storage (disk, drum, tape) appears to give the only feasible solution. The mere transfer of the active structures to one end of the list store is still fairly difficult, particularly when sublists are shared.

An alternative to garbage collection is the *reference counter* approach. We recommend it for systems that keep structure and symbols separate (see Section 10h). In this approach an empty list has a reference count of 0 to start with. A nonempty list starts with a count of 1. Whenever the list is made a sublist of another list the reference count is increased by 1; when the list ceases to be the sublist of a list the count is decreased by 1. In terms of Figure 10.23, the reference counts of N(1), N(2), ..., N(5) are, respectively, 2, 4, 3, 2, 4; that of SYMBOL is 1. A field can be set aside in the list name for the reference counter. Whenever an erasure is called for, or a list is removed from a structure, the appropriate reference count is adjusted. A list is returned to available space when its count drops to zero. Looking at Figures 10.22 and 10.23, let us remove arc ⟨5, 4⟩. This means removing the second element of N(5) (it becomes part of the available space list), and decreasing the reference count in N(4) by 1. Next, let node 4 be removed; i.e., let N(4) be erased. Since the reference count in N(4) now becomes 0, list N(4) is returned to available space. At the same time counts in N(1) and N(2) are decreased. Next consider what happens when the two operations are performed in reverse order. Now the erasure of N(4) results merely in a decrease of the reference count. Next, when ⟨5, 4⟩ is removed, the reference count in N(4) becomes 0, return of N(4) to available space takes place, and the reference counts of N(1) and N(2) are decremented at this stage.

In our example the reference counter approach seems to lead to an unnatural situation: Why should the removal of a node result in the removal of arcs originating from the node when the indegree of the node is zero, but not when it is nonzero? This perplexing state of affairs has nothing to do with reference counters, but is a consequence of the peculiar interpretation we have given to "removal of a node." When a node is removed we must remove *all* arcs incident with it. Therefore, node 4 is not properly removed before the removal of arc ⟨5, 4⟩, and it is meaningless to comment on the situation before this has taken place.

The problem of reentrant structures is much more serious. One should be able to return the whole digraph to available space by erasing `N(1)`, `N(2)`, ..., `N(5)` in turn. Actually what happens is that the counts in the list names decrease by 1 in each instance, without any one decreasing to zero! Consider the erasure of `N(1)`. Since the reference count in `N(1)` decreases only to 1, the list is not returned to available space. We can, however, revise our approach, and decrease the reference counts in `N(2)` and `N(3)` at this point, instead of waiting until the return of `N(1)` to available space. Then all five counts ultimately reduce to zero, but they do so at rather awkward times. The count in `N(1)` becomes zero while `N(4)` is being erased: We move along list `N(4)`, see the reference to `N(1)` in the first element, and decrement the counter of `N(1)`. Since the count is now zero, processing of `N(4)` has to be interrupted so that list `N(1)` can be returned to available space. This type of interrupt is fairly easy to deal with, but things become rather chaotic when a list refers to itself, as `N(5)` does. When we come to the processing of `N(5)` its counter holds the value 2. This is reduced to 1, and processing of the list begins. The first element of `N(5)` contains a reference to `N(5)`, and the count is decreased again. Now it becomes 0, and `N(5)` should be returned to available space, but if we do so at once, then `N(2)` and `N(4)` are left up in the air. The solution of the problem requires some delicate programming.

One of our aims in separating data and structural information was to eliminate markers. Since garbage collection depends on the use of markers, we recommend the reference counter technique for "separated" lists, even when they form reentrant structures, despite the greater complexity of this scheme. An exception is provided by lists in which the elements are all of the same size, but they rarely are in systems of the type considered here. Then markers for a garbage collector can be arranged in external tables of bits, the *n*th bit in a table being associated with the *n*th element in the list store.

10k. PL/I-Type Data Structures

The *structures* of Cobol and PL/I are more complicated than arrays. The PL/I declaration

```
DECLARE 1U(2), 2V, 2W, 3Y, 3Z(2,2), 2X;
```

generates a data structure that has the form

```
U(1)  ┌V                        U(2)  ┌V
      │W  ┌Y                          │W  ┌Y
      │   │Z(1,1)                     │   │Z(1,1)
      │   │Z(1,2)                     │   │Z(1,2)
      │   │Z(2,1)                     │   │Z(2,1)
      │   └Z(2,2)                     │   └Z(2,2)
      └X                              └X
```

A prefix in the declaration indicates the *level* of the variable in the structure; numbers in parentheses indicate that the variable represents an array of the given dimensions. The first V is referenced by writing U(1).V, or, if no ambiguity can arise, simply V(1). The second Z(1,2) is referenced by U(2).W.Z(1,2), or by U(2).Z(1,2), or by W(2).Z(1,2), or by Z(2,1,2), and this list does not exhaust all the possible forms. Structure U contains 14 *simple* data elements: V(1), Y(1), Z(1,1,1), Z(1,1,2), Z(1,2,1), Z(1,2,2), X(1), V(2), Y(2), Z(2,1,1), Z(2,1,2), Z(2,2,1), Z(2,2,2), X(2). Names U and W refer to the data collectively: U names the main structure, W a substructure of U.

Let us now generate a second structure of identical form by

```
DECLARE 1UU(2), 2VV, 2WW, 3Y, 3ZZ(2,2), 2XX;
```

Since Y now appears in two declarations, every reference to it must be *qualified*; i.e., we must specify the structure in which the particular Y we want resides.

The "simple" PL/I statement,

```
U = U + UU;
```

is equivalent to

```
U(1) = U(1) + UU(1);
U(2) = U(2) + UU(2);
```

The second of these is, in turn, equivalent to the three statements

```
V(2) = V(2) + VV(2);
W(2) = W(2) + WW(2);
X(2) = X(2) + XX(2);
```

and W(2) = W(2) + WW(2) is equivalent to

```
U.Y(2)   = U.Y(2)   + UU.Y(2)   ;
Z(2,1,1) = Z(2,1,1) + ZZ(2,1,1);
Z(2,1,2) = Z(2,1,2) + ZZ(2,1,2);
Z(2,2,1) = Z(2,2,1) + ZZ(2,2,1);
Z(2,2,2) = Z(2,2,2) + ZZ(2,2,2);
```

The collective assignment statement U = U + UU is equivalent to 14 normal assignment statements in all.

These structures differ from list structures in that their structural features cannot be altered once they have been declared. Nevertheless, a compiler has to use a system of pointers in setting up the structures so as to achieve efficiency in the compilation of statements that refer to the structures. Representation of structures by trees can greatly assist one in coming to an

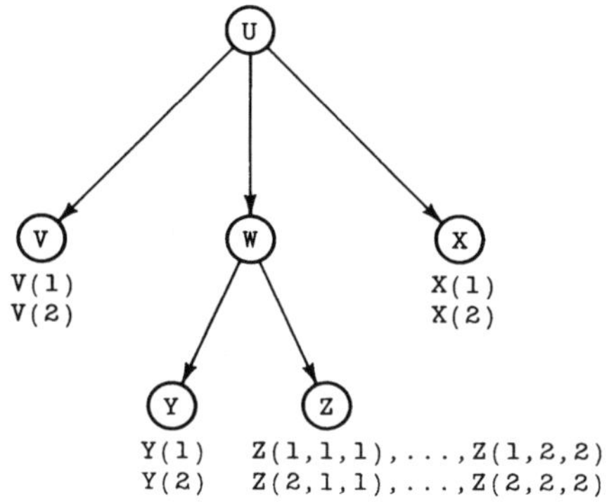

Figure 10.24

understanding of the concepts involved. The labeled tree of Figure 10.24 represents structure U.

Let us consider how a compiler would deal with the two declarations of our example. We need two tables: a *symbol table* and a *structure table*. An entry in the symbol table consists of a symbol and a pointer to an entry in the structure table. In compiling the declaration of U an entry is generated in the symbol table for every node of the tree corresponding to U. When the declaration of UU is compiled, entries are generated for symbols that are not already in the symbol table, i.e., for UU, VV, WW, ZZ, and XX, but not for Y.

The structure table contains an entry for *every* node. In our example there are 11 entries in the symbol table, but 12 in the structure table. This means that all entries in the structure table corresponding to the one entry in the symbol table must be chained by means of pointers. Another set of pointers is needed to conserve the structural features established by the declarations. Moreover, in the case of terminal nodes, the structure table must contain information relating to the memory locations reserved for storage of the data elements associated with these nodes. If a single data element is associated with a node, then the information is simply the address of the location assigned to this datum. If the data elements form an array (in our example they do so at every terminal node), dimensions of the array may have to be stored in addition to the address of its first element. Storage assignment does not interest us here, and we shall ignore all parts of the structure table that relate to this aspect of the problem.

For our example the tables are as shown in Figure 10.25. The first column in STRUCT contains pointers linking entries that correspond to the same symbol. For a given node the second column contains a pointer to the right-most terminal node in the structure that can be reached from this node. In

Symbol table SYMBOL

1	U	1
2	V	2
3	W	3
4	Y	10
5	Z	5
6	X	6
7	UU	7
8	VV	8
9	WW	9
10	ZZ	11
11	XX	12

Structure table STRUCT

1	0	6	
2	0	2	
3	0	5	
4	0	4	
5	0	5	Information
6	0	6	relating to
7	0	12	storage
8	0	8	assignment
9	0	11	
10	4	10	
11	0	11	
12	0	12	

Figure 10.25

the case of a terminal node this is a reference to itself. We shall now see that this information is sufficient for the location of a given terminal node.

ALGORITHM 10.7 Given specification $E_1.E_2.---.E_N$, the algorithm finds the entry in the structure table STRUCT corresponding to this specification. If the specification is ambiguous or improperly written, then an error return is taken (Step 17). An auxiliary array NREF of size N is used to store row numbers of STRUCT.

1. Set J = 1, FIRST = .TRUE..
2. Look up E_J in SYMBOL—assuming that E_J = SYMBOL(K,1), set NREF(J) = SYMBOL(K,2).
3. Set J = J + 1.
4. If J $\leqq$ N, go to 2.
5. Set J = 1.
6. Set K = NREF(J). If K = 0, go to 14.
7. If J = N, go to 12.
8. If K $\geqq$ NREF(J+1), set NREF(J) = STRUCT(K,1) and go to 5.
9. Set KK = NREF(J+1). If KK = 0, go to 18.
10. If STRUCT(KK,2) > STRUCT(K,2), set NREF(J+1) = STRUCT(KK,1) and go to 5.
11. Set J = J + 1 and go to 6.
12. If FIRST = .FALSE., go to 17.
13. Set J = 1, FIRST = .FALSE., INDEX = NREF(N).
14. Set K = NREF(J).
15. If STRUCT(K,1) = 0, go to 19.

16. Set NREF(J) = STRUCT(K,1), J = J + 1. If J ≦N, go to 14; else go to 5.
17. Return—error condition.
18. If FIRST = .TRUE., go to 17.
19. Return—INDEX contains the number of the required row of STRUCT.

Comments. Array NREF is initiated in Steps 1 to 4. Steps 5 to 11 locate a region of STRUCT that the given specification refers to. Changes are made to NREF if the row numbers stored in NREF are not in strictly ascending order (Step 8), or if two row numbers in NREF refer to two different structures (Step 10). If the given specification does not refer to any declared structure, then K or KK becomes zero, and an error return is taken. If no inconsistency is detected, a check is carried out in Steps 13 to 16, and if the check indicates a possibility that there might be a second structure that could be described by the given specification, then Steps 5 to 11 are executed again. This time an inconsistency indicates that the given specification is unambiguous; if J attains the value N (Step 7), then there are at least two structures to which the given specification might refer, and the error return is taken.

Next let us consider how we would locate all the simple data elements denoted by U.W, say. A.10.7 would be used to find the entry corresponding to U.W in STRUCT. In terms of Figure 10.25, the relevant entry would be in row 3. Note that STRUCT(3,2) = 5. This value is used to delimit the relevant section of STRUCT; we look for a row K farther down the table such that STRUCT(K,2) is 5, and use the information relating to storage assignments in rows 3 to K to gain access to the data. Here K = 5, and, since only rows 4 and 5 refer to terminal nodes, the information contained in these rows would be used.

Notes

[Kn68] is a good general reference that covers the material of this chapter in greater detail. A general work to consult on lists and their processing is [Fo67]. [Im69] describes list storage schemes employed for internal representation of data in a number of list and string processing systems. [Gr67] surveys representations for graphic data. [Wi64, Wi65] contain examples of applications of list structures; further references to examples can be easily extracted from [Sa66a], an excellent annotated and indexed bibliography of 297 items on the use of computers for nonnumerical mathematics.

An assembler language implementation of a stack is described in [Ba62a];

also [Ro63, Ay63] deal with implementation of a stack by means of programming. Hardware implementation of stacks in the KDF9 and B5000 computers is described in [Ha62b] and [Ca63b], respectively (see also [Sa62] for the drawing of a stack machine that was patented as early as 1957). More recent hardware implementations of stacks are reported in [Sp67, Ha68]. Abstract automata based on the stack have an extensive literature; [Ev63, Gi67, Sc67a, Kn67b] constitute a small sample.

The use of a stack in the compilation of arithmetic expressions is described in the classic [Ba59b]. Algol permits recursive procedure calls, i.e., a procedure may call itself, and the structure of the language is such that recursive programming can be used extensively in the compilation of an Algol program. The storage problems associated with recursive programming are elegantly solved by the use of a stack (see [Di60]), and the main reason for the stacks in the KDF9 and B5000 is that the designers envisaged Algol as the major programming language to be used on these computers. [Go67c] is a good introduction to recursive methods for the solution of problems in general. For a more detailed presentation of applications of recursive programming see [Ba68a]. Backtrack programming, which is a close relative of recursive programming, is described in [Go65, Fl67] (see also [Pe66] for a specific example). [Ha62c] is a reference for Section 10b.

List processing begins with the work of Newell, Shaw, and Simon on their "logic theory machine," see [Ne57]. Threaded lists were introduced by Perlis and Thornton [Pe60], and symmetric lists by Weizenbaum [We63]. [Co64] is a survey of the organization and use of list structures with multiword elements. Storage allocation when list elements are restricted in size to 2^n ($n = 0, 1, 2, 3$) words is described in [Kn65]. For a list storage scheme with multiword elements of arbitrarily variable size see [Ro67b]. Formats of multiword list elements composed of IBM System/360 32-bit words are described in [Ma67b, Th67]. Under these formats pointers occupy 3 or 4 bytes (24 or 32 bits). The suggestion of Section 10e that a directory of base addresses could be used in System/360 list processing may be new. In string processing it is sometimes advantageous to make a distinction between a logical element (a string) and a list element, where it is permissible to have a logical element distributed over several list elements. In [Wo65] a formula is derived for the optimal number of words in multiword list elements of constant size in terms of the average number of words in a logical element. A string storage scheme with list elements of variable size is described in [Be65].

Garbage collection based on the marking of the accessible list elements was first described by McCarthy [Mc60]; the reversal-of-pointers technique of A.10.6 is taken from [Sc67b]. A garbage collection scheme for the case when lists occupy two levels of storage is described in [Co67a]. Reference counters were introduced by Collins [Co60]. The inadequacy of the basic technique when it comes to recovery of erased reentrant list structures was pointed out in [Mc63b]. Nevertheless, the technique has been used quite effectively in two Fortran-embedded list processing languages, see [We63, Co66]. A hybrid scheme that uses a marking procedure in addition to the counter technique to get around the difficulty of reentrant list structures is described in [We69]. Alternatively, one can use the recursive scheme outlined at the end of Section 10j; note that the effectiveness of this scheme does *not* depend on the separation of structural information from symbolic data. Compaction of storage blocks of variable size is discussed in [Ha67c].

Facilities for structuring data along the lines of Section 10k are provided in Cobol and PL/I (see [Sa68c], pp. 330–381 for a description of Cobol and pp. 540–582 for PL/I). The storage mapping function of Section 9b is generalized to these data structures in [Ho62, De66]. [Kn68] contains a very useful section on them (pp. 423–434, see in particular p. 432 where the system of tables of Figure 10.25 is ascribed to Dahm).

Exercises

10.1 (a) Consider the Fortran implementation of a stack as described in Section 10a. What does the stack contain after execution of the following program segment?

```
      CALL SKOPEN
      DO  5  K = 1,10
  5   CALL PUSH (K)
      DO  6  K = 1,5
  6   NEX(K) = IPOP (IT)
      CALL PUSH (IT)
      DO  7  K = 1,5
  7   CALL PUSH (NEX(K))
```

10.2 (a) Consider the Fortran implementation of a stack as described in Section 10a. Assuming that 20 numbers have been stored in the stack, write a program that removes the element that is just below the topmost element in the stack, and every second element thereafter. (This leaves 10 elements in the stack at the end of the operation.)

10.3 (a) Write PUSH and IPOP routines for a *first in–first out* store. (An IPOP routine that shifts every item that remains in the store to an adjacent location after an item has been extracted is not acceptable.)

10.4 (b) Implement A.10.3 as a Fortran program and apply it to the tables of Figure 5.15.

10.5 (b) Develop an algorithm that converts an infixed arithmetic statement to prefix form.

10.6 (b) Combine A.10.1 and A.10.2 into a single algorithm that uses two stacks (an operator stack and an operand stack) concurrently.

10.7 (b) Develop an algorithm that converts a representative set of K-formulas (see Section 3e) to a list of arcs and a list of isolated nodes. Implement the algorithm as a Fortran program and apply it to the set

$$\{**a***a*bc**dcbef,\ **bf**e**bad*ac\}.$$

Use the program to show that

$$*a*a*bc*****dcbef$$

is not a K-formula.

10.8 (b) Develop an algorithm that tests a Boolean form for well-formedness (see D.2.8 and its example).

10.9 (d) Elements in the first 5 rows of a 250 × 2 array `LIST` (and only these locations) are to be used for list names, and the remaining elements of `LIST` are to function as a list store. Taking `LIST(1,1)` for the name of the list of available space, write a routine `INITIO` that initializes `LIST`.

10.10 (d) Assuming that a list store is arranged as described in Exercise 10.9, write the following subprograms:

(i) `KOUNT`—to find the number of elements in a given list.
(ii) `LAST`—to find the value of the last element in a given list.
(iii) `NTHOFF`—to delete the last element in a given list.
(iv) `KTHOFF`—to delete the kth element in a given list.
(v) `KTHADD`—to introduce a given datum into a given list so that the new element becomes the kth element of the list, the old kth element becomes the $(k + 1)$th element, etc.
(vi) `ALLOFF`—to delete an entire given list.
(vii) `MCHOFF`—to delete every element in a given list that matches a given value.

10.11 (d) Assuming that a list store is arranged as described in Exercise 10.9, make appropriate changes to subprograms `PUSH` and `IPOP` of Section 10d so that they will function for this arrangement. What is the appearance of the array `LIST` after execution of the following program segment?

```
CALL INITIO
DO 5 K = 1,10
DO 5 J = 1,2
```

```
5   CALL PUSH (LIST(J,2),K+J)
    DO  6  K = 1,5
    DO  6  J = 1,2
6   NEX(J,K) = IPOP (LIST(J,2))
    CALL PUSH (LIST(1,2),NEX(1,5))
    CALL PUSH (LIST(2,2),NEX(2,5))
    DO  7  J = 1,2
    DO  7  K = 1,5
7   CALL PUSH (LIST(J,2),NEX(J,K))
```

10.12 (d) Assuming that a list store is arranged as described in Exercise 10.9, write PUSH and IPOP routines for the implementation of *first in–first out* stores as lists. (You may make calls to the routines of Exercise 10.10.) The efficiency of the FIFO stores can be greatly improved by making each "list name" a pair of pointers. To where should the two pointers point?

10.13 (f) Develop an algorithm that generates a list structure from a parenthesized string.

10.14 (f) Represent the tree of Figure 3.18 and the digraph of Figure 3.1 by list structures.

10.15 (f) Develop an algorithm for the output in sorted form of the information contained in a sort tree that has the representation of Figure 10.15.

10.16 (f) Design list representations of the following structures: (i) binary trees in which only terminal nodes are labeled, (ii) weighted digraphs, (iii) weighted labeled digraphs, (iv) trees that have the form shown in Figure 6.5, allowing for the sharing of sublists.

10.17 (g) Redraw the list structures of Exercise 10.14 using (i) threaded list representation, (ii) symmetric list representation.

10.18 (g) In the structure of Figure 10.21 only terminal nodes carry labels. Redraw the structure of Figure 10.20 to make it conform to this convention, and generate the corresponding symmetric list representation. Find general formulas for the numbers of list elements required for symmetric list representation of parenthesized strings under the two conventions. Find also a general formula for the number of list elements required for threaded list representation of a parenthesized string.

10.19 (g) Design algorithms for processing of symmetric lists that correspond to subprograms KOUNT, LAST, NTHOFF, KTHOFF, and MCHOFF of Exercise 10.10.

10.20 (g) Let a FIFO store be represented by a symmetric list. Develop PUSH and IPOP procedures for this representation.

10.21 (j) Repeat Exercise 10.15 using now the reversal-of-pointers technique to keep track of the traversal.

10.22 (j) A reference counter technique for recovering reentrant list structures is outlined at the end of Section 10j. Expand the outline to an explicit detailed algorithm.

10.23 (k) Draw a tree that represents the data structure declared by the PL/I statement

```
DECLARE 1X, 2Y, 3Z(3), 3YY(2), 2X, 3Y;
```

10.24 (k) Develop an algorithm for the creation of table SYMBOL and the first two columns of table STRUCT (see Figure 10.25) from declarations of data structures.

CHAPTER 11

Organization of Files

11a. Records and Files

Let us start with a quote: "Data management problems are what the Internal Revenue Service has when it receives 65 million income tax returns every April 15." The returns are records in an extremely large file. Another example of a file is a telephone directory in which the triples {*name*, *address*, *telephone number*} are the records. The information kept by a computer installation on the jobs it has run, which may be a set of quintuples of the form {*job number*, *name*, *department*, *phone*, *cumulative total of money spent on computer use*}, is another example. The bibliography at the end of this book is a file of modest size; its records are the quintuples {*code*, *author*(*s*), *title*, *publishing data*, *reference data*}. A *file*, then, is a set of n-tuples, the n-tuples are *records* in the file, and we shall call the coordinates of an n-tuple the *fields* of a record. Large or small, files have to be organized and managed, and the purpose of this chapter is to provide some insight in the problems associated with these activities.

A considerable effort has gone into the formalization of file processing concepts, but, although we must be ready to recognize the benefits that can derive from a formal approach, we should be cautious not to become too pedantic. The formal theories of files propounded so far do not appear to satisfy significantly the requirement set down in another quote: "The mathematical system serving as a model must yield theorems whose interpretation affords some deeper insight or knowledge." For this reason, apart from giving a sound definition of the objects that we shall talk about, we shall adopt an informal approach.

DEFINITION 11.1 A *file* is a collection of data, each datum consisting of three elements:

(a) a *unit*—an entity (object, person, concept, etc.) that may be considered, for data processing purposes, in terms of a finite number of properties;
(b) a *property*—a characteristic to which measures can be assigned;
(c) a *measure*—a value capable of being expressed in a finite number of information units (bits, characters, etc.).

The set of data in a file associated with the one unit is a *unit record* or simply a *record*.

The values are the fields of a record, and in most files only values are explicitly represented. Such files are called *homogeneous*. It is then essential to have the values ordered so that they can be correlated with the appropriate properties as and when required. Let $P = \{p_1, p_2, \ldots, p_n\}$ be a set of properties, and let V_i be the set of values that property p_i can take. A homogeneous file on P is a subset of $V_1 \times V_2 \times \cdots \times V_n$, and the members of the subset are the records. A V_i may have to include a special null value in case p_i is inapplicable to a given unit, or its measure is unknown. A case in point: In the bibliography of this book references that carry a Zz code have anonymous authors, and, if we consider the bibliography as a subset of $V_1 \times V_2 \times V_3 \times V_4 \times V_5$, then V_2 must contain the special null value, which is assigned to the Zz entries.

The decision of whether or not to make a file homogeneous is determined by the characteristics of the unit records and by the purpose behind the setting up of the file, i.e., by what is to be done with the file afterwards. In many instances homogeneous design is natural. For example, the population census of a country produces a unit record for each resident of the country. Here $P = \{$*name*, *address*, *date of birth*, *place of birth*, $\ldots\}$. The file can be stored on magnetic tapes. Since the properties for each unit record are the same, there is no need to give an explicit representation of the properties in any one record. The file is homogeneous, with each record occupying a block of fixed size on a tape. The representation of the measures within a record adheres to a fixed format, so that a particular measure, say *date of birth*, occupies a well-defined field in each block.

Next consider a fact retrieval system based on a file of Nobel Prize winners, with the records having the format Name/date of birth/date of death/Peace/

Literature/Chemistry/Physics/Medicine/Economics. A record in the file might be

Watson, James Dewey/1928/ ∅ / ∅ / ∅ / ∅ / ∅ /1962/ ∅

On the other hand, the record could be stored in a nonhomogeneous file as

Watson, James Dewey/B1928/M1962

Although the homogeneous file would take up more space, retrieval of, say, all winners of the Medicine Prize after 1950 would be faster, and this consideration could lead one to the adoption of the homogeneous design.

A file may be designed so that every record in the file occupies the same amount of storage space, or the size of individual records may be permitted to vary. We speak then of *fixed length records* and *variable length records*, respectively. A homogeneous file generally consists of fixed length records. In a nonhomogeneous file the amount of information that makes up a record may greatly vary from record to record, and the greater storage efficiency achieved with variable length records may outweigh the benefits deriving from the greater simplicity of a file in which all records have the same length.

A file must give reasonably fast access to a specified item of information contained in it. This calls for organization. The most common organization of records in a file is the *sequential* organization. Although there are exceptions (for example, the sequential order in a file of incoming messages in a message-switching system may be determined by the order of arrival of the messages), generally the records in a sequential file are in lexicographic order of the values in some particular field or fields within the records. The fields that determine the sequential order are called *keys*. In a telephone directory the key is the property *name*. In our bibliography the key is the code field; since the names of the authors have been used in the construction of the codes, the names are also very nearly in lexicographic order, but not quite.

When the key of a number of records has the same value, the use of a single key may be inadequate, and a second key may have to be used to establish the sequence of records for which the first key has the same value. In some bibliographies the field *name of author* is the first key, and *publication year* the second.

The records are put into sequential order by *sorting* on the key or keys. Some sorting procedures will be given in Section 11d. When the values of a key are in lexicographic order, *logarithmic search* (also called *binary search*) can be used to gain access to a record specified by a given value of the key. This value is called the *search key*.

Logarithmic search is based on the following observation. Assume that A is a vector of n elements, and that the elements are in lexicographic order, i.e., that $a_k \leqq a_{k+1}$ for $k = 1, 2, \ldots, n - 1$. Further assume that a given datum d, if it is in A, is one of $a_i, a_{i+1}, \ldots, a_j$. Compute $k = \lfloor (i+j)/2 \rfloor$, where $\lfloor x \rfloor$ denotes the largest integer not greater than x, and compare d with a_k. If $d = a_k$, the search is over; if $d < a_k$, then d can only be one of $a_i, \ldots, a_{k-1}$; if $d > a_k$, it can only be one of $a_{k+1}, \ldots, a_j$. Logarithmic search starts with $i = 1$ and $j = n$, and the search region is successively reduced. The total number of comparisons is at most $\lfloor \log_2 n \rfloor + 1$; it is $\lfloor \log_2 n \rfloor$ if one can assume that d is in A. By contrast, if the elements of A were unordered, one would have to go down A, comparing d against each element in turn. If d were in A, the expected number of comparisons would be $n/2$; if it were not, then one would have to go through all n elements before one could be sure.

ALGORITHM 11.1 A program for finding a given datum KEY in the first column of a matrix **M** by logarithmic search. It is assumed that the first column of **M** contains integers in ascending order. SEARCH returns the number of the row in which KEY is found; it returns 0 if KEY is not in the first column of **M**.

```
      INTEGER FUNCTION SEARCH (KEY,M,N,K)
      DIMENSION M(N,K)
      INTEGER HI
      LO = 1
      HI = N
      SEARCH = 0
 5    IF (LO.EQ.HI .AND. KEY.NE.M(LO,1)) RETURN
      MID = (LO+HI)/2
      IF (KEY - M(MID,1)) 6,10,7
 6    HI = MID
      GO TO 5
 7    LO = MID + 1
      GO TO 5
10    SEARCH = MID
      RETURN
      END
```

11b. Indexed Files

The registers kept by motor registration authorities are files. Assume that records in the file of one such authority are divided into the fields *name/address/registration number/date of expiry/model of car*. The file might serve a variety of purposes. The primary purpose should be to provide information for sending out renewal notices, but the file might be used in crime detection as well. As far as the primary purpose is concerned, the appropriate sort key is *date of expiry*. However, sorted this way, the file is not properly ordered for finding the name and address corresponding to a given registration number—an application that could arise in crime detection.

One solution is to arrange the file in 12 *blocks*, one for each month of expiry, and to have each block sorted on the *registration number*. The block starting addresses are kept in an external directory of 12 entries. Then, given a registration number, one applies logarithmic search to each block in turn. Assuming that the blocks are approximately equal in size and that the total number of records in the file is n, the expected number of comparisons for locating a record is approximately $6 \log_2 (n/12)$.

A more drastic solution would be to keep two complete files: one sorted on *date of expiry*, the other on *registration number*. This design would squander memory; the only information that we really need in the second file is the set of registration numbers and pointers to records in the first file. Let us, therefore, generate the second file as follows: (a) from each record in the first file copy the registration number and attach to it the address of the record, and (b) sort these ordered pairs on the registration numbers. The second file is now an *index* of the first. An index can be prepared using any of the fields. A file that has a set of indexes associated with it is known as an *indexed file*.

Let us turn again to document retrieval, which was introduced in Section 8c. Assume that we have a set of documents, and that the documents have index terms associated with them. The index terms might be:

Document 1—memory protection, monitor, operating system, time-sharing;
Document 2—integer programming, operations research, zero–one variables;
Document 3—data structures, file handling, paging, storage allocation, storage organization, time-sharing;
⋮

An indexed document file is set up by storing the bibliographic information in order of document numbers (preferably with all entries made the same size), and by creating a sorted list of index terms in which each term is followed by identifying numbers of all documents with which it is associated. Let us call the two components of the indexed file document register and index, respectively. A user might make a request for documents that carry all of the following index terms: memory protection, operating system, time-sharing. The three terms would be looked up in the index, and the intersection of the three sets of document numbers computed. The resulting set of document numbers then identifies all documents satisfying the request, and the bibliographic information relating to these documents is extracted from the document register.

The nature of the document file differs somewhat from that of the motor registration file. The basis of the indexed motor registration file is a sequential file containing all the information that is stored in the system, and this sequential file remains intact throughout; the indexes are *additions* to it. Searches are made in the sequential file and in the indexes. Sequential files to which indexes are added are called *indexed sequential files*. Imagine that the indexed document file, too, is generated from a sequential file, one in which each record consists of bibliographic data and the index terms, i.e., of more than one field. Now, however, each record is reduced to a single field, and *all* other data are placed in the index; i.e., a separation of the data takes place. Searches are made in the index alone. A file of this type is called *inverted*. The distinction between indexed sequential and inverted files can become very blurred. For instance, if it is considered worthwhile to supply the user of the document retrieval system with additional information that would help him come to a finer appreciation of the relevance of the retrieved documents, then the full sets of index terms associated with the retrieved documents should be incorporated in the output, and this means that the index terms should be part of the register entries. Then the document register carries too much information for the file to be a pure inverted file. On the other hand, since there is no direct interrogation of the register, the file is not indexed sequential.

The process of making changes to a file is called *updating* of the file. The document file has to be updated when new documents come to hand, and the ease or difficulty of updating is determined by the original design of the file. No complex system is perfect, and sooner or later an existing system ought to be scrapped and replaced by a new system designed on the basis of operational experience with the existing system. Therefore, one can decide on the greatest number of documents that the file will be permitted to contain, taking the view that when this number is in danger of being exceeded the time will have come to redesign the system anyway. One may also be able to decide in advance on a set of permissible index terms, and, having a bound on the number of documents, one would have a fair idea of how many locations should be assigned to each index term for storing document numbers. We shall call these locations the *reference vector* of the index term. With this design updating of both the document file and the index is trivial. When a new document is brought into the system, the bibliographic data are simply appended to the existing document register. The index terms associated with the document are looked up in the index, and, for each index term, the number of the new document is inserted in the first free location of its reference vector.

If the set of index terms cannot be fixed in advance, then the document register is still easy to update, but updating of the index becomes difficult. Assume that an index term for which there is no entry in the index is associated with the new document. One cannot just append the entry for the new index term to the end of the index, because then the index is no longer ordered and logarithmic search becomes inapplicable. Therefore, storage has to be found for the new entry at its rightful place within the index, and this involves shifting of all entries that come after it. A file or a component of a file organized sequentially in lexicographic order can be efficiently searched, but the efficiency of updating is very poor.

Efficiency of updating improves when the index is set up as a list structure. Consider the index of the document file arranged as a binary sort tree. The tree is stored as a list structure (Figure 10.15 shows an example of a list structure corresponding to a sort tree) in which an element consists of left and right pointers, the index term, and locations for storing document numbers. Search in a *balanced* binary sort tree is as efficient as binary search in the corresponding sequentially organized sorted index, but, since the sort tree is unlikely to be balanced, we can expect some loss of search efficiency. Because of the left and right pointers there is some loss of storage efficiency as well.

A change in the design of the list structure eliminates the need to estimate in advance the number of documents that will be indexed by a particular index term. Instead of representing a node in the sort tree by an element, let us represent it by a list. We now need two types of list elements. An element of the first type carries left and right pointers and an index term as before, but, in place of the locations for storing document numbers, there is a pointer to a list of these numbers. Document numbers are stored in elements of the second type. These elements consist of two fields: one for a document number, the other for a pointer to the next element containing a document number. The document numbers are stored in ascending order. Figure 11.1 shows the list corresponding to a node of the sort tree.

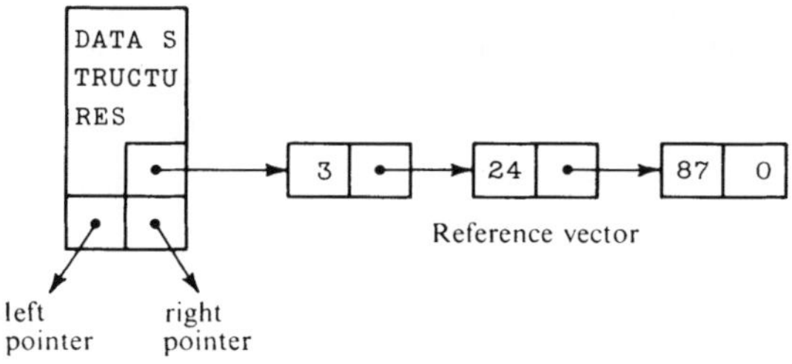

Figure 11.1

11c. Scatter Storage Techniques

Logarithmic search locates a record in a sorted sequential file fairly rapidly, but even better search times are achieved when the design of a file is based on scatter storage techniques. Moreover, updating efficiency matches the search efficiency. Although a scatter file occupies more storage than a sequential file containing the same amount of information, the additional locations required may be fewer than the locations taken up by pointers in a sort tree.

Let us take the compilation of Fortran programs for the context of our discussion. A compiler has to keep track of locations assigned to variables in the program that is being compiled; i.e., a file with records of the form ⟨*variable name*, *location*⟩ has to be maintained. If A is the set of permissible Fortran variable names, then $|A|$ is approximately equal to 1.6×10^9 (Exercise 1.28), but in a particular program only a comparatively small number of possible names will be encountered. Let T be a vector in which we store names as they are encountered. (We also need a vector in which to store the assigned locations, but this vector is irrelevant to our discussion.) Consider a function f: $A \to I$, where I is the set of subscripts of the elements of T, and the restriction $f \mid N$, where N is the set of variable names actually appearing in the program. Now, $f \mid N$ cannot be one to one, i.e., cannot associate all names in N with unique subscripts (otherwise, for T of any reasonable size, the computation of values of the function would take so long that the purpose for having the function would be lost), but it is possible to define functions f such that the number of elements in the range of $f \mid N$ is not much smaller than $|N|$, i.e., the restriction is close to being one to one, and the number of elements in the range of f is not much greater than $|N|$, i.e., the storage requirements are modest. Moreover, the subscripts can be computed very rapidly.

In general terms, we take a record, and apply a function f, called the *scatter function* or *hash function*, to a key in the record to arrive at an address in a scatter file storage area. Several different scatter functions have been devised. We shall consider the three most commonly used. To simplify our discussion we shall continue with the example of the Fortran compiler, and describe the operations in terms of variable names and subscripts defining locations in the table T (which is our *scatter table* or *hash table*).

a. Square the name and extract n bits from the middle of the result. The value of this field (the *hash address*) defines a location in T. Since the middle bits of the square of a word depend on every bit in the word, the probability that different variable names will give rise to different addresses is nearly the same for LOCA and LOCB, XARRAY and YARRAY, ABCDEF and ZYXWUV. Unless one employs a scaling procedure, the size of T must be 2^n, where n is the number of bits extracted. This can prove rather annoying: If a table of, say, 2048 locations is too small, then the table having the next possible size of 4096 may be much too large.

b. Compute the hash address by separating the name into n-bit fields, adding the fields, and taking the n lowest bits of the sum. Whether this method is preferable to squaring depends entirely on the speed with which the computer performs the operations. A variant of this method is to split the name in half, and to apply the fast logical *exclusive or* operation on the two halves, e.g., from GORSE (represented by the octal integer 274651622560_8 in the IBM 7040 character code) and HORSE (304651622560_8) derive 456331_8 and 526331_8, respectively. The n bits are extracted from this result. Unfortunately, nearly equal names will tend to give the same hash address. Here the same hash address is obtained if one takes the low bits. If the high bits were taken, then the same hash address would be obtained for names differing only in the last character. Splitting the name into three or more parts and applying the *exclusive or* to all parts is no solution because the probability that the hash address contains more zeros than ones grows with the number of parts.

c. Divide the name by the size of the table and take the remainder for the hash address. The method is more effective if the size of the table is an odd number. Otherwise the datum and the hash address have the same lowest bit. When a variable name does not occupy an entire word, the spare character positions are normally filled with blanks. All short names will then give rise to even hash addresses, and the result will be a heavier use of the even locations in the table. The division method has two advantages: One does not have to go below the level of a computer word in the computation of the hash address, and the size of the table is not restricted to powers of 2.

In nearly all practical situations one has to deal with multifield records. Quite often the key is not as important as the other fields. For example, in symbol tables of compilers and assemblers the addresses assigned to variables, rather than the variable names, are the important data, and the names may actually be discarded at some stage of the process. It would be to our advantage to discard the names as early as possible, and store in T the locations assigned to variables instead of their names. However, if we discard keys while the file is being added to, then we cannot deal with situations in which a hash address refers to a location that is already occupied. If the computed hash address refers to an empty location, or the location is found to be occupied, but the key agrees with the stored key, the job is done. If, however, the key and the stored key differ, then we have a *collision*. We shall now consider a number of methods for dealing with collisions.

The simplest technique is to look for a free location (or for agreement of the key with a stored key) in the neighborhood of the location to which the hash address points. One examines location after location in the forward direction until the job is done. If the end of the table is reached, the search shifts to the head of the table. This linear search procedure leads to clustering of entries and is rather inefficient as a consequence.

The clustering effect is reduced by making the search procedure cover the table in jumps of varying magnitude. With a table of size N, if collision occurs at the kth location, compute a pseudorandom integer r on the interval $[1, N-1]$, and make a new try at the $(k+r)$th location, or at the location given by the remainder of $(k+r)/N$ in case $k+r > N$. The pseudorandom number generator must produce every integer on $[1, N-1]$ exactly once. The table is full when repetition sets in. The same sequence of integers r must be produced for every new key; i.e., every insertion or look-up operation involving collisions must start with a new initialization of the pseudorandom number generator. The generator may, for example, generate the sequence 1, 6, 31, 13, ... Then, if the hash address is 18, say, and a collision occurs, one tries location 19. If this probe results in a collision as well, one tries 24, then 49, and so on. If the table size is 161 and the hash address 153, say, collisions would make one try in turn locations 154, 159, 23, and so forth.

The efficiency of the random search procedure depends only on the ratio of occupied locations t to total number of locations N. This ratio is known as the *load factor*, and we write $\alpha = t/N$. The expected number of probes necessary to enter the $(k+1)$th record is given by

$$\begin{aligned} A_k &= 1 + \sum_{i=1}^{k} P(k, i)/P(N, i) \\ &= 1 + k/(N-k+1) \\ &= 1/(1 - k/N + 1) \\ &\approx 1/(1-\alpha). \end{aligned}$$

The average number of probes E that must be made to look up an item is equal to the average of $A_0, A_1, \ldots, A_{k-1}$, and this is (approximately)

$$E = \frac{1}{\alpha}\int_0^\alpha \frac{dx}{1-x} = -\left(\frac{1}{\alpha}\right)\ln(1-\alpha).$$

Some representative values:

α:	0.5	0.75	0.9	0.95
E:	1.39	1.83	2·56	3.15

Compare these with the average number of probes in linear search: approximately 5.5 when $\alpha = 0.9$ and 10.5 when $\alpha = 0.95$.

A third approach is to set up all records having the same hash address as a list. The hash address is computed, and, if the location indicated by this address is empty, the new record is inserted there. If the hash address points to the first element of a list, the sequence of pointers is followed until the keys agree or the end of the list is reached. In the latter case a free location is found by some means or other, the record is inserted in this location, and the location joined to the list. A difficulty arises when the hash address points to a location occupied by an element that is not the *first* element of a list, i.e., belongs to the list corresponding to some other hash address. In this case the old entry must be moved to a new location so that the record can become the first element of a new list. The relocation involves changing the pointer in the list element preceding the element that has to be moved, and there arises the problem of finding this element (Exercise 11.8). The average number of probes is significantly smaller than for the other schemes:

α:	0.5	0.75	0.9	0.95
E:	1.25	1.38	1.45	1.48

but these values do not tell the whole story. Relocation of elements can be quite costly in terms of time, the actual overheads depending greatly on the skill of the programmer. Moreover, additional space is taken up by the pointers.

A modification of the method eliminates relocation. The total storage region is divided into two segments. One is the scatter storage area itself; the other is an overflow area, arranged as a list of available space. Only the first element of a list is located in the scatter storage area. All other elements go into the overflow area. Then a hash address must always point either to a free location or to the first element of a list. Consider the second alternative, and assume that the key of the new record does not agree with the stored key. If the pointer is zero, then an element is taken from the list of available space, the new record is inserted in this overflow element, and a pointer to the element put in the pointer field of the element in the scatter storage area. If the pointer is not zero, then one follows pointers until the end of the list is reached (assuming that the keys do not agree anywhere along the way). Then the record is stored in an element taken from the list of available space, and this element is added to the list.

11d. Sorting

Although file organization is becoming more dependent on scatter storage techniques, the importance of sorting is not diminishing. Files are maintained for the sake of providing us with information, and we expect to receive the information that we request in what we call a "natural" order. Therefore, irrespective of how files are organized inside a computer, records of an output file must be in lexicographic order of the values of some key or another.

The algebra of sets provides another context in which we depend on order. Set operations, such as union or intersection, take much longer when the operations are carried out on sets with unordered elements than when the elements are ordered. Assume that vector `A` contains the `N` elements of a set A, and that `B` contains the `M` elements of set B. A.11.2 is a procedure that merges elements of `A` and `B` into `C`. It is easy to modify the procedure so that it finds $A \cup B$, i.e., removes duplicates.

ALGORITHM 11.2 A procedure for merging numbers stored in order of magnitude in arrays A and B.

```
      SUBROUTINE MERGE (A,N,B,M,NPLUSM)
      DIMENSION A(N), B(M), C(NPLUSM)
      I = 1
      J = 1
      K = 1
    1 C(K) = A(I)
      IF (B(J).LT.C(K)) GO TO 3
      I = I + 1
      K = K + 1
      IF (I.LE.N) GO TO 1
      DO 2  II = J,M
      C(K) = B(II)
    2 K = K + 1
      RETURN
    3 C(K) = B(J)
      IF (A(I).LT.C(K)) GO TO 1
      J = J + 1
      K = K + 1
      IF (J.LE.M) GO TO 3
      DO 4  II = I,N
      C(K) = A(II)
    4 K = K + 1
      RETURN
      END
```

ALGORITHM 11.3 Let arrays A and B contain *distinct* elements of sets A and B in order of magnitude. Subroutine DIFF finds in C the K elements of $A - B$. Note that

```
      CALL DIFF (A,N,B,M,A,K)
```

is permitted.

```
      SUBROUTINE DIFF (A,N,B,M,C,K)
      DIMENSION A(N), B(M), C(N)
      J = 1
      K = 0
      DO 15  I = 1,N
    3 IF (A(I) - B(J))  10, 15, 5
```

```
    5  J = J + 1
       IF (J - M)  3, 3, 17
   10  K = K + 1
       C(K) = A(I)
   15  CONTINUE
       RETURN
   17  DO 20  J = I,N
       K = K + 1
   20  C(K) = A(J)
       RETURN
       END
```

Perhaps *selection sort* is the sorting method that is easiest to understand. Assume that an array $[a_1, a_2, \ldots, a_n]$ is to be sorted. One selects the largest of $a_1, a_2, \ldots, a_n$ and interchanges this number with a_n, then selects the second largest number in a scan of elements $a_1, a_2, \ldots, a_{n-1}$ and interchanges it with a_{n-1}, and so on. This method requires $(n-1)+(n-2)+\cdots+1 = \frac{1}{2}(n^2-n)$ comparisons. Next we consider *exchange sort.* In principle this method does not greatly differ from selection sort. In the first pass of the exchange sort one compares a_i with a_{i+1} for $i = 1, 2, \ldots, n-1$, and exchanges the elements whenever $a_i > a_{i+1}$. This shifts the largest element into position a_n, and a_n is not involved in subsequent comparisons. The exchanges of the second pass shift the second largest element into a_{n-1}, and so on. The process is stopped when no exchanges have been made in a pass. This can reduce the number of comparisons below the maximal value of $\frac{1}{2}(n^2-n)$ in certain cases.

ALGORITHM 11.4 Exchange sort.

```
       SUBROUTINE EXSORT (A,N)
       INTEGER A(N), TEMP
       LOGICAL FINISH
       DO  6  K = 2,N
       FINISH = .TRUE.
       LP = N - K + 1
       DO  5  L = 1,LP
       IF (A(L).LE.A(L+1)) GO TO 5
       TEMP = A(L+1)
       A(L+1) = A(L)
       A(L) = TEMP
       FINISH = .FALSE.
    5  CONTINUE
```

```
6   IF (FINISH) RETURN
    RETURN
    END
```

Array A has been declared INTEGER because the comparisons of A(L) with A(L + 1) may take less time in integer than in real arithmetic. Assume, however, that we make the call

```
CALL EXSORT (ARRAY, 100)
```

where ARRAY is REAL. Most systems do not check whether the actual argument ARRAY agrees in type with the dummy argument A. Moreover, representation of real numbers in many computers is such that, for real a and b, $a \leqq b$ if and only if $a \leqq b$ when a and b are *interpreted as integers*. Such a representation permits the use of the one subroutine for sorting both integer and real arrays.

Example

The array [3, 2, 1, 6, 4, 5] is sorted in three passes (no exchanges are made in the third pass). On the other hand, sorting of [2, 3, 4, 5, 6, 1] requires the maximal five passes to get the 1 into the first location.

We shall now describe a sorting procedure based on the merging operation of A.11.2. There are fewer comparisons than with selection or exchange, but $2n$ locations are now needed to sort n numbers. The numbers are paired, and placed into "arrays" of length 2, the smaller number of a pair becoming the first element of the "array." Next pairs of "arrays" of length 2 are merged to produce "arrays" of length 4, pairs of these "arrays" are merged to produce "arrays" of length 8, and so on. The method is most effective when $n = 2^m$, but usually n is not an exact power of 2. For all n satisfying $2^{m-1} < n \leqq 2^m$ the number of passes is m, i.e., the number of passes is given by $\lceil \log_2 n \rceil$ where $\lceil x \rceil$ denotes the least integer not smaller than x. The method is illustrated by Figure 11.2. Actually one uses only the two arrays A and B, both of size n, and the "arrays" of the discussion above correspond to sequences of elements in A or B.

The procedure that we have described is known as a *two-way merge* because pairs of arrays are merged. If, instead of merging two arrays, we produced the new arrays by merging q arrays at a time, then we would have a *q-way merge*. The number of passes in a q-way merge is $\lceil \log_q n \rceil$, and the total number of comparisons is less than $(q - 1)n\lceil \log_q n \rceil$.

When n is about 1000, two-way merge makes about 10,000 comparisons. Exchange sort makes about 50 times as many, but is minimal as far as storage

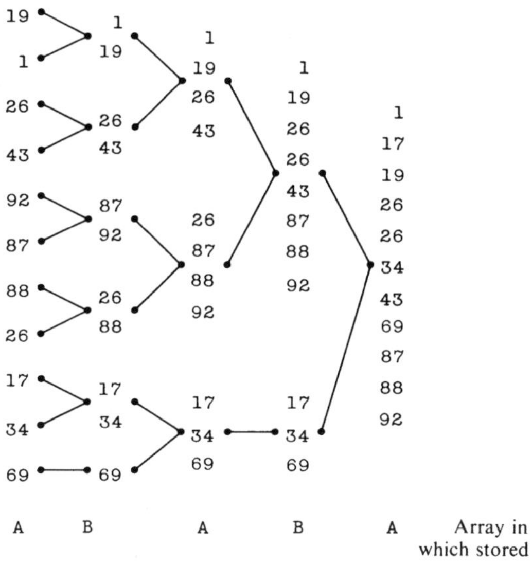

Figure 11.2

requirements are concerned. Our objective is to find a sorting procedure that combines the speed of two-way merge with minimal storage requirements. One minimal storage procedure that makes the same number of comparisons as the two-way merge is the *merge–exchange sort* (or *Shell sort*). Let $[a_1, a_2, \ldots, a_n]$ be the array to be sorted. Define the numbers

$$\begin{aligned} d_1 &= 2^{\lfloor \log_2 n \rfloor} - 1; \\ d_k &= \lfloor \tfrac{1}{2} d_{k-1} \rfloor, \qquad k = 2, \ldots, m-1; \\ d_m &= 1. \end{aligned}$$

In the first stage of the process strings $a_i\, a_{i+d_1}$ $(i = 1, 2, \ldots, n - d_1)$ are sorted by exchange, and, in general, in the kth stage strings $a_i\, a_{i+d_k}\, a_{i+2d_k} \cdots$ $(i = 1, 2, \ldots, d_k - 1)$ are sorted by exchange. For the array of Figure 11.2, since $n = 11$, have $d_1 = 7$ and $m = 3$. The first stage sorts strings $a_1 a_8$, $a_2 a_9$, $a_3 a_{10}$, $a_4 a_{11}$. We find, however, that in our example these strings are already in order. In the second stage, since $d_2 = \lfloor 7/2 \rfloor = 3$, exchange is applied to strings $a_1 a_4 a_7 a_{10}$, $a_2 a_5 a_8 a_{11}$, $a_3 a_6 a_9$. For example, string 19 43 88 34 becomes 19 34 43 88. At the end of the second stage the array is in rather good order: [19, 1, 17, 34, 26, 26, 43, 69, 87, 88, 92]. Therefore, the sort is completed with very few exchanges in the third stage.

ALGORITHM 11.5 Merge–exchange sort. (For reasons of efficiency the sequence of operations is not exactly as described above—there is some overlap in the sorting of the strings in stages 2, 3, ..., $m - 1$.)

```
      SUBROUTINE SHELL (A,N)
      INTEGER A(N), TEMP, D
      D = 1
    1 D = 2 * D
      IF (D.LE.N) GO TO 1
    2 D = (D-1)/2
      IF (D.EQ.0) RETURN
      ITOP = N - D
      DO  4  I = 1,ITOP
      J = I
    3 L = J + D
      IF (A(L).GE.A(J)) GO TO 4
      TEMP = A(J)
      A(J) = A(L)
      A(L) = TEMP
      J = J - D
      IF (J.GT.0) GO TO 3
    4 CONTINUE
      GO TO 2
      END
```

Evaluation of the procedures has to be approached with great caution. When the numbers of comparisons differ greatly, as they do with ordinary exchange and merge–exchange, then it is obvious which of the two is the better method. This is not at all obvious when one compares merge–exchange with the tree sort (A.5.3), although in the worst case tree sort requires twice as many comparisons as merge–exchange. In addition to the number of comparisons, efficiency depends on the quality of coding and the complexity of the programs; a much larger proportion of the total execution time is spent sequencing the actual computation steps in a complicated program than in a program having a simple structure. Efficiency depends also on the language in which the program is written, on the compiler, and on the peculiarities of the computer. A test on Algol versions of A.5.3 and A.11.5 run on the CDC 1604A has shown that with $n = 10{,}000$ merge–exchange takes 1.33 times as long as the tree sort for integer arrays, but 1.38 times as long for real arrays. The difference in these ratios should be due to causes unrelated to the algorithms themselves.

Throughout this book we are emphasizing that a high-level language, such as Fortran, which was designed primarily for coding numerical algorithms, can be applied to problems of a nonnumerical nature. We have seen that Fortran can deal with such problems quite effectively, provided the data of the problems can be translated to numbers. Difficulty begins when the data cannot be interpreted as numbers. Consider alphabetic data read into the computer under A-conversion. In IBM 360 machines the trouble is minimal. Characters are stored in 8-bit fields (bytes), and 4 such fields make up a word. For example, we would have the representation,

BATE	11000010	11000001	11100011	11000101
ABET	11000001	11000010	11000101	11100011
BAT	11000010	11000001	11100011	01000000

where the code 01000000 in the last word represents a blank. Negative integers are represented by a 1 in the leading bit position and the 2's complement of the integer in the remaining 31 bit positions. Interpreted as integers our data become

−111101	00111110	00011100	00111011
−111110	00111101	00111010	00011101
−111101	00111110	00011100	11000000

Clearly, when these data are sorted as integers, we get the proper order

```
ABET
BAT
BATE
```

But, if the words to be sorted contain both upper and lower case letters, then the words beginning with lower case letters precede *all* words beginning with upper case letters, and we would get, for example, a "sorted" list

abet
bat
bate
Adam

The position is more difficult with some other codes. For example, the code used for internal representation of alphabetic data in the IBM 7040 uses 6 bits (or 2 octal digits) to represent a character, and there are 6 characters to a word. With this code, interpreting the representations as integers, we have

GORSE	274651622560_8
HORSE	304651622560_8
MORSE	-44651622560_8
MORSEL	-44651622543_8
MORSES	-44651622562_8

and an impossible "sorted" list is obtained:

```
MORSES
MORSE
MORSEL
GORSE
HORSE
```

Other than going to assembler language, there is no solution. A Fortran program interprets the 36-bit word (6 characters) as a number composed of a sign bit and 35 other bits, and the sign bit causes part of the trouble. Assembler language provides "logical" comparison, which, in effect, treats all 36 bits as a positive integer, and this facility should be used whenever the Fortran procedure would make an arithmetic comparison. Then the sorted list is

```
GORSE
HORSE
MORSEL
MORSE
MORSES
```

Only MORSE is out of place in this list. If, moreover, blanks are translated to zeros before sorting, and translated back to blanks afterward, the proper order is obtained.

Some assembler languages give access to individual characters, and this facility is used in *radix sort*, which is a popular sorting procedure for mechanical card sorters, but is not widely used on computers because of very high storage requirements. In this method, to sort words over an alphabet of k characters, $k + 1$ "bins" are used. If d is the length of the longest word in the set to be sorted, then every word in the set is assumed to be d characters long; the extra characters are assumed to be blanks at the right-hand end of the word. The dth character is examined first, and the word is assigned to the bin associated with that character. Then the words are collected and distributed again on the $(d - 1)$th character, and so on. Table 11.1 shows how the procedure works. Storage requirements come to $(k + 2)n$, but there are only nd comparisons. The $(k + 1)$-way comparisons can be executed very rapidly by making use of assembler language indexing techniques. Storage requirements can be significantly reduced if the bins are set up as lists.

A refinement of the method, which requires only $2n$ storage locations, can be implemented when individual bits are accessible. Examine the lowest bit position: Going through the array from top to bottom first transfer all items with a 0 in this position to an auxiliary array; then transfer all items

TABLE 11.1

RADIX SORT

	A	B	C	D	E	G	H	blank
BEACH CAB ACE ADAGE BEAD					ADAGE		BEACH	CAB ACE BEAD
ADAGE BEACH CAB ACE BEAD			BEACH	BEAD		ADAGE		CAB ACE
BEACH BEAD ADAGE CAB ACE	BEACH BEAD ADAGE	CAB			ACE			
BEACH BEAD ADAGE CAB ACE	CAB		ACE	ADAGE	BEACH BEAD			
CAB ACE ADAGE BEACH BEAD	ACE ADAGE	BEACH BEAD	CAB					

with a 1 in this bit position. Next examine the second lowest bit position, again transferring first all items with a 0, then items with a 1 in this position (these transfers are back to the original array), and so on. Figure 11.3 illustrates the procedure with 5-bit numbers. Although the method involves bn transfers, where b is the number of bits in an item, and $2bn$ individual bits are examined, the method, which we call *binary radix sort*, can be quite fast.

Another method that has greater than minimal storage requirements is *sorting by address calculation*. In order to be able to apply this method one has to know the range of the numbers to be sorted, and—for efficient performance—something must be known about their distribution. To illustrate the *principle* of the method assume that 90 distinct unordered integers in the

17	10001	01100	01100	11000	10001	00110	6
9	01001	00110	11000	10001	10001	01001	9
12	01100	11000	10001	01001	10011	01100	12
6	00110	10001	01001	10001	00110	10001	17
24	11000	01001	10001	10011	11000	10001	17
19	10011	10011	00110	01100	01001	10011	19
17	10001	10001	10011	00110	01100	11000	24
Pass	1	2	3	4	5		

Figure 11.3

range [1, 100] occupy array **N**, and that a second array **M**, of dimension 100, is available. The numbers can be sorted by the following procedure:

```
      DO  4  K = 1,100
   4  M(K) = 0
      DO  5  K = 1,90
      I = N(K)
   5  M(I) = I
      I = 0
      DO 10  K = 1,90
   7  I = I + 1
      IF (M(I).EQ.0) GO TO 7
  10  N(K) = M(I)
```

Of course, the sort is not nearly as neat in a practical situation. First, the calculation of the address, i.e., of subscript I of the location in **M** that an item having value t is to occupy, can take considerable time. If n numbers have uniform distribution and range in value from k_1 to k_2, and **M** has dimension D, then an estimate of I is given by $(t - k_1)c$, where $c = \text{D}/(k_2 - k_1)$. The formula would be more complicated for nonuniformly distributed items. Secondly, the same address may be computed for more than one item, e.g., when there is duplication. Then we may find **M(I)** already occupied, and another location has to be found for the new item. All items in **M** at any particular stage must be in their correct order. Therefore, insertion of the new item may involve some shifts. To avoid having to shift too many items, D has to be made greater than n. When $\text{D} = 1.5n$, then, assuming that the formula for the address calculation reflects the distribution of the data reasonably well, the method requires about $1.33n$ comparisons. One must, of course, include address calculations in time estimates.

Let us describe sorting as follows: In sorting of a file we compare values of a *key* and interchange *records*. This description applies equally well to a file of multifield records and a file in which the key is the entire record. Despite the identity of the processes in principle, sorting of multifield records may create its own particular problems. Consider matrix Q of Section 9c. It is a very simple file; its columns are records. If the first element in a column is taken for the key, then the file is already sorted. We may, however, decide to sort on the second element of a column, and, to make things more difficult, on the first element within each group of columns having the same second element. But the file is already sorted on the first element, and this should be taken into account in selecting the sorting procedure. We want a procedure that keeps records having the same second element in their existing relative order. Then, at the end of the sort on the second element, the records are in order on the secondary key as well. A single pass radix sort, with five bins for the five values of the key, is obviously one such method.

Shifting of records during a sort is inconvenient when the file consists of long records. The situation is aggravated when the records are of varying length, or the file is too large to fit in core memory. The usual procedure then is to extract values of the key, and to have a record identifier associated with each value (the identifier may be the address of the first location occupied by a record). The ordered pairs ⟨*key*, *identifier*⟩ are sorted, and after the sort the sequence of record identifiers defines the ordered file; i.e., the sorted ordered pairs constitute an index of the file. If we can afford the space for permanent retention of the index, then we have now an indexed file, and there is no need for physical rearrangement of the records. If the file is in core memory, a vector of identifiers is all that we require. The identifiers are used to gain access to values of the key in the file itself when a comparison has to be made, and only the identifiers are ever exchanged. This variant is slower, and the records have to be rearranged at the end of the sort.

11e. Files and Secondary Storage

The properties of the medium on which a file is stored affect the organization of the file and the procedures used in its processing. If a sequential file is stored in core memory, then logarithmic search is the proper procedure for gaining access to a record, but imagine applying logarithmic search to a file stored on magnetic tape! The tape is spun forward 1000 feet, say, back 500 feet, back 250 feet, again forward 125 feet, and so on. Certainly a fascinating spectacle, but that is about all it would be.

When a file is not in core memory the *page* of a file becomes an important concept. Records are regarded as grouped into a sequence of pages, the size of a page depending on the characteristics of the storage medium. In the case of a magnetic tape the pages are tape blocks. With this sequential storage medium the only way of getting at the nth page is by going through the preceding $n - 1$ pages, and one might as well have the search guided by data on these pages. Therefore, instead of performing logarithmic search on the file, one reads the pages in sequence, inspecting the value of the key in the last record of a page. If the search key exceeds this value, the next page is read; if not, then logarithmic search is applied to the page currently in core memory.

Magnetic disks (and drums) call for a different strategy. Let us interpret the tracks as pages. Nearly all of the time required to access particular records is spent in gaining access to the pages to which they belong, but each page can be accessed independently. The best search time is achieved if one can identify the page that holds the record indicated by a given search key without direct reference to other pages of the file. Therefore, a directory, which holds the value of the key of the first record of every page, is kept in core storage. Then, given a search key, the appropriate page of the file is identified by something similar to logarithmic search in the directory. This page is brought into core memory, and logarithmic search takes over. If it is impracticable to hold the directory in core storage at all times, then it can be made part of the first page of the file. Now a record is accessed by bringing at most two pages into core memory.

Next, let us investigate the effects of the use of secondary storage on the design of the document retrieval system of Section 11b. We base our design on core memory, a disk, and magnetic tape storage. Assume that the document register is stored on tape, and the index distributed over core memory and the disk: index terms in core memory (either as a sort tree or as a scatter table), reference vectors on the disk. We shall see that the design illustrated by Figure 11.1 can be retained only in part. The main requirement is that any reference vector should be stored on just the one track, i.e., that each track be associated with a fixed set of index terms. This prevents us from setting up the tracks reserved for the reference vectors as a single list store. We can, of course, arrange each track as a separate list store. Note that half the available space in the list store is then taken up by pointers. This means that overflow of a track could occur earlier than under a scheme in which the track is divided into blocks, each block assigned to a particular index term, and the reference vector stored sequentially in this block. Assume that the latter scheme is adopted. Then the pointer from the index term to the reference vector consists of a track number and of the location of the head of the block within the track. The blocks must contain back references to the index terms. If overflow occurs, then one of the reference vectors has to be moved to a different track, and the location of the index term must then be known so that the pointer associated with the index term can be updated.

If the entire file is stored in core memory, then the system can respond to a retrieval request with the output of bibliographic data as soon as the relevant documents have been identified. If the document register is on magnetic tape, then output is best postponed until the relevant documents have been identified for a number of requests. The output for all these requests is then generated in a single tape pass.

Under time-sharing and multiprocessing the entire random access memory (core, disks, drums) is sometimes divided into pages. Pages belonging to the one program may be located anywhere. When a datum is referenced by the program, and the datum is not already in core memory, the page that contains the datum is brought into core memory by the operating system (and a different page—not necessarily belonging to one's program—is transferred from core memory to disk or drum). This is the *virtual memory* concept. Although the programmer is not concerned with the mechanism for the transfer of pages between different levels of storage (all memory *looks* to him as being on the one level), he should use programming techniques that minimize the number of page interchanges. For example, if a scatter table occupies more than one page, then the programmer should aim at resolving collisions within a page; i.e., if the first probe leads to a particular page, then subsequent probes should be confined to this same page. Linear search is then more effective than either randomized search or the use of lists.

Clearly, then, organization of files, and search and updating techniques must be made to suit the properties of the storage devices. Another factor to be taken into consideration is frequency of use. It would be very poor policy to devote a large amount of time to the optimization of search efficiency if the file is not likely to be much looked at. In some applications the ease with which items can be deleted is another important factor—this is one aspect of updating that we have not considered at all.

Notes

File organization has a very extensive literature, and a representative selection of sources for additional reading would take up a disproportionately large part of our bibliography. Instead of extending the bibliography by too many pages, we recommend the detailed surveys in the *Annual Review of Information Science and Technology* [Cl66b, Mi67b, Sh68] as guides for further study. However, a novice should first read [Ga66], which is a case study describing the design of a file and the implementation of interrogation facilities for users of the file, in order to gain some appreciation of what a file designer has to consider in real life.

An alternative to binary search is Fibonacci search, described in [Fe60]. Fibonacci search avoids division (cf. the statement following statement in 5 A.11.1); it may, therefore, be more efficient than A.11.1. Scatter storage techniques are surveyed in [Mo68a]; the division method for arriving at a hash address is described in [Ma68].

[Ba62b] contains algorithms in the form of flowcharts for set operations on sets stored as lists. A survey of sorting procedures can be found in [Go67d]. [Fl69], a book on sorting, has a major fault that the author expresses the algorithms in his own private assembler language. [Ba68b] is a very interesting paper dealing with hardware solution of the sorting problem. The ratios of times given near the end of the paragraph following A.11.5 have been calculated using data taken from [Bl66].

It is well known that natural language texts are highly redundant (an indication of this is our ability to understand abbreviations). [Ba60] is a classic paper giving a systematic abbreviation procedure. A survey of a variety of text compression procedures can be found in Chapter 3 of [Bo63c]. [Ma67c, Sc67c] are more recent publications dealing with the compression of data.

The two quotes in Section 11a are, respectively, due to Dodd [*Computing Surveys* **1,** p. 117 (1969)] and Lees [*Language* **35,** p. 300 (1959)].

Exercises

11.1 (a) Is the structure defined by the following PL/I statement a file?

```
DECLARE 1U(50), 2V, 2W, 3Y, 3Z(2,2), 2X;
```

If you have decided that the structure is a file, how many records does it contain? What are the properties (or fields)? How many measures does a record in this file contain?

11.2 (a) Why is statement 6 of A.11.1 `HI = MID` rather than `HI = MID - 1`?

11.3 (a) Fibonacci search is based on the Fibonacci numbers c_k, which are defined as follows:

$$c_0 = 0; \quad c_1 = 1; \quad c_k = c_{k-1} + c_{k-2} \qquad (k \geqq 2).$$

Assume that A is a vector of lexicographically ordered elements, and that a value d, if it is in A, belongs to an interval of c_k elements beginning with a_i. Then one starts the Fibonacci search process by comparing d with $a_{i+c_{k-1}}$. If the values match, then the search is finished. If $d < a_{i+c_{k-1}}$, then d can only be in the interval of c_{k-1} elements beginning with a_i, and one repeats the process with k set to $k - 1$. If $d > a_{i+c_{k-1}}$, then d can only be in the interval of $c_k - c_{k-1}$ $(=c_{k-2})$ elements beginning with $a_{i+c_{k-1}}$, and the process is repeated with k set to $k - 2$ and i set to $i + c_{k-1}$. Implement the Fibonacci search algorithm as a Fortran program.

11.4 (b) What is the purpose of storing the document numbers in the reference vector of Figure 11.1 in ascending order?

11.5 (b) Describe the process of updating an entry in the index of a document file with reference to the scheme illustrated by Figure 11.1. Assuming that the number of a document that is being added to the file is always greater than the number of a document already in the file, suggest improvements to the design of the index entry.

11.6 (b) Assume that the index of a document file is arranged as a sort tree with individual entries having the form shown in Figure 11.1. Give a full description of the procedure for amending the index when a document is deleted from the file.

11.7 (c) Write a Fortran implementation of a scatter table using the division method (Method c) for computing hash addresses, and linear search for resolving collisions. (You are free to select any format you wish for the data that this table is to cater for.)

11.8 (c) Assume that collisions in a scatter table are resolved by storing all records that have the same hash address as a list, and that one does not use an "overflow area." Then one must have a procedure for locating the predecessor of a given element in the list (see the second-last paragraph of Section 11c). What format would you choose for the lists that would give access to predecessors? Justify your scheme. Describe an algorithm for locating the predecessor of a given element in your scheme.

11.9 (c) Do Exercise 11.7 again, using now the scheme described in the final paragraph of Section 11c to resolve collisions.

11.10 (c) Design a scatter table to cater for variable length records.

11.11 (c) Design a scheme for storing sparse matrices in a scatter table. (Points to watch: What is to be the argument of the hash function? Is the hash function going to produce reasonably random distribution of hash addresses?)

11.12 (d) Modify A.11.2 so that it finds the set union of `A` and `B`.

11.13 (d) Consider a symmetric list of n elements in which the first m elements, which are in lexicographic order, and the remaining $n - m$ elements, which are also in lexicographic order, represent two sets. Develop an algorithm (as a flowchart) that replaces the list by a new list representing the intersection of the two sets.

11.14 (d) Consider two symmetric lists `A` and `B`. The elements in each list are distinct, and they are in lexicographic order. Design a procedure that replaces list `A` with the set difference `A - B` and list `B` with the set difference `B - A`.

11.15 (d) Assume that you are given a vector of integers in which the first k elements (k is unknown) are in ascending order. Devise a strategy for sorting the vector that makes use of this.

11.16 (d) Consider A.5.3 and the sorting algorithms of Section 11d. Which of the sorting methods keep records that have the same key in their existing relative order?

11.17 (e) The records of a file are so ordered that the values of a key, which are integers, are in ascending order. The file is considered as a sequence of pages, and each page occupies a track on a magnetic disk (the track numbers are in no particular order). A directory is kept in core storage. This is a two-column matrix in which the first element of row k contains the value of the key of the first record on the kth page. The second element contains the number of the track on which the kth page is located. Devise an algorithm for identifying the track that contains the record corresponding to a given search key.

CHAPTER 12

Programming Languages for Information Structures

12a. List Processing Languages

In Section 10d we pointed out the difference between symbolic and numerical computation. It may be more appropriate to speak of *formatted* and *unformatted* computation. In formatted computation data are of known size and structure, and do not undergo structural changes during processing. Numerical computation and the processing of formatted files, i.e., files that have fixed structure, belong to this type of computation. In unformatted computation the size and structure of the data are unpredictable. List and string processing languages are designed to facilitate the processing of such data. The difference between these two types of languages is a matter of emphasis. List processing languages are more concerned with structuring of data into trees or digraphs, and the processing of such structures; string processing languages stress facilities for manipulating unstructured lists, i.e., strings. The differences are not as important as the feature common to all list and string processing languages: The data on which the languages operate are stored dynamically as lists.

Although some languages will be described in detail, no language is going to be specified to the level of detail required by a programmer in that language. Our aim is to define list and string *processing* in general by describing the facilities made available by a number of specific languages. This approach is taken in the hope that two secondary objectives will be achieved: The reader will become sufficiently knowledgeable to select a language that fits a given problem reasonably well, and our discursive descriptions will enable him to approach the appropriate programming manuals with some feeling of familiarity.

The best known examples of list processing languages are IPL-V (Information Processing Language—version 5) and LISP (LISt Processor). IPL-V is a low-level language. This means that the programmer is handicapped in much the same way a machine language programmer is as compared to a Fortran or Algol programmer. For example, return of erased lists to the available space list is the programmer's responsibility. In particular, he has to make sure that a shared list is not returned too early. Further, transfer of control to a subroutine involves the programmer in a considerable amount of detailed housekeeping, but recursive definition and use of subroutines is then no more difficult than their standard definition and use. A feature of IPL-V that a programmer trained in the use of modern assembler languages finds particularly irksome is the anachronistic avoidance of mnemonics. We should recognize the historical importance of IPL (the earliest member of the IPL family dates as far back as the earliest Fortran), but the language can now be considered effectively obsolescent.

LISP is perhaps the most unusual programming language in existence. It is a logician's language. Therefore, while it is a delight to some, others find its unconventionality harder to take. We shall first investigate the general philosophy of computation on which LISP is based, in the hope that these preliminaries will lead to a better understanding of the language and of its significance. By necessity our exposition will be rather brief. At this point it would be advisable to read again the part of Section 2b that deals with the difference between a function and a form, and the part of Section 2c dealing with statements.

Consider statements $s_1, s_2, \ldots, s_n$, and define the form

$$(s_1 \to e_1, s_2 \to e_2, \ldots, s_n \to e_n).$$

This is a *conditional form.* The *value* of the form is the value of the e corresponding to the first s that has the value T (is true). If all the statements have value F (are false), then the value of the form is undefined. Furthermore, the form is undefined if an undefined s is encountered before a true s, or if the s corresponding to the first true s is undefined. Take

$$(4 < 3 \to 7, 2 > 3 \to 8, 2 < 3 \to 9, 4 < 5 \to 7).$$

Here the first true s is $2 < 3$ and, therefore, the value of the form is 9. Propositional forms containing connectives can be expressed as conditional forms. Thus

$$p \wedge q = (p \to q, \text{T} \to \text{F});$$

i.e., the form takes the truth value of q if p is true, but if p is false, then the next statement is examined. If this statement is reached at all, then we want the conditional form to take the value F. To ensure this we want the statement to be always true, and the simplest way of achieving this is to make it the constant T. The two other primary connectives can be defined as follows:

$$p \vee q = (p \rightarrow \mathtt{T},\ \mathtt{T} \rightarrow q),$$
$$\neg p = (p \rightarrow \mathtt{F},\ \mathtt{T} \rightarrow \mathtt{T}).$$

Then, for example, the propositional form $\neg (p \wedge q)$ can be written as the composite form

$$((p \rightarrow q,\ \mathtt{T} \rightarrow \mathtt{F}) \rightarrow \mathtt{F},\ \mathtt{T} \rightarrow \mathtt{T}).$$

Next we define the factorial function:

$$n! = (n = 0 \rightarrow 1,\ \mathtt{T} \rightarrow n\cdot(n-1)!).$$

This is a recursive definition; e.g., we have

$$\begin{aligned} 2! &= (2 = 0 \rightarrow 1,\ \mathtt{T} \rightarrow 2\cdot(2-1)!) \\ &= 2\cdot 1! \\ &= 2\cdot(1 = 0 \rightarrow 1,\ \mathtt{T} \rightarrow 1\cdot(1-1)!) \\ &= 2\cdot 1\cdot 0! \\ &= 2\cdot 1\cdot(0 = 0 \rightarrow 1,\ \mathtt{T} \rightarrow 0\cdot(0-1)!) \\ &= 2\cdot 1\cdot 1 \\ &= 2. \end{aligned}$$

This example shows that even though $(0 - 1)!$ is undefined the form defining $0!$ has a value.

Let us see how the formalism works in practice. One starts with a few undefined functions and predicates (i.e., functions having {F, T} for their range), the *basic* functions appropriate to the problem area. Then, in order to evaluate a particular function, a definition of the function has to be set up, and the function applied to particular arguments. The function is defined by means of functional composition, starting with some or all of the basic functions. Unless the function that one wants computed is very simple, the definition is a conditional form, which may itself contain conditional forms. To compute a function one puts down the form *def*(arguments), where *def* stands for the function. In the pure formalism this form is an independent single expression; i.e., in building up the function from other functions all these other functions except the basic functions have to be defined within *def*. This creates a few difficulties.

Suppose a function is defined by an expression e, which is some form in variables $x_1, x_2, \ldots, x_n$. In particular, consider the function defined by the form $x^2 + y$. Now, if one wants to apply the function to the arguments 2 and

3, say, one cannot simply write $(x^2 + y)(2, 3)$. This form is ambiguous; there is nothing to indicate whether one wants $2^2 + 3$ or $3^2 + 2$ evaluated. The ambiguity is resolved by use of Church's lambda notation, which converts the *form* $x^2 + y$ to a *function*. We normally deal with this type of ambiguity by having a separate definition $f(x, y) = x^2 + y$, and then, on encountering $f(2,3)$, we refer to the definition, which tells us that 2 is to take the place of x and 3 the place of y. The λ-notation is similar, but avoids the separation of the definition and the arguments. If we want the interpretation $2^2 + 3$, the form $x^2 + y$ is converted to the function $\lambda((x, y), x^2 + y)$, and then $\lambda((x, y), x^2 + y)(2, 3) = 7$; if the other evaluation is intended, then the form is converted to the λ-expression $\lambda((y, x), x^2 + y)$, or to $\lambda((u, v), v^2 + u)$; since the variables are dummies, they can be replaced by others in a consistent manner.

We say that λ *binds* the variables. Binding of variables alone is, however, insufficient in recursive definitions. Consider

$$factorial(n) = (n = 0 \rightarrow 1,\ \mathrm{T} \rightarrow n \cdot factorial(n - 1)).$$

The expression

$$\lambda((n), (n = 0 \rightarrow 1,\ \mathrm{T} \rightarrow n \cdot factorial(n - 1)))$$

is inadequate because we do not know whether the *factorial* here refers to the function that is being defined or to some other function in the definition of which the present definition might be contained. The name of the function has to be bound as well, and this is done by the *label* notation as follows:

$$label(factorial, \lambda((n), (n = 0 \rightarrow 1,\ \mathrm{T} \rightarrow n \cdot factorial(n - 1)))).$$

We could, of course, equally well write

$$label(h, \lambda((x), (x = 0 \rightarrow 1,\ \mathrm{T} \rightarrow x \cdot h(x - 1)))).$$

This concludes our discussion of the pure formalism. It can be shown that all number theoretical functions and predicates can be defined in the formalism by starting with just the predicate of equality and the successor function *succ*, where $succ(n) = n + 1$. This means that the formalism is as powerful a basis for the theory of computability as that of Turing machines (see Section 4c). However, our purpose is not to discuss the theory of computability. Rather, as far as we are concerned, the main appeal of the formalism is that it can be applied to any problem area simply by taking a new set of basic functions. In the case of operations on structured symbolic data the result is LISP, but pure LISP, i.e., a system in which every nonbasic function has to be defined anew each time it is to be used, is impracticable. Therefore, LISP as a programming language deviates from the pure formalism. For one thing, it has a `DEFINE` facility for separating definitions from applications of functions, i.e., there is a subroutine structure. For another, the programming language provides arithmetic facilities that do not derive from the basic functions. Pure LISP is a declarative language, and a program written in pure LISP is simply a series of functional forms. Algol and Fortran, on the other hand, are imperative languages; their programs consist of commands. Although use of conditional forms obviates a real need for sequential execution and branching, sometimes there is a greater convenience in being able to define a computation by a sequence of commands. This facility is also provided by the programming language. In what follows we shall consider the IBM 7090 implementation of LISP.

LISP operates on data in the form of *S-expressions* (Symbolic expressions), defined recursively as follows:

(a) Every *atomic symbol* or *atom*, i.e., an unstructured string, is an S-expression. In particular, the null symbol, `NIL` or `( )`, is an S-expression.

(b) If α_1 and α_2 are S-expressions, then $(\alpha_1.\alpha_2)$ is an S-expression.

S-expressions can be pictured as binary trees with labeled terminal nodes. For example, the S-expression corresponding to the tree of Figure 5.21 (disregarding labels on arcs and nonterminal nodes) is `(A1.((A2.A3).A1))`. An abbreviation for linear lists is permitted. If one has, say, a list of four atoms `A`, `B`, `C`, `D`, the list can be written as `(A B C D)`, which is an abbreviation of the S-expression `(A.(B.(C.(D.NIL))))`. Expressions `A` and `(A)` are different: `A` is an atom, `(A)` is a linear list whose expanded form is `(A.NIL)`. The abbreviated list notation and the dot notation can be used within the same expression, e.g., `(A B(C.D))` is equivalent to `(A.(B.((C.D).NIL)))`.

An atomic symbol is stored as a *property list*, which contains the external representation of the symbol and other information, all of which is accessible to the programmer. An S-expression $(\alpha_1.\alpha_2)$ is represented by a list element; in the IBM 7090 this is one machine word. The list element has two pointer fields: The address field contains a pointer to the representation of α_1, the decrement field to the representation of α_2. If, for example, α_1 is an atom, then the pointer refers to the property list of this atom; if α_1 is not an atom, then it is an S-expression of the form $(\beta_1.\beta_2)$, which is represented by another list element, and the pointer indicates this element. The representation of `(X.(Y.(Z.NIL)))` consists of three elements, representing (X.---), (Y.---), and `(Z.NIL)`. Assume that `X`, `Y`, and `Z` are atoms. Then the pointer in the address field of each element indicates a property list, and the decrement field points to the next element. In the case of `(Z.NIL)` the decrement field contains an end of list marker standing for `NIL`. `(X Y Z)` has, of course, exactly the same representation as `(X.(Y.(Z.NIL)))`.

There are five basic functions: *car*, *cdr*, *cons*, *eq*, and *atom*. The value of *car*$((\alpha_1.\alpha_2))$ is α_1, that of *cdr*$((\alpha_1.\alpha_2))$ is α_2. These names derive from the machine representation; *car* stands for "Contents of Address field of Register," and *cdr* for "Contents of Decrement field of Register." The value of *cons*(α_1, α_2) is $(\alpha_1.\alpha_2)$; i.e., given S-expressions α_1 and α_2, *cons* constructs the S-expression $(\alpha_1.\alpha_2)$. The predicate $eq(a_1, a_2)$ has the value `T` if the *atoms* a_1 and a_2 are equal, and `F` if they differ. The value of the predicate *atom*(α) is `T` if α is an atom, and `F` if it is not. In an IBM 7090 program only uppercase letters can be used, and $f(x_1, x_2, \ldots, x_n)$ is actually written `(F X1 X2 --- XN)`. The conditional form is written

$$(\texttt{COND}\ (s_1\ e_1)(s_2\ e_2) \text{ - - - } (s_n\ e_n)).$$

Consider a function that extracts the final atom embedded in an S-expression, e.g., `A1` in the case of `(A1.((A2.A3).A1))`. If we would not wish to give a special name to the function and would have the definition followed at once by a specific argument, then we would write

```
(LABEL LAST(LAMBDA(X)(COND((ATOM X)X)(T(LAST(CDR X))))))
```

In LISP anything in parentheses is a list, and the mass of parentheses that annoy one looking at a LISP program is there to help the LISP system set up the program itself in the form of list structures. Since the functions are then S-expressions, they can be modified by the program. This flexibility can be a great asset to an experienced programmer. On the other hand, the LISP system has to be an interpreter rather than a compiler; the machine code to compute a function cannot be put together prior to execution because then the final form of the function is unknown. With some restrictions, however, LISP can be compiled, and LISP compilers exist. Compiled programs are much faster than interpreted programs.

Apart from the basic functions, which are written in machine code, all other LISP functions can be defined in LISP. In principle all of these functions could be left to the programmer to define, but in practice this would be too great an imposition on the programmer, and the LISP system provides a great number of predefined functions. Even the LISP interpreter can be written in LISP. The interpreter is a universal function `EVAL`, universal in the sense that, given a description of any function and a set of arguments, it computes the value of the function for these arguments.

There is a difference between arguments used in the definition of functions and the specific arguments for which the function is to be evaluated. In familiar terms, the difference is exemplified by $f(x, y)$ and $f(2, 3)$—the 2 and 3 do not *represent* anything further; they stand for *themselves*. The LISP function `QUOTE` indicates that an expression stands for itself rather than for something to be interpreted; e.g., to get the first atom of `(A B C)`, we would write

```
(CAR(QUOTE(A B C)))
```

Let us now set up a definition of the function that extracts the final atom of an S-expression by means of the DEFINE function:

```
(DEFINE(QUOTE((LAST(LAMBDA(X)(COND((ATOM X)X)
                                   (T(LAST(CDR X)))))))))
```

DEFINE is supplied with the name of the function that is being defined—here it is the atom LAST—and the λ-expression that specifies the function. The evaluation of DEFINE consists of storing the λ-expression as a list structure and setting up a property list for the name. The property list contains an indicator saying that LAST is a function and a pointer to the place where the list structure specifying the function is to be found. Since we are storing the actual name LAST, and the actual definition of LAST, i.e., since there is no interpreting to be done here, we have to use QUOTE. The DEFINE function, because it labels a function anyway, makes the explicit LABEL notation superfluous.

Normally a LISP program consists of the definitions of functions by means of the DEFINE function, followed by applications of the functions to specific arguments. This is a sequence of S-expressions. EVAL looks at each S-expression in turn and evaluates it. Assume that the interpreter has dealt with all the definitions, and that it now receives as input the expression

```
(LAST(QUOTE(A1.((A2.A3).A1))))
```

Having access to the definition of LAST, EVAL uses the definition to evaluate the function for the given argument. The result is the atom A1.

However, the extent to which QUOTE has to be used is unbearable; QUOTE has to be applied to virtually everything in the program. In order to eliminate most of the explicit references to QUOTE, EVAL is combined with QUOTE; the actual IBM 7090 LISP interpreter is the function EVALQUOTE. Then the form of the definitions and references to actual arguments changes. Their changed form is shown in the following complete LISP program.

```
DEFINE(((NULL(LAMBDA(X)
             (COND((ATOM X)(EQ X NIL))(T F))))
        (MEMBER(LAMBDA(A X)
             (COND((NULL X)F)((EQ A(CAR X))T)
             (T(MEMBER A(CDR X))))))
        (UNION(LAMBDA(X Y)
             (COND((NULL X)Y)((MEMBER(CAR X)Y)
             (UNION(CDR X)Y))
             (T(CONS(CAR X)(UNION(CDR X)Y))))))
```

```
        (SETPROD(LAMBDA(X Y)
              (COND((NULL X)NIL)((MEMBER(CAR X)Y)
              (CONS(CAR X)(SETPROD(CDR X)Y)))
              (T(SETPROD(CDR X)Y)))))
      ))
SETPROD((A1 A2 A3)(A1 A3 A5))
UNION((X Y Z)(U V W X))
STOP
```

The one DEFINE is used to define four functions. The predicate NULL is defined first. Its value is T if X is empty, i.e., if it is NIL. For example, CDR((A)) is empty. MEMBER is another predicate. The value of MEMBER(A X) is T if A is a member of the linear list X. True values result with MEMBER(B(A B C)) and with MEMBER((X.Y)(C.(X.Y))), but the value of MEMBER(X((X))) is F. NULL is used in the definition of MEMBER, and NULL and MEMBER are both used in definitions of UNION and SETPROD, which are functions for finding, respectively, the set union and set intersection of two sets represented by lists. The result of the evaluation of SETPROD is (A1 A3), the result of evaluation of UNION is (Y Z U V W X). Actually functions NULL and MEMBER did not have to be defined; they are included in the set of predefined functions provided by the system.

Predefined function CSET can be used to provide a list structure with an identifying name. Thus, after evaluation of CSET(XX(A B C)), the list (A B C) can be referenced by the name XX. Functions RPLACA and RPLACD can be used to make radical changes in list structures: (RPLACA X Y) replaces the address field of the element representing X by Y; (RPLACD X Y) performs a similar replacement in the decrement field. For example, if one wants to convert (A B C) into (A(B C)), which is an abbreviation for (A.((B.(C.NIL)).NIL)), one defines a function ALTER and uses it as follows:

```
DEFINE(((ALTER(LAMBDA(X)
              (RPLACD(RPLACA(CDR X)(CONS(CAR(CDR X))
                                  (CDR(CDR X))))NIL))
      )))
CSET(XX(A B C))
ALTER(XX)
```

This is an interesting example of a function used for its *side effects*: Evaluated for XX, ALTER has the value ((B C)), possibly of no interest to the programmer, but RPLACA and RPLACD have changed the structure of XX to

`(A(B C))`. It is also an example of a complicated definition that is not a conditional form. The main use of `RPLACD` is to create circular lists. The LISP system provides many other functions for operating on list structures. They are used for the building up of list structures, extraction or replacement of specified elements, pairwise association of elements of two linear lists having the same number of elements, etc. For example, `(SUBST X Y Z)` substitutes `X` for all occurrences of the S-expression `Y` in S-expression `Z`; `SUBST(X1 A1(A1.((A2.A3).A1)))` produces `(X1.((A2.A3).X1))`, and the result of `SUBST(X1(A2.A3)(A1.((A2.A3).A1)))` is `(A1.(X1.A1))`.

The LISP arithmetic functions do not derive from the basic functions. Nevertheless, they conform to the general format of LISP functions. They are, therefore, rather unconventional; e.g., for $(X + Y + Z)W$ one writes `(TIMES(PLUS X Y Z)W)`.

Storage reclamation is by means of garbage collection; the garbage collector, which uses the technique of marking accessible elements, and returning unmarked elements to the list of available space in a second pass, is called into action when in the process of evaluating `CONS` all available space is found to be used up.

The elegance and flexibility of LISP have made it a tool for many serious investigations in the field of symbol manipulation. However, the basic format of a list element is fixed, and this can lead to inefficiency; too much space may be used up by pointers, and the mechanism for accessing data can be very slow. Greater efficiency may be achieved if the programmer is given facilities for specifying formats of list elements that correspond more closely to the form of the data he has to deal with. The result is a lower level language, and the improvements in storage efficiency and execution times are achieved at the cost of increased programming time. Arguments in favor of the use of high level languages in all circumstances concentrate on this aspect, neglecting to point out that a high level language is not always as easy to use as one is made to believe. Unless the processor for the language contains elaborate optimizing procedures, *gross* inefficiencies can arise if programmers neglect to familiarize themselves with the implementation in *great* detail, and this familiarization process also takes time.

The language L^6 has been selected for our example of a low level list processing language because of its wide implementation. L^6 stands for the six occurrences of L in "Bell Telephone Laboratories' Low-Level Linked List Language." Because of the very precise control that the programmer has over allocation of storage L^6 is a highly efficient language when used by an experienced programmer. The main storage unit is a *block* of 2^n words (in the IBM 7094 implementation n ranges from 0 to 7). The programmer imposes structure on blocks by defining *fields* within them. A field is defined by specifying its name, its word in a block (0–127 in the 7094 implementation), and the leftmost and rightmost bits of the field (0–35). The name of a field is a single digit or letter. This means that as many as 36 fields may be declared. Fields may be overlapped, or nested one within another; they may be redefined during execution. Entry into the system is through a set of 26 *base fields* or *bugs*, identified by single letters. The bugs contain pointers to blocks. One can imagine that a bug sits on the block to which it points, and hops to a different block when the pointer is changed. The choice of the term is quite apt; we can readily picture the 26 bugs hopping all over the place during execution of a program.

Besides acting as pointers, bugs can perform other functions. Consider two L^6 instructions, each made up of three operations:

```
THEN(0,DA,3,17)(0,DB,21,35)(1,DC,0,35)
THEN(2,DD,0,35)(3,DE, B,35)(0,DF,0,35)
```

The first two operations of the first instruction define the decrement and address fields of the first word (word 0) of a block as fields A and B, respectively. The next operation defines the second word (word 1) as C. The first operation of the second instruction defines word 2 as D. The B in (3,DE, B,35) refers to a bug, and the leftmost limit of field E is specified by the current contents of bug B. The field definitions apply to *every* block in the system. Now, in some blocks we might wish to use word 0 for storing pointers; in others we might wish to store a symbolic datum in this word. Therefore, in the final operation, word 0 is defined as field F as well. When the word contains a symbolic datum, we refer to it by the name F; when it contains pointers, we refer to the pointers by A and B. In a block of 2 words fields D and E are, of course, undefined; in a block of one word A, B, and F alone are defined.

If only 36 fields may be defined, and a field occupies at most one word, what is the use of a block of 128 words? Consider first how a datum is accessed. Assume that bug B sits on some block (block 1), that field A of the block points to another block (block 2), that field B of this block points to yet another block (block 3), and that field A of this block points to the block (block 4) in which we wish to access field C. The "address" of this field is BABAC. Normally the BABA points to word 0 of block 4. However, if we add *n* to the contents of field A in block 3, then all fields in the block accessed by pointer BABA are looked for *n* words away from their normal positions. By successive incrementation of the pointer field A of block 3, the programmer can scan through block 4, applying the same field name to different words in the block.

L^6 instructions are conditional or unconditional. There are four conditional instructions

$$\left.\begin{array}{l}\texttt{IFNONE}\\ \texttt{IFANY}\\ \texttt{IFALL}\\ \texttt{IFNALL}\end{array}\right\}(t1)(t2)\cdots(tn)\quad \texttt{THEN}\quad (op1)(op2)\cdots(opm)\ \texttt{LABEL}$$

The test expressions (t1), (t2), . . . , (t*n*) have logical values. If none is true, then `IFNONE` is satisfied. If at least one is true, then `IFANY` is satisfied. If all are true, then `IFALL` is satisfied. If not all are true, then `IFNALL` is satisfied. Now, if the complete conditional is satisfied, then operations (op1), (op2), . . . ,(op*m*) are performed, and control passes to the instruction labeled `LABEL` (if the list of operations is empty, then there is just the transfer of control). Otherwise control passes to the next instruction in sequence. Examples of test expressions: `(BABA,P,XA)` is true if `BABA` and `XA` point to the same block; `(XQ55,G,89)` is true if the decimal integer in the field specified by `XQ55` exceeds 89.

The unconditional instruction has the form

$$\texttt{THEN}\ (op1)(op2)\cdots(opm)$$

The field definitions given above are examples of unconditional instructions. The list of operations may be followed by a label, in which case control passes to the instruction indicated by the label after the operations have been performed. A simple transfer instruction is written

```
THEN PLACE
```

where `PLACE` is the label of the statement to which control is to pass.

Operations can be grouped into seven classes. The first class consists of operations for setting up the list store, defining fields, getting a block, and releasing a block to free storage. For example, operation `(BABA,GT,32)` gets a block of 32 words from free storage and inserts a pointer to this block in the field indicated by `BABA` (in terms of our example above this would be field `A` of block 3 of that example). Operation `(BABA,GT,32,X)` performs all the above and, in addition, saves the old pointer in the field indicated by the last argument (in terms of our example bug `X` is placed on block 4, and the field that we were wishing to access by writing `BABAC` can now be accessed by writing `XC`). The second class of operations permits one to copy fields and blocks, and interchange field contents.

Arithmetic operations belong to the third class. Assume that we wish to access every second word in the block indicated by `BABA`, and that we start with `BABA` pointing to word 0 of the block. Then incrementation of `BABA` is performed by (BABA, A, 2), an addition operation. If the original contents of BABA had to be saved, we could have used operation (BABB, E, BABA) to copy the pointer into the field specified by BABB. Afterwards the pointer could be restored by means of (BABA, E, BABB). L^6 caters for addition, subtraction, multiplication, and division of integers.

"Logical" operations, which belong to the fourth class, are applied bitwise. For example, if fields XY2 and XY3 contain, respectively, 0110101 and 1100110, then the or-operation (XY2 O XY3) produces 1110111 in XY2, and-operation (XY2 N XY3) produces 0100100, the exclusive or-operation (XY2 X XY3) produces 1010011, and complementation (XY2 C XY2) changes the contents of the field specified by XY2 to 1001010. Operations of the fifth class are also bit operations. This class comprises operations for left and right shifts, counting of 0-bits and 1-bits in a field, and the determination of the position of the rightmost 0-bit or 1-bit in a field.

The final two classes contain operations for input and output, and for writing of subroutines. Subroutines may be used recursively. For this reason a system stack is provided. Two other stacks are provided for the programmer —one for saving definitions of fields, the other for saving contents of fields. For example, the pointer BABA, which we were incrementing in an example above, could have been saved prior to its incrementation by means of the field contents pushdown operation (S, FC, BABA), and restored by means of (R, FC, BABA).

In the implementation of L^6 the structure of the word in the computer on which the language is implemented is taken into account. For example, the fact that address and decrement fields are directly accessible on the IBM 7094 is used to give very rapid access to L^6 fields that coincide with these fields. Therefore, the 7094 L^6 programmer should locate his pointers in these fields, defining a pointer field A by (0, DA, 3, 17) or (0, DA, 21, 35), and not by (0, DA, 0, 14) or (0, DA, 10, 24), say.

12b. String Processing Languages

The data on which a string processing program operates are commonly stored as lists consisting of linked elements, and the lists may be built up into structures, but the user of a string processing language need not concern himself with the internal representation of his data. In contrast to list processing, which consists mainly of building and traversal of structures, the basic operation of string processing is the very simple substitution of D.4.13. Deletion and replacement of substrings, rearrangement of substrings into new patterns, and the concatenation of strings are all instances of substitution.

A string processing program is similar to a Markov algorithm, but there are many refinements. The proper context for the study of Markov algorithms is the theory of computability; there one wants a simplest possible characterization of recursively solvable problems, and Markov algorithms provide one such characterization. In the theory of computability, where it is sufficient to know that an algorithm for solving a problem can be constructed, the simplicity is an advantage. In practical computing, where one actually has to produce the algorithm, this same simplicity becomes a handicap. Even the pure Markov formalism could not function in a computer environment without input and output facilities; other facilities, which are not strictly essential in the sense that effective string processing routines could be written without them, are superimposed on the Markov formalism to make a programmer's life easier. It is important to realize that string processing languages derive from Markov algorithms, but lists of differences between the Markov formalism and actual string processing languages would have little relevance. Therefore, we shall make no further mention of Markov algorithms here.

Comit is the oldest string processing language; it is as old as Fortran. Despite its several faults, the numerous original concepts it embodies make it an outstanding intellectual achievement. This is why Comit will be discussed here in rather great detail in preference to the more recent and more widely used Snobol. Snobol is also a fine achievement, but its designers had Comit to give them a starting point. The designers of Comit started from nothing.

The operational area of Comit is the *workspace*, which contains a single string. The string is regarded as a sequence of *constituents*, which may be characters or groups of characters. In the external representation of a string plus signs are used to separate consituents. The minus sign stands for a Comit blank; this means that true blanks can be freely used to improve the readability of a program. Some examples:

```
HIS - OR - HERS                              (12.1)
HIS + -OR + -HERS                            (12.2)
H + I + S + - + O + R + - + H + E + R + S    (12.3)
```

Here we have 1, 3, and 11 constituents, respectively.

Material in the workspace can be manipulated by means of the Comit *rule*, which, for the time being, we take to be simply a device for effecting a canonical substitution (see D.4.13). It contains a *left half* and a *right half*. The left half of the rule specifies an existing pattern that is to be looked for in the workspace; the right half specifies the desired change. In terms of the notation of D.4.13, the left and right halves of a rule are β and β', respectively (the

σ is not specified because there is only one string on which one may operate, the string in the workspace).

The specifiers that can be used in a left half are of four types: (i) an actual symbol, e.g., -HERS; (ii) $n, to denote a specific number of contiguous constituents, e.g., $3 for three constituents; (iii) $, to denote an indefinite number of constituents; (iv) an integer *m*, which denotes the *m*th specifier in the left half. For example, let us apply the left half

S + $2 + $1 + $ + 3 (12.4)

to string (12.3). First a search is made in the string for the symbol S, which is found. Then the two constituents that follow the S are counted off (they are - and O), then another constituent is counted off (it is R), then an indefinite number of constituents is passed over in a search for a symbol that corresponds to the third specifier, i.e., another R is looked for, and found. Let us call the substring corresponding to the *i*th specifier the *i*th *component* of the pattern. Here we have five components:

S

- + O

R

- + H + E

R

Note that left half (12.4) does not match strings (12.1) and (12.2), but that it would match

S + P + A + R + R + O + W (12.5)

and the five components corresponding to the five specifiers would now be

S

P + A

R

nothing

R

The complete substring specified by the left half of a rule is replaced by a substring specified by the right half of this rule. Specifiers in the right half are either actual symbols or integers denoting the components of the pattern specified by the left half. Symbolic specifiers are used to effect insertion or replacement of symbols; numerical specifiers to effect rearrangement of the components. If the pattern established by the left half has n components, but not all of the n component numbers appear in the right half, then the components whose numbers do not appear in the right half are deleted. For example, assume that (12.5) is to be changed to

```
R + O + W
```

and that left half (12.4) has been applied to the string. Then any one of the following right halves will do the job: R, or 5, or 3. If we wished to change string (12.5) to

```
A + R + R + O + W
```

then the following would be acceptable right halves:

```
A + 5 + 3

A + 3 + 3
```

The string cannot be changed into, say,

```
S + P + A + R + R + I + N + G                (12.6)
```

because this would require replacement of consituents O and W, which are not part of the substring specified by left half (12.4). Left and right halves of a rule are separated by an equal sign, e.g.,

```
$ + R + 2 + $ = 1 + 2 + 3 + O + W
```

which would change (12.6) into (12.5).

In addition to the two halves, a rule may carry an identifying name (label). It may also contain one or more routing instruction and a branching indicator. The routing instructions specify what is to be done with the workspace or some constituents of it after the operation specified by the right half has been completed; the branching indicator specifies the next rule that is to be executed. The order of the parts of a complete rule is, rule name—left half—right half—routing instructions—branching indicator.

Comit provides for a maximum of 128 strings: the string in the workspace and strings that are stored in 127 numbered auxiliary storage areas called *shelves*. For an example of possible transfers, one may exchange the contents of the workspace with those of a designated shelf. The shelves can be used as fairly sophisticated pushdown stores. Although data can be transferred componentwise into the workspace from only the left end of a shelf, data from the workspace may be pushed down at either end of a shelf. This means that a shelf can function as a stack or as a queue. These data transfers are effected by routing instructions. Input and output is another function of routing instructions. In particular, since an input–output device can be a magnetic tape unit, Comit has a backing store, but all transfers to and from the backing store are the programmer's responsibility. Another set of routing instructions expands or compresses designated constituents, e.g., converts words to characters or characters to words.

Although some parts of a rule are optional (e.g., one may have a rule with a left half and no right half), the rule name and the branching indicator are obligatory. The rule name is a sequence of characters that serves to identify the rule if a reference is made to the rule, or, if there is no explicit reference to the rule, it may be an asterisk (*). If no substring of the string in the workspace matches the left half of a rule, then the rest of the rule is ignored, and control passes to the next rule in sequence. If there is a match, then the rule to which control next passes is determined by the branching indicator—if it is an asterisk, then control still passes to the next rule in sequence; if it is an explicit rule name, then the rule that is being pointed to is applied next. The "next" rule may be the same rule that is being applied. Consider the rules

```
CUT    $ + -THE  =  1  // *Q17 1            CUT
*      $         =     // *Q17 1,    *X17   *
```

The first rule has the name `CUT`. The left half of this rule searches the string in the workspace for the word `THE`. The word is deleted, and all the text that precedes it is shifted out of the workspace and pushed down at the right-hand end of shelf 17 by means of the routing instruction `*Q17 1`, where `*Q` is the operation code and the `1` indicates that the first component is to be pushed down. Since the branching indicator corresponds to the name of the rule, control is not passed out of the rule until every `THE` has been deleted from the string, i.e., until the left half fails. Control then passes to the next rule in sequence. Shelf 17 now contains all of the original string up to the final `THE`, but stripped of every `THE`, and the workspace contains the text that follows the final `THE`. The first routing instruction of the second rule now appends the contents of the workspace to what is already on shelf 17, and the second routing instruction exchanges the now empty contents of the workspace with the contents of shelf 17, i.e., the modified string is shifted back into the workspace.

One of the most interesting features of Comit is that a rule may consist of a number of *subrules*. Assume that the workspace contains a coded message (in the cryptographic sense) and that a dummy is to be inserted after every fifth letter. Letters A, E, I, O, U are to be used for dummies, with E to have twice the frequency of the other dummies. The following program segment will do the job:

```
*                           // INSERT A                 *
INSERT A    $5  =   1 + A   // *Q15 1 2,  INSERT E1     INSERT
       E1           1 + E   // *Q15 1 2,  INSERT I      INSERT
       I            1 + I   // *Q15 1 2,  INSERT O      INSERT
       O            1 + O   // *Q15 1 2,  INSERT U      INSERT
       U            1 + U   // *Q15 1 2,  INSERT E2     INSERT
       E2           1 + E   // *Q15 1 2,  INSERT A      INSERT
*           $   =           // *Q15 1,  *X15            *
```

Lines 2–7 constitute rule INSERT, which consists of subrules A, E1, I, O, U, E2, and must have only one left half serving all subrules. The selection of the subrule that is to be applied is the function of the Comit *dispatcher*. The dispatcher is told by an appropriate routing instruction in the first rule (note that this rule contains neither a left nor a right half) that subrule A is to be applied when control passes to rule INSERT. Within INSERT each subrule provides for a resetting of the dispatcher, so that the subrules are applied cyclically in the order A, E1, I, O, U, E2, A, E1, ... If the dispatcher is not set, i.e., if we omit the first rule and the dispatcher instructions in all subrules of rule INSERT, then the choice of the subrule is made at random under the control of a built-in pseudorandom number generator. In the present example one might well wish to avoid cyclic repetition of the dummies. The random selection process should then be used.

Now let us consider how one would encode the message in the first place. Assume that we use a simple letter-for-letter substitution code. The code can be set up as a *list rule*, which functions as a dictionary. A list rule consists of a number of subrules, which have the general form

l.half = r.half // routing section branching indicator

The name of the list is carried by the first subrule; it must have a hyphen for its first character to distinguish the rule as a list rule. Assume that the message to be encoded has been *stacked* on shelf 15 (its final character is the leftmost symbol on the shelf). The following program segment will take characters off shelf 15 one at a time, translate them, and *stack* the translated characters on shelf 16 (i.e., put the characters of the encoded message in their proper sequence).

```
ENCODE     $ =    // *N15 1, *L1    CODE
-CODE      A = F  // *S16 1         ENCODE
           B = X  // *S16 1         ENCODE
           C = D  // *S16 1         ENCODE
           :
           Z = Q  // *S16 1         ENCODE
```

The left half of the first rule identifies all of the workspace as a single constituent. The `*N15 1` instruction fetches the leftmost constituent from shelf 15 and substitutes it for the first component of the pattern established by the left half (here it is the entire workspace). The `*L1` instruction prepares the character for the dictionary lookup (actually, no preparation is needed here, but a `*L` instruction must still be given—the precise specification of this instruction need not concern us here), and control then passes into `CODE`. The subrule whose left half corresponds to the given character is located, and the right half of this subrule is pushed down on shelf (stack) 16 by means of the `*S16 1` instruction. Control then passes back to `ENCODE`. Finally shelf 15 is empty: `CODE` is entered with a null string, which, of course, fails to match any left half in the dictionary. When this happens, control automatically passes to the rule that follows the list rule. The Comit system automatically sorts all subrules belonging to a list rule into alphabetical order of left halves; binary search is used in the lookup.

Although we have spoken of the workspace and the shelves as if they were fixed regions in store, there are no actual transfers of data. Properly one should speak of a string called workspace and strings called shelf 1, shelf 2, and so on. Consider, for example, exchange of workspace with a shelf. The data that constitute the two strings are *not* shifted about; the exchange is effected simply by interchange of pointers.

With some ingenuity one can create reentrant data structures, but the basic structure of Comit is a three-level tree. Tree structures are created by the Comit *subscript* facility. There are two types of subscripts, numerical and logical. A single numerical subscript and any number of logical subscripts may be attached to a constituent. A numerical subscript is an integer in the range 0 to 32767. A logical subscript consists of a subscript name optionally followed by as many as 36 subscript values. The same subscript name may be assigned to more than one constituent, but the total number of elements in the union of all the sets of values associated with the one subscript name may not exceed 36. A subscripted constituent is denoted by the constituent symbol, then a slash, and then the subscripts. For example, if constituent OBJECT carries the numerical subscript 54, and logical subscripts COLOR with value RED, and SHAPE with values ROUND and HOLLOW, then one writes

```
OBJECT/.54, COLOR RED, SHAPE ROUND HOLLOW
```

for the constituent. One can picture a subscripted constituent as being the root of a tree (level 0). The names of logical subscripts then occupy nodes on level 1 of the tree, and their values are on level 2. Figure 12.1 shows the tree structure corresponding to the constituent of our example. Note that the order of writing down the subscripts and their values is immaterial, e.g.,

```
OBJECT/.54, SHAPE HOLLOW ROUND, COLOR RED
```

is an equivalent description of the constituent. This means that the tree is unordered. A constituent is matched by a specifier in the left half of a rule that gives a full description of the constituent or by one in which some or all of the subscripts and some or all of the values of a logical subscript are missing.

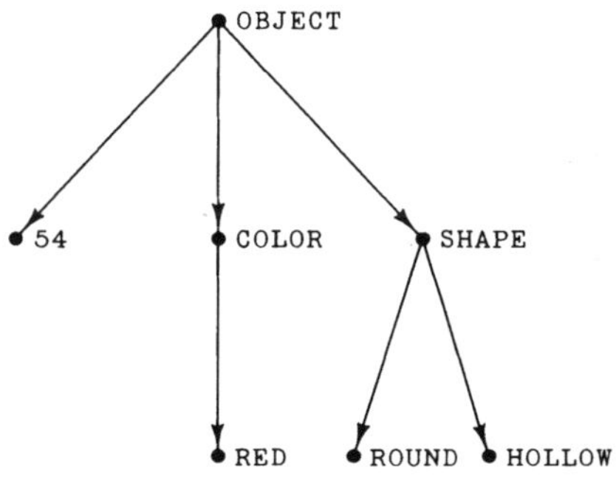

Figure 12.1

The constituent to which a numerical subscript is attached functions more or less as a variable name, with the subscript functioning as an integer value of the variable. Numerical subscripts are used in loop control and for indirect addressing, and arithmetic operations may be performed on them in a rudimentary fashion—moreover, this is the only facility for arithmetic that Comit provides. Let us look at an example. Assume that the first 127 constituents of the workspace are to be provided with subscripts 1, 2, . . . , 127 and then stored on the left of shelves 1, 2, ..., 127, respectively. The following program segment will accomplish this:

```
*                   $1 = $1/.0                          *
LOOP  $1/.L127 + $1 = 1/.I1 + 2/.*1 // *S*1 1      LOOP
*                   $1                  // *S*1 1      *
```

The first rule provides the first constituent with the numerical subscript 0. The left half of the second rule specifies two constituents. The first specifier matches the first constituent of the workspace for as long as its numerical subscript is less than 127 (L*n*. stands for any subscript less than *n*). The `.I1` in the right half causes the subscript to be increased by 1 (`.I`*n* causes addition of *n* to the subscript), and the `2/.*1` then assigns the incremented subscript carried by the first constituent to the second constituent as well (`*`*n* is an indirect reference to the numerical subscript carried by the *n*th component of the specified pattern). Finally the routing instruction moves the first constituent out to the shelf designated by the current numerical subscript of this constituent (the `*1` is another indirect reference). Control stays in rule `LOOP` until the numerical subscript carried by the first constituent attains the value 127 (in terms of the original workspace contents we are then looking at the 127th constituent). Then the left half of the rule does not match the workspace, and control passes to the third rule, which stacks the 127th constituent of the original workspace contents on shelf 127.

The values of logical subscripts form sets, and set operations can be performed on these sets. Only set union and complementation are provided, but, since these operations constitute a functionally complete set (see Example 2 of D.2.11), they suffice. Complementation requires the universal set to be explicitly defined, and, for complementation to be practicable, the universal set must be kept reasonably small. In Comit this is achieved by having each logical subscript name give rise to its own algebra. The universal set of this algebra is the full set of all the values that are ever associated with the subscript in the entire program. The requirement that the universal set be of a manageable size is the reason why the total number of subscript values may not exceed 36. These values are collected into a list by the compiler. Interaction between different algebras, of which finding the union of a set of values of one subscript and a set of values of a different subscript would be an example, is not provided for. Otherwise the compiler would find the generation of the universal sets too difficult a task.

In common with most existing programming languages Comit suffers from the preoccupation with numbers in the early days of computing, which led to the mistaken belief that a small character set would prove sufficient for all applications. (Another reason for restricting character sets was the smallness of computer memories. This influenced the adoption of a 6-bit code, but the 6-bit code still provides for at least 63 characters, which the older card punches do not; a 48-character set is standard for the punches.) In Comit the shortage of characters was overcome by the introduction of a "second font" of *double characters*, which are normal characters preceded by asterisks. Thus, `*A` counts as a single character, and it differs from `A`. Some characters, because of their special significance in Comit rules, e.g., the digits, `$`, `+`, and the `*` itself, cannot be used as data under their normal guise, and must be converted to double characters: `*0`, `*1`, ..., `*9`, `*$`, `*+`, `**`. When data are read from cards the conversion is performed automatically. Moreover, words may be expanded into single character constituents as part of the reading operation. For example, a card with the string `HASTINGS = 1066` in columns 1–15 can be made to arrive in the workspace as

```
H + A + S + T + I + N + G + S + - +  *= + - +
                                     *1 + *0 + *6 + *6
```

In summary, the most interesting features of Comit are multiple subrules, particularly when a subrule is selected in a random manner, logarithmic search in a list rule, and the logical subscripts when they are interpreted as set algebras. On the debit side there is the extreme awkwardness of arithmetic facilities for integers and the complete absence of floating point arithmetic, which may discriminate against the use of Comit in a situation for which the language would be otherwise most suitable. Text analysis involving some statistical computations is a case in point. The privileged status given to the workspace is a source of inconvenience. One would prefer to have strings identified by names, with no limit on the number of strings, and the ability to operate on any string. However, a Comit programmer soon becomes so proficient at using routing instructions for transfers to and from shelves that the inability to name strings ceases to nettle him. More serious is the limitation that the absence of named strings sets on further developments of the language. One might wish, for example, to provide a greater variety of pattern specifiers for left halves of rules. Assume that one wants to match *any* vowel in the workspace. It would be convenient to set up the five vowels in a named string, and use the string name in the left half of the rule with the understanding that a match is established when a workspace constituent matches *any* constituent of the named string. It is not that this "set of alternatives" type of specifier could not be implemented under the existing formalism (a feasible solution would be to store the vowels on a shelf and permit reference to be made to this shelf in the left half), but the solution of this and related problems would introduce so much artifice that it would be better to design a completely new language.

Snobol is such a language. Several versions of the language exist, the latest being Snobol 4. Each new version has been an improvement over its predecessor. Some features of Comit find no counterpart in Snobol, the more significant being the dictionary feature (list rule), explicit operations of set algebra, and the notion of a constituent. In the case of Snobol 3 one can still argue that the inconvenience due to the lack of these features balances out the improvements of Snobol over Comit, but this argument is definitely inapplicable to Snobol 4, which represents a significant advance with respect to both Snobol 3 and Comit.

Consider the problem of matching any vowel. In Snobol 4 one may set up a pattern of *alternatives*:

```
VOWEL = 'A' | 'E' | 'I' | 'O' | 'U'
```

If, then, one wishes to delete every vowel from a particular string TEXT, say, one writes

```
VOWOUT   TEXT VOWEL =              : S(VOWOUT)
```

The first symbol is a statement label. The two symbols that precede the equal sign indicate a string and a pattern that is to be looked for in the string. Since there is no "right half," the action is one of deletion. The field following the colon contains branching indicators; here we want to stay in rule VOWOUT if there has been a successful match, and have control transferred to the next rule in sequence when the pattern fails to match a substring of TEXT, i.e., when there are no more vowels left in the string. The symbol S is to be read "transfer on success" (a symbol F can be used as well—it means "transfer on failure").

Patterns may be combined. Thus, if we were looking for a two-vowel sequence, we could set up a pattern

```
DVOWEL = VOWEL VOWEL
```

DVOWEL would match any one of AA, AE, AI, ..., UO, UU. Alternatively, a pattern can be set up by means of the Snobol built-in function SPAN:

```
VOW.STRING = SPAN('AEIOU')
```

Pattern VOW.STRING will match a sequence of vowels of any length. Snobol has excellent facilities for setting up very complex patterns, but the ease with which patterns can be set up may deceive the user; a pattern that takes very little time to construct may give rise to prohibitively long search times.

Other noteworthy features of Snobol 4 are facilities for real (floating point) arithmetic and structuring of data. The simplest data structure above the level of a string is an array. For example, in

```
OBJECT = ARRAY(100)
```

the Snobol function `ARRAY` is used to create a vector of size 100 and to assign the name `OBJECT` to the vector. Very complicated reentrant list structures can be created by means of the Snobol `DATA` function. There are no explicit facilities for the traversal of these structures, but it should be noted that such facilities are not provided by L^6 either. Since the Snobol programmer should experience no difficulty in the writing of his own list processing routines, Snobol 4 is potentially as effective for processing list structures as it is for processing strings.

12c. Extension of General Purpose Languages

The addition of list processing routines to Snobol 4 would be an example of embedding: a list processing "language" would be embedded in Snobol 4. We have already discussed embedding in Section 9c, where an example of embedding in Fortran was given. Embedding extends the power of a language, and two types of extension can be recognized. In the first type the syntax of a language is added to; in the second type the language is made semantically richer. Syntactic extension is outside the scope of this book, and the meaning of semantic extension has already been covered in Section 9c.

Here we shall restrict ourselves to embedding in Fortran. A particularly simple example is the addition of a stack. We could consider the three stack routines of Section 10a a stack "language," but somehow we feel that this is too trivial an extension to justify being called a language. Indeed, if embedding consists merely of the addition of subprograms, there is no hard and fast rule that can tell us just how extensive the semantic enrichment must be before we can speak of a new language. This is a matter of opinion, and we would call a system of subprograms a new language if there is general agreement that this system constitutes a language. A system that has achieved the status of a language by virtue of general agreement is SLIP, a set of list processing routines embedded in Fortran.

The basic data structure of SLIP is the symmetric list (for examples see Figures 10.19 to 10.21). A SLIP run is started by a call of the form

```
      CALL INITAS (SPACE,N)
```

where INITAS is the SLIP initiating routine, SPACE is an array declared to contain at least N elements, and N is the size of the storage area that INITAS is to set up for SLIP use, e.g.,

```
DIMENSION LOCS(10000)
CALL INITAS (LOCS,10000)
```

causes the 10,000 elements of LOCS to be strung together into a list of available space of 5000 two-word list elements. Elements are taken off this list of available space when required, and they are returned when there is no further need for them by a variant of the reference counter technique discussed in Section 10j.

A list is created by taking elements from the list of available space. The list normally has a Fortran name containing a pointer to the first list element. This first element is a list *header*. The first word of every pair of words making up a list element contains a marker field (the ID or identifier field) and two pointers: the backward (or left) and forward (or right) pointers. The second word contains a datum. The type of the datum is indicated by the value of ID. If an element is a list header, then ID = 2, and the datum word is divided into three fields: a two-bit marker field, which is available to the user, a pointer field that may contain the address of a description list (to be defined further on), and the reference counter, which indicates the number of lists of which this list is a sublist. An *empty list* consists only of a header. If a list is not empty, then it has elements additional to the header. The datum words of such elements contain either "normal" data or pointers to other lists, these possibilities being distinguished by ID having value 0 or 1, respectively.

The SLIP list structure corresponding to the structure of Figure 10.21 (it would differ from the structure of Figure 10.21 by having list headers) could be created by the following program segment:

```
CALL LIST (TREE)
CALL LIST (SUBT)
CALL NEWTOP (1HA,TREE)
CALL NEWBOT (SUBT,TREE)
CALL NEWBOT (1HB,TREE)
CALL NEWBOT (1HC,TREE)
CALL NEWTOP (1HF,SUBT)
CALL NEWTOP (1HE,SUBT)
CALL NEWTOP (1HD,SUBT)
```

Subroutine LIST creates an empty list, i.e., generates a header. Note here that

TREE and SUBT are Fortran variable names and that they receive the addresses of the two headers as their values. Subroutine NEWTOP adds an element to the head of a list, subroutine NEWBOT adds it to the bottom of the list. In calls to these routines Hollerith data are used everywhere except in statement 4, which makes SUBT a sublist of TREE. The system automatically differentiates between storing of "normal" data and creation of sublists; i.e., it automatically provides the added list element with the appropriate identifier.

Some Fortran systems permit a subprogram to be used both as subroutine and function. The original SLIP implementation was under such a system, and in this implementation LIST can be used as a function. Then it creates an empty list as before, but, being a function, it now has a value, namely the address of the header. Statements 1 and 3 can then be combined into

```
51   CALL NEWTOP (1HA, LIST(TREE))
```

Moreover, it is then possible to set up list structures in which sublists do not have explicit Fortran names, by using the literal 9 for the argument of LIST. Here this means that statements 2 and 4 can be replaced by

```
52   CALL NEWTOP (LIST(9), TREE)
```

which creates a nameless sublist and sets up a pointer to this sublist in the main list, but there is now some difficulty in the building up of the sublist. One possible approach is to use the function BOT, which has the datum stored in the last element of a list for its value (a companion function TOP has the datum in the topmost element for its value). After execution of statements 51 and 52 in our new program the last element of TREE contains the pointer to the sublist, and one can therefore write statements 7–9 in the form

```
     CALL NEWTOP (datum, BOT(TREE))
```

Finally statements 5 and 6 are placed unchanged at the end of the new program segment. NEWTOP and NEWBOT, when used as functions, have for their values the addresses of the elements they have inserted in a list, and the pair of routines NXTLFT(D,C) and NXTRGT(D,C) insert an element containing D to the left or right, respectively, of the element indicated by C. This means that statements 5 and 6 could be combined into

```
     CALL NXTRGT (1HC, NEWBOT(1HB,TREE))
```

if Fortran functions could take Hollerith arguments, but this they do not.

All facilities of Fortran are, of course, available to the SLIP programmer. To give an example, we rewrite our program segment as follows:

```
      DIMENSION DEF(3)
      DATA A,B,C,DEF /1HA,1HB,1HC,1HD,1HE,1HF/
      CALL NEWTOP (A, LIST(TREE))
      CALL NEWTOP (LIST(9), TREE)
      DO  3 K = 1,3
   3  CALL NEWBOT (DEF(K), BOT(TREE))
      CALL NXTRGT (C, NEWBOT(B,TREE))
```

In addition to the routines already mentioned, there are 12 other routines for manipulating data on lists. For example, while BOT (or TOP) simply returns a datum, a similar function POPBOT (or POPTOP) returns the datum, but also removes the element in which the datum was stored from the list and returns it to the list of available space.

Although the structures created by the two program segments given above both correspond to Figure 10.21, they are different. The LIST(9) form leaves the reference counter of the created list as zero, but LIST(SUBT) makes it one. Therefore, after this list has been made a sublist of TREE, the reference counts of the sublist are 1 or 2, depending on how it was created. Routine IRALST is used to decrease the value of a reference counter by one, and to erase the list if the new value is zero. Therefore, CALL IRALST(TREE) would erase the entire structure in one instance, but would leave SUBT as an independent active list in the other.

Traversal of a list, or of a list structure, is an important processing activity, and SLIP provides a good set of routines for this activity. One uses sequencers and readers. A *sequencer* is a Fortran variable to which the function SEQRDR assigns the address of the header of a specified list, e.g.,

```
      S = SEQRDR (TREE)
```

inserts in S the address of the header of list TREE. The function SEQLR(S,IND) advances sequencer S so that it points to the next element down the list (advances it to the right), sets indicator IND to $+1$ if the new element is a header, to 0 if it is a pointer to a sublist, and to -1 otherwise. The value of the function is the datum stored in the new element. A similar advance to the left is made by the companion routine SEQLL. Note that SEQLR advances the sequencer back to the header when it has reached the last element of the list. This is a consequence of the symmetry of SLIP lists. SEQLL initially moves the sequencer from the header to the last element, and then advances it up the list back to the header. Therefore, if we were required to count the number of sublists that list TREE has, we could do it by means of the following program segment in which SEQLL could be used equally well in place of SEQLR.

```
      STREE = SEQRDR (TREE)
      N = 0
    5 DUMMY = SEQLR (STREE,J)
      IF (J.EQ.0) N = N+1
      IF (J.LE.0) GO TO 5
```

Exit from the loop takes place when the sequencer has been advanced back to the header of TREE.

The L in SEQLR stands for *linear*. SEQSR (and its companion SEQSL) is similar to SEQLR (SEQLL) except that the advance is *structural.* This means that if we had the program segment

```
      ST = SEQRDR (T)
    5 DUMMY = SEQSR (ST,J)
      GO TO 5
```

the sequencer would be advanced to the first sublist of T (call it T1), would be moved down T1 until it encountered the first sublist of T1 (call it T11), would be moved down T11 until it met a pointer to a sublist of T11, and so on until, finally, it would find itself in a list without sublists. Then it would keep on cycling round this list.

Sequencing routines contain no mechanism for returning the sequencer to a list higher up in the structure, e.g., from T11 to T1 of the example above. The programmer could, of course, keep a record of the points at which descent into a sublist took place, and arrange for returns on the basis of this record. Fortunately he does not have to do this: SLIP provides a second set of traversal routines, which keep the record automatically. This record is called a *reader*; it is a stack created and maintained by the SLIP system. The programmer does, however, have access to readers and may modify them. Eight routines are provided for this purpose, but they will not be discussed here. Another difference between sequencer and reader controlled advance is this: While a sequencer merely advances to an adjoining element, a reader searches a list or a list structure until an element of a specified type (the target) is found. The reader is then said to point to the target. There are three types of targets: an E-target is an element with ID = 0, an N-target is an element with ID = 1, and a W-target is an element with ID = 0 or 1.

The reader K of a list T is created by

```
      K = LRDROV (T)
```

and initially K points to the header of T. List traversal under reader control is effected by *advance routines*, which all have ADV for the first three characters in their names. The other characters in the name of an advance routine indicate the type of advance that this routine is to bring about. The fourth character is L or S, depending on whether the search is linear or structural. The fifth character is E, N, or S; it indicates the type of target. The sixth character, which is L or R, indicates direction: L for left, R for right. An advance routine is a function, and it has two arguments: It has to be supplied with the name of a reader, and it sets an indicator. Its own value is the datum stored in the target (or 0 if the search terminates without the target being found).

For example, ADVSNR(K, IND) traces round the list structure served by reader K in a rightward (clockwise) direction, searching for a pointer to a sublist. The search terminates when the target has been found, or when the header of the list associated with K is reached. In the first case IND is given a nonzero value; in the second case IND is made zero. The other five structural advance routines are similar. The six linear advance routines do not descend into sublists. In other respects they are identical with the corresponding structural advance routines.

To see how an advance routine would be used in practice consider Figure 12.2. The figure represents a SLIP list structure corresponding to the sort tree of Figure 5.15. The representation is very sketchy in that the symmetry of the lists is not even hinted at. Let the structure have the name TREE. The following program segment transfers numbers from TREE to an array N in ascending order:

```
      INTEGER ADVSEL
      K = LRDROV (TREE)
      DO 50  J = 1,12
   50 N(J) = ADVSEL (K,IND)
```

On exit from the loop we have no further use for the reader. But, since it still points to an element in a subtree, the stack is not empty. We should, therefore, erase the reader explicitly as follows:

```
      CALL IRARDR (K)
```

With any SLIP list one may associate a description list. A description list is not a sublist in that a pointer to the description list is found in the header of the list and a list may have no more than one description list. A description list consists of pairs of elements. The first element of a pair

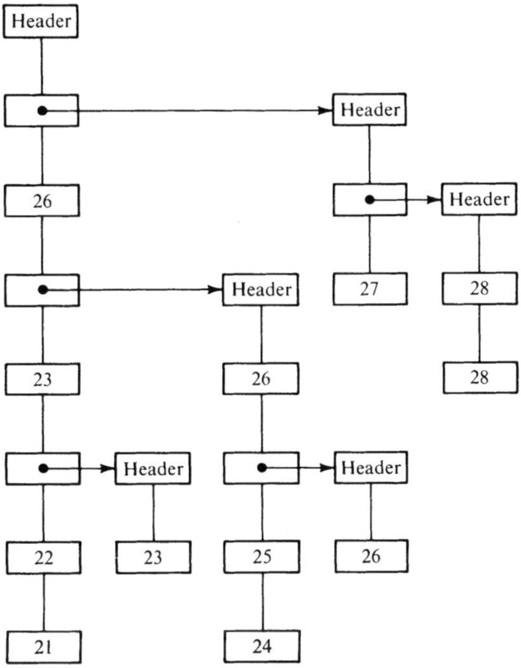

Figure 12.2

contains an *attribute*, the second contains a *value* of the attribute. Description lists can be used in similar fashion to Comit subscripts. Let us define two of the seven SLIP routines for dealing with description lists, and then look at an example of their use. Routine `NEWVAL(A,V,L)` searches the description list of list `L` for the attribute `A`. If it is found, the existing value is replaced by `V`; if not, then `A` and `V` are added to the description list. If `L` has no description list, one is created of `A` and `V`. Function `ITSVAL(A,L)` returns the value corresponding to attribute `A` in the description list of `L` (or zero, if there is no attribute `A`).

Assume now that we have 100 objects, represented by lists with names `OBJ(1),OBJ(2),...,OBJ(100)`. The list `OBJ(54)` could have been associated with a description list as follows:

```
CALL NEWVAL (5HSHAPE, 5HROUND, OBJ(54))
CALL NEWVAL (5HCOLOR, 3HRED, OBJ(54))
```

Assume that each list `OBJ(K)` has a description list, and that we are to search the description lists for the attribute `SHAPE` having value `ROUND`. All lists whose description lists do contain this attribute–value pair are to be made sublists of a new list ROUNDS. We can proceed as follows:

```
      DATA  SH,IR /5HSHAPE, 5HROUND/
      CALL LIST (ROUNDS)
      DO 10  N = 1,100
   10 IF (ITSVAL(SH,OBJ(N)).EQ.IR)
     1      CALL NEWBOT(OBJ(N),ROUNDS)
```

Weizenbaum's original paper on SLIP described more than a hundred routines. Although some of these routines merely support other routines, i.e., a user would not call on them directly, the majority are there for direct use. We have discussed only a limited number of these routines. Of the remaining routines the most interesting are VISIT and TERM, which permit recursion in Fortran. Besides creating the list of available space, INITAS creates 100 stacks W(1), W(2), ..., W(100). These stacks, called *public lists*, may be used to store information for the purposes of recursion. We shall explain recursion in SLIP by examining a particular program, a function for computing the factorial of a number.

```
      INTEGER FUNCTION  FACT (N)
      INTEGER TOP
C   THE COMMON STATEMENT IS A SLIP TECHNICALITY
      COMMON AVSL, W(100)
    7 ASSIGN 10 TO NSTAT
    8 FACT = VISIT(NSTAT,PARM1(N))
    9 RETURN
   10 IF (TOP(W(1)).EQ.1) GO TO 20
   17 CALL VISIT(NSTAT,PARM1(TOP(W(1))-1))
   18 NN = NN * TOP(W(1))
   19 CALL TERM(NN,RESTOR(1))
   20 NN = 1
   21 CALL TERM(NN,RESTOR(1))
      END
```

PARM1 belongs to a class of functions of the general form

PARM*m*(P1,P2, ..., PM),

which push down parameters P1, P2, ..., PM in stacks W(1), W(2), ..., W(M), respectively. RESTOR(M) removes the top element from each of stacks W(1), W(2), ..., W(M), and discards it. The value of RESTOR is immaterial. VISIT and TERM must be considered as a pair; in a sense TERM actually sets the value of function VISIT. VISIT(J,PARM*m*

(P1, P2, ..., PM)) pushes down the address of the location following the command that effected the transfer to VISIT in a stack of return addresses (to which only VISIT has access), transfers control to the statement whose statement number has been assigned to J, and execution of statements then proceeds in normal manner until a reference to TERM is reached. Then TERM(K,RESTOR(M)) transfers control back to VISIT, the value of VISIT is made K, and normal execution is resumed from the location indicated by the address popped up from the stack of return addresses. The second argument of VISIT and TERM is ignored in these routines. Since, for example, RESTOR(M) is executed before control passes into TERM, it could have been possible to design SLIP in such a way that one would have had to write the statement pair

```
CALL RESTOR (M)
CALL TERM (K)
```

instead of the single CALL TERM(K,RESTOR(M)). The existing design was adopted to reduce the number of statements in programs. In finding FACT(3) the order of execution of the statements in FACT is 7, 8a (transfer to 10), 10, 17a (transfer to 10), 10, 17a (transfer to 10), 10, 20, 21, 17b (since VISIT used as a subroutine nothing further done here), 18 (NN made 2), 17b, 18 (NN made 6), 19, 8b (current value of NN given to VISIT and this value assigned to FACT), 9.

The entire SLIP system, with the exception of VISIT and TERM, can be written in Fortran. This applies even to the primitive support routines for accessing parts of a computer word. Consider the 36-bit IBM 7090 word. The first word of a SLIP list element can have the following format in a 7090 implementation: bits 4–18 for the left link, bits 19–20 for ID, bits 21–35 for the right link, with the sign bit and bits 1–3 remaining unused (and set to zero). Let us call this word IW. Then the left link is given by IW/131072, the right link by MOD(IW,32768), and ID by MOD(IW,131072)/32768, where MOD is the Fortran remaindering function. However, while the loss of efficiency due to Fortran coding can be tolerated for most SLIP routines, a crippling increase in execution times would result if the primitives were coded in Fortran rather than in assembler language.

To an Algol programmer SLIP recursion may look very clumsy. Nevertheless, it is a highly ingenious solution of a very difficult problem. It illustrates the wide range of extensions that embedding can give even to Fortran, which, because of its rigid conventions, does not easily accept embedding. It is actually possible to set up the equivalent of a Comit rule in the form

```
IF (FOUND(...)) CALL ALTER (...)
```

where the arguments of the logical function FOUND are used to specify a search pattern, and the arguments of subroutine ALTER indicate desired changes to the pattern. As one would expect, FOUND and ALTER have to be coded in assembler language.

Let us list the advantages of embedding over the use of independent special purpose languages.

1. *Efficiency.* A special purpose language can be considered as made up of two components: the facilities specific to the problem area in which the language is used and general facilities, such as input and output, arithmetic capability, and sequencing of the computational process. A general purpose language provides all these facilities, and, since many people are involved in the design and improvement of the language, these facilities are generally excellent. Moreover, by being able to spread the cost over a large number of users, the compiler can be optimized to produce highly efficient machine code. The designers of an independent special purpose language have to provide the general facilities and a compiler or interpreter, and they lack the resources to do as good a job. With embedding the general facilities already exist, and all available resources can be directed to the implementation of just those facilities that relate to the special problem area.

2. *Computing power.* One way of looking at SLIP is to consider it as two languages: a list processing language, and a language for recursion. If a problem does not call for recursion, then there is no need to include the VISIT and TERM routines in the program. On the other hand, if the problem involves processing of strings as well as lists, then one may develop an embedded string processing language and use it together with SLIP in one's program. It is possible to conceive a large set of embedded languages from which a number of languages can be selected for use according to needs in the one program. This highly flexible approach has not been given much thought in the past because of storage limitations, but, with the advent of large memories and the virtual memory concept, it has now become feasible.

3. *Ease of learning.* An experienced Fortran programmer finds the Fortran embedded SLIP much easier to learn than, say, LISP. However, users of special purpose languages are often inexperienced in programming. Therefore, the aim of designers of independent special purpose languages has been to make their languages as "natural" for their prospective users as possible. For example, the form of the Comit rule reflects the notational convention of linguists, and a linguist certainly has less difficulty understanding

```
*    N = VP + NP    *
```

than the equivalent of this rule in an embedded language, where it might take the form

IF (*logical function*) *subroutine call*

However, in a realistic Comit or Snobol program very few rules are as simple. The programmer has input and output, data transfers, and sequencing added to his purely linguistic concerns, and the rules he has to write become very complex. It takes a considerable amount of study to learn to write

```
LOOP  $1/.L127 + $1 = 1/.I1 + 2/.*1 // *S*1 1  LOOP
```

which is a fair example of a nontrivial Comit rule. This rule could hardly be said to be "natural" for the linguist. Indeed, it takes no more of the linguist's time to become experienced in the use of a general purpose language and an embedded language than it takes to learn the *proper* use of Comit or Snobol.

Notes

List and string processing languages are surveyed in [Ab68] and in [Bo64b, Ra68]. Chapter 6 (pp. 382–470) of [Sa68c] is devoted to list and string processing languages; a very good feature of this book is the clear way the significant contributions made to technology in the design and implementation of a language are pointed out. A description of the data structures in a number of list and string languages and of their internal representation can be found in [Im69].

IPL-V is described in [Ne60]; [Ne64] is the IPL-V manual. The LISP philosophy is set out by McCarthy in [Mc60, Mc63c]. The latter paper deals with LISP in the context of the theory of computability. [Mc62] is the manual. The LISP literature is quite extensive: [Be64b] is a book that alone contains 20 papers on LISP, ranging from exposition to implementation and applications; [Ba65, Co65, Tr66] deal with the embedding of LISP in Algol; [Bo67e, Bo68, Co67b] discuss the problems associated with the implementation of LISP in two-level storage systems; [Fe69] deals with garbage collection when LISP is implemented in a virtual memory system (in which the available memory is practically infinite, and garbage collection is, therefore, a matter more of efficiency than necessity). LISP has been extended to make it a general purpose language; the result, LISP2, is described in [Ab66]. Addition of Fortran-like features to LISP is the subject matter of [Kn69]. Modification in a different direction, towards incremental computation, is dealt with in [Sa68d]. In addition to the papers dealing with applications of LISP in [Be64b] there is [Ra64]. L^6 is defined in [Kn66].

[Zz62b] introduces one to Comit programming; [Zz62c] is the manual. A number of papers describe applications: [Yn62, Da65, Sa66b, Li67]. Perhaps the most ambitious program written in Comit is an early version of the

General Inquirer; the General Inquirer is described in [St62, Du65], and in [St66], which is a book of 650 pages. The first version of Snobol is described in [Fa64], and Snobol 3 is described in [Fa66]; the latter paper contains three complete programs as examples. [Gr68] is the manual of Snobol 4; it contains six complete Snobol programs in an appendix and there are several more complete programs given in the text itself. See also [De67] for a good introduction to Snobol 3.

The rationale of embedding is spelled out in [Bo64c]. For examples of embedding see [Ba64b, Mo64, Co66, Sa68b]. The ingenious system of subroutines for semi-analytic differentiation described in [We64] constitutes a particularly simple but effective example of embedding. FLPL, the first well-known Fortran-embedded list processor, is described in [Ge60]. For applications of this language see [Da62, Ta62]. Weizenbaum's SLIP is defined in [We63]; [Sm67] is a good introduction. Applications of SLIP are described in [La65b, Ra68b]. Weizenbaum's own application papers, [We66b, We67], do not give programming details, but they should be read for a discussion of the type of problems SLIP was designed for and for Weizenbaum's elegant style. [Sv66] is the listing of an Algol-embedded list processor that is simpler than the Algol-embedded LISP systems referred to above.

[Ax67, De67, Sm67] constitute a comparison study: They contain, respectively, LISP, Snobol, and SLIP programs for an expression recognition problem. Annotated sample programs can be found also in [Ra68a].

Embedding of the type that enriches the syntax of a programming language is too specialized a topic for discussion in this book. Nevertheless, a few bibliographic pointers are being provided. A very intelligent discussion of the whole problem can be found in [Ne68b]; this paper should be read thoroughly. A solution that is much more attractive to a user than doing his own embedding is to persuade computer manufacturers to incorporate the desired additional features in the language itself. Although there has been an instance where this approach has been successful (see [La67b] for a description of the list processing facilities added to the original specifications of PL/I), it is not a general solution; manufacturers do not have the resources necessary to produce a language that would answer the needs of every user, and, even if they did, no strong case has been made for the desirability of such a language. The user generally has to fend for himself. One approach is to put a program that is written in an extended language through a preprocessor that converts statements of the embedded language to statements in the host language; an example of this is described in [Do66]. A more general solution is to provide the language with definition facilities (also called macro facilities). The first implementation of a definition facility that enables the user to extend the vocabulary of a higher-level language appears to have been made in 1963; this is the MAD definition facility described in [Ar69b]. For a description of macro facilities in assembler languages see [Ke69] and pp. 130–144 of [We68]. Macro processing in general is surveyed in [Br69], but the reader is advised to familiarize himself with the basic concepts in Chapter 3 of [We68] (p. 145 to as far as can be managed) before tackling this survey.

One of the hardest problems confronting the designer of a list or string processing language is how to deal with null objects. A very valuable discussion of this problem can be found on pp. 185–186 of [Mo68b].

There are no exercises, due to two reasons. First, our description of the languages has not been sufficiently detailed to justify the setting of programming exercises based on the material provided in the text. The same reason excludes exercises on comparison and evaluation of the languages. Second, the material given in this chapter should be supplemented by detailed study of a particular language, and by programming in that language; the programming assignments will, of course, be determined by the language that is selected. A few suggestions:

(a) An output routine that prints a given body of text, right adjusted by insertion of blanks (cf. [Be69b]).

(b) As (a), but including a hyphenation procedure (see [Ri65]).

(c) A recognizer of Boolean forms (see Exercise 10.8).

(d) The procedure of A.6.8 for finding cycles in a digraph reformulated as a list processing problem.

(e) Conversion routines for converting a tree in which only terminal nodes are labeled to an equivalent tree in which every node carries a label, and vice versa (cf. Figures 10.21 and 10.20).

(f) Conversion procedures for converting an unbalanced binary sort tree to an equivalent balanced sort tree, and a procedure for setting up a sort tree that generates the tree in balanced form by continuously rearranging it while it is being built up.

(g) A radix sort procedure in which the bins are structured as lists.

Solutions to Selected Exercises

Chapter 1

1.1 Only (iv) presents a clear case of equality. In (ii) there is equality if members of B are interpreted as binary numbers, and if it is assumed that there is no basic difference between binary and decimal numbers. If, however, the elements are interpreted as symbols; i.e., if the base of the number system is considered an important distinguishing feature, then A and B are not equal.

1.2 The basis of the inductive proof is the same as part (i) of the proof in the text. For the induction step assume that $\mathscr{P}(A)$ has 2^{n-1} members when A has $n-1$ members, and consider a set $B = A \cup \{x\}$. $\mathscr{P}(B)$ has all elements of $\mathscr{P}(A)$ and, in addition, each element C of $\mathscr{P}(A)$ gives rise to a new element $C \cup \{x\}$. Hence $\mathscr{P}(B)$ has $2^{n-1} + 2^{n-1} = 2^n$ elements.

1.4 Let $a \in A \cap C$. Then $a \in A$ and $a \in C$. But $a \in A$ implies $a \in B$ (since $A \subseteq B$) and, similarly, $a \in C$ implies $a \in D$. In other words, $a \in B \cap D$. Since $a \in A \cap C$ implies $a \in B \cap D$, $A \cap C \subseteq B \cap D$.

1.6 See Section 11d for hints.

1.8 Let $x \in A$. Since $\mathscr{X}$ and $\mathscr{Y}$ are partitions, there exist precisely one $X_i \in \mathscr{X}$ such that $x \in X_i$ and precisely one $Y_j \in \mathscr{Y}$ such that $x \in Y_j$. Then $x \in X_i \cap Y_j$; i.e., each member of A is a member of some set in $\mathscr{Z}$. Next assume $x \in X_i \cap Y_j$ and $x \in X_m \cap Y_n$. Then $x \in X_i$ and $x \in X_m$, implying $i = m$, and $x \in Y_j$ and $x \in Y_n$, implying $j = n$. Hence the sets comprising $\mathscr{Z}$ are disjoint.

1.10 Example 2 of Th.1.4 gives proof of $\overline{A_1 \cap A_2} = \overline{A_1} \cup \overline{A_2}$. The proof is completed by induction. Take $J = \{1, 2, \ldots, n-1\}$ and assume $\overline{\bigcap_{J\in J} A_J} = \bigcup_{J\in J} \overline{A_J}$. Then

$$\begin{aligned}\overline{(\bigcap_{J\in J} A_J) \cap A_n} &= \overline{(\bigcap_{J\in J} A_J)} \cup \overline{A_n}\\ &= (\bigcup_{J\in J} \overline{A_J}) \cup \overline{A_n}\\ &= \bigcup_{I\in I} \overline{A_I}\end{aligned}$$

The proof of the other part is similar.

1.12 No. If $A = U$, then $\{A, \bar{A}\} = \{U, \varnothing\}$. If $A = \varnothing$, then $\varnothing \in \{A, \bar{A}\}$. By D.1.9, a partition may not have $\varnothing$ for a member. (However, in some circumstances it may be convenient to change D.1.9 to allow partitions having null sets for members.)

1.14 (i) By D.1.11 and D.1.12, $A + A = (A \cap \bar{A}) \cup (A \cap \bar{A})$ and hence, by Th.1.4, $A + A = \varnothing \cup \varnothing$. But $\varnothing \cup \varnothing$ has no members. It is therefore identical with $\varnothing$.

(ii) Since $\varnothing$ has no members, $x \in A + \varnothing$ if and only if $x \in A$.

(iii) By Part (i) of Exercise 1.13,

$$A + B = (A \cup B) \cap (\bar{A} \cup \bar{B}) \text{ and } \bar{A} + \bar{B} = (\bar{A} \cup \bar{B}) \cap (\bar{\bar{A}} \cup \bar{\bar{B}}).$$

But $x \in S$ if and only if $x \notin \bar{S}$, and $x \notin \bar{S}$ if and only if $x \in \bar{\bar{S}}$. Hence $S = \bar{\bar{S}}$. Consequently $\bar{A} + \bar{B} = (\bar{A} \cup \bar{B}) \cap (A \cup B)$, and it is an easy matter to prove $(\bar{A} \cup \bar{B}) \cap (A \cup B) = (A \cup B) \cap (\bar{A} \cup \bar{B})$.

(iv) The defining formula of $A + B$ (see the example of D.1.12) shows that we can write $(A + B) + B$ as $A + (B + B)$. But $B + B = \varnothing$. Hence $(A + B) + B = A + \varnothing = A$.

1.15
$$\begin{aligned}X &= X \cap U && \text{(Th.1.5, Part 4B)}\\ &= X \cap (A \cup \bar{A}) && \text{(5A)}\\ &= (X \cap A) \cup (X \cap \bar{A}) && \text{(3B)}\end{aligned}$$

Similarly, $Y = (Y \cap A) \cup (Y \cap \bar{A})$. Consequently, if $X \cap A = Y \cap A$ and $X \cap \bar{A} = Y \cap \bar{A}$, then $X = Y$.

1.16 By Th.1.6 this is equivalent to proving that $A = A \cap B$ and $C = C \cap D$ imply $A \cap C = (A \cap C) \cap (B \cap D)$. Assuming $A = A \cap B$ and $C = C \cap D$, we have $A \cap C = (A \cap B) \cap (C \cap D)$. Rearrangement gives $A \cap C = (A \cap C) \cap (B \cap D)$.

1.18 See A.11.3 for hints regarding the modification. Since $M + N = (M \cup N) - (M \cap N)$, it is not necessary to have a routine that finds $M + N$ directly. A first call to our routine gives $M \cup N$ and $M \cap N$. A second call, with $M \cup N$ and $M \cap N$ as arguments, gives $(M \cup N) - (M \cap N)$.

1.19 Assuming the principle of duality, we need only prove Part 1 of the theorem. Assume $a \oplus b = a$ for all a. Then, in particular, $0 \oplus b = 0$. But

$$\begin{aligned}0 \oplus b &= b \oplus 0 && \text{(1A)}\\ &= b. && \text{(4A)}\end{aligned}$$

Hence $b = 0$.

1.20 (ii)
$$\begin{aligned}a \oplus (a * b) &= (a * 1) \oplus (a * b) && \text{(4B)}\\ &= a * (1 \oplus b) && \text{(3B)}\\ &= a * 1 && \text{(1A and Part 2 of Th.1.10)}\\ &= a. && \text{(4B)}\end{aligned}$$

1.21 For the proof that $x * z = y * z$ and $x * z' = y * z'$ imply $x = y$ use the method of Exercise 1.15. Then

$$\begin{aligned}(a \oplus (b \oplus c)) * a &= a * (a \oplus (b \oplus c)) && \text{(1B)}\\ &= a; && \text{(Th.1.11, Part 3)}\end{aligned}$$

$$
\begin{aligned}
((a \oplus b) \oplus c) * a &= a * ((a \oplus b) \oplus c) && \text{(1B)}\\
&= (a * (a \oplus b)) \oplus (a * c) && \text{(3B)}\\
&= a \oplus (a * c) && \text{(Th.1.11, Part 3)}\\
&= a; && \text{(Th.1.10, Part 3)}\\
(a \oplus (b \oplus c)) * a' &= (a' * a) \oplus (a' * (b \oplus c)) && \text{(1B, 3B)}\\
&= a' * (b \oplus c) && \text{(1B, 5B, 1A, 4A)}\\
&= (a' * b) \oplus (a' * c) && \text{(3B)}\\
&= (a' * (b \oplus a)) \oplus (a' * c) && \text{(4A, 5B, 1B, 3B)}\\
&= ((a \oplus b) \oplus c) * a'. && \text{(1A, 3B, 1B)}
\end{aligned}
$$

$a * (b * c) = (a * b) * c$ is given by the principle of duality. At this point, however, the more observant readers might have noticed that the "proof" is invalid. Our proof of Part 3 of Th.1.10 (Exercise 1.20) makes use of Part 2 of Th.1.10, and this in turn uses Axiom 2A. Hence we have to prove $a \oplus 1 = 1$ again without resorting to Axiom 2A:

$$
\begin{aligned}
a \oplus 1 &= (a \oplus 1) * 1 && \text{(4B)}\\
&= (a \oplus 1) * (a \oplus a') && \text{(5A)}\\
&= a \oplus (1 * a') && \text{(3A)}\\
&= a \oplus a' && \text{(1B, 4B)}\\
&= 1. && \text{(5A)}
\end{aligned}
$$

1.23 The Venn diagrams should suggest that

$$(W \cup X) + (Y \cup Z) \subseteq (W + Y) \cup (X + Z).$$

Show that this condition is equivalent to

$$((W \cup X) + (Y \cup Z)) \cap \overline{(W + Y) \cup (X + Z)} = \varnothing.$$

The rest is tedious algebraic manipulation. (See also Exercise 2.11.)

1.25 Under this definition $\langle a, a, a\rangle = \{\{a\}\} = \langle a, a\rangle$.

1.28 There are 26 letters from which to choose the first character. The rest of the name is an n-sample of a set of 36 elements, where $n \in \{0, 1, \ldots, 5\}$. Hence the number of different names is $26 \sum_{i=0}^{5} 36^i = 1{,}617{,}038{,}306$.

1.30 Let us be given a set of integers stored as array **N**, say **N** $= \langle 2, 3, 9, 11, 16\rangle$. We are required to find M-samples of this set as array SAMPLE. Instead of dealing with **N** itself, we can find—in array REF—M-samples of the set of subscripts $\{1, 2, 3, 4, 5\}$. Elements of REF are used to extract those elements of **N** that make up SAMPLE. Thus, if we require 3-samples, we generate, in turn, $\langle 1, 1, 1\rangle, \langle 1, 1, 2\rangle, \ldots, \langle 2, 5, 3\rangle, \ldots, \langle 5, 5, 5\rangle$ in REF, and in each instance find the corresponding 3-sample $\langle 2, 2, 2\rangle, \langle 2, 2, 3\rangle, \ldots, \langle 3, 16, 9\rangle, \ldots, \langle 16, 16, 16\rangle$ in SAMPLE by means of the program segment

```
      DO 222  K = 1,M
      J = REF(K)
  222 SAMPLE(K) = N(J)
```

There is no point in giving a complete solution here; you should test your subroutine by running it on a computer.

1.32 The subroutine performs properly when NZ = 0 but not when M = NZ. Immediately after statement 102 insert

```
      IF (M.LT.KA) RETURN
```

1.33 You are either convinced or not convinced.

Chapter 2

2.1 (i) and (ii).

2.5 The number of one-to-one functions is equal to $n!$, the number of n-permutations of the elements of the set. This is also the number of onto functions. The number of into functions is n^n (by Th.2.2).

2.8 All except (ii). In (ii) the number of opening parentheses exceeds the number of closing parentheses. Since, for example, $(1 \oplus 1) * 0 \neq 1 \oplus (1 * 0)$, the function corresponding to form (i) cannot be unambiguously defined without rules of precedence for the operations. In arithmetic normal evaluation is from left to right with precedence ( ), unary −, × and /, + and −. In Boolean evaluation the precedence is ( ), ′, *, ⊕.

2.9 The Boolean function is $\{\langle 0,0,0,0\rangle, \langle 0,0,1,0\rangle, \langle 0,1,0,1\rangle, \langle 0,1,1,0\rangle, \langle 1,0,0,1\rangle, \langle 1,0,1,1\rangle, \langle 1,1,0,1\rangle, \langle 1,1,1,1\rangle\}$.

2.11 Let $h(W, X, Y, Z) = ((W \cup X) + (Y \cup Z)) \cap \overline{(W + Y) \cup (X + Z)}$. One can show that $v(h(m_1, m_2, m_3, m_4)) = \varnothing$ for all 4-samples $\langle m_1, m_2, m_3, m_4\rangle$ of $\{\varnothing, U\}$ by means of a valuation table.

2.12
$$\begin{aligned} x_1 * (x_1' \oplus x_2) * x_2' &= ((x_1 * x_1') \oplus (x_1 * x_2)) * x_2' \\ &= (x_1 * x_2) * x_2' \\ &= 0. \end{aligned}$$

The simplified circuit is an open circuit without any switches.

2.14 In the bridge circuit a signal entering from the left can go along the arm bearing switch x_1, or along the arm bearing switch x_1'. If it passes through x_1, it can next go through x_2', or through x_2 and x_3'. If it passes through x_2', then it cannot continue along the arm bearing switch x_2 because this switch is closed when x_2' is open. Hence the signal continues through x_3 and x_4', or through x_3' and x_4. If, on the other hand, the signal passed through x_1, x_2, and x_3', then it can only go through x_4' next. Thus we have the possible paths $x_1—x_2'—x_3—x_4'$, $x_1—x_2'—x_3'—x_4$, $x_1—x_2—x_3'—x_4'$. An open x_1 gives rise to the paths $x_1'—x_2—x_3'—x_4$, $x_1'—x_2—x_3—x_4'$, $x_1'—x_2'—x_3—x_4$. But these are the six parallel paths of Figure 2.3.

2.15 We introduce the notation $f(x_1, x_2) = x_1 \downarrow x_2$. Function f is known as the dagger function. We have

(i) $x_1' = x_1 \downarrow x_1$;
(ii) $x_1 \oplus x_2 = (x_1 \downarrow x_2)' = (x_1 \downarrow x_2) \downarrow (x_1 \downarrow x_2)$;
(iii) $x_1 * x_2 = x_1' \downarrow x_2' = (x_1 \downarrow x_1) \downarrow (x_2 \downarrow x_2)$.

Hence any Boolean function can be represented by a circuit consisting entirely of NOR-gates. Then, since $x_1 | x_2 = (x_1 * x_2)'$, we have

(iv) $x_1 | x_2 = ((x_1 \downarrow x_1) \downarrow (x_2 \downarrow x_2)) \downarrow ((x_1 \downarrow x_1) \downarrow (x_2 \downarrow x_2))$.

2.18 The definition can be derived from D.2.8 with obvious changes of symbols. It is optional whether or not we permit symbols → and ↔. If we do, the definition ends with

(7) If α and β are statement forms, then $\alpha \to \beta$ is a statement form.
(8) If α and β are statement forms, then $\alpha \leftrightarrow \beta$ is a statement form.
(9) Only expressions given by (1)–(8) are statement forms.

Although we have not been considering statements containing logical constants F and T in the text (such as $p \vee \mathrm{T}$ or $p \wedge \neg p \leftrightarrow \mathrm{F}$), any familiarity with logical expressions in Fortran or Algol should convince one that provision for such statements must be made in the definition.

2.20 Statements (i), (ii), and (iii) are shown to be tautologies by means of truth tables. Since (ii) is a biconditional tautology, one can substitute $\neg p \to \neg q$ for $q \to p$ in (iii): Statement (iv) results.

2.22 Here one should *not* use truth tables or a computer program. Consider the first implication. This is an argument with premises (1) $p \to q$, (2) $q \to r$, (3) $r \to p$, and conclusion $p \leftrightarrow r$. By laws of detachment and syllogism, (1) and (2) give (4) $p \to r$. Statements (4) and (3) can be combined into (5) $(p \to r) \wedge (r \to p)$. But (5) is the definition of $p \leftrightarrow r$. One proves the other two implications in a similar fashion.

2.23 If $\to$ is to be associative, then $(p \to q) \to r$ must be equivalent to $p \to (q \to r)$. Since, for example, $(\mathrm{F} \to \mathrm{F}) \to \mathrm{F}$ reduces to F, but $\mathrm{F} \to (\mathrm{F} \to \mathrm{F})$ to T, operation $\to$ is not associative. By means of a truth table one can show that $(p \leftrightarrow q) \leftrightarrow r$ is equivalent to $p \leftrightarrow (q \leftrightarrow r)$, i.e., that the biconditional is associative.

2.25

(1)	$b \to i \wedge e$,	(Premise)
(2)	s,	(Premise)
(3)	$s \to m$,	(Premise)
(4)	$m \to \neg i \wedge e$,	(Premise)
(5)	m,	(2 and 3)
(6)	$\neg i \wedge e$,	(5 and 4)
(7)	$\neg i$,	(6—$p \wedge q \to p$ is a tautology)
(8)	$\neg i \vee \neg e$,	(7—$p \to p \vee q$ is a tautology)
(9)	$\neg (i \wedge e)$,	(8—De Morgan's law)
(10)	$\neg (i \wedge e) \to \neg b$,	(Contrapositive of 1)
(11)	$\neg b$.	(9 and 10)

2.27 Including the null relation, there are $|\mathscr{P}(A \times B)| = 2^{mn}$ relations, irrespective of whether $A \subseteq B$ holds.

2.29 Yes, in some superset of $\{a, b, c, \langle a, b \rangle, \langle b, c \rangle\}$.

2.32 Denote the relation by R. Since R is nonempty, there exists some $\langle a, b \rangle \in R$. But then $\langle b, a \rangle \in R$ (by symmetry), and $\langle a, b \rangle \in R$ and $\langle b, a \rangle \in R$ imply $\langle a, a \rangle \in R$ (by transitivity). Hence the relation cannot be irreflexive. Next assume that R is transitive, symmetric, and reflexive in some set A, and consider a set $A \cup \{x\}$ where $x \notin A$. Then relation R is still transitive and symmetric in $A \cup \{a\}$, but $\langle x, x \rangle \notin R$.

2.33 (i) The relation is not transitive.

(ii) Denote the relation by R. Clearly $\langle x_1, x_2 \rangle \; R \; \langle x_1, x_2 \rangle$. Next let $\langle x_1, x_2 \rangle \; R \; \langle x_3, x_4 \rangle$. Then $x_1 + x_4 = x_2 + x_3$, giving $x_3 + x_2 = x_4 + x_1$ on rearrangement. But then $\langle x_3, x_4 \rangle \; R \; \langle x_1, x_2 \rangle$. Finally assume $\langle x_1, x_2 \rangle \; R \; \langle x_3, x_4 \rangle$ and $\langle x_3, x_4 \rangle \; R \; \langle x_5, x_6 \rangle$. We have

$$x_1 + x_4 = x_2 + x_3,$$
$$x_3 + x_6 = x_4 + x_5.$$

Add: $$x_1 + x_4 + x_3 + x_6 = x_2 + x_3 + x_4 + x_5.$$

Subtract $x_3 + x_4$ from both sides:

$$x_1 + x_6 = x_2 + x_5.$$

Hence $\langle x_1, x_2 \rangle \; R \; \langle x_5, x_6 \rangle$.

(iii) Again denote the relation by R and assume $\langle x_1, x_2 \rangle \; R \; \langle x_1, x_2 \rangle$. Then, by definition, $x_1 + x_1 = x_2 + x_2$, which is only true if $x_1 = x_2$. Hence R is not reflexive.

2.35 It is sufficient to show that $S(n, m)$ is the number of ways a set of n elements can be partitioned into m disjoint nonempty subsets. This we do by induction. Clearly, if $n = 1$, there is one partition for $m = 1$ and no partition for $m > 1$. But $S(1, 1) = 1$, and it is easily shown that $S(1, m) = 0$ for $m > 1$. For the induction step assume that $S(n-1, m)$ is the number of ways a set of $n-1$ elements can be partitioned into m subsets. Since m is arbitrary, the assumption also gives $S(n-1, m-1)$ for the number of ways the $n-1$ elements can be partitioned into $m-1$ subsets. Denote the set of $n-1$ elements by A_{n-1}, and let $A_n = A_{n-1} \cup \{a_n\}$. Partitions of A_n into m subsets are of two types: (i) the set $\{a_n\}$ and a partition of A_{n-1} having $m-1$ members; (ii) a partition of A_{n-1} having m members, but with a_n merged into one of the members. There are $S(n-1, m-1)$ partitions of the first type and $mS(n-1, m)$ of the second. But $S(n-1, m-1) + mS(n-1, m) = S(n, m)$.

2.37 Let R be an equivalence relation that is not the identity relation. Then there exists some $\langle a, b\rangle \in R$ such that $a \neq b$. Since R is symmetric, also $\langle b, a\rangle \in R$. Now assume that R is a partial ordering, i.e., that it is antisymmetric. Then $\langle a, b\rangle \in R$ and $\langle b, a\rangle \in R$ imply $a = b$, and we have a contradiction.

2.39 Any equivalence relation except the identity relation.

2.41 Denote the partial ordering by $\leqq$ and assume that there exist two least elements, b and b'. Then $b \leqq a$ and $b' \leqq a$ for any element a in the set. In particular $b \leqq b'$ and $b' \leqq b$; i.e., $b < b'$ or $b = b'$, and $b' < b$ or $b' = b$. But $b < b'$ and $b' < b$ imply $b < b$, which is impossible. Hence $b = b'$. Similarly one shows that there is at most one greatest element. Assume that there exists a least element b and that b is not minimal. Then there exists some a such that $a < b$. But, by D.2.24, $b \leqq a$; i.e., $b < a$ or $b = a$. If $b < a$, then $b < a$ and $a < b$ imply $b < b$. If $b = a$, $a < b$ becomes $b < b$. Hence a least element is minimal. Next assume that there exists a least element b and an arbitrary minimal element c. Since b is a least element, $b \leqq c$. Since c is minimal, $b < c$ does not hold. Hence $b = c$. The proof for a greatest element is similar.

2.45 None of the sets has a least element.

(i) Minimal elements: 2, 3, 5, 7. Greatest (and maximal) element: 210.
(ii) Minimal elements: 2, 3, 5. Greatest (and maximal) element: 1080.
(iii) Minimal elements: 2, 3, 5. Greatest (and maximal) element: 300.

2.48 One of the order-isomorphic sets is {1, 2, 3, 5, 6, 7, 10, 14, 15, 21, 35, 900, 1764, 4900, 11,025, 85,766, 121,000,000}. The diagram turned through 180° represents the partially ordered set $\langle \mathscr{P}(\{a, b, c, d\}), \supseteq\rangle$. In general, there is no definite relation between sets represented by a diagram and the diagram turned through 90°.

2.50 (i) No. (ii) No.

2.52 Assume $a * b = a$. Then $a = \inf\{a, b\}$; i.e., a is a lower bound of $\{a, b\}$. Hence, by D.2.27, $a \leqq b$. Next assume $a \leqq b$. Clearly $b \leqq b$ as well, and b is an upper bound of $\{a, b\}$. Assume $b \neq \sup\{a, b\}$; i.e., assume $c = \sup\{a, b\}$ such that $b \neq c$. Then $c \leqq b$ (c is the supremum) and $b \leqq c$ (c is an upper bound), and, since $b \neq c$, $c < b$ and $b < c$. But this is impossible. Hence $b = \sup\{a, b\}$. Finally assume $a \oplus b = b$. This means that b is an upper bound of $\{a, b\}$, i.e., that $a \leqq b$. Since $a \leqq a$ as well, a is a lower bound of $\{a, b\}$. Just as the assumption $b \neq \sup\{a, b\}$ led to the impossible $b < b$ above, assumption $a \neq \inf\{a, b\}$ leads to $a < a$. Hence $a = \inf\{a, b\} = a * b$.

2.54 By D.2.27 and D.2.24, $\inf\{a, b\} \leqq a \leqq \sup\{a, b\}$, or, by D.2.29, $a * b \leqq a \leqq a \oplus b$. The equivalences of Exercise 2.52 then give

$$(a * b) \oplus a = a = a * (a \oplus b).$$

2.59 Take a system $\langle A, \oplus, *\rangle$ with

(i) $a \oplus b = b \oplus a$, $\quad a * b = b * a$;
(ii) $a \oplus (b \oplus c) = (a \oplus b) \oplus c$, $\quad a * (b * c) = (a * b) * c$;
(iii) $a \oplus (a * b) = a$, $\quad a * (a \oplus b) = a$

as axioms, and define relation $\leqq$ such that $a \leqq b$ if and only if $a \oplus b = b$ (or $a * b = a$). We can then prove theorems

$$a \oplus a = a,$$
$$a \leqq b \text{ and } b \leqq a \text{ imply } a = b,$$

etc. Axioms (i) and (ii) form part of the set of axioms of Boolean algebra. In Boolean algebra Statements (iii) can be derived as theorems, but Statements (i) and (ii) do not suffice for this purpose (see Exercise 1.20). Hence Statements (iii) are axioms here. (Distributive complemented lattices are Boolean algebras.)

2.60 (i) Let array LAT contain the elements of X. Then

```
      SUBROUTINE SETUP (LAT, LESS, N)
      LOGICAL TEST
      DIMENSION LAT(N), LESS(N,N)
      DO  5  I = 1,N
      DO  5  J = 1,N
      LESS(I,J) = 0
    5 IF (TEST(LAT(I),LAT(J))) LESS(I,J) = 1
      RETURN
      END
```

Because of the restriction on the order in which elements of the set may be stored in LAT we need not define LESS(I, J) when J is less than I, and the DO 5 J =1,N can be changed to DO 5 J =I,N. For the lattice of Example 1 of D.2.29 TEST can be written as follows:

```
      LOGICAL FUNCTION TEST (N,M)
      TEST = .FALSE.
      IF (MOD(M,N).EQ.0) TEST = .TRUE.
      RETURN
      END
```

Subroutine TABLE generates the addition–multiplication table in array LESS, and in so doing overwrites its previous contents. Some attention must therefore be given to the order in which the entries are computed. Entries of the addition table must be created first. The computation is based on the observation that if $x_i \leqq x_k$ and $x_j \leqq x_k$, then x_k is an upper bound of $\{x_i, x_j\}$. Hence the least value of K such that LESS (I , K) = LESS (J , K) = 1 gives the least upper bound.

```
      SUBROUTINE TABLE (LAT,LESS,N)
      DIMENSION LAT(N), LESS(N,N)
      DO 10  I = 2,N
      II = I - 1
      DO 10  J = 1,II
      DO  5  K = I,N
    5 IF (LESS(I,K)*LESS(J,K).EQ.1) GO TO 10
   10 LESS(I,J) = LAT(K)
      DO 20  I = 1,N
      II = N + 1 - I
      DO 20  J = 1,I
```

```
      JJ = N + 1 - J
      DO 15  K = I,N
      KK = N + 1 - K
15    IF (LESS(KK,II)*LESS(KK,JJ).EQ.1) GO TO 20
20    LESS(II,JJ) = LAT(KK)
      RETURN
      END
```

2.62 (i) Set B is closed under $\oplus$, $*$, and $'$. Hence $x \in B$ and $y \in B$ imply, in turn, $y' \in B$, $x * y' \in B$, $x' \in B$, $x' * y \in B$, $x + y \in B$.

(ii) The most obvious examples are addition and multiplication. Set N is not closed under subtraction (e.g., $5 - 12 \notin N$), but one can have an operation $\dot{-}$ (proper subtraction), defined by

$$\begin{aligned} x \dot{-} y &= x - y, \qquad x \geqq y, \\ &= 0, \qquad x < y, \end{aligned}$$

under which N is closed. Out of the multitude of other operations under which N is closed we select a one-argument operation S (successor function), defined by $S(n) = n + 1$.

2.63 The similar algebras are (i), (iii), and (v). Since $\oplus$ and $*$ are both two-argument operations, each operation in a corresponding pair from (i) and (iii) has the same number of arguments. Two algebras need not have the same number of elements for similarity. Hence, even though $|B| \neq |\{0, 1\}|$, (i) and (v) are similar. Algebra (ii) does not have as many operations as the other algebras, and $'$ in (iv) does not have as many arguments as the corresponding operation ($\oplus$ or $*$) in any of the other algebras.

2.64 If X is a sublattice of a distributive lattice L, then, since the identities of D.2.30 hold for any $a, b, c \in L$, they hold, in particular, for any $a, b, c \in X$ ($X \subseteq L$). Hence X is distributive. If Y is a homomorphic image of L under some homomorphism $h: L \to Y$, and x, y, z are arbitrary elements of Y, then, by definition of homomorphism, there may be found in L elements a, b, c such that $h(a) = x$, $h(b) = y$, $h(c) = z$. Consequently, if L is distributive, we have

$$\begin{aligned} x * (y \oplus z) &= h(a) * (h(b) \oplus h(c)) \\ &= h(a) * h(b \oplus c) \\ &= h(a * (b \oplus c)) \\ &= h((a * b) \oplus (a * c)) \\ &= h(a * b) \oplus h(a * c) \\ &= (h(a) * h(b)) \oplus (h(a) * h(c)) \\ &= (x * y) \oplus (x * z), \end{aligned}$$

and, by Exercise 2.55, the other distributive law holds as well. Hence Y is distributive.

Chapter 3

3.2 A digraph $\langle A, R \rangle$ is bipartite if there exists a partition $\{X, Y\}$ of A such that, for all $\langle a, b \rangle \in R$, $a \in X$ and $b \in Y$.

3.5 Simple cycles: $(a_1, a_2, a_6, a_5, a_1)$, (a_2, a_6, a_2), (a_2, a_6, a_3, a_2), (a_9, a_9). Nonsimple cycle: $(a_1, a_2, a_6, a_3, a_2, a_6, a_5, a_1)$. Note that, for example, node sequences $(a_2, a_6, a_2, a_6, a_2)$ and (a_2, a_6, a_2) correspond to the same cycle.

3.6 The first part is obvious. The second part can be established by means of a counter-example: Consider paths (a, c, d, b) and (b, d, a) in a digraph

$$\langle \{a, b, c, d\}, \{\langle a, c\rangle, \langle b, c\rangle, \langle c, d\rangle, \langle d, a\rangle, \langle d, b\rangle\}\rangle;$$

there exists no simple cycle $(a, \ldots, b, \ldots, a)$.

3.9 The subroutine should be based on Corollary 2 of Th.3.6. At this point it does not matter how efficient the subroutine is (A.6.1 is an efficient algorithm for finding the path matrix). Whether or not the subroutine accepts a variable adjacency matrix for its input depends, of course, on how it is written.

3.13 The antisymmetry test is changed to a symmetry test:

```
C3   SYMMETRY TEST
       IF (MX(I,J).NE.MX(J,I)) RETURN
```

3.18 Let **M** be the adjacency matrix. "Symmetrize" it, possibly as follows:

```
       DO  1  I = 1,N
       DO  1  J = 1,N
    1  IF (M(I,J).EQ.1)  M(J,I) = 1
```

Now find **MP**, the path matrix corresponding to the "symmetrized" adjacency matrix and in **MP** set all diagonal elements to 1:

```
       DO 50  J = 1,N
   50  MP(J,J) = 1
```

MP is then the matrix *C*.

3.20 By the corollary of Th.3.8 the strong components of a digraph are the connected components of its cycle digraph. But connectedness is an equivalence relation. Hence, if the procedure of Exercise 3.18 is applied to the adjacency matrix of the cycle digraph, and the resulting matrix made the input to ISEQUI as modified in Exercise 3.15, then the equivalence classes printed by ISEQUI define the strong components of the digraph.

3.22 The minimal sets contain, respectively, one and four K-formulas.

3.25

```
       LOGICAL FUNCTION ISKFOR (KFORM,M)
       INTEGER  KFORM(M)
C   THE M SYMBOLS OF THE FORMULA ARE SUPPLIED IN ARRAY
C   KFORM - THE K-OPERATOR IS REPRESENTED BY 0.
       ISKFOR = .FALSE.
       K = 0
       N = 0
       J = 0
    5  J = J+1
       IF (J.GT.M) RETURN
       IF (KFORM(J).EQ.0) GO TO 10
       N = N+1
       GO TO 15
```

```
10  K = K+1
15  IF (K.GE.N) GO TO 5
    IF (J.EQ.M) ISKFOR = .TRUE.
    RETURN
    END
```

3.27 For example, a digraph $\langle\{1, 2, 3, 4, 5\}, \{\langle 1, 2\rangle, \langle 3, 4\rangle, \langle 4, 5\rangle, \langle 5, 3\rangle\}\rangle$.

3.28 By Th.3.3, a node belonging to a node base must have zero indegree, or must lie on a cycle. The node base of a disconnected digraph contains at least two nodes. Therefore, an acyclic digraph with exactly one node having indegree 0 must be connected. Next we show that a connected digraph without slings in which exactly one node has indegree 0 and every other node has indegree 1 is acyclic. (The requirement that there be no slings excludes arcs that do not contribute to indegrees of nodes.) A connected digraph with no arcs satisfies the conditions and is acyclic; a connected digraph with one arc also satisfies the conditions and is acyclic. This is the base of an induction proof. Assume that a digraph with n arcs that satisfies the conditions is acyclic. Since all nodes but one have indegree 1, and the remaining node has indegree 0, there are $n + 1$ nodes in the digraph. Add an $(n + 1)$th arc. The arc originates at one of these $n + 1$ nodes (otherwise there are now two nodes with indegree 0), and it does not terminate at any of them (otherwise now there is no node with indegree 0, or there is a node with indegree 2). Alternatively, the arc does not originate at any of the $n + 1$ nodes, but then it terminates at the node having indegree 0 (otherwise there are now two nodes with indegree 0). In either case the new arc does not close a cycle.

3.30 Work with complements of the digraph. Then it is easy to see that the digraphs are isomorphic. The correspondence of the nodes is

a	b	c	d	e	f
6	3	5	2	4	1

(For practice one should establish the isomorphism of D_1 and D_2 directly as well.)

3.31 The definition of a complete digraph (D.3.15) is of the form $p \to q$, where p is "$\langle a, b\rangle \notin R$" and q is "$\langle b, a\rangle \in R$", and statement $p \to q$ is false if and only if p is true and q is false. With respect to a node pair $\langle a, a\rangle$ this means that $\langle a, a\rangle \notin R$ makes $p \to q$ false, i.e., that a complete digraph must have a sling on every node. Therefore, if $D = \langle X, R\rangle$ is complete symmetric, then R is the universal relation in X, and the adjacency matrix of D is invariant under permutations of rows and columns, i.e., for all D', D and D' are isomorphic. (Actually, we have a much stronger result: D and D' are identical.)

3.34 We have to show that a connected graph contains no circuits (p) if and only if it becomes disconnected when any one of its edges is removed (q). Assume p, and prove q: In a graph that contains no circuits an edge $\{a, b\}$ is a unique chain between a and b, and after removal of this edge there is no longer a chain between the nodes, i.e., the graph is disconnected. Next assume q and prove p. If removal of an arbitrary edge $\{a, b\}$ has disconnected the graph, then there are now nodes c and d in the graph such that no chain $(c, \ldots, d)$ exists. A chain $(c, \ldots, d)$ must have existed before the removal of the edge, and the removed edge must have belonged to the chain. But, if there had existed a circuit containing $\{a, b\}$, then a chain $(c, \ldots, d)$ would still exist. Therefore, no circuit containing $\{a, b\}$ could have existed, and since $\{a, b\}$ is arbitrary, no circuit at all could have existed.

Chapter 4

4.1 Consider arbitrary matrices

$$A=\begin{bmatrix} a & b \\ c & d \end{bmatrix}, \quad B=\begin{bmatrix} s & t \\ u & v \end{bmatrix}, \quad C=\begin{bmatrix} w & x \\ y & z \end{bmatrix}.$$

Then

$$A.(B.C)=(A.B).C=\begin{bmatrix} asw+aty+buw+bvy & asx+atz+bux+bvz \\ csw+cty+duw+dvy & csx+ctz+dux+dvz \end{bmatrix}.$$

The identity element is

$$\begin{bmatrix} 1 & 0 \\ 0 & 1 \end{bmatrix}.$$

4.3 An equilateral triangle ABC can be rotated through 120°, 240°, and 360° in the plane. Let the midpoints of sides AB, BC, CA be denoted by X, Y, and Z, respectively. The triangle can be rotated about axes AY, BZ, and CX. The six rotations form a group, where the rotation through 360° is the identity. Setting up of the "multiplication" table is rather trivial. For the pentagon there are 5 rotations in the plane, and 5 rotations about axes drawn from a vertex to the midpoint of the side opposite it.

4.5 Give a permutation

$$p=\begin{pmatrix} a_1 & a_2 & \cdots & a_n \\ b_1 & b_2 & \cdots & b_n \end{pmatrix},$$

construct a digraph $D=\langle A, R\rangle=\langle\{a_1, \ldots, a_n\}, \{\langle a_1, b_1\rangle, \ldots, \langle a_n,b_n\rangle\}\rangle$. We shall prove that there exist strong components $\langle A_i, R_i\rangle$ of D such that the R_i constitute a partition of R and each R_i is a simple cycle. Then each digraph cycle defines a permutation cycle and the product of the permutation cycles is p. Since $\{a_1, \ldots, a_n\}=\{b_1, \ldots, b_n\}$, exactly one arc originates from and exactly one arc terminates at each node. This means that every node a_i lies on a path $(\ldots, a_i, \ldots)$. Consider the longest possible path through a_i. If arc $\langle a_i, a_i\rangle$ exists, then, since this arc can be the only one incident with node a_i, the longest path is the simple cycle (a_i, a_i) and $\langle\{a_i\}, \{\langle a_i, a_i\rangle\}\rangle$ is a strong component. Otherwise denote the longest path through a_i by $(c_1, \ldots, c_{k-1}, c_k)$. It must be a simple cycle: Since an arc $\langle c_j, c_1\rangle$ exists, but the path is not $(c_j, c_1, \ldots, c_k)$, the arc appears elsewhere in the path, and, since $c_1, \ldots, c_{k-1}$ are distinct (otherwise more than one arc would originate from a node), the path is $(c_1, \ldots, c_j, c_1)$. No arc that does not belong to the cycle can be incident with a node on the cycle. Hence the nodes and arcs that define the cycle constitute a strong component and a node that belongs to the strong component cannot belong to any other strong component. Moreover, there is a one-to-one correspondence between nodes $a_i \in A$ and arcs $\langle a_i, \ldots\rangle \in R$, and, if $a_i \in A_i$, then $\langle a_i, \ldots\rangle$ belongs to R_i and to R_i alone. But every $a_i \in A$ belongs to some A_i. Consequently the R_i constitute a partition of R.

4.7 They alternate.

4.9 Obviously the set is closed under *max*; since $0 \leq ab \leq b$ for $0 \leq a \leq 1$ and $b \geq 0$, the set is closed under multiplication. Multiplication and *max* are associative, and $a\cdot 1=1\cdot a=a$ and $max(b, 0)=max(0, b)=b$ when $b \geq 0$. Hence $\langle[0, 1], max, 0\rangle$ and $\langle[0, 1], \cdot, 1\rangle$ are semigroups. Also, $max(a, b)=max(b, a)$, $max(a, 1)=1$ when $a \leq 1$, and $a\cdot 0=0\cdot a=0$ for all a. It remains to be shown that $a\cdot max(b, c)=max(ab, ac)$ and $max(b, c)\cdot a=max(ba, ca)$, but, since multiplication and *max* are commutative, it suffices to show that $a\cdot max(b, c)=max(ab, ac)$ when $b \geq c$. If $b \geq c$, then $max(b, c)=b$ and $ab \geq ac$ for $a \geq 0$. Consequently $a\cdot max(b, c)=ab$ and $max(ab, ac)=ab$.

4.11 The domain of f_c is not V^*. Hence f_c and $f_c | B$ cannot be operations in V^*.

4.13 Any Markov algorithm that contains a production having the null string for its antecedent.

4.15 The problem is overlap: Removal of all occurrences of *ara* from *ararat* should leave just the *t*. Write the specified substring as $\alpha\beta$, where α is the first character of the substring. Markers s, u, v, w are used. The algorithm is

1. $s\alpha\beta \rightarrow \alpha us\beta$
2. $sx \rightarrow xs$ $\quad (x \in V)$
3. $\alpha u \rightarrow v\alpha$
4. $v\alpha\beta \rightarrow w$
5. $vw \rightarrow \Lambda$
6. $vx \rightarrow v$ $\quad (x \in V \cup V')$
7. $w \rightarrow \Lambda$
8. $s \rightarrow .\Lambda$
9. $\Lambda \rightarrow s$

Consider, for example, the removal of $\alpha\beta = abca$ from *ababcaca*. We are left with a new *abca*, which we do not wish to delete. Marker s prevents deletions of this type. Production 1 marks *all* substrings $\alpha\beta$; Production 3 makes a marker substitution and shifts each marker to the start of its substring. Production 4 deletes nonoverlapped substrings, and, after their deletion, markers v and w enclose parts of overlapped substrings that remain to be deleted. This deletion is effected by Production 6.

4.20 Some discussion of this is to be found in Section 12b.

4.21 Strings (i) and (iv) are sentences. String (ii) is not because *The* $\notin V$ (only *the* $\in V$), and string (iii) is not because *an* $\notin V$.

4.23 (ii) $G = \langle\{a, b\}, \{S\}, P, S\rangle$, where P is $\langle S\rangle ::= b \,|\, a\langle S\rangle a$. G is Type 2.

(v) The grammar derives from D.2.8. It is Type 2. A possible set of production rules is

$$\begin{aligned}
\langle\text{var}\rangle &::= x_1 \,|\, x_2 \,|\cdots|\, x_n \\
\langle\text{cons}\rangle &::= 0 \,|\, 1 \\
\langle\text{unop}\rangle &::= {}' \\
\langle\text{binop}\rangle &::= \oplus \,|\, * \\
\langle\text{prim}\rangle &::= \langle\text{var}\rangle \,|\, \langle\text{cons}\rangle \\
\langle\text{form}\rangle &::= \langle\text{prim}\rangle \,|\, (\langle\text{form}\rangle) \,|\, \langle\text{form}\rangle\langle\text{unop}\rangle \,|\, \langle\text{form}\rangle\langle\text{binop}\rangle\langle\text{form}\rangle
\end{aligned}$$

4.24 (ii) $\{a^n bc^n \,|\, n \geqq 0\}$.

(iv) $\{\alpha aca\beta \,|\, \alpha \in \{a, adb^n d\}^*, n \geqq 1$; if $\alpha = s_1 s_2 \cdots s_k$, then $\beta = s_k \cdots s_2 s_1\}$.

4.26 Denote the symbols by $a_1, a_2, \ldots, a_i$. Assume that $\frac{1}{2}n(n+1)+1$ rules are sufficient to produce all permutations of n symbols. There is an initiating rule $S \rightarrow A_1 A_2 \cdots A_n$, n rules of the form $A_t \rightarrow a_t$, and the remaining $\frac{1}{2}n(n-1)$ rules, which have the form $A_s A_t \rightarrow A_t A_s$, transpose the metalinguistic variables. To get a particular permutation, say $a_3 a_2 a_1 a_4$ when $i = 4$, one first produces string $A_3 A_2 A_1 A_4$ by means of transpositions, and then replaces the variables with terminal symbols. When a_{n+1} is added, the initiating rule becomes $S \rightarrow A_1 \cdots A_{n+1}$, and, to generate a particular permutation, one permutes symbols $A_1, \ldots, A_n$ and then sends A_{n+1} to its required position in the string (see the discussion preceding A.1.5). This requires n additional transposition rules

$$A_t A_{n+1} \rightarrow A_{n+1} A_t \ (t = 1, \ldots, n),$$

and we need the further rule $A_{n+1} \rightarrow a_{n+1}$; i.e., the total number of rules becomes $\frac{1}{2}n(n+1) + 1 + (n+1) = \frac{1}{2}(n+1)(n+2) + 1$, as required. To establish the base of the induction proof consider $P_2 = \{S \to A_1A_2,\ A_1A_2 \to A_2A_1,\ A_1 \to a_1,\ A_2 \to a_2\}$. These rules produce the two permutations a_1a_2 and a_2a_1, and $|P_2| = 4 = \frac{1}{2}i(i+1)+1$ when $i = 2$.

Chapter 5

5.1 There are two phases. (i) The check on indegrees is made by computing the sum of elements for each column of the adjacency matrix. This should come to 0 for one column (store its number in NROOT), and be 1 for every other column. (ii) The digraph can be shown to be acyclic, or to be connected and without slings. It is easier to establish acyclicity: Find the path matrix and test its diagonal elements; they should all be 0. Digraph D is a directed tree if both tests are passed, and then it has a root, which is identified by NROOT.

5.3 The string *abab* can be produced in two different ways:

$$\begin{aligned} S &\Rightarrow abSb \\ &\Rightarrow abab \end{aligned} \qquad \text{or} \qquad \begin{aligned} S &\Rightarrow aAb \\ &\Rightarrow abSb \\ &\Rightarrow abab \end{aligned}$$

$L(G) = \{(ab)^n ab^n \mid n \geqq 0\}$. An unambiguous grammar is $G_1 = \langle\{a, b\}, \{S\}, P, S\rangle$, where P is

$$\langle S\rangle ::= ab\langle S\rangle b \mid a.$$

5.5 One Type 3 grammar that generates the language has the productions

$$\begin{aligned} \langle S\rangle &::= \langle A\rangle \mid \langle B\rangle b \mid b \\ \langle A\rangle &::= ab \mid \langle A\rangle ab \\ \langle B\rangle &::= bc \mid \langle B\rangle bc \mid a \mid \langle B\rangle a \end{aligned}$$

A simple algorithm for converting a syntactic chart to a state graph does not appear to exist. One might get some guidance in the design of the automaton from the syntactic chart, but this would be a purely subjective matter.

5.7 The grammar might have the following productions:

$$\begin{aligned} \langle S\rangle \ &::= (\langle\text{var}\rangle = \langle\text{expr}\rangle) \mid (\langle\text{var}\rangle = \langle\text{var}\rangle) \\ \langle\text{expr}\rangle &::= (\langle\text{var}\rangle\langle\text{op}\rangle\langle\text{var}\rangle) \mid (\langle\text{expr}\rangle\langle\text{op}\rangle\langle\text{expr}\rangle) \mid (\langle\text{var}\rangle\langle\text{op}\rangle\langle\text{expr}\rangle) \mid \\ &\qquad (\langle\text{expr}\rangle\langle\text{op}\rangle\langle\text{var}\rangle) \\ \langle\text{var}\rangle &::= a \mid b \mid c \mid \cdots \mid z \\ \langle\text{op}\rangle &::= + \mid - \mid / \mid \times \end{aligned}$$

Strictly speaking, the definitions of ⟨var⟩ and ⟨op⟩ are invalid in a parenthesis grammar. These productions could be corrected to

$$\begin{aligned} \langle\text{var}\rangle &::= (a) \mid (b) \mid (c) \mid \cdots \mid (z) \\ \langle\text{op}\rangle &::= (+) \mid (-) \mid (/) \mid (\times) \end{aligned}$$

but since we prefer $(a = (b + c))$ to $((a)(=)((b)(+)(c)))$, say, we would be reluctant to write the productions this way. The alternative is to omit the last two definitions altogether. Then, in the definition of $\langle S\rangle$, we would have to replace $(\langle\text{var}\rangle = \langle\text{expr}\rangle)$ with the 26 alternatives $(a = \langle\text{expr}\rangle) \mid (b = \langle\text{expr}\rangle) \mid \cdots \mid (z = \langle\text{expr}\rangle)$, and make similar replacements elsewhere. However, our grammar is equivalent to this parenthesis grammar, and this is our justification for writing it the way we have written it. (A grammar that is not a parenthesis grammar, but is equivalent to one, is called a concealed parenthesis grammar.)

5.9 The average number of comparisons for the next word coming into the dictionary is (i) unbalanced—4·79, balanced—3·92; (ii) unbalanced—5·66, balanced—4.10.

5.10

		Array P	
the	1	2	4
only	2	5	3
problem	3	0	13
to	4	0	8
be	5	10	6
faced	6	22	7
in	7	9	11
using	8	14	0
foregoing	9	16	0
algorithm	10	12	17
obtain	11	19	21
a	12	0	0
solution	13	15	24
traveling	14	0	0
salesman	15	0	0
for	16	0	0
an	17	0	18
arbitrarily	18	0	0
large	19	23	20
number	20	0	0
of	21	0	0
cities	22	0	0
is	23	0	0
storage	24	0	0

5.12 Call the two trees A and B, respectively, and assume that B is to be merged into A. Their roots are *the* and *assuming*. All words that precede *the* and *assuming* in the two trees are reached by taking the left arc out of the root of the appropriate tree, and, since *assuming* precedes *the*, all such words in B (including *assuming*) can be sent into A through the left arc. Moreover, in going down the leftmost path in A, we reach in succession *only*, *be*, and *algorithm*. We find that *assuming* precedes the first two of these words, but that it does not precede *algorithm*. Therefore, all words in B that are reached by the left arc out of the root (i.e., *all*, *an*, *and*, and also *assuming* itself) can enter tree A at node *algorithm*. This discussion should be enough of a hint as to how an efficient merging algorithm can be designed.

5.14 Extended-entry decision tables:

Input 1	1	1	1	1	0	0	0	0
2	1	1	0	0	1	1	0	0
3	1	0	1	0	1	0	1	0
Action (Part ii)	A	A	A	B	A	B	B	B
(Part iv)	B	A	A	A	A	A	A	B

Limited-entry decision tables:

Input 1	1	1	1	0	0	0
2	1	0	0	1	1	0
3	–	1	0	1	0	–
Action (Part ii)	A	A	B	A	B	B

Input 1	1	1	1	0	0	0
2	1	1	0	1	0	0
3	1	0	–	–	1	0
Action (Part iv)	B	A	A	A	A	B

Note, however, that if we rearrange the second limited-entry table as follows

Input 1	1	0	1	0	1	0
2	1	0	1	0	0	1
3	1	0	0	1	–	–
Action (Part iv)	B	B	A	A	A	A

then, in searching the table for a column that corresponds to a given data vector, only the first two columns have to be examined. If the data vector does not match either of these columns, then the action must be A.

5.16 The limited-entry tables are

	0.30	0.25	0.25	0.20
q_1	N	Y	Y	Y
q_2	–	–	Y	N
q_3	–	Y	N	N
	A1	A1	A3	A2

	0.35	0.25	0.20	0.20
q_1	–	Y	N	Y
q_2	–	Y	–	N
q_3	Y	N	N	N
	A1	A3	A1	A2

The average number of comparisons in the two cases is

$$0.30 + 2 \times 0.25 + 3 \times 0.25 + 4 \times 0.20 = 2.35,$$
$$0.35 + 2 \times 0.25 + 3 \times 0.20 + 4 \times 0.20 = 2.25.$$

This analysis shows that the second table is more efficient. Note again that a further simplification is possible (see solution to Exercise 5.14):

	0.25	0.20	0.55
q_1	Y	Y	–
q_2	Y	N	–
q_3	N	N	–
	A3	A2	A1

Under this scheme the average number of comparisons is

$$0.25 + 2 \times (0.20 + 0.55) = 1.75.$$

5.18 The sequence of evaluation of conditions in the two trees is q_1—q_3—q_2 or q_3—q_1—q_2 respectively, and the average costs are, respectively,

$$1.5 + 0.50 \times 2 \quad + 0.25 \times 4 = 3.50,$$
$$2 \quad + 0.50 \times 1.5 + 0.25 \times 4 = 3.75.$$

The first procedure is more efficient.

Chapter 6

6.1 See Exercise 3.20 and its solution. In finding the cycle digraph by A.3.3 one should now use A.6.1 to compute the path matrix.

6.3 If indices j and k are interchanged, then the entire algorithm has to be iterated $\log_2 n$ times. A study of the second part of the proof of Th.6.1 ($p_{ik} = 1$ implies $x^*_{ik} = 1$) will tell you what happens if j and i are interchanged.

6.5 In Step 6 make a special case of the condition $(x^*_{ij} + x^*_{jk}) = x^*_{ik}$. If this condition holds, then set $v^*_{ik} = v^*_{ik} + v^*_{ij}v^*_{jk}$, were + denotes juxtaposition. Note that either or both of v^*_{ij} and v^*_{jk} may be composed of juxtaposed strings; see the paragraph preceding A.6.3 for the distribution laws.

6.8 The algorithm depends on the triangle inequality for distances:

$$d_{kt} \leqq d_{kJ} + d_{Jt}.$$

The triangle inequality does not hold for negative distances. Whereas a minimization process can be based on the triangle inequality, there is no corresponding principle on which to base a maximization process.

6.10 Take the path matrix of A_k, adjoin to it the $(k + 1)$th row and column of X, and call this matrix **M**. First compute the $(k + 1)$th column:

```
      KP = K + 1
      DO  5  J = 1,K
      IF (M(J,KP).EQ.1) GO TO 5
      DO  3  I = 1,K
  3   IF (M(J,I)*M(I,KP).EQ.1) GO TO 4
      GO TO 5
  4   M(J,KP) = 1
  5   CONTINUE
```

A possible alternative is

```
      KP = K + 1
      DO  5  J = 1,K
      IF (M(J,KP).EQ.0) GO TO 5
      DO  3  I = 1,K
  3   IF (M(I,J).EQ.1) M(I,KP) = 1
  5   CONTINUE
```

Next deal with the $(k + 1)$th row in a similar fashion:

```
      DO 10  J = 1,KP
      IF (M(KP,J).EQ.1) GO TO 10
      DO  8  I = 1,KP
  8   IF (M(KP,I)*M(I,J).EQ.1) GO TO 9
      GO TO 10
  9   M(KP,J) = 1
 10   CONTINUE
```

It is a very common mistake to omit the final part:

```
       DO 15   I = 1,K
       DO 15   J = 1,K
       IF (M(I,J).EQ.1) GO TO 15
       M(I,J) = M(I,KP)*M(KP,J)
   15  CONTINUE
```

6.12 The ith term in the sum over i gives the number of cycles of length i. When $n = 5$ the terms have values 10, 20, 30, and 24, respectively, and their sum is 84. Denote the formula by $N(n)$. The problem is to make the formula separable for an induction proof. Some rather involved rearrangements finally result in

$$N(n) = \sum_{i=2}^{n}\left(\sum_{j=2}^{i}\left(\prod_{k=1}^{j-1}(i-k)\right)\right),$$

and then

$$N(n) = N(n-1) + \sum_{j=2}^{n} \prod_{k=1}^{j-1} (n-k),$$

to which an induction proof can be applied.

6.14 The simple cycles are (a, b, c, d, e, a), (a, c, d, e, a), (a, d, e, a), (b, c, d, e, b), (d, e, d). The K-formula as written defines cycle (a, b, c, d, e, a). Consider cycle (a, d, e, a). Switching of the parenthesized substrings in

$$***a(*b*c*d***eabd)(c)(d)$$

produces

$$***a(d)*b*c(*d***eabd)c$$

and interchange of the parenthesized substrings in this formula gives

$$***a*d***eabd*b*cdc$$

6.16 With arc $\langle 8, 1\rangle$ added to the digraph, A.6.8 finds that (1, 3, 2, 5, 4, 7, 6, 8, 1) is the longest simple cycle through nodes 1 and 8. Hence the longest simple path is (1, 3, 2, 5, 4, 7, 6, 8).

6.19 In the adjacency matrix of the network show that the elements of exactly one column and of exactly one row are all zero.

6.21 Of the five critical paths in the network of Figure 6.11 there are just two left: (1, 3, 5, 7, 8, 9) and (1, 4, 3, 5, 7, 8, 9).

6.23 Smith can be away for 11 days.

6.26 Let D be the matrix of path lengths and let $P = (1, \ldots, n)$ be the longest path found by the program of Exercise 6.9. There is more than one longest path from 1 to n if there exists some $k(k = 2, \ldots, n-1)$ such that $d_{1k} + d_{kn} = d_{1n}$ and k does not lie on P. Moreover, the existence of such a k is a necessary condition for the existence of more than one longest path. This requires some explanation. Referring to the network of Exercise 6.21, what if the program had found (1, 4, 3, 5, 7, 8, 9) for the longest path? In general, if subpath $(i, \ldots, j)$ of $P = (1, \ldots, i, \ldots, j, \ldots, n)$ is equal in length to arc $\langle i, j\rangle$, then the existence of longest path $(1, \ldots, i, j, \ldots, n)$ is not detected. However, if the program of Exercise 6.9 is written sensibly, then this situation does not arise; i.e., a properly written program selects $(1, \ldots, i, j, \ldots, n)$ in preference to $(1, \ldots, i, \ldots, j, \ldots, n)$ for the longest path.

6.28 A.6.9 merely tells whether the network contains cycles. A.3.3 identifies all arcs that belong to cycles; i.e., it enables the designer of the network to locate the cause of error with much greater efficiency.

Chapter 7

7.3 Presume that A.6.1 is used to find the path matrix in the program of Exercise 7.2 and take a hint from Exercise 6.3. A further note: Elements in the lowest $k+1$ rows of the adjacency matrix are all zero; hence iteration on i in Steps 2 and 3 may be limited to $i=1, 2, \ldots, n-k-1$ alone.

7.5 Consider the digraph $\langle\{a, b, c, d, e, f, g\}, \{\langle a, b\rangle, \langle a, d\rangle, \langle b, c\rangle, \langle c, d\rangle, \langle c, f\rangle, \langle d, e\rangle, \langle e, b\rangle, \langle e, g\rangle, \langle f, g\rangle\}\rangle$, in which (a, d, e, b, c, f, g) is a forward path by our new definition. Therefore, according to the new definition $\{\langle e, b\rangle\}$ is not a return path. Now take $P=(a, b, c, d, e, g)$ and $Q=(a, b, c, d, e, b, c, d, e, g)$. We have that P is a simple path, $P \subset Q$, and Q does not contain a path that is parallel to a subpath of P. Hence $(Q-P)=\{\langle e, b\rangle\}$ is a return path in terms of the definition of Section 7c.

7.6 No to both questions. Add arc $\langle b, d\rangle$ to the digraph of the solution to Exercise 7.5. The number of cycles increases from 1 to 2, but the number of return paths and final arcs remains 1.

7.8 Consider a digraph $D=\langle A, R\rangle$. For every arc $\langle a_i, a_j\rangle \in R$ generate sets

$$B_i=\{b_i \mid a_i \text{ is reachable from } b_i\},$$
$$B_j=\{b_j \mid b_j \text{ is reachable from } a_j\}.$$

Arc $\langle a_i, a_j\rangle$ is a separator if and only if (1) $\{B_i, B_j\}$ is a partition of A, and (2) $\langle a_i, a_j\rangle$ is the only arc in R such that $a^i \in B_i$, $a_j \in B_j$. Sets B_i and B_j can be generated very rapidly from column i and row j of the path matrix of D; note that every node a_i is reachable from itself, even though p_{ii} in the path matrix may be zero. Condition (2) is tested in the adjacency matrix. As an added exercise you should prove that the procedure works.

Chapter 8

8.1 9.7 and 1.0, respectively.

8.3 Arcs $\langle 1, 2\rangle$ and $\langle 3, 6\rangle$ (c_{12} increased to 2, or c_{36} increased to 5).

8.5 For every capacitated node a add a dummy node a'. Also add dummy arc $\langle a, a'\rangle$, and assign the capacity of node a to this arc. Arcs that terminated at a still terminate there, but arcs that originated from a now originate from a'.

8.7 The union of the five sets {CS1, CS13, CS248}, {CS1, CS31}, {CS1, CS248, CS31}, {CS1, CS13}, and {CS1, CS13, CS31} has only four elements. Therefore, by Th.8.3, it is impossible to set up a one-to-one correspondence between faculty members and courses.

Chapter 9

9.1 $loc(5, 6, 7) = loc(1, 1, 1) + 456,$
$loc(10, 9, 8) = loc(1, 1, 1) + 987,$
$loc(1, 5, 9) = loc(1, 1, 1) + 48.$

9.3 Assume that bit positions are numbered 0, 1, 2, . . . , 35. Then we write

```
      SUBROUTINE PLACE (LOGIC,LDIM,I,J,K,LWD,LBIT)
      DIMENSION LOGIC(LDIM)
      LOC = LOGIC(2)*LOGIC(3)*(I-1) + LOGIC(3)*(J-1)
     1      + (K-1)
      LWD = LOC/36 + 1
      LBIT = MOD(LOC,36)
      RETURN
      END
```

If MN is stored instead of M in `LOGIC(2)`, then we can write

```
      LOC = LOGIC(2)*(I-1) + LOGIC(3)*(J-1) + (K-1)
```

Note, however, that an identical improvement in efficiency (reduction of the number of multiplications by one) can be achieved under the existing scheme by nesting:

```
      LOC = LOGIC(3)*(LOGIC(2)*(I-1) + (J-1)) + (K-1)
```

The routine does not require dimension L. Therefore `LOGIC(1)` can be made to contain $(LMN-1)$, and this quantity may then be used in a check as to whether the specified a_{ijk} is within array bounds:

```
      IF (LOC.GT.LOGIC(1)) GO TO 999
```

where the transfer is to an error exit.

9.6 The representation of X^3 is

$$\mathrm{QQ} = \begin{bmatrix} 1 & 1 & 1 & 2 & 2 & 2 & 2 & 3 & 3 & 3 & 5 & 6 & 6 & 6 & 6 & 9 \\ 2 & 3 & 5 & 1 & 2 & 6 & 7 & 2 & 3 & 5 & 6 & 2 & 3 & 5 & 6 & 9 \\ 1 & 1 & 1 & 1 & 1 & 1 & 1 & 1 & 1 & 1 & 1 & 2 & 1 & 1 & 1 & 1 \\ 5 & 9 & 10 & 0 & 8 & 11 & 0 & 12 & 13 & 14 & 15 & 0 & 0 & 0 & 0 & 0 \end{bmatrix}$$

$$\mathrm{NC} = [4 \quad 1 \quad 2 \quad 0 \quad 3 \quad 6 \quad 7 \quad 0 \quad 16]$$

$$\mathrm{NR} = [1 \quad 4 \quad 8 \quad 11 \quad 11 \quad 12 \quad 16 \quad 16 \quad 16 \quad 17]$$

Chapter 10

10.1 The contents of the stack are

1 2 3 4 5 6 10 9 8 7 6

(6 is the topmost datum).

10.2 Use common sense: Treat the stack storage area as a normal Fortran array.

10.3 Let N be the array used for the FIFO store, and let its size be 100. Two pointers are needed. Let us call them `IH` and `IL`. Pointer `IH` indicates the element that was last stored; `IL` points to the element that is next to be extracted. In `IPOP` one should first evaluate

`(IL.GT.IH)`. If this expression is true, then there is underflow; otherwise `N(IL)` is extracted, and `IL` incremented by 1. In PUSH one should first test `IH`. If `(IH.EQ.100)` is true, then a further test has to be made. If `(IL.EQ.1)` is also true, then there is overflow; otherwise all elements in the FIFO store are shifted down, and `IL` and `IH` adjusted, perhaps as follows:

```
      IH = 101 - IL
      DO 20  I = 1,IH
      N(I) = N(IL)
20    IL = IL + 1
      IL = 1
```

Then `IH` is incremented by 1, and the datum placed in `N(IH)`.

10.7 Each K-formula must be provided with an end-of-formula marker to avoid ambiguity; otherwise a string $**abcd$, for example, could be interpreted as a single (invalid) formula $**abcd$ or as a pair of K-formulas $\{**abc, d\}$. Let \$ be the end-of-formula marker, and use & to indicate the end of the string. The set on which the algorithm is to be tested is then written

$$**a***a*bc**dcbef\$**bf**e**bad*ac\$\&$$

Let A be the set of arcs, and N the set of isolated nodes. A stack is used to process the string, which we write as $s_1s_2s_3\cdots$, in the following manner:

1. Set $i = 1$.
2. If $s_i = *$, then go to 8.
3. Add s_i to N.
4. Set $i = i + 1$.
5. If $s_i \neq \$$, go to 17.
6. Set $i = i + 1$.
7. If $s_i = \&$, stop; else go to 2.
8. Push s_i.
9. Set $i = i + 1$.
10. If $s_i = \$$, go to 17.
11. If $s_i = *$, push s_i and go to 9.
12. Set $u = s_i$.
13. If stack empty, go to 4.
14. Pop up v.
15. If $v = *$, push u and go to 9.
16. Add $\langle v, u\rangle$ to A, set $u = v$, go to 13.
17. Error exit; the string is not a representative set of K-formulas.

Steps 1–8 process the first symbol in a K-formula and the markers. A special case is made here of a K-formula that consists of just one node symbol (the K-formula of an isolated node). To gain an understanding of the rest of the algorithm take a specific K-formula and look at how the algorithm processes it. In converting the algorithm to a program one must first choose a representation for the string that is to be processed. The string can be written into a linear array, one symbol to an element. The K-operator can be represented by 0, node symbols by positive integers, and markers by negative integers.

10.9

```
      SUBROUTINE INITIO
      COMMON LIST(250,2)
      DO  5  K = 6,249
```

```
 5   LIST(  K,2) = K + 1
     LIST(250,2) = 0
     DO  6  K = 1,5
     DO  6  J = 1,2
 6   LIST(K,J) = 0
     LIST(1,1) = 6
     RETURN
     END
```

10.12 We assume that `LIST(1,1)` is the name of the list of available space. Let both elements in the Kth row of `LIST` be the name of FIFO store K (K = 2, 3, 4, 5 in the scheme of Exercise 10.9). Make `LIST(K,1)` point to the last element and `LIST(K,2)` to the first element. Under this arrangement the routines may be written as follows:

```
     SUBROUTINE PUSH (K,IN)
     COMMON LIST(250,2)
     IF (LIST(1,1).EQ.0) GO TO 999
     IA = LIST(1,1)
     IB = LIST(K,1)
     IF (IB.EQ.0) IB = K
     LIST( 1,1) = LIST(IA,2)
     LIST(IA,1) = IN
     LIST(IA,2) = 0
     LIST( K,1) = IA
     LIST(IB,2) = IA
     RETURN
999  WRITE(6,100)
     CALL EXIT
100  FORMAT(1H1,28HLAVS EXHAUSTED - EXIT CALLED)
     END

     FUNCTION IPOP (K)
     COMMON LIST(250,2)
     IB = LIST(K,2)
     IF (IB.EQ.0) GO TO 999
     IPOP = LIST(IB,1)
     LIST(K,2) = LIST(IB,2)
     IF (LIST(K,2).EQ.0) LIST(K,1) = 0
     LIST(IB, 2) = LIST(1,1)
     LIST( 1,1) = IB
     RETURN
999  WRITE(6,100) K
     CALL EXIT
100  FORMAT(1H1,18HUNDERFLOW IN STACK,I2)
     END
```

10.15 The algorithm is as given in Figure 10.2. The K of Figure 10.2 is interpreted as the address of a list element. Then $D(K)$ is the datum field of this element, and $P(K, 1)$ and $P(K, 2)$ are left and right pointers, respectively. There is just one change: Instead of setting K to 1, the initial value of K should be picked up from the name of the list structure.

10.18 To get our representations straight take it that the parenthesized strings corresponding to Figures 10.20 and 10.21 are *a*(*b*(*d*)*c*(*efg*)) and *a*(*def*)*bc*, respectively. First consider the representation in which the tree has all nodes labeled. Then both the threaded list (as in Figure 10.16) and the symmetric list (as in Figure 10.20) have $2n - 1$ elements, where n is the number of alphabetic symbols in the string. Next consider the representation in which only terminal nodes of the tree are labeled. Then the symmetric list (as in Figure 10.21) contains $n + m$ elements, where n is the number of alphabetic symbols and m is the number of opening parentheses in the string.

Chapter 11

11.1 The structure is a file containing 50 records. The properties are V, Y, Z(1,1) Z(1,2), Z(2,1), Z(2,2), and X. A record contains seven measures. Note, however, that there are no hard and fast rules. For example, an interpretation under which Z is a single property having a 2×2 matrix for its measure is just as consistent with D.11.1.

11.4 Descending order would do as well. The reason for order is that it takes less time to find the set union or set intersection of reference vectors when their elements are ordered. (For example, if we were searching for documents described by index terms, "data structures" *and* "operating systems," then we would require the documents defined by the intersection of the reference vectors associated with these two terms.)

11.6 What follows is merely an analysis rather than a full solution. The procedure for deleting a document is determined by the design of the file. The index terms associated with documents were either retained in the document register, or the entries in the register were stripped of index terms when the index was being constructed. In the former case one searches for the appropriate index terms in the sort tree and deletes the document number in the reference vector of each such index term; in the latter case *all* reference vectors must be examined. If the number of the document that is being deleted is the only document number in a reference vector, then we are left with an empty vector, and the entire node should be removed from the sort tree. However, there is no great harm in leaving the index entry in the tree; a zero in the field that would point to the reference vector tells that no document has this particular index term associated with it. We can go even further and confine the deletion procedure to the document register alone. Provide each entry in the register with a marker. Normally the marker shows that the document is *active*. Deletion is merely the process of changing the marker to the *inactive* setting. The retrieval system responds to a retrieval request by generating a set of document numbers, and printing document register entries corresponding to these numbers, but output of inactive entries is suppressed.

11.8 There are two solutions: (a) Use symmetric (two-way) lists, or (b) have the last element of a simple (one-way) list pointing back to the first element. Under Scheme (b) we have what are called *circular* lists. The sequence of pointers is followed right round the list; the predecessor is the element that has an address equal to the hash address in its pointer field. Under Scheme (a) the predecessor is located at once, but more pointers have to be changed whenever a new element is added to a symmetric list. Hence, in view of the fact that the lists contain very few elements (for a load factor as high as 0.95 the average number of probes is still only 1.48), Scheme (b) has the better overall efficiency. Moreover, Scheme (a) requires more storage space (but not always—consider records that are one computer word long in an IBM 7090 implementation; one must use a second word for pointers and markers, and here it makes no difference whether this word contains one or two pointers).

11.10 Have a scatter table HASH and a separate record table FILE. Make FILE a list store of the form described in the final paragraph of Section 10i. Initially it is a single block of available space. As records are entered, blocks of the size required for their storage are chopped off the block of available space and become multiword list elements. The first word in each of these list elements indicates the size of the block and contains a pointer field; the record occupies the rest of the block. All records associated with hash address N are linked in a list, and HASH(N) is the name of the list. A record is processed as follows: (a) Store the record in FILE; since there is no erasing, records are stored sequentially in order of input. (b) Compute the hash address. Let it be N. (c) If HASH(N) = 0, create a list by inserting in HASH(N) a pointer to the block that holds the record. If HASH(N) ≠ 0, then a list already exists. Look for the last block in this list and insert the pointer in its pointer field; i.e., add the block that is being processed to the list.

11.14 The procedure consists of tracing through the two lists, deleting elements that are common to both. Assume that neither list is empty, and that locations AA and BB contain addresses of first elements of lists A and B. Denote the forward pointer in element N of list X by $X_N(F)$, and the datum stored in this element by $X_N(D)$; e.g., $A_{AA}(F)$ denotes the forward pointer in the first element of list A.

1. Set $J = AA$ and $K = BB$.
2. If $A_J(D) < B_K(D)$, go to 5; if $A_J(D) > B_K(D)$, go to 6.
3. Set $JTEMP = A_J(F)$ and $KTEMP = B_K(F)$; delete elements A_J and B_K; set $J = JTEMP$ and $K = KTEMP$.
4. If $J = AA$, stop; else go to 7.
5. Set $J = A_J(F)$. If $J = AA$, stop; else go to 2.
6. Set $K = B_K(F)$.
7. If $K = BB$, stop; else go to 2.

(Your solution should contain an explicit procedure for deletion of an element in a symmetric list.)

11.17 Use a variant of A.11.1. Let array **M** be the directory, and change the first five executable statements of SEARCH to

```
      LO = 1
      HI = N
  5   IF (LO.NE.HI) GO TO 51
      SEARCH = LO
      IF (KEY.LT.M(LO,1)) SEARCH = LO - 1
      RETURN
 51   MID = (LO + HI)/2
```

The value of SEARCH is the number of the row in the directory that contains the appropriate track number. It is possible that the file does not contain a record corresponding to the given key. In particular, SEARCH returns the value 0 if the given key is smaller than the key of the first record on the first page of the file.

Bibliography

Parentheses at the end of an entry enclose the following items of information: (a) the number of references cited in the work (usually omitted in the case of a book), (b) the number of its review in *Computing Reviews*, and (c) the chapter or chapters to which it is relevant. The fact that some very important and not too recent papers do not carry a review number is a warning that a thorough literature search should not be confined to a search through *Computing Reviews*.

The following abbreviations, some of which may be non-standard, are used in titles of periodicals:

AFIPS	American Federation of Information Processing Societies
BIT	Nordisk Tidskrift for Informationsbehandling
CACM	Communications of the Association for Computing Machinery
FJCC	Fall Joint Computer Conference
ICC	International Computation Centre
IEEE	Institute of Electrical and Electronics Engineers
JACM	Journal of the Association for Computing Machinery
JCC	Joint Computer Conference
SIAM	Society for Industrial and Applied Mathematics
SJCC	Spring Joint Computer Conference
RIRO	Revue Française d'Informatique et de Recherche Opérationelle

Ab64 Abramowitz, M., and Stegun, I. A., eds., *Handbook of Mathematical Functions with Formulas, Graphs, and Mathematical Tables.* U.S. Govt. Printing Office, Washington, D.C., 1964. xiv + 1046 pp. (CR 10100—errata are published in *Math. Comp.*; Chapters 1, 2.)

Ab66 Abrahams, P. W., *et al.*, The LISP2 programming language and system. *Proc. AFIPS* **29** (*FJCC, 1966*), 661–676. (8 refs.; CR 11934; Chapter 12.)

Ab68 Abrahams, P. W., Symbol manipulation languages. *Advances in Comput.* **9,** 51–111 (1968). (41 refs.; CR 17684; Chapter 12.)

Al69 Allen, F. E., Program optimization. *Ann. Rev. Automatic Programming* **5,** 239–307 (1969). (15 refs.; Chapter 7.)

An65 Anderson, H. E., Automated plotting of flowcharts on a small computer. *CACM* **8,** 38–39 (1965). (3 refs.; Chapter 7.)

Ar69a Arona, S. R., and Dent, W. T., Randomized binary search technique. *CACM* **12,** 77–80 (1969). (2 refs.; CR 17257; Chapter 5.)

Ar69b Arden, B. W., Galler, B. A., and Graham, R. M., The MAD definition facility. *CACM* **12,** 432–439 (1969). (9 refs.; CR 18041; Chapter 12.)

Ax67 Axsom, L. E., An expression recognition routine in LISP 1.5. See Ro67a, pp. 481–489. (4 refs.; CR 16236; Chapter 12.)

Ay63 Ayers, J. A., Recursive programming in Fortran II. *CACM* **6,** 667–668 (1963). (No refs.; Chapter 10.)

Ba59a Backus, J. W., The syntax and semantics of the proposed international algebraic language of the Zurich ACM-GAMM conference. *Proc. Internat. Conf. Inf. Proc., Paris, 1959*, pp. 125–132. (No refs.; CR 3158; Chapter 4.)

Ba59b Samelson, K., and Bauer, F. L., Sequential formula translation. *CACM* **3,** 76–83 (1960) (reprinted in Ro67a, pp. 206–220). Earlier publ. in German: Bauer, F. L., and Samelson, K., *Elektron. Rechenanlagen* **1,** 176–182 (1959). (14 refs.; CR 219; Chapter 10.)

Ba60 Barrett, J. A., and Grems, M., Abbreviating words systematically. *CACM* **3,** 323–324 (1960). (1 ref.; CR 221; Chapter 11.)

Ba62a Baecker, H. D., Implementing a stack. *CACM* **5,** 505–507 (1962). (2 refs.; Chapter 10.)

Ba62b Banerji, R. B., The description list of concepts. *CACM* **5,** 426–432 (1962). (6 refs.; CR 3595; Chapter 11.)

Ba63 Barnett, M. P., Indexing and the Λ-notation. *CACM* **6,** 740–745 (1963). (4 refs.; CR 5668; Chapter 9.)

Ba64a Bach, E., *An Introduction to Transformational Grammars.* Holt, New York, 1964. xii + 205 pp. (Selected bibliography of 86 items; Chapter 4.)

Ba64b Bailey, M. J., Barnett, M. P., and Burleson, P. B., Symbol manipulation in Fortran—SASP I routines. *CACM* **7,** 339–346 (1964). (6 refs.; CR 6694; Chapter 12.)

Ba65 Barnes, J. G., A KDF9 Algol list-processing scheme. *Comput. J.* **8,** 113–119 (1965–1966). (7 refs.; CR 8994; Chapter 12.)

Ba67a Basu, S. K., On computation in programming languages. *ICC Bull.* **6,** 1–26 (1967) (note that the paper is incomplete as printed—Figure 1 is missing). (6 refs.; Chapter 3.)

Ba67b Battersby, A., *Network Analysis for Planning and Scheduling*, 2nd ed. Macmillan, London, 1967. x + 414 pp. (Chapter 6.)

Ba68a Barron, D. W., *Recursive Techniques in Programming.* Macdonald, London, 1968, viii + 64 pp. (CR 17273; Chapter 10.)

Ba68b Batcher, K. E., Sorting networks and their applications. *Proc. AFIPS* **32** (*SJCC, 1968*), 307–314. (3 refs.; Chapter 11.)

Ba69a Baer, S. L., Matrice de connexion minimale d'une matrice de précédence donnée. *RIRO* **3,** No. 16, 65–73 (1969). (6 refs.; Chapter 6.)

Ba69b Bayes, A. J., A network representation of serial code. *Austral. Comput. J.* **1,** 246–250 (1967–1969). (5 refs.; CR 19924; Chapter 7.)

Be58 Berge, C., *Théorie des Graphes et ses Applications.* Dunod, Paris, 1958 (English transl.: *The Theory of Graphs and its Applications,* Methuen, London, 1962. x + 247 pp.). (Chapter 3.)

Be62 Berge, C., and Ghouila-Houri, A., *Programmes, Jeux et Réseaux de Transport.* Dunod, Paris, 1962 (English transl.: *Programming, Games and Transportation Networks.* Methuen, London, 1965. x + 260 pp.). (CR 8348; Chapters 3, 6, 8.)

Be64a Beckenbach, E. F., Network flow problems. In *Applied Combinatorial Mathematics* (E. F. Beckenbach, ed.), Chapter 12, pp. 348–365. Wiley, New York, 1964. (9 refs.; Chapter 8.)

Be64b Berkeley, E. C., and Bobrow, D. G., eds., *The Programming Language LISP: Its Operation and Applications.* M.I.T. Press, Cambridge, Massachusetts, 1966 (originally published in 1964 by Information Internat.). x + 382 pp. (CR 10865; Chapter 12.)

Be65 Berztiss, A. T., A note on storage of strings. *CACM* **8,** 512–513 (1965). (3 refs.; CR 9713; Chapter 10.)

Be66 Bellman, R., Kalaba, R., and Za'deh, L., Abstraction and pattern classification. *J. Math. Anal. Appl.* **13,** 1–7 (1966). (2 refs.; Chapter 1.)

Be68a Berztiss, A. T., A note on segmentation of computer programs. *Information and Control* **12,** 21–22 (1968). (4 refs.; Chapter 3.)

Be68b Bellmore, M., and Nemhauser, G. L., The traveling salesman problem: A survey. *Operations Res.* **16,** 538–558 (1968). (37 refs.; Chapter 6.)

Be69a Berztiss, A. T., and Watkins, R. P., Directed graphs and automatic flowcharting. *Proc. 4th Austral. Comput. Conf., Adelaide, 1969,* pp. 495–499. (20 refs.; Chapters 3, 7.)

Be69b Berns, G. M., Description of FORMAT, a text-processing program. *CACM* **12,** 141–146 (1969). (13 refs.; CR 16833; Chapter 12.)

Bi41 Birkhoff, G., and MacLane, S., *A Survey of Modern Algebra.* Macmillan, New York, 1941 (3rd ed., 1965, x + 437 pp.). (Chapter 4.)

Bi62 Bigelow, C. G., Bibliography on project planning and control by network analysis: 1959–1961. *Operations Res.* **10,** 728–731 (1962). (68 refs.; Chapter 6.)

Bl66 Blair, C. R., Certification of Algorithm 271. *CACM* **9,** 354 (1966). (No refs.; Chapter 11.)

Bo60 Booth, A. D., and Colin, A. J. T., On the efficiency of a new method of dictionary construction. *Information and Control* **3,** 327–334 (1960). (2 refs.; Chapter 5.)

Bo63a Bobrow, D. G., Syntactic analysis of English by computer—A survey. *Proc. AFIPS* **24** (*FJCC, 1963*), 365–387. (54 refs.; CR 8838; Chapter 5.)

Bo63b Boyell, R. L., and Ruston, H., Hybrid techniques for real-time radar simulation. *Proc. AFIPS* **24** (*FJCC, 1963*), 445–458. (1 ref.; Chapter 5.)

Bo63c Bourne, C. P., *Methods of Information Handling.* Wiley, New York, 1963. xiv + 241 pp. (Good sets of references throughout; Chapter 11.)

Bo64a Böhm, C., and Santolini, A., A quasi-decision algorithm for the P-equivalence of two matrices. *ICC Bull.* **3,** 57–69 (1964). (1 ref.; CR 6986; Chapter 3.)

Bo64b Bobrow, D. G., and Raphael, B., A comparison of list-processing computer languages. *CACM* **7,** 231–240 (1964) (reprinted with 35 additional references in Ro67a, pp. 490–511). (13 refs.; CR 9728; Chapter 12.)

Bo64c Bobrow, D. G., and Weizenbaum, J., List processing and extension of language facility by embedding. *IEEE Trans. Electronic Computers* **EC-13,** 395–400 (1964). (6 refs.; Chapter 12.)

Bo67a Booth, T. L., *Sequential Machines and Automata Theory.* Wiley, New York, 1967. xvi + 592 pp. (CR 15238; Chapter 4.)

Bo67b Bobrow, D. G., Fraser, J. B., and Quillian, M. R., Automated language processing. *Ann. Rev. Inf. Sci. Tech.* **2,** 161–186 (1967). (121 refs.; Chapter 5.)

Bo67c Boehm, B. W., Tabular representation of multivariate functions—with applications in topological modeling. *Proc. 22nd ACM Nat. Conf., 1967*, pp. 403–415. (26 refs.; CR 12795; Chapter 5.)

Bo67d Boothroyd, J., Algorithm 22: Shortest path between start node and end node of a network. *Comput. J.* **10,** 306–307 (1967–1968). (2 refs.; Chapter 6.)

Bo67e Bobrow, D. G., and Murphy, D. L., Structure of a LISP system using two-level storage. *CACM* **10,** 155–159 (1967). (6 refs.; CR 12701; Chapter 12.)

Bo68 Bobrow, D. G., and Murphy, D. L., A note on efficiency of a LISP computation in a paged machine. *CACM* **11,** 558, 560 (1968). (2 refs.; CR 15546; Chapter 12.)

Br59 de la Briandais, R., File searching using variable length keys. *Proc. Western JCC, 1959*, pp. 295–298. (2 refs.; Chapter 5.)

Br69 Brown, P. J., A survey of macro processors. *Ann. Rev. Automatic Programming* **6,** 37–88 (1969). (39 refs.; Chapter 12.)

Bu58 Burge, W. H., Sorting, trees, and measures of order. *Information and Control* **1,** 181–197 (1958). (6 refs.; Chapter 5.)

Bu65 Busacker, R. G., and Saaty, T. L., *Finite Graphs and Networks: An Introduction with Applications.* McGraw-Hill, New York, 1965. xiv + 294 pp. (Chapters 3, 6, 8.)

Ca63a Caporaso, S., A composition method for normal Markov algorithms. *ICC Bull.* **2,** 195–204 (1963). (1 ref.; CR 6075—includes correction; Chapter 4.)

Ca63b Carlson, C. B., The mechanization of a push-down stack. *Proc. AFIPS* **24** (*FJCC, 1963*), 243–250. (5 refs.; Chapter 10.)

Ca67 Carlson, A. R., Concept frequency in political text: An application of a total indexing method of automated context analysis. *Behavioral Sci.* **12,** 68–72 (1967). (9 refs.; CR 11897; Chapter 8.)

Ca69 Cannon, J. J., Computers in group theory: A survey. *CACM* **12,** 3–12 (1969). (78 refs.; CR 17617; Chapter 4.)

Ch59 Chomsky, N., On certain formal properties of grammars. *Information and Control* **2,** 137–167 (1959) (reprinted in *Readings in Mathematical Psychology* (R. D. Luce, R. R. Bush, and E. Galanter, eds.), Vol. 2, pp. 125–155. Wiley, New York, 1965). (10 refs.; Chapter 4.)

Ch63 Chomsky, N., Formal properties of grammars. In *Handbook of Mathematical Psychology* (R. D. Luce, R. R. Bush, and E. Galanter, eds.), Vol. 2, pp. 323–418. Wiley, New York, 1963. (78 refs.; Chapter 4.)

Ch66 Chen, Y. C., and Wing, O., Some properties of cycle-free directed graphs and the identification of the longest path. *J. Franklin Inst.* **281,** 293–301 (1966). (6 refs.; CR 10574; Chapter 6.)

Ch67 Chapin, N., Parsing of decision tables. *CACM* **10,** 507–510, 512 (1967). (4 refs.; CR 13316; Chapter 5.)

Cl62 Clippinger, R. F., Information algebra. *Comput. J.* **5,** 180–183 (1962–1963). (No refs.; CR 4164; Chapter 2.)

Cl64 Clampett, H. A., Randomized binary searching with tree structures. *CACM* **7,** 163–165 (1964). (1 ref. in text; CR 5957; Chapter 5.)

Cl66a Clark, E. R., On the automatic simplification of source-language programs. *Proc. 21st ACM Nat. Conf., 1966*, pp. 313–319 [or *CACM* **10**, 160–165 (1967)]. (3 refs.; CR 11212; Chapter 7.)

Cl66b Climenson, W. D., File organization and search techniques. *Ann. Rev. Inf. Sci. Tech.* **1**, 107–135 (1966). (95 refs.; CR 11900; Chapter 11.)

Co60 Collins, G. E., A method for overlapping and erasure of lists. *CACM* **3**, 655–657 (1960). (5 refs.; Chapter 10.)

Co64 Comfort, W. T., Multiword list items. *CACM* **7**, 357–362 (1964). (11 refs.; CR 6686; Chapter 10.)

Co65 Cohen, J., and Nguyen-Huu-Dung, Définition de procédures LISP en Algol. *Rev. Française Traitement Information Chiffres* **8**, 271–293 (1965). (8 refs.; CR 10851; Chapter 12.)

Co66 Collins, G. E., PM, a system for polynomial manipulation. *CACM* **9**, 578–589 (1966). (14 refs.; CR 10884; Chapters 10, 12.)

Co67a Cohen, J., and Trilling, L., Remarks on "garbage collection" using a two-level storage. *BIT* **7**, 22–30 (1967). (8 refs.; CR 12758; Chapter 10.)

Co67b Cohen, J., The use of fast and slow memories in list-processing languages. *CACM* **10**, 82–86 (1967). (4 refs.; Chapter 12.)

Cu63 Curry, H. B., *Foundations of Mathematical Logic*. McGraw-Hill, New York, 1963. xii + 408 pp. (Chapter 4.)

Da58 Davis, M., *Computability and Unsolvability*. McGraw-Hill, New York, 1958. xxvi + 210 pp. (Chapter 4.)

Da62 Darnaut, P., and Sandier, G., Utilisation de FLPL dans la résolution d'un problème d'ordonnancement. *Symbolic Languages in Data Processing*, pp. 731–739. Gordon and Breach, New York, 1962. (2 refs.; Chapter 12.)

Da65 Darlington, J. L., Machine methods for proving logical arguments expressed in English. *Mach. Transl.* **8**, 41–67 (1964–1965). (11 refs.; CR 11832; Chapter 12.)

De64 Demoucron, G., Malgrange, Y., and Pertuiset, R., Graphes planaires: Reconnaissance et construction de représentations planaires topologiques. *Rev. Française de Rech. Opérationelle* **8**, No. 30, 33–47 (1964). (2 refs.; Chapter 3.)

De66 Deuel, P., On a storage mapping function for data structures. *CACM* **9**, 344–347 (1966). (3 refs.; Chapter 10.)

De67 Desaulets, E. J., and Smith, D. K., An introduction to the string processing language Snobol. See Ro67a, pp. 419–454. (9 refs.; CR 16235; Chapter 12.)

Di60 Dijkstra, E. W., Recursive programming. *Numer. Math.* **2**, 312–318 (1960) (reprinted in Ro67a, pp. 221–227). (4 refs.; CR 17275; Chapter 10.)

Do66 Dodd, G. G., APL—A language for associative data handling in PL/I. *Proc. AFIPS* **29** (*FJCC, 1966*), 677–684. (13 refs.; CR 12753; Chapter 12.)

Du65 Dunphy, D. C., Stone, P. J., and Smith, M. S., The General Inquirer: Further developments in a computer system for content analysis of verbal data in the social sciences. *Behavioral Sci.* **10**, 468–480 (1965). (18 refs.; CR 11180; Chapter 12.)

El66 Elmaghraby, S. E., *The Design of Production Systems*. Reinhold, New York, 1966. xviii + 491 pp. (Chapter 6.)

Ev63 Evey, R. J., Application of pushdown-store machines. *Proc. AFIPS* **24** (*FJCC, 1963*), 215–227. (19 refs.; Chapter 10.)

Fa63 Fang, J., *Abstract Algebra*. Schaum, New York, 1963. xii + 339 pp. (Chapter 4.)

Fa64 Farber, D. J., Griswold, R. E., and Polonsky, I. P., Snobol, a string manipulation language. *JACM* **11**, 21–30 (1964). (5 refs.; CR 6940—points out errors; Chapter 12.)

Fa66 Farber, D. J., Griswold, R. E., and Polonsky, I. P., The Snobol 3 programming language. *Bell System Tech. J.* **45,** 895–944 (1966). (8 refs.; CR 11942; Chapter 12.)

Fe60 Ferguson, D. E., Fibonaccian searching. *CACM* **3,** 648 (1960). (No refs.; Chapter 11.)

Fe68 Feldman, J., and Gries, D., Translator writing systems. *CACM* **11,** 77–113 (1968). (226 refs.; CR 14729; Chapter 5.)

Fe69 Fenichel, R. R., and Yochelson, J. C., A LISP garbage-collector for virtual-memory computer systems. *CACM* **12,** 611–612 (1969). (7 refs.; Chapter 12.)

Fi66 Fischer, D. L., Data, documentation and decision tables. *CACM* **9,** 26–31 (1966). (1 ref.; CR 9615; Chapter 5.)

Fi68 Fisher, A. C., Liebman, J. S., and Nemhauser, G. L., Computer construction of project networks. *CACM* **11,** 493–497 (1968). (7 refs.; CR 18221; Chapter 6.)

Fl62 Floyd, R. W., Algorithm 97: Shortest path. *CACM* **5,** 345 (1962). (1 ref.; Chapter 6.)

Fl64 Floyd, R. W., Algorithm 245: Treesort 3. *CACM* **7,** 701 (1964). (2 refs.; Chapter 5.)

Fl67 Floyd, R. W., Nondeterministic algorithms. *JACM* **14,** 636–644 (1967). (4 refs.; CR 14587; Chapters 6, 10.)

Fl69 Flores, I., *Computer Sorting.* Prentice-Hall, Engelwood Cliffs, New Jersey, 1969. xii + 237 pp. (CR 16053; Chapter 11.)

Fo62 Ford, L. R., and Fulkerson, D. R., *Flow in networks.* Princeton Univ. Press, Princeton, New Jersey, 1962. xiv + 194 pp. (CR 4845; Chapter 8.)

Fo67 Foster, J. M., *List Processing.* Macdonald, London, 1967. vi + 54 pp. (33 refs.; CR 13446; Chapter 10.)

Fr58 Fraenkel, A. A., and Bar-Hillel, Y., *Foundations of Set Theory.* North-Holland Publ., Amsterdam, 1958. x + 415 pp. (Chapter 1.)

Fr60 Fredkin, E., Trie memory. *CACM* **3,** 490–499 (1960). (5 refs.; CR 475; Chapter 5.)

Fr66 Fraenkel, A. A., *Abstract Set Theory*, 3rd ed. North-Holland Publ., Amsterdam, 1966. viii + 295 pp. (Chapter 1.)

Ga64 Garwick, J. V., Data storage in compilers. *BIT* **4,** 137–140 (1964). (No refs.; CR 7258; Chapter 9.)

Ga66 Gatto, J. J., Application of computer-based retrieval concepts to a marketing information dissemination system. *Proc. AFIPS* **28** (*SJCC, 1966*), 285–295. (3 refs.; CR 10763; Chapter 11.)

Ge60 Gelernter, H., Hansen, J. R., and Gerberich, C. L., A Fortran-compiled list-processing language. *JACM* **7,** 87–101 (1960). (6 refs.; CR 142; Chapter 12.)

Gi60 Gill, A., Analysis of nets by numerical methods. *JACM* **7,** 251–254 (1960). (1 ref.; Chapter 6.)

Gi66 Ginsburg, S., *The Mathematical Theory of Context-Free Languages.* McGraw-Hill, New York, 1966. xiv + 232 pp. (CR 12079; Chapter 4.)

Gi67 Ginsburg, S., Greibach, S. A., and Harrison, M. A., Stack automata and compiling. *JACM* **14,** 172–201 (1967). (19 refs.; Chapter 10.)

Gi69 Gibbs, N. E., A cycle generating algorithm for finite undirected linear graphs. *JACM* **16,** 564–568 (1969). (6 refs.; CR19833; Chapter 3.)

Gl64 Glushkov, V. M., *Introduction to Cybernetics.* Academic Press, New York, 1966. x + 322 pp. (Original Russian edition: *Vvedeniye v Kibernetiku.* Acad. Sc. Ukr. S.S.R., Kiev, 1964.) (CR 11012; Chapter 4.)

Go63 Gorn, S., Processors for infinite codes of the Shannon-Fano type. In *Mathematical Theory of Automata* (J. Fox, ed.), pp. 223–240. Polytech. Inst. of Brooklyn, Brooklyn, New York, 1963. (7 refs.; Chapter 3.)

Go65 Golomb, S. W., and Baumert, L. D., Backtrack programming. *JACM* **12,** 516–524 (1965). (13 refs.; CR 9442; Chapter 10.)

Go67a Goguen, J. A., L-fuzzy sets. *J. Math. Anal. Appl.* **18,** 145–174 (1967). (7 refs.; Chapter 1.)

Go67b Gotlieb, C. C., and Corneil, D. G., Algorithms for finding a fundamental set of cycles for an undirected linear graph. *CACM* **10,** 780–783 (1967). (6 refs.; CR 14355; Chapter 3.)

Go67c Goldstein, A. J., Recursive techniques in problem solving. *Proc. AFIPS* **30** (*SJCC, 1967*), 325–329. (3 refs.; CR 17027; Chapter 10.)

Go67d Goetz, M. A., Sorting and merging. In *Digital Computer User's Handbook* (M. Klerer and G. A. Korn, eds.), Pt. 1, Chapter 1.10, pp. 292–320. McGraw-Hill, New York, 1967. (57 refs.; Chapter 11.)

Gr64 Grossman, I., and Magnus, W., *Groups and Their Graphs.* Random House (Singer), New York, 1964. viii + 195 pp. (CR 8003; Chapter 4.)

Gr65 Griffiths, T. V., and Petrick, S. R., On the relative efficiencies of context-free grammar recognizers. *CACM* **8,** 289–300 (1965) [see also *CACM* **8,** 594 (1965)]. (38 refs.; CR 7999; Chapter 5.)

Gr67 Gray, J. C., Compound data structure for computer aided design: A survey. *Proc. 22nd ACM Nat. Conf., 1967,* pp. 355–365. (17 refs. and glossary; Chapter 10).

Gr68 Griswold, R. E., Poage, J. F., and Polonsky, I. P., *The Snobol 4 Programming Language.* Prentice-Hall, Englewood Cliffs, New Jersey, 1968. x + 221 pp. (CR 17858; Chapter 12.)

Gu62 Gumin, H., Digital computers, mathematical logic and principal limitations of computability. *Proc. IFIP Congr., Munich, 1962,* pp. 29–32. (10 refs.; CR 10154; Chapter 4.)

Ha59 Haibt, L. M., A program to draw multilevel flow charts. *Proc. Western JCC, 1959,* 131–137. (No refs.; CR 27; Chapter 7.)

Ha60 Halmos, P. R., *Naïve Set Theory.* Van Nostrand, Princeton, New Jersey, 1960. viii + 104 pp. (Chapter 1.)

Ha62a Hardgrave, W. W., and Nemhauser, G. L., On the relation between the traveling-salesman and the longest-path problems. *Operations Res.* **10,** 647–657 (1962). (10 refs.; Chapter 6.)

Ha62b Haley, A. C. D., The KDF9 computer system. *Proc. AFIPS* **22** (*FJCC, 1962*), 108–120 [or *Proc. 2nd Austral. Comput. Conf., Melbourne, 1963,* Paper C.1 (22 pages)]. (11 refs.; CR 5466; Chapter 10.)

Ha62c Hamblin, C. L., Translation to and from Polish notation. *Comput. J.* **5,** 210–213 (1962–1963). (6 refs.; CR 3864; Chapter 10.)

Ha65a Harrison, M. A., *Introduction to Switching and Automata Theory.* McGraw-Hill, New York, 1965. xviii + 499 pp. (Extensive bibliography; CR 9109; Chapter 2.)

Ha65b Harary, F., Norman, R. Z., and Cartwright, D., *Structural Models: An Introduction to the Theory of Directed Graphs.* Wiley, New York, 1965. x + 415 pp. (CR 8421; Chapters 2, 8.)

Ha66 Hamburger, P., On an automated method of symbolically analyzing times of computer programs. *Proc. 21st ACM Nat. Conf., 1966,* pp. 321–330. (5 refs.; CR 11052; Chapter 7.)

Ha67a Hall, M., *Combinatorial Theory.* Ginn (Blaisdell), Boston, Massachusetts, 1967. x + 310 pp. (Chapter 1.)

Ha67b Hays, D. G., *Introduction to Computational Linguistics.* Amer. Elsevier, New York, 1967. xvi + 231 pp. (CR 15139; Chapter 5.)

Ha67c Haddon, B. K., and Waite, W. M., A compaction procedure for variable length storage elements. *Comput. J.* **10,** 162–165 (1967–1968). (8 refs.; CR 13547; Chapter 10.)

Ha68 Hauck, E. A., and Dent, B. A., Burroughs' B6500/B7500 stack mechanism. *Proc. AFIPS* **32** (*SJCC, 1968*), 245–251. (3 refs.; Chapter 10.)

Ha69 Harary, F., *Graph Theory*. Addison-Wesley, Reading, Massachusetts, 1969. x + 274 pp. (CR19472; Chapter 3.)

He62 Hellerman, H., Addressing multidimensional arrays. *CACM* **5,** 205–207 (1962). (3 refs.; CR 2619; Chapter 9.)

He68 Herrmann, R. L., Selection and implementation of a ternary switching algebra. *Proc. AFIPS* **32** (*SJCC, 1968*), 283–290. (9 refs.; Chapter 2.)

Hi62a Hibbard, T. N., Some combinatorial properties of certain trees with application to searching and sorting. *JACM* **9,** 13–28 (1962). (9 refs.; Chapter 5.)

Hi62b Hill, U., Langmaack, H., Schwarz, H. R., and Seegmüller, G., Efficient handling of subscripted variables in Algol 60. *Symbolic Languages in Data Processing*, pp. 331–340. Gordon and Breach, New York, 1962. (1 ref.; Chapter 9.)

Ho62 Hoffman, S. A., Data structures that generalize rectangular arrays. *Proc. AFIPS* **21** (*SJCC, 1962*), 325–333. (3 refs.; Chapter 10.)

Ho63a Hopley, J., Algorithm 152: Nexcom. *CACM* **6,** 385 (1963). (No refs.; Chapter 1.)

Ho63b Hoffmann, T. R., Assembly line balancing with a precedence matrix. *Management Sci.* **9,** 551–562 (1962–1963). (16 refs.; Chapter 6.)

Ho66 Hohn, F. E., *Applied Boolean Algebra*, 2nd ed. Macmillan, New York, 1966. xiv + 273 pp. (CR 12078; Chapter 2.)

Ho67 Howden, W. E., A program for the construction of PERT flowcharts. *Comput. J.* **10,** 278–281 (1967–1968). (2 refs.; CR 13792; Chapter 6.)

Hu68 Hu, T. C., A decomposition algorithm for shortest paths in a network. *Operations Res.* **16,** 91–102 (1968). (9 refs.; Chapter 6.)

Im69 D'Imperio, M. E., Data structures and their representation in storage. *Annual Rev. Automatic Programming* **5,** 1–75 (1969). (34 refs.; CR 18275; Chapters 10, 12.)

In62 Ingerman, P. Z., Algorithm 141: Path matrix. *CACM* **5,** 556 (1962). (1 ref.; Chapter 6.)

In66 Ingerman, P. Z., *A Syntax-Oriented Translator*. Academic Press, New York, 1966. x + 131 pp. (95 refs.; CR 11509; Chapter 5.)

Ka60 Karp, R. M., A note on the application of graph theory to digital computer programming. *Information and Control* **3,** 179–190 (1960). (16 refs.; CR 3563; Chapter 7.)

Ka62 Kahn, A. B., Topological sorting of large networks. *CACM* **5,** 558–562 (1962). (2 refs.; CR 4359; Chapter 6.)

Ka63 Kahn, A. B., Skeletal structure of PERT and CPA computer programs. *CACM* **6,** 473–479 (1963). (22 refs.; CR 4835; Chapter 6.)

Ka64 Kaufmann, A., *Méthodes et Modèles de la Recherche Opèrationelle*, Vol. 2. Dunod, Paris, 1964 (English transl.: *Graphs, Dynamic Programming and Finite Games*. Academic Press, New York, 1967. xviii + 484 pp.). (Chapter 3.)

Ka65 Kanner, H., Kosinski, P., and Robinson, C. L., The structure of yet another Algol compiler. *CACM* **8,** 427–438 (1965) (reprinted in Ro67a, pp. 228–252). (13 refs.; CR 15194; Chapter 5.)

Ka66 Karp, R. M., and Miller, R. E., Properties of a model of parallel computations: Determinacy, termination, queueing. *SIAM J. Appl. Math.* **14,** 1390–1411 (1966). (7 refs.; Chapter 7.)

Ke69 Kent, W., Assembler-language macroprogramming. *Comput. Surveys* **1,** 183–196 (1969). (2 refs.; CR 18369; Chapter 12.)

Ki66 Kirkpatrick, T. I., and Clark, N. R., PERT as an aid to logical design. *IBM J. Res. Develop.* **10,** 135–141 (1966). (5 refs.; CR 10753; Chapter 6.)

Ki67 King, P. J. H., Decision tables. *Comput. J.* **10,** 135–142 (1967–1968). (29 refs.; CR 12948—includes a correction; Chapter 5.)

Ki68 King, P. J. H., Ambiguity in limited entry decision tables. *CACM* **11,** 680–685 (1968). (9 refs.; CR 15973; Chapter 5.)

Kl67 Klein, M. M., Scheduling project networks. *CACM* **10,** 225–231 (1967). (7 refs.; CR 12275; Chapter 6.)

Kn63 Knuth, D. E., Computer-drawn flowcharts. *CACM* **6,** 555–563 (1963). (7 refs.; Chapter 7.)

Kn65 Knowlton, K. C., A fast storage allocator. *CACM* **8,** 623–625 (1965). (2 refs.; Chapter 10.)

Kn66 Knowlton, K. C., A programmer's description of L^6. *CACM* **9,** 616–625 (1966). (9 refs.; CR 11513; Chapter 12.)

Kn67a Knuth, D., A characterization of parenthesis languages. *Information and Control* **11,** 269–289 (1967). (3 refs.; CR 14350; Chapter 5.)

Kn67b Knuth, D. E., and Bigelow, R. H., Programming languages for automata. *JACM* **14,** 615–635 (1967). (6 refs.; CR 13675; Chapter 10.)

Kn68 Knuth, D. E., *The Art of Computer Programming*, Vol. 1, *Fundamental Algorithms*. Addison-Wesley, Reading, Massachusetts, 1968. xx + 634 pp. (CR 14505; Chapters 1, 5, 7, 10.)

Kn69 Knowlton, P., An algebraic extension to LISP. *Proc. AFIPS* **35** (*FJCC, 1969*), 169–178. (1 ref.; CR18881; Chapter 12.)

Ko66 Korfhage, R. R., *Logic and Algorithms*. Wiley, New York, 1966. xii + 194 pp. (CR 11339; Chapters 1, 4.)

Kr64 Krider, L., A flow analysis algorithm. *JACM* **11,** 429–436 (1964). (No refs.; CR 6954; Chapters 3, 6, 7.)

Kr68 Kral, J., One way of estimating frequencies of jumps in a program. *CACM* **11,** 475–480 (1968). (6 refs.; CR 16713; Chapter 7.)

Ku62 Kurtzberg, J., Algorithm 94: Combination. *CACM* **5,** 344 (1962). (No refs.; Chapter 1.)

Ku67 Kurtukov, A., On optimal arrangement of graphs. *ICC Bull.* **6,** 143–159 (1967). (2 refs.; Chapter 7.)

La61 Lasser, D. J., Topological ordering of a list of randomly numbered elements of a network. *CACM* **4,** 167–168 (1961). (1 ref.; CR 1192; Chapter 6.)

La64 Landweber, P. S., Decision problems of phrase-structure grammars. *IEEE Trans. Electronic Computers* **EC–13,** 354–362 (1964). (11 refs.; Chapter 4.)

La65a Lass, S. E., PERT Time calculation without topological ordering. *CACM* **8,** 172–174 (1965). (No refs.; Chapter 6.)

La65b Lapidus, A., and Goldstein, M., Some experiments in algebraic manipulation by computer. *CACM* **8,** 501–508 (1965). (13 refs.; Chapter 12.)

La67a Langdon, G. G., An algorithm for generating permutations. *CACM* **10,** 298–299 (1967). (No refs.; Chapter 1.)

La67b Lawson, H. W., PL/I list processing. *CACM* **10,** 358–367 (1967). (5 refs.; Chapter 12.)

Le49 Lederman, W., *Introduction to the Theory of Finite Groups*. Oliver & Boyd, Edinburgh, 1949 (3rd ed., 1957). x + 170 pp. (Chapter 4.)

Le66 Lerda-Olberg, S., Bibliography on network-based project planning and scheduling: 1962–1965. *Operations Res.* **13,** 925–931 (1966). (109 refs.; Chapter 6.)

Li64 Lipschutz, S., *Theory and Problems of Set Theory and Related Topics*. Schaum, New York, 1964. vi + 233 pp. (Chapter 1.)

Li67 Libbey, M. A., The use of second order descriptors for document retrieval. *Amer. Doc.* **18,** 10–20 (1967). (6 refs.; CR 13133; Chapter 12.)

Lo64 Lockyer, K. G., *An Introduction to Critical Path Analysis.* Pitman, London, 1964. viii + 111 pp. (CR 16373; Chapter 6.)

Lo66 Lockyer, K. G., *Critical Path Analysis; Problems and Solutions.* Pitman, London, 1966. 118 pp. (CR 16318; Chapter 6.)

Lo69a Lowry, E. S., and Medlock, C. W., Object code optimization. *CACM* **12,** 13–22 (1969). (7 refs.; CR 17066; Chapter 7.)

Lo69b Lowe, T. C., Analysis of Boolean program models for time-shared, paged environments. *CACM* **12,** 199–205 (1969). (15 refs.; CR 17862; Chapter 7.)

Ma51 Markov, A. A., The theory of algorithms. *Amer. Math. Soc. Transl.* [2], **15,** 1–14 (1960) [from *Trudy Mat. Inst. Steklov* **38,** 176–189 (1951)]. (14 refs.; Chapter 4.)

Ma54 Markov, A. A., *Theory of Algorithms.* Israel Progr. for Sci. Transl., Jerusalem, 1961. viii + 444 pp. (Original Russian ed.: *Teoriya Algorifmov.* Acad. Sc. USSR., Moscow, 1954.) (Chapter 4.)

Ma60 Marimont, R. B., Applications of graphs and Boolean matrices to computer programming. *SIAM Rev.* **2,** 259–268 (1960). (9 refs.; Chapter 7.)

Ma67a Martin, D., and Estrin, G., Models of computations and systems—evaluation of vertex probabilities in graph models of computations. *JACM* **14,** 281–299 (1967). (7 refs.; CR 12864; Chapter 7.)

Ma67b Madnick, S. E., String processing techniques. *CACM* **10,** 420–424 (1967). (10 refs.; Chapter 10.)

Ma67c Marron, B. A., and deMaine, P. A. D., Automatic data compression. *CACM* **10,** 711–715 (1967). (7 refs.; CR 13852; Chapter 11.)

Ma68 Maurer, W. D., An improved hash code for scatter storage. *CACM* **11,** 35–38 (1968). (2 refs.; Chapter 11.)

Mc60 McCarthy, J., Recursive functions of symbolic expressions and their computation by machine, Pt. I. *CACM* **3,** 184–195 (1960) (reprinted in Ro67a, pp. 455–480). (5 refs.; CR 479; Chapters 10, 12.)

Mc61 McNaughton, R., The theory of automata, a survey. *Advances in Comput.* **2,** 379–421 (1961). (119 refs.; CR 3920; Chapter 4.)

Mc62 McCarthy, J., Abrahams, P. W., Edwards, D. J., Hart, T. P., and Levin, M.I., *LISP 1.5 Programmer's Manual.* M.I.T. Press, Cambridge, Massachusetts, 1962. vi + 99 pp. (CR 5689; Chapter 12.)

Mc63a McGee, W. C., The formulation of data processing problems for computers. *Advances in Comput.* **4,** 1–52 (1963). (82 refs.; Chapter 2.)

Mc63b McBeth, J. H., On the reference counter method. *CACM* **6,** 575 (1963). (3 refs.; Chapter 10.)

Mc63c McCarthy, J., A basis for a mathematical theory of computation. In *Computer Programming and Formal Systems* (P. Braffort and D. Hirschberg, eds.), pp. 33–70. North-Holland Publ., Amsterdam, 1963 (an earlier version: *Proc. Western JCC, 1961*, pp. 225–238). (14 refs.; CR 4544 and 7994; Chapter 12.)

Mc67 McNaughton, R., Parenthesis grammars. *JACM* **14,** 490–500 (1967). (3 refs.; CR 13260; Chapter 5.)

Mc68 McDaniel, H., *An Introduction to Decision Logic Tables.* Wiley, New York, 1968. ix + 96 pp. (CR 15397; Chapter 5.)

Mc69 McIlroy, M. D., Algorithm 354: Generator of spanning trees. *CACM* **12,** 511 (1969). (4 refs.; Chapter 3.)

Me67 Mealy, G. H., Another look at data. *Proc. AFIPS* **31** (*FJCC, 1967*), 525–534. (6 refs.; CR 14726; Chapter 2.)

Mi67a Minsky, M. L., *Computation: Finite and Infinite Machines.* Prentice-Hall, Englewood Cliffs, New Jersey, 1967. xviii + 317 pp. (Chapter 4.)

Mi67b Minker, J., and Sable, J., File organization and data management. *Ann. Rev. Inf. Sci. Tech.* **2,** 123–160 (1967). (135 refs.; Chapter 11.)

Mo63 Moshman, J., Johnson, J., and Larsen, M., RAMPS—a technique for *R*ecursive *A*llocation and *M*ulti-*P*roject *S*cheduling. *Proc. AFIPS* **23** (*SJCC, 1963*), 17–27. (12 refs.; CR 5599; Chapter 6.)

Mo64 Moraff, N., *B*usiness and *E*ngineering *E*nriched *F*ortran (BEEF). *Proc. 19th ACM Nat. Conf., 1964*, Paper D1.4, 7 pp.) (3 refs.; Chapter 12.)

Mo67 Montalbano, M., High-speed calculation of the critical paths of large networks. *IBM Systems J.* **6,** 163–191 (1967). (6 refs.; Chapter 6.)

Mo68a Morris, R., Scatter storage techniques. *CACM* **11,** 38–44 (1968). (5 refs.; Chapter 11.)

Mo68b Mooers, C. N., How some fundamental problems are treated in the design of the TRAC language. In *Symbol Manipulation Languages and Techniques* (D. G. Bobrow, ed.), pp. 178–190. North-Holland Publ., Amsterdam, 1968. (7 refs.; CR 15199; Chapter 12.)

Mo69 Moyles, D. M., and Thompson, G. L., An algorithm for finding a minimal equivalent graph of a digraph. *JACM* **16,** 455–460 (1969). (4 refs.; CR 19127; Chapter 6.)

Na63 Naur, P., ed., Revised report on the algorithmic language Algol 60. *CACM* **6,** 1–17 (1963) (reprinted in Ro67a, pp. 79–117, and in various other publications). (No refs.; CR 4540; Chapter 4.)

Ne57 Newell, A., and Shaw, J. C., Programming the logic theory machine. *Proc. Western JCC, 1957*, pp. 230–240. (2 refs.; Chapter 10.)

Ne60 Newell, A., and Tonge, F. M., An introduction to Information Processing Language V. *CACM* **3,** 205–211 (1960) (reprinted in Ro67a, pp. 362–374). (14 refs.; CR 15977; Chapter 12.)

Ne64 Newell, A., *et al., Information Processing Language-V Manual*, 2nd ed. Prentice-Hall, Englewood Cliffs, New Jersey, 1964. xxxvi + 267 pp. (CR 1931 and 6668; Chapter 12.)

Ne68a Nelson, R. J., *Introduction to Automata.* Wiley, New York, 1968. xiv + 400 pp. (CR 14096; Chapter 4.)

Ne68b Newey, M. C., An efficient system for extendible languages. *Proc. AFIPS* **33** (*FJCC, 1968*), 1339–1347. (15 refs.; Chapter 12.)

Ni66 Nicholson, T. A. J., Finding the shortest route between two points in a network. *Comput. J.* **9,** 275–280 (1966–1967). (3 refs.; CR 12877; Chapter 6.)

Oe57 Oettinger, A., Account identification for automatic data processing. *JACM* **4,** 245–253 (1957). (5 refs.; Chapter 5.)

Pa64 Pandit, S. N. N., Some observations on the longest path problem. *Operations Res.* **12,** 361–364 (1964). (12 refs.; Chapter 6.)

Pa69a Paton, K., An algorithm for finding a fundamental set of cycles of a graph. *CACM* **12,** 514–518 (1969). (4 refs.; CR 18332; Chapter 3.)

Pa69b Patt, Y. N., Variable length tree structures having minimum average search time. *CACM* **12,** 72–76 (1969). (2 refs.; CR 16843; Chapter 5.)

Pe60 Perlis, A. J., and Thornton, C., Symbol manipulation by threaded lists. *CACM* **3,** 195–204 (1960). (4 refs.; CR 214; Chapter 10.)

Pe66 Penny, S. J., and Burkhard, J. H., Multidimensional correlation lattices as an aid to three-dimensional pattern reconstruction. *Proc. AFIPS* **28** (*SJCC, 1966*), 449–455. (3 refs.; CR 10394; Chapter 10.)

Po66 Ponstein, J., Self-avoiding paths and the adjacency matrix of a graph. *SIAM J. Appl. Math.* **14,** 600–609 (1966). (3 refs.; Chapter 6.)

Qu67 Quillian, M. R., Word concepts: A theory and simulation of some basic semantic capabilities. *Behavioral Sci.* **12,** 410–430 (1967). (37 refs.; CR 14875; Chapter 8.)

Ra64 Raphael, B., A computer program which "understands". *Proc. AFIPS* **26** (*FJCC, 1964*), 577–589. (21 refs.; CR 7207; Chapter 12.)

Ra66a Ramamoorthy, C. V., Analysis of graphs by connectivity considerations. *JACM* **13,** 211–222 (1966). (14 refs.; Chapter 7.)

Ra66b Ramamoorthy, C. V., The analytic design of a dynamic look ahead and program segmenting system for multiprogrammed computers. *Proc. 21st ACM Nat. Conf., 1966*, pp. 229–239. (7 refs.; CR 11502; Chapter 7.)

Ra67 Ramamoorthy, C. V., A structural theory of machine diagnosis. *Proc. AFIPS* **30** (*SJCC, 1967*), 743–756. (13 refs.; CR 12903; Chapter 7.)

Ra68a Raphael, B., Bobrow, D. G., Fein, L., and Young, J. W., A brief survey of computer languages for symbolic and algebraic manipulation. In *Symbol Manipulation Languages and Techniques* (D. G. Bobrow, ed.), pp. 1–54. North-Holland Publ., Amsterdam, 1968. (36 refs.; CR 15196; Chapter 12.)

Ra68b Ramani, S., SLIP operations on trees and their relevance to problems of linguistic interest. In *Symbol Manipulation Languages and Techniques* (D. G. Bobrow, ed.), pp. 312–339. North-Holland Publ., Amsterdam, 1968. (10 refs.; CR 15341; Chapter 12.)

Re66 Reinwald, L. T., and Soland, R. M., Conversion of limited-entry decision tables to optimal computer programs I: Minimum average processing time. *JACM* **13,** 339–358 (1966). (17 refs.; Chapter 5.)

Re67 Reinwald, L. T., and Soland, R. M., Conversion of limited-entry decision tables to optimal computer programs II: Minimum storage requirement. *JACM* **14,** 742–755 (1967). (7 refs.; Chapter 5.)

Re68 Reiter, R., Scheduling parallel computations. *JACM* **15,** 590–599 (1968). (8 refs.; CR 17864; Chapter 7.)

Ri58 Riordan, J., *An Introduction to Combinatorial Analysis.* Wiley, New York, 1958. xii + 244 pp. (Chapter 1.)

Ri65 Rich, R. P., and Stone, A. G., Method for hyphenating at the end of a printed line. *CACM* **8,** 444–445 (1965). (No refs.; CR 8546; Chapter 12.)

Ri67 Richards, R. K., *Electronic Digital Components and Circuits.* Van Nostrand, Princeton, New Jersey, 1967. xii + 526 pp. (About 1000 refs.; CR 14809; Chapter 9.)

Ro59 Roy, B., Transitivité et connexité. *C. R. Acad. Sci.* **249,** 216–218 (1959). (Chapter 6.)

Ro63 Rotenberg, N., and Opler, A., Variable width stacks. *CACM* **6,** 608–610 (1963). (No refs.; CR 5358; Chapter 10.)

Ro66 Roberts, S. M., and Flores, B., Systematic generation of Hamiltonian circuits. *CACM* **9,** 690–694 (1966). (2 refs.; Chapter 6.)

Ro67a Rosen, S., ed., *Programming Systems and Languages.* McGraw-Hill, New York, 1967. xvi + 734 pp. (Contains reprints of Ba59b, Bo64b, Di60, Ka65, Mc60, Na63, Ne60, Yn62 and original papers Ax67, De67, Sm67; CR 15975.)

Ro67b Ross, D. T., The AED free storage package. *CACM* **10,** 481–492 (1967). (6 refs.; CR 13437; Chapter 10.)

Ro68 Robert, P., and Ferland, J., Généralisation de l'algorithme de Warshall. *RIRO* **2,** No. 7, 13–25 (1968). (6 refs.; CR 18698; Chapter 8.)

Ru65 Rutherford, D. E., *Introduction to Lattice Theory.* Oliver & Boyd, Edinburgh, 1965. x + 117 pp. (CR 8591; Chapter 2.)

Ry63 Ryser, H. J., *Combinatorial Mathematics.* Math. Assoc. of Amer., 1963. xvi + 154 pp. (CR 7371; Chapter 1.)

Sa62 Samelson, K., Programming languages and their processing. *Proc. IFIP Congr., Munich, 1962*, pp. 487–492. (18 refs.; CR 7262; Chapter 10.)

Sa64 Salton, G., and Sussenguth, E. H., Some flexible information retrieval systems using structure matching procedures. *Proc. AFIPS* **25** (*SJCC, 1964*), 587–597. (17 refs.; CR 6916; Chapters 3, 8.)

Sa66a Sammet, J. E., An annotated descriptor based bibliography on the use of computers for non-numerical mathematics. *Comput. Rev.* **7,** No. 4, B.1–B.29 (1966); later, expanded version in *Symbol Manipulation Languages and Techniques* (D. G. Bobrow, ed.), pp. 358–484. North-Holland Publ., Amsterdam, 1968. (About 300 and 380 references, respectively; CR 15211; Chapter 10.)

Sa66b Satterthwait, A. C., Programming languages for computational linguistics. *Advances in Comput.* **7,** 209–238 (1966). (14 refs.; CR 11429; Chapter 12.)

Sa67 Sager, N., Syntactic analysis of natural language. *Advances in Comput.* **8,** 153–188 (1967). (54 refs.; CR 14435; Chapter 5.)

Sa68a Salton, G., *Automatic Information Organization and Retrieval.* McGraw-Hill, New York, 1968. xiv + 514 pp. (CR 16841; Chapter 8.)

Sa68b Sakoda, J. M., DYSTAL: *dy*namic *st*orage *a*llocation *l*anguage in Fortran. In *Symbol Manipulation Languages and Techniques* (D. G. Bobrow, ed.), pp. 302–311. North-Holland Publ., Amsterdam, 1968. (5 refs.; CR 15202; Chapters 9, 12.)

Sa68c Sammet, J. E., *Programming Languages: History and Fundamentals.* Prentice-Hall, Englewood Cliffs, New Jersey, 1968. xxx + 785 pp. (CR 17682 and 17854; Chapters 10, 12.)

Sa68d Sandewall, E. J., LISP A: A LISP-like system for incremental computing. *Proc. AFIPS* **32** (*SJCC, 1968*), 375–384. (8 refs.; CR 15403; Chapter 12.)

Sc58 Scott, A. E., Automatic preparation of flowchart listings. *JACM* **5,** 57–66 (1958). (1 ref.; Chapter 7.)

Sc63 Scidmore, A. K., and Weinberg, B. L., Storage and search properties of a tree-organized memory system. *CACM* **6,** 28–31 (1963). (3 refs.; CR 4780; Chapter 5.)

Sc64 Schurmann, A., The application of graphs to the analysis of distribution of loops in a program. *Information and Control* **7,** 275–282 (1964). (5 refs.; CR 8016; Chapter 7.)

Sc67a Schneider, V., Syntax-checking and parsing of context-free languages by pushdown-store automata. *Proc. AFIPS* **30** (*SJCC, 1967*), 685–690. (8 refs.; CR 13263; Chapter 10.)

Sc67b Schorr, H., and Waite, W. M., An efficient machine-independent procedure for garbage collection in various list structures. *CACM* **10,** 501–506 (1967). (13 refs.; CR 13179; Chapter 10.)

Sc67c Schwartz, E. S., and Kleiboemer, A. J., A language element for compression coding. *Information and Control* **10,** 315–333 (1967). (16 refs.; CR 12895; Chapter 11.)

Sc68 Schurmann, A., GAN, a system for generating and analyzing activity networks. *CACM* **11,** 675–679 (1968). (3 refs.; Chapter 6.)

Sh66 Sherman, P. M., Flowtrace, a computer program for flowcharting programs. *CACM* **9,** 845–854 (1966). (11 refs.; CR 11921; Chapter 7.)

Sh68 Shoffner, R. M., The organization, maintenance and search of machine files. *Ann. Rev. Inf. Sci. Tech.* **3,** 137–167 (1968). (94 refs.; CR 16551; Chapter 11.)

Si64 Sikorski, R., *Boolean Algebras*, 2nd ed. Springer, New York, 1964. x + 237 pp. (Chapter 1.)

Si65 Simões Pereira, J. M. S., On Boolean matrix equation $M' = \bigvee_{i=1}^{d} M^i$. *JACM* **12,** 376–382 (1965). (3 refs.; CR 9805; Chapter 6.)

Si66a Simmons, R. F., Automated language processing. *Ann. Rev. Inf. Sci. Tech.* **1,** 137–169 (1966). (116 refs.; CR 11097; Chapter 5.)

Si66b Simmons, R. F., Storage and retrieval of aspects of meaning in directed graph structures. *CACM* **9**, 211–214 (1966). (5 refs.; CR 10043; Chapter 8.)

Si66c Simmons, R. F., Burger, J. F., and Long, R. E., An approach toward answering English questions from text. *Proc. AFIPS* **29** (*FJCC, 1966*), 357–363. (19 refs.; CR 12655; Chapter 8.)

Sm67 Smith, D. K., An introduction to the list-processing language SLIP. See Ro67a, pp. 393–418. (7 refs.; CR 16427; Chapter 12.)

Sp67 Spain, R. J., Marino, M. J., and Jauvtis, H. I., DTPL push down memory. *Proc. AFIPS* **30** (*SJCC, 1967*), 491–498. (4 refs.; CR 12535; Chapter 10.)

St61 Stoll, R. R., *Sets, Logic, and Axiomatic Theories.* Freeman, San Francisco, California, 1961. x + 206 pp. (Chapter 1.)

St62 Stone, P. J., Bales, R. F., Zvi Namenwirth, J., and Ogilvie, D. M., The General Inquirer: A computer system for content analysis and retrieval based on the sentence as a unit of information. *Behavioral Sci.* **7**, 484–498 (1962). (6 refs.; Chapter 12.)

St63 Stoll, R. R., *Set Theory and Logic.* Freeman, San Francisco, California, 1963. xiv + 474 pp. (CR 5428; Chapter 1.)

St65 Stockham, T. G., Some methods of graphical debugging. *Proc. IBM Sci. Comput. Symp. Man-Machine Communication, White Plains, May 1965*, pp. 57–71. IBM Data Proc. Div., White Plains, New York, no date. (11 refs.; Chapter 7.)

St66 Stone, P. J., Dunphy, D. C., Smith, M. S., Ogilvie, D. M., *et al.*, *The General Inquirer.* M.I.T. Press, Cambridge, Massachusetts, 1966. xx + 651 pp. (General bibliography on content analysis, specialized bibliographies follow each application chapter; CR 12245; Chapter 12.)

Su60 Suppes, P., *Axiomatic Set Theory.* Van Nostrand, Princeton, New Jersey, 1960. xii + 265 pp. (Chapter 1.)

Su63 Sussenguth, E. H., Use of tree structures for processing files. *CACM* **6**, 272–279 (1963). (6 refs.; Chapter 5.)

Su65 Sussenguth, E. H., A graph-theoretical algorithm for matching chemical structures. *J. Chem. Doc.* **5**, 36–43 (1965). (3 refs.; CR 7812; Chapter 8.)

Sv66 Svejgaard, B., Contribution No. 17: List processing. *BIT* **6**, 164–175 (1966). (6 refs.; Chapter 12.)

Sz63 Szasz, G., *Introduction to Lattice Theory*, 3rd ed. Academic Press, New York, 1963. 229 pp. (Chapter 2.)

Ta62 Tabory, R., Premiers elements d'un langage de programmation pour le traitement en ordinateur des graphes. *Symbolic Languages in Data Processing*, pp. 717–730. Gordon and Breach, New York, 1962. (4 refs.; Chapter 12.)

Ta67 Tate, F. A., Handling chemical compounds in information systems. *Ann. Rev. Inf. Sci. Tech.* **2**, 285–309 (1967). (179 refs.; Chapter 8.)

Th67 Thomas, E. M., GRASP—a *gra*phic *s*ervice *p*rogram. *Proc. 22nd ACM Nat. Conf., 1967*, pp. 395–402. (4 refs.; Chapter 10.)

Tr62 Trotter, H. F., Algorithm 115: Perm. *CACM* **5**, 434–435 (1962). (1 ref.; Chapter 1.)

Tr66 Trundle, R. W. L., LITHP—an Algol list processor. *Comput. J.* **9**, 167–172 (1966–1967). (3 refs.; Chapter 12.)

Tu68 Turner, J., Generalized matrix functions and the graph isomorphism problem. *SIAM J. Appl. Math.* **16**, 520–526 (1968). (8 refs.; Chapter 3.)

Un64 Unger, S. H., GIT—a heuristic program for testing pairs of directed line graphs for isomorphism. *CACM* **7**, 26–34 (1964). (8 refs.; CR 5749; Chapter 3.)

Ve66 Veinott, C. G., Programming decision tables in Fortran, Cobol or Algol. *CACM* **9**, 31–35 (1966). (2 refs.; CR 9724; Chapter 5.)

Wa62 Warshall, S., A theorem on Boolean matrices. *JACM* **9,** 11–12 (1962). (2 refs.; Chapter 6.)

We63 Weizenbaum, J., Symmetric list processor. *CACM* **6,** 524–544 (1963). (4 refs.; CR 5023; Chapters 10, 12.)

We64 Wengert, R. E., A simple automatic derivative evaluation program. *CACM* **7,** 463–464 (1964). (No refs.; CR 6698; Chapter 12.)

We66a Welch, J. T., A mechanical analysis of the cyclic structure of undirected linear graphs. *JACM* **13,** 205–210 (1966). (6 refs.; CR 10573; Chapter 3.)

We66b Weizenbaum, J., ELIZA—a computer program for the study of natural language communication between man and machine. *CACM* **9,** 36–45 (1966). (6 refs.; CR 9655; Chapter 12.)

We67 Weizenbaum, J., Contextual understanding by computers. *CACM* **10,** 474–480 (1967). (6 refs.; CR 13062; Chapter 12.)

We68 Wegner, P., *Programming Languages, Information Structures, and Machine Organization.* McGraw-Hill, New York, 1968. xx + 401 pp. (CR 16228; Chapter 12.)

We69 Weizenbaum, J., Recovery of reentrant list structures in SLIP. *CACM* **12,** 370–372 (1969). (3 refs.; CR 18051; Chapter 10.)

Wi60 Windley, P. F., Trees, forests and rearranging. *Comput. J.* **3,** 84–88 (1960–1961). (5 refs.; Chapter 5.)

Wi64 Wiseman, N. E., Application of list-processing methods to the design of interconnections for a fast logic system. *Comput. J.*, **6,** 321–327 (1963–1964). (11 refs.; CR 6256; Chapter 10.)

Wi65 Wilkes, M. V., Lists and why they are useful. *Comput. J.* **7,** 278–281 (1964–1965) [or *Proc. 19th ACM Nat. Conf., 1964*, Paper F1 (5 pp.)]. (4 refs.; CR 6945; Chapter 10.)

Wo65 Wolman, E., A fixed optimum cell-size for records of various lengths. *JACM* **12,** 53–70 (1965). (No refs.; CR 8218; Chapter 10.)

Yn62 Yngve, V. H., Comit as an IR language. *CACM* **5,** 19–28 (1962) (reprinted in Ro67a, pp. 375–392). (5 refs.; Chapter 12.)

Za65a Zadeh, L. A., Fuzzy sets. *Information and Control* **8,** 338–353 (1965). (3 refs.; Chapter 1.)

Za65b Zadeh, L. A., Fuzzy sets and systems. *Proc. Symp. on System Theory*, pp. 29–37. Polytech. Inst. of Brooklyn, Brooklyn, New York, 1965. (6 refs.; Chapter 1.)

Za68 Zadeh, L. A., Fuzzy algorithms. *Information and Control* **12,** 94–102 (1968). (7 refs.; Chapter 1.)

Zz62a An information algebra—phase 1 report, language structure group of the CODASYL development committee. *CACM* **5,** 190–204 (1962). (No refs.; CR 2621; Chapter 2.)

Zz62b *An Introduction to Comit Programming*, revised ed. M.I.T. Press, Cambridge, Massachusetts, 1962. ii + 60 pp. (Chapter 12.)

Zz62c *Comit Programmer's Reference Manual*, corrected ed. M.I.T. Press, Cambridge, Massachusetts, 1962. vi + 61 pp. (Chapter 12.)

Index

Printed in Dunstable, United Kingdom

91336867R00257